Arthropod Management and Landscape Considerations in Large-scale Agroecosystems

Arthropod Management and Landscape Considerations in Large-scale Agroecosystems

Edited by

Michael J. Brewer

Professor and Regents Fellow, Texas A&M AgriLife Research, Texas A&M University System

Gary L. Hein

Professor Emeritus, Department of Entomology, Doctor of Plant Health Program, University of Nebraska – Lincoln

CABI

CABI is a trading name of CAB International

CABI
Nosworthy Way
Wallingford
Oxfordshire OX10 8DE
UK

Tel: +44 (0)1491 832111
E-mail: info@cabi.org
Website: www.cabi.org

CABI
200 Portland Street
Boston
MA 02114
USA

Tel: +1 (617)682-9015
E-mail: cabi-nao@cabi.org

A catalogue record for this book is available from the British Library, London, UK.

Library of Congress Cataloging-in-Publication Data

Names: Brewer, Michael J. (Michael John), 1959- editor. | Hein, Gary Lee, editor.
Title: Arthropod management and landscape considerations in large-scale agroecosystems / Michael J. Brewer, Gary L. Hein
Description: Boston, MA : CAB International, [2024] | Includes bibliographical references and index | Summary: "Focusing on the worlds major plant-based food crops this important book reviews arthropod pest management practice in large-scale agroecosystems from the field to the landscape scale. These major challenges are illustrated with case studies throughout"-- Provided by publisher
Identifiers: LCCN 2024020736 (print) | LCCN 2024020737 (ebook) | ISBN 9781800622753 (hardback) | ISBN 9781800622760 (ebk) | ISBN 9781800622777 (epub)
Subjects: LCSH: Arthropod pests--Control--Environmental aspects. | Agricultural pests--Control--Environmental aspects. | Food crops--Protection. | Agricultural ecology.
Classification: LCC SB951.5 .A77 2024 (print) | LCC SB951.5 (ebook) | DDC 632/.951--dc23/eng/20240617
LC record available at https://lccn.loc.gov/2024020736
LC ebook record available at https://lccn.loc.gov/2024020737

ISBN-13: 9781800622753 (hardback)
9781800622760 (ePDF)
9781800622777 (ePub)

DOI: 10.1079/9781800622777.0000

Commissioning Editor: Ward Cooper
Editorial Assistant: Emma McCann
Production Editor: Rosie Hayden

Typeset by Straive, Pondicherry, India

Contents

Contributors

Kristen Baum, Department of Integrative Biology, Oklahoma State University, 501 Life Sciences West, Stillwater, OK 74078, USA; kristen.baum@okstate.edu

Mahendra Bhandari, Texas A&M AgriLife Research, 10345 Hwy 44, Corpus Christi, TX 78406, USA; Mahendra.Bhandari@ag.tamu.edu

Michael J. Brewer, Texas A&M AgriLife Research, 10345 Hwy 44, Corpus Christi, TX 78406, USA; brewermj@tamu.edu

Haley Butler, Department of Entomology and Plant Pathology, Oklahoma State University, 127 Noble Research Center, Stillwater, OK 74078, USA; haley.butler@okstate.edu

Hannalene du Plessis, IPM Program, Unit for Environmental Sciences and Management, Private Bag X6001, North-West University, Potchefstroom 2520, South Africa; hannalene.duplessis@nwu.ac.za

Norman C. Elliott, US Department of Agriculture, Agricultural Research Service, Peanut and Small Grains Research Unit, 1301 N Western Rd, Stillwater, OK 74075, USA; norman.elliott@usda.gov

Peter C. Ellsworth, Department of Entomology, University of Arizona, Maricopa Agricultural Center, 37860 W. Smith-Enke Rd, Maricopa, AZ 85138, USA; peterell@cals.arizona.edu

Sarah Elzay, Department of Biological Sciences, Albertson Hall 302, Fort Hays State University, Hays, KS 67601, USA; sdelzay@fhsu.edu

Isaac L. Esquivel, Department of Entomology & Nematology, University of Florida, North Florida Research and Education Center, 155 Research Rd, Quincy, FL 32351, USA; Isaac.Esquivel@ufl.edu

Ashleigh M. Faris, Department of Entomology and Plant Pathology, Oklahoma State University, 127 Noble Research Center, Stillwater, OK 74078, USA; ashleigh.faris@okstate.edu

Aaron J. Gassmann, Department of Plant Pathology, Entomology and Microbiology, Iowa State University, 1344 ATRB, 2213 Pammel Dr, Ames, IA 50011, USA; aaronjg@iastate.edu

Kristopher Giles, Department of Entomology and Plant Pathology, Oklahoma State University, 127 Noble Research Center, Stillwater, OK 74078, USA; kris.giles@okstate.edu

Maura Hall, Department of Plant Pathology, Entomology and Microbiology, Iowa State University, 1344 ATRB, 2213 Pammel Dr, Ames, IA 50011, USA; mjhall@iastate.edu

Gary L. Hein, Department of Entomology, University of Nebraska, 103 Entomology Hall, Lincoln, NE 68583, USA; ghein1@unl.edu

Louis S. Hesler, US Department of Agriculture, Agricultural Research Service, Integrated Cropping Systems Unit, 2923 Medary Ave, Brookings, SD 57006, USA; louis.hesler@usda.gov

Anders S. Huseth, Department of Entomology and Plant Pathology and the North Carolina Plant Sciences Initiative, North Carolina State University, Campus Box 7613, Raleigh, NC 27695, USA; ashuseth@ncsu.edu

William D. Hutchison, Department of Entomology, University of Minnesota, 1980 Folwell Ave, St Paul, MN 55108, USA; hutch002@umn.edu

Ian MacRae, Northwest Research and Outreach Center, University of Minnesota, 2900 University Ave, Crookston, MN 56716, USA; imacrae@umn.edu

Robert L. Meagher, Jr, US Department of Agriculture, Agricultural Research Service, Insect Behavior and Biocontrol Research Unit, 1700 S.W. 23RD Dr, Gainesville, FL 32608, USA; Rob.Meagher@usda.gov

Lance J. Meinke, Department of Entomology, University of Nebraska, 103 Entomology Hall, Lincoln, NE 68583, USA; lmeinke1@unl.edu

Steven E. Naranjo (Emeritus), US Department of Agriculture, Agricultural Research Service, Arid-Land Agricultural Research Center, 21881 North Cardon Ln, Maricopa, AZ 85138, USA; senaranjo@cox.net and steve.naranjo@usda.gov

Matthew E. O'Neal, Department of Plant Pathology, Entomology and Microbiology, Iowa State University, 1344 ATRB, 2213 Pammel Dr, Ames, IA 50011, USA; oneal@iastate.edu

Shannon L. Osborne, US Department of Agriculture, Agricultural Research Service, Integrated Cropping Systems Unit, 2923 Medary Ave, Brookings, SD 57006, USA; shannon.osborne@usda.gov

Pankaj Pal, Texas A&M AgriLife Research, 10345 Hwy 44, Corpus Christi, TX 78406, USA; pankaj.pal.03@outlook.com

Katherine A. Parys, US Department of Agriculture, Agricultural Research Service, Pollinator Health in Southern Crop Ecosystems Research Unit, 141 Experiment Station Rd, Stoneville, MS 38776, USA; Katherine.Parys@usda.gov

Dominic D. Reisig, Department of Entomology and Plant Pathology, Vernon James Research and Extension Center, North Carolina State University, 207 Research Station Rd, Plymouth, NC 27962, USA; ddreisig@ncsu.edu

Tom Royer, Department of Entomology and Plant Pathology, Oklahoma State University, 127 Noble Research Center, Stillwater, OK 74078, USA; tom.royer@okstate.edu

Nina Rudin, Department of Entomology and Plant Pathology, Oklahoma State University, 127 Noble Research Center, Stillwater, OK 74078, USA; nrudin@okstate.edu

Thomas W. Sappington, US Department of Agriculture, Agricultural Research Service, Corn Insects and Crop Genetics Research Unit, 819 Wallace Rd, Ames, IA 50011, USA; tom.sappington@usda.gov

Gregory A. Sword, Department of Entomology, Texas A&M University, 370 Olsen Blvd, TAMU 2475, College Station, TX 77843, USA; Gregory.Sword@ag.tamu.edu

Ashley E. Tessnow, Department of Entomology, Texas A&M University, 370 Olsen Blvd, TAMU 2475, College Station, TX 77843, USA; Ashley.Tessnow@ag.tamu.edu

Paul A. Umina, School of BioSciences, The University of Melbourne, Grattan St, Parkville, Victoria 3010, Australia; pumina@unimelb.edu.au

Johnnie van den Berg, IPM Program, Unit for Environmental Sciences and Management, Private Bag X6001, North-West University, Potchefstroom 2520, South Africa; johnnie.vandenberg@nwu.ac.za

Foreword

It has long been known that arthropod pest occurrence and intensity are influenced by the agroecosystems in which crops are grown. Consequently, populations of specific pests and pest complexes vary greatly at geographic scales that span fields, farms production systems, regions, and continents. These patterns are strongly influenced by the availability and location of alternate crop and non-crop hosts and habitats for pest and beneficial species, and by the ability of these species to move among them. In the coming decades, agricultural production systems are expected to change significantly as new technologies are adopted and crop distributions and production practices change in response to global warming, shifting pest/pathogen distributions, and the need to preserve environmental quality, ecosystem services and biodiversity. These changes are expected to alter landscape-scale interactions affecting pest populations in important ways.

Understanding of these relationships has grown dramatically in recent decades in tandem with advances in information technology, geospatial tools and analytics, and molecular techniques. *Arthropod Management and Landscape Considerations in Large-scale Agroecosystems* provides an important and timely overview of the conceptual framework linking landscape ecology to integrated pest management (IPM), as well as numerous case studies that explore important landscape–pest and beneficial insect interactions in relation to pest management. Importantly, it illustrates that we now have the capability to elucidate and understand these relationships at multiple spatial scales, and to apply that understanding to inform the coming change in agricultural production systems and practices and enhance the development and implementation of sustainable crop and pest management systems.

George G. Kennedy
William Neal Reynolds Distinguished Professor of Agriculture, Department of Entomology and Plant Pathology, North Carolina State University, Raleigh, NC, USA

Foreword

Agroecology encompasses patterns and processes occurring within the crop plant, among plants within a field, between fields and natural habitats within landscape mosaics, and among landscapes within regions. Although this spatial continuum has long been appreciated, theory and practice of integrated pest management (IPM) occurs primarily at the farm or field level, first because these are the typical management units, and second because landscape-scale processes and patterns are complex and difficult to study and understand, let alone manage. None the less, work on the ecology and management of agricultural insect pests, natural enemies, and pollinators at the landscape scale has continued to advance. In recent decades, geographic information systems, remote sensing, and molecular tools have enabled greater understanding and more sophisticated applications for managing insects while considering the landscapes beyond the farm. Despite this growth in interest, there have been no edited collections on the theme for more than a decade. *Arthropod Management and Landscape Considerations in Large-scale Agroecosystems* superbly addresses this need.

Co-editors Drs Brewer and Hein, each with extensive experience in landscape-scale studies of arthropod pests, have assembled contributions by leading experts working at local, landscape, regional, and intercontinental scales on key pests and pollinator management in production systems in the USA, Africa, and Australia. Two chapters, one from each co-editor, set the stage with overviews of aspects of pest management in a landscape context. The 12 remaining chapters provide deep dives, with thorough reviews of foundational and ongoing work on pest and pollinator management beyond the farm, using case studies, highlights of recent developments, and recommendations for future work. Principal challenges to incorporating larger spatial scales into pest and pollinator management are addressed throughout. These include data acquisition and modeling, detecting and accounting for variability among and within species, regional variations in how landscape heterogeneity affects pests and natural enemies, and the need for effective communication and cooperation among producers and managers in shared landscapes to enable effective areawide pest management.

Arthropod Management and Landscape Considerations in Large-scale Agroecosystems is a primer for those of us working on mobile pests and pollinators in heterogeneous landscapes and regions, those seeking an introduction, and those initiating new research in this area of research and application. As climates change, and along with them entire production landscapes, so will the ecology of the insects affecting the agricultural systems in these landscapes. The volume will be a valuable resource for agricultural entomologists working to meet the challenges arising from these changes.

Sanford D. Eigenbrode
Distinguished Professor, Entomology,
University of Idaho, Moscow, ID, USA

Foreword

The science and practice of integrated pest management (IPM) has evolved in the last 60+ years, with remarkable breakthroughs in the fields of biology, ecology, insect–plant interactions, phenology, and predictive modeling, which together have helped improve decision support aids for making conscious pest management resolutions. From the field-scale perspective, information is available to help us understand how pests and beneficial arthropods respond to changes driven by pest management practices and how these are affected by environmental factors; however, there is less information about approaching issues at large scale and how they may affect long-term pest management planning. Therefore, by addressing fundamental pest dynamics and control approaches in a landscape context, this book covers a series of topics that can help researchers, educators, pest management practitioners, students, and, potentially, future decision makers, learn how spatial scale influences our understanding of pest and beneficial dynamics and the selection of targeted decision support tools. This book presents a series of successful case studies that are skillfully described, including a well-documented cotton IPM program in Arizona (Chapter 3), several aphid and biological control systems in wheat, canola, and other crops (Chapters 5, 6, 7, 8), the billion-dollar pest "western corn rootworm" in corn (Chapter 9), the worldwide consequence of the fall armyworm (Chapters 4 and 14), and corn earworm (Chapter 10); the role of bees is well characterized in Chapter 12, including how pesticide exposure can influence the fate of bees' ability to seek resources. All chapters lead to Chapter 13 where data acquisition and integration concepts are highlighted as the basis for contemporary decision support tools development.

Arthropod Management and Landscape Considerations in Large-scale Agroecosystems is highly recommended for applied ecologists, pest management researchers and practitioners, and advanced undergraduate and graduate students. The book is a valuable resource that can enlighten the reader's perception of how regional approaches' successes and failures can lead to broader approaches for long-term sustainable results.

Silvia I. Rondon
Professor of Entomology,
Specialist and Director of the Oregon IPM Center,
Corvallis, OR, USA

https://blogs.oregonstate.edu/iaep/sample-page/
https://agsci.oregonstate.edu/oipmc

Introduction

On behalf of all chapter contributors, we are pleased to present *Arthropod Management and Landscape Considerations in Large-scale Agroecosystems*. Understanding pest and beneficial (i.e. natural enemies, pollinators) arthropod dynamics, developing integrated pest management (IPM) tactics, establishing decision support aids, and stewarding management and environmental resources have advanced during the 60+ years of IPM as a discipline and practice. For large-scale agricultural crops, the literature is particularly rich in understanding how pest and beneficial arthropods respond to and interact with agronomic and environmental features within cropland and in developing pest management tactics applied at the local (farm) level. Yet for crops in large production systems, pest population increases and decreases, both gradual and abrupt, vary considerably from the field scale up through the local and regional landscape scales. Understanding pest and beneficial arthropod population dynamics and the influence of landscape related factors across the various spatial scales is critical to developing more sustainable and effective IPM programs. As applied to large-scale agroecosystems, the consideration and integration of ecosystem services and trophic interactions across the various spatial scales into pest management approaches are the main themes of this book.

Fourteen chapters are segmented into three book sections that facilitate the content flow. The two chapters of the first section present general concepts of arthropod–plant dynamics and pest management services (naturally occurring and managed inputs) from within-field, local, regional, to continental landscape-scale perspectives. Seven case studies across the two chapters enrich the concepts and illustrate opportunities and challenges for incorporating landscape considerations into arthropod management in large-scale agroecosystems. The ten chapters of the second section highlight specific research approaches and findings that improve management of pest and beneficial arthropods, including pollinators, and stewardship of pest management technologies and pest management as an ecosystem service in the context of ecological, agricultural, and socio-economic components of large agroecosystems. The last section contains two chapters that review research and applications of information systems and molecular tools. Emphasis is placed on future development and use of these tools to improve our understanding and management of arthropod pests, leading to more sustainable IPM from local (farm) to regional (growing region) levels and ability to adapt to changes in the agricultural and environmental landscape.

Arthropod Management and Landscape Considerations in Large-scale Agroecosystems is of special interest to applied ecologists, pest management researchers and practitioners, and advanced undergraduate and graduate students. Each chapter is independent, but please consider the book as a whole in its presentation that reflects on spatial relationships important to the development, stewardship, and advancement of more sustainable IPM programs in large-scale agroecosystems.

Michael J. Brewer
Professor and Regents Fellow,
Texas A&M AgriLife Research,
Texas A&M University System,
Corpus Christi, TX, USA

Gary L. Hein
Professor Emeritus, Department of Entomology,
Doctor of Plant Health Program,
University of Nebraska – Lincoln,
Lincoln, NE, USA

February 29, 2024

1 Influence of Spatial Scale on Landscape Effects Important to Arthropod Management

Gary L. Hein,[1,*] Thomas W. Sappington,[2] and Ian MacRae[3]

[1]Department of Entomology, University of Nebraska – Lincoln, Lincoln, NE, USA; [2]US Department of Agriculture, Agricultural Research Service, Corn Insects and Crop Genetics Research Unit, Ames, IA, USA; [3]Northwest Research and Outreach Center, University of Minnesota, Crookston, MN, USA

Abstract

The development and expansion of efficient spatial technologies and geostatistical analysis capabilities over the last few decades has enabled a dramatic increase in spatial landscape research. The knowledge gained from this research in large-scale agroecosystems impacts implementation of integrated pest management (IPM) programs across four spatial scales: (i) intrafield, (ii) local landscape, (iii) regional, and (iv) continental. Understanding the variability and interactions across these agroecological landscape scales can increase management options to encompass wider impacts from mobile pest and beneficial insects. IPM programs implemented by farmers and other managers have largely been applied uniformly at the field level. However, management of pest insects with clear and predictable intrafield spatial distributions may become more efficient through site-specific applications. The makeup of local landscapes surrounding fields directly affects the dynamics of field populations of pests, natural enemies, and pollinators, and this impact will expand considerations for managing local landscapes. Beyond the local landscape (i.e. regional landscape), longer range movement by pests and natural enemies and their interactions with crop and non-crop habitats can dramatically influence pest risk potential, natural enemy prevalence, and management itself. Finally, continental-scale migration of pests can contribute significantly to risk assessment important for IPM program development and implementation. Case studies are presented illustrating the incorporation of improved spatial understanding into the development and implementation of IPM programs. Increasingly, sustainability of large-scale crop production systems will be enhanced by farm-level IPM decisions that account for local, regional, and continental landscape interactions. The success of future IPM programs will depend on better understanding these broader ecological interactions to enhance the prediction of agroecosystem services across all spatial scales and increase overall effectiveness of landscape services and economic and environmental resilience.

Introduction

Developments in spatial technologies through the last few decades have dramatically impacted research and application of agricultural production practices. These technologies have expanded the sphere of research questions that can be addressed and the resulting knowledge and understanding of production systems. Perhaps most importantly, they have enabled

*Corresponding author: ghein1@unl.edu

 Arthropod Management and Landscape Considerations in Large-Scale Agroecosystems (eds M.J. Brewer and G.L. Hein)
DOI: 10.1079/9781800622777.0001

the application of improved management processes. Primary among these technologies are global positioning systems (GPS) and geographic information systems (GIS). With expanded access and reliability of these technologies and increased geostatistical analysis capabilities, researchers and farm managers have begun to quantify and understand spatial variability in the interactions between soil, water, pests, and other environmental stressors that influence plant growth in agricultural production systems.

Historically, the management target for integrated pest management (IPM) programs has been the field or groups of fields (e.g. farm). However, spatial technologies have enabled the study of ecological interactions involving arthropod pests, natural enemies, and pollinators that extend well beyond the field level. This includes the spatial details of intrafield interactions but also encompasses spatial interactions ranging across the local landscape to a regional or even continental scale. These technologies have enabled approaches to address spatially related ecological questions and develop applications critical to improving the efficiency and effectiveness of management programs.

Spatial variability and resulting spatial interactions occur across a continuum from individual plants to well beyond an individual field or even a farm. The extensive dispersal capabilities of many insects dramatically expand the spatial dimensions important for population management. The influence and interactions of population regulation mechanisms vary across spatial scales. In spatial research efforts, this continuum of spatial scales can be divided into four levels (Fig. 1.1): (i) intrafield scale, (ii) local landscape scale, (iii) regional scale, and (iv) continental scale. Improved understanding of intrafield spatial variability has enabled development of site-specific management (i.e. precision agriculture) within fields when proper equipment and information systems are available. Understanding variability and interactions across the agroecological landscape surrounding a field (local landscape scale) can expand management options to encompass

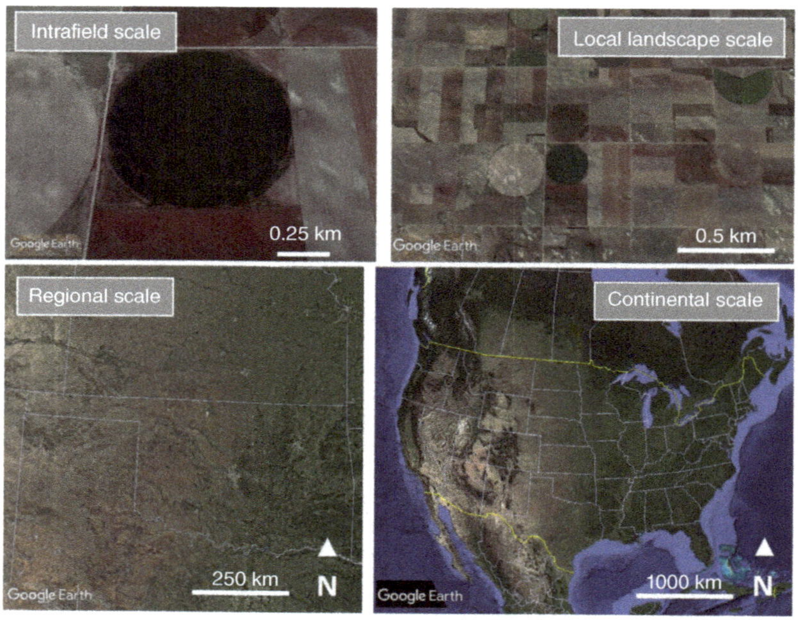

Fig. 1.1. Ecological and integrated pest management (IPM) research has been carried out across widely varying spatial scales ranging from the intrafield scale up to the continental scale. Farmers generally target their IPM decisions at the field or farm scale, but impacts on those decisions often result from insect and landscape interactions across all these spatial scales. (Landscape image credits: Google Earth.) (Original artwork by G. Hein.)

wider interactions and impacts of mobile pests, natural enemies, and pollinators. Beyond the local landscape, longer range movement by pests and natural enemies and their interactions with crop and non-crop habitats across the larger agroecosystem can dramatically influence pest risk potential, natural enemy prevalence, and management itself. This can occur at the regional scale (e.g. across watersheds or states with similar agroecosystem characteristics), but even continental-scale movement of pests often contributes significantly to local pest risk experienced by farmers.

One of the greatest challenges accompanying access to more accurate spatial data at increasing spatial scales is to bring it to bear on IPM decisions by farmers. Increased understanding of ecological systems at larger spatial scales must translate to quantifiable impact at the field level to enable more efficient and sustainable decision making. Understanding pest and natural enemy dynamics at larger spatial scales also enables broader management options of perhaps greater economic impact while enhancing the resilience of the larger agroecosystem. Ecological interactions at spatial scales beyond the field are particularly relevant to large-scale agroecosystems because of the widespread presence of the target crop and its pests. However, pest and natural enemy dynamics vary across the broader agroecosystem due to differences in weather, diversity of the non-crop habitat, and diversity in cropping systems that include the target crop.

The objective of this chapter is to provide a broad overview of the use of spatial data at these four spatial scales to improve IPM programs. The impacts of these spatial relationships are particularly important because of the broader spatial interactions and considerations that occur with large-scale agricultural systems. Case studies will be provided, and we identify chapters in this book illustrating how knowledge at these scales can improve IPM in practice. Finally, we discuss incorporation of spatial dynamics and interactions at various scales into local and regional IPM programs.

Intrafield Scale

Recent advances in GIS and GPS, coupled with geostatistical analyses, have enabled development of improved methods to address the variability of parameters that impact crop production on a site-specific basis within fields (i.e. precision agriculture). Early programs involved fertility management, usually relying on separating fields into zones of similar physical characteristics such as soil type, slope, etc. (Doerge, 1999; Khosla *et al.*, 2002). Delineation of management zones became more sophisticated by incorporating other physical factors such as organic matter and electrical conductivity, improving application of these technologies. Soil data have also been merged with yield mapping to better estimate plant response (Adamchuk *et al*, 2010). More recently, remote sensing has been used to continuously monitor plant condition, and variable rate fertilizer applications have been developed to optimize plant response (Sishodia *et al.*, 2020). Increasingly expensive inputs have spurred the ongoing improvement of technology for targeted and variable rate application of pesticides, irrigation, and planting. The greatest use of these precision technologies for crop protection has been in weed management (Lowenberg-DeBoer and Erickson, 2019; Sishodia *et al.*, 2020).

Historically, arthropod pests in large-scale crops were managed on a scale encompassing the entire field. However, the variable nature of within-field spatial distribution patterns of many arthropod pests has long been recognized as non-random. Few ecological studies addressed this variability until advances in GPS/GIS technologies and geostatistics enabled more efficient and accurate spatial characterization of these often-complex interactions (Liebhold *et al.*, 1993). Liebhold *et al.* (1993) emphasized the need to analyze insect population processes functioning through space (i.e. landscape ecology). Park *et al.* (2007) identified three steps to implementing site-specific management of insect pests. The first is to characterize the spatial structure of the pest population. Pests with limited dispersal capacity and that show spatial aggregation across a field are the best prospects for site-specific management. Border gradients resulting from biotic (e.g. host, dispersal differences) and abiotic (e.g. soil, water, light, effects on plant physiology, wind) factors are the most common spatial effect seen for insect populations (van Helden, 2010). Additionally, gradients must be reasonably stable to enable a useful site-specific management response; however, spatial distributions of typically mobile insects are often temporally dynamic, presenting a significant challenge (Fleischer *et al.*, 1999).

The second step in implementing site-specific management is to use the spatial structure data to generate a spatial distribution map that can be used to develop a response or prescription map (Park *et al.*, 2007). Finally, the prescription map is used to apply a site-specific management response. Types of response could include variable application of pesticide, fertility, or irrigation, or the planting of crop varieties with differential pest resistance characteristics; however, the primary examples thus far have focused on variable pesticide application. Early examples of the application of these technologies in IPM, summarized by Park *et al.* (2007) and van Helden (2010), involved establishing spatial distribution of the pest, carrying out spatially variable insecticide applications, and in some cases, determining the spatial distribution of natural enemy populations.

Where pest populations are aggregated, the cost of sampling to establish accurate spatial distribution maps is a major limitation to implementing site-specific management in IPM (Fleischer *et al.*, 1999; Park *et al.*, 2007). An approach to address the costs of mapping spatial aggregation is to delineate management zones based on more readily quantified factors that influence insect populations. Deleon *et al.* (2017) used pest sampling maps in cotton to establish treatment zones that were largely limited to field borders. Temple (2007) used historical yield and production data to establish high-risk zones in cotton. Davidson *et al.* (2007) correlated insect population levels across high, medium, and low corn production zones. Some of these examples demonstrated potential for reduction in insecticide applications and economic savings, but there are limited examples of highly effective site-specific IPM in large-scale agricultural systems.

Case Study 1.1 illustrates one successful effort at site-specific IPM at the intrafield scale. However, even with expanded interest beginning more than two decades ago, successful examples in large-scale agricultural systems are limited. There are several limitations to further advances in IPM precision agriculture. The detailed spatial distribution of numerous insect pests has not been determined, or it may not express the level of aggregation necessary for the practical application of spatial management approaches. The challenge of establishing spatial distribution maps is often costly, and aggregated pest distributions may be temporally unstable (Fleischer *et al.*, 1999). Other factors impacting practicality of site-specific management are crop value and input costs. Input and application costs influence the potential economic savings from site-specific management; however, realizing a profitable return is more difficult with low- or moderate-value crops. Park *et al.* (2007) and van Helden (2010) identified limitations of site-specific application technology as an important constraint, but this has been extensively addressed in the last decade through development of variable rate technology for pesticide and fertilizer application, irrigation, and planting (Lowenberg-DeBoer and Erickson, 2019; Sishodia *et al.*, 2020; Gerhards *et al.*, 2022). Further advances in application technology are likely, along with reduced cost of this technology. Continued work is needed to address the limitations in implementing site-specific management to enable increased adoption of these practices and to realize potential benefits.

Local Landscape Scale

Ecological interactions of pests and natural enemies in a target field with its surrounding landscape are not hard to envision; however, efforts to quantify these often-complex interactions are more problematic. Research through the 1990s to the present has been extensive. Our understanding of these interactions has increased dramatically and generated some guidelines on managing the local landscape to increase services provided to agricultural production (Brewer and Goodell, 2012; Schellhorn *et al.*, 2015; Zhao *et al.*, 2016). The spatial scale of the local landscape in most of this research includes the area within about 1–2 km of the target field (Schellhorn *et al.*, 2015). Much of the research dealing with the local landscape focuses on natural enemy populations and their interactions with pests, but emphasis has increased on the value and needs of pollinators (Lundin *et al.*, 2021; see Chapters 11 and 12, this volume).

Case Study 1.1: Aphid management in seed potatoes

The economic sustainability of commercial potato production relies on disease-free seed potatoes, and state agencies test seed lots visually for disease within season and through post-season grow-outs of randomly selected tubers. The threshold for losing certification is very low, typically 0.5% infected plants. The greatest threat to certified seed production in the United States (US) is an ongoing pandemic of two potato viruses, potato leafroll virus (PLRV) and potato virus Y (PVY). Both can impact tuber quality and plant vigor, and both are transmitted by aphids (Stevenson *et al.*, 2001). The most efficient vector is *Myzus persicae* (Sulzer) (Hemiptera: Aphididae), the green peach aphid (MacGillivray, 1981) (Fig. 1.2).

PLRV is a persistent virus, with the vector remaining viruliferous for life. After being acquired by the vector, the virus must cross the gut lining and make its way to the salivary glands before it can be transmitted to an uninfected plant. The interval between virus acquisition and transmissibility (latency period) ranges from 6 to 24 h (Sugawara *et al.*, 1974). Consequently, the spread of PLRV can be managed with the application of standard insecticides (DiFonzo *et al.*, 1996; Ragsdale *et al.*, 2001).

Conversely, PVY is a non-persistent virus, wherein the vector does not remain viruliferous after transmitting the virus. PVY is vectored by more than 65 aphid species (Radcliffe and Ragsdale, 2002; Verbeek *et al.*, 2010; Pelletier *et al.*, 2012). Virions attach to the aphid's mouth parts during feeding on an infected plant, and within seconds to minutes the virus can be transferred to an uninfected plant, leaving the vector non-viruliferous (DiFonzo, 1996). This rapid transmission precludes management with standard insecticides. Weekly application of crop oils, though expensive, is effective in decreasing PVY spread (Suranyi, 2000). Insecticides targeting aphid probing and feeding can also manage spread of PVY, but they require multiple applications of various chemistries per season which tends to be expensive.

Overwintering of *M. persicae* is precluded in the Northern Great Plains by winter temperatures and lack of primary hosts. The species overwinters in more southerly states (US) and immigrates into the Northern Great Plains on low-level jet winds. Zhu *et al.* (2006) found that low-level jet wind events were correlated with virus disease severity. *Myzus persicae* immigrate into the region early in the growing season, but they likely first colonize host plants other than potato (Zhu *et al.*, 2006). Seasonal movement of *M. persicae* and other vector species into potato fields was monitored by a regional aphid trapping network, *Aphid Alert*, and this monitoring indicated that most aphid immigration into potato fields began in mid July and peaked in mid August (Suranyi *et al.*, 1999; Radcliffe *et al.*, 2008).

Researchers have observed an apparent edge effect in the establishment of winged aphids entering a field. Because of the amount of cultivation required in potato production, areas at the end of rows are kept bare to facilitate the turning of equipment. This results in a stark contrast of bare ground to a green canopy, attracting colonizing and non-colonizing aphid species near edges of potato fields (DiFonzo *et al.*, 1996; Suranyi, 2000) (Fig. 1.2). In addition, open areas as small as ~1 m diameter within fields bounded by canopy are attractive to aphids (Davis *et al.*, 2009; Davis *et al.*, 2015). The original observations led to planting non-PVY-host border crops as a PVY management tactic (DiFonzo, 1996; DiFonzo *et al.*, 1996). If the spatiotemporal parameters of the preferred colonization area at the edge of fields could be defined, recommendations for targeted application of border treatments could ease the economic impact of PVY management.

Large-scale field trials were conducted over three growing seasons (2000–2002) to establish the within-field distribution of *M. persicae* in seed potato fields (18–49 ha) in the Red River Valley of Minnesota and North Dakota, US (Carroll, 2005). *Myzus persicae* presence and density were determined in a uniform grid of ~300 sample sites with stratified edges per field. Density of *M. persicae* was recorded on each georeferenced (via GPS) sample plant twice weekly from mid July through August. Spatial analyses indicated significantly more alate and apterous aphids within 18.3 m of the field's edge and few if any away from the edge. Subsequent research (Carroll, 2005) indicated this edge effect deteriorated over time as aphids continued to colonize the field. But for 2 weeks following initial colonization, *M. persicae* remain concentrated within the 18-m edge, approximately the width of one pass of an aerial application of an insecticide.

In a proof-of-concept trial in 2003, efficacy of targeted border treatments to control aphid vectors of PVY was tested in 27 seed potato fields (all >8 ha) (Carroll *et al.*, 2009). Seasonal flight of aphid vector species was monitored by the *Aphid Alert* network (Radcliffe and Ragsdale, 2002), and fields were scouted weekly after initial detection. When a field was colonized by *M. persicae*, the grower cooperator was informed, and the field edge was sprayed within 1 day. Post-application counts were conducted within 3–7 days. Participating growers were interviewed after receiving seed certification test results.

Continued

Case Study 1.1: Continued.

Fig. 1.2. Potato field showing the high contrast between growing potatoes and field margins that provides attraction to immigrating *Myzus persicae*. Inset: winged and wingless forms of *M. persicae*. (Photo credits: I. MacRae; inset, J. Kalisch, University of Nebraska.)

Analyses indicated this technique controlled *M. persicae* colonization and population growth within the test fields, with costs of targeted applications significantly (~93%) lower than those of whole-field treatments (Carroll *et al.*, 2009). Seed lots from all participating growers passed summer disease inspection. Results (reported from growers) from post-season testing of the fields by seed certification agencies were mixed. Some reported better than average results, some no change, but two reported some fields were worse than expected. One of these two growers indicated the result may have been due to failure to continue management until vine kill. Importantly, however, all participating growers intended to continue using border treatments in the future.

Through publication and outreach, the findings of this research stimulated increased use of border treatments to manage aphid vectors of viral diseases in the potato growing areas of Minnesota and North Dakota. One of the participants in the original study continued border applications for well over 10 years until they discontinued seed production. The technique may be enhanced when combined with emerging scouting technologies.

Remote sensing of canopy reflectance has been demonstrated to identify aphid feeding damage within fields (e.g. Marston *et al.*, 2020). Open areas within fields, a point source for aphid colonization, are easily identified using unmanned aerial systems (UAS) (MacRae *et al.*, 2021), and successful identification of PVY within fields using remote sensing has been demonstrated (Griffel *et al.*, 2018; Polder *et al.*, 2019). The costs associated with obtaining remotely sensed data to identify spatiotemporal distributions are also becoming more economical. The use of UAS to obtain reflectance data provides the capability of obtaining immediate high-resolution data, but UAS have limited flight time along with regulatory requirements for their commercial use. The use of cameras for pest identification mounted on variable rate and targeted applicators is being actively developed, and some are now commercially available. However, with a spray width of a single nozzle of 0.5 m and an applicator speed of 5.6 m/s, a treatment pixel is likely to be measured in meters. This means these systems are likely compatible with high-resolution satellite data. The use of satellite imagery provides greater potential spatial coverage and has been shown to be as effective in identifying aphid damage as that obtained by UAS (Ribeiro *et al.*, 2024). Overall, the mapping of predictable vector populations has potential to increase the applicability of precision agriculture in seed potato production.

From this research Zhao *et al.* (2016) extracted four regulating mechanisms that connect landscape patterns and pest management impacts:

1. The spatial arrangement of plants in the landscape that provide food or shelter can impact natural enemy populations.
2. The temporal dynamics of the plants that provide diverse habitats in the landscape can meet seasonal requirements of beneficial arthropod populations.
3. Landscape connectivity (i.e. between non-crop and crop areas) can increase biocontrol services in the landscape.
4. A reduction in non-crop habitat can have a negative impact on natural enemy populations.

The effects of these factors on pest and natural enemy populations can be important, but are quite variable (Karp *et al.*, 2018). Haan *et al.* (2020) reinforced the influence of many of these characteristics on natural enemy abundance and pest suppression. However, they indicated that predicting effects will require better understanding of interactions among trophic levels and between arthropod dispersal capabilities and landscape spatial scale. Extensive efforts have been made in understanding these relationships in large-scale agroecosystems, and this is a focus of several chapters in this book (see Chapters 3–6 and 8–12, this volume). Karp *et al.* (2018) also indicated the potential for "disservices" from increased pest pressure arising from local landscapes. It is important to understand the potential contribution of neighboring crop and non-crop areas to pest risk. Case Study 1.2 demonstrates the importance of understanding dispersal capabilities of pests in risk assessment and management among local landscapes.

Regional Landscape Scale

Factors impacting pest, natural enemy, and pollinator dynamics usually occur on a continuum across spatial scales. Regional-scale research involves larger areas that could be encompassed in an areawide management effort or perhaps beyond (Koul *et al.*, 2008; Brewer and Elliott, 2023; see Chapter 2, this volume). The primary delimiter for this scale would be the commonality of crops, pests, and environmental conditions within an agroecosystem. Many factors causing local-scale effects will likely impact the larger regional scale as well. However, the increased distance from the managed crop will make specific interactions more difficult to document and to estimate their effects at the farm level. Importantly, the larger scale makes impacts more difficult for farmers/managers to envision, and thus, affects their decision-making process. Decision making is also affected by the increasing complexity of ownership and land management objectives across these larger spatial scales.

Research at the regional scale has predominately addressed the impact of crop and non-crop diversity on pest and natural enemy populations. The impact of regional landscape diversity on pest management has been documented in various systems, including cotton in Australia (Wilson *et al.*, 2018; see Chapter 2, this volume), sorghum aphid in sorghum (Brewer *et al.*, 2022; see Chapter 8, this volume), and cereal aphids in wheat (see Chapters 5 and 6, this volume). One benefit from a regional pest management effort lies in supplementing existing natural control capacity with regional deployment of IPM practices. Use of resistant varieties, crop rotation, planting dates, residue management, or biological control efforts (including classical releases) at the regional level can supplement and in some cases synergize natural controls such as pest suppression by natural enemies (Brewer and Goodell, 2012; Zhao *et al.*, 2016; see Chapter 2, this volume).

Pest and natural enemy movement at the regional scale are an important consideration in IPM programs (see Chapters 3, 4, 7, 9, 10, and 14, this volume). Flight capacity that can displace a pest well beyond the local landscape creates challenges in predicting movement patterns and dealing with its effects on populations. Pest dispersal capacity may interact with geographic location of the agroecosystem. As an example, the army cutworm, *Euxoa auxiliaris* (Grote) (Lepidoptera: Noctuidae), is an important pest that can affect regrowth or establishment of early season crops (e.g. alfalfa, wheat, sugarbeet) across the Western Great Plains of North America. This insect is tied geographically to the Rocky Mountains where it migrates to spend the summer at cooler temperatures, feeding on flowering plants at higher elevations (Burton *et al.*, 1980). In the fall, moths return to the grasslands and crops of the Western Great Plains to lay eggs

Case Study 1.2: Wheat curl mite dispersal and local disease spread

The wheat curl mite, *Aceria tosichella* (Keifer) (Acariformes: Eriophyidae), is a serious pest of wheat throughout the Western Great Plains of the US, and it is an invasive pest that has spread to several other areas in the world (Skoracka *et al.*, 2014). Direct damage from the mite does occur, but the mite's major impact results from its ability to transmit four viruses to wheat worldwide. Three of these viruses (wheat streak mosaic virus, High Plains wheat mosaic virus, and Triticum mosaic virus) form a virus complex that causes serious yield loss in wheat across the Great Plains (Hein *et al.*, 2023a, b). All three viruses are common throughout the Great Plains, but wheat streak mosaic virus was the first one identified (in the 1950s), and it is the most prevalent of the three (Burrows *et al.*, 2009; Byamukama *et al.*, 2013). Annual yield loss resulting from this mite–virus disease complex for Kansas alone has been estimated at 1.5%, amounting to about US$20 million yearly (Hollandbeck *et al.*, 2021). However, the most serious impact from the wheat streak mosaic disease complex is often seen with local losses reaching 100% in epidemic areas (Hein *et al.*, 2023a).

As with most arthropod-transmitted viral diseases, management is tied very closely to the biology and ecology of the vector. The wheat curl mite is tiny (~250 mm) (Fig. 1.3), and it has a very short life cycle (7–10 days depending on temperature) resulting in a very high reproductive rate. Mite dispersal capabilities are critical to virus spread. The mite can disperse via walking, but due to its size its walking range is very limited (i.e. plant-to-plant movement only when plants are touching); therefore, it relies primarily on wind for significant spatial dispersal across the landscape (Majer *et al.*, 2021). This wind-borne dispersal is passive as the mite possesses no means of propelling itself in the air.

Fig. 1.3. Extensive volunteer wheat growing through hail-damaged wheat stubble and at extreme risk of being infested with wheat curl mites and infected with viruses. (A) Mite adults and egg. (B) Volunteer wheat germinating from seeds shelled out by hail. (Photo credits: G. Hein.)

Continued

Case Study 1.2: Continued.

Mite populations through the year are variable. In normal (healthy) winter wheat through the fall, winter, and into the spring, mite populations will be present at very low levels. However, through the spring the high reproduction rate of the mite enables mite populations to build to their highest levels of the year just before harvest, easily reaching 1–3 billion mites/ha (Byamukama *et al.*, 2015; McMechan and Hein, 2017). These populations do possess the ability to transmit viruses, thus virus inoculum is present every year at this time (Byamukama *et al.*, 2015). Mites will survive less than 24 h at summer temperatures and humidities if they are not on a living host (Wosula *et al.*, 2015); therefore, mite dispersal shows an extensive peak as mites seek alternate hosts. In western Nebraska, this peak has been detected in sampling up to 30 km away from the nearest wheat field (G.L. Hein, unpublished data). After harvest, dispersal activity drops off dramatically.

During this preharvest dispersal event in the Great Plains where wheat production is prevalent, mites are widespread and move onto living summer hosts that allow them to bridge the summer period (i.e. the "green bridge"). There have been around 90 grass species listed as hosts for the mite, but most of these are of little economic concern due to low prevalence in the field, limited mite reproductive capacity, asynchrony of seasonal occurrence with the green bridge, and inability to host viruses (McMechan, 2016). Grasses present through the summer serve as hosts, with the most common being volunteer wheat, corn, and a few other forage and weed grasses. However, volunteer wheat when present is the optimal host for mites and viruses to survive this summer bridge to the fall wheat crop. Thus, preharvest volunteer wheat presents the greatest risk for development of this serious mite–virus disease complex. Preharvest volunteer wheat can be present as a weed in summer crops, but the most problematic source for this volunteer results when serious hail occurs within the 3-week period before wheat harvest (Staples and Allington, 1956; McMechan, 2016; Chen *et al.*, 2024). When damaged by hail during this window of risk, wheat seeds are shelled to the ground and can quickly germinate to establish volunteer and become infested by mites carrying virus (Fig. 1.3).

The presence of a green bridge for mites and viruses is the primary risk determinant for this disease complex. Thus, the primary focus of management of this disease complex is control of green bridge hosts, primarily volunteer wheat (Thomas *et al.*, 2004; Hein *et al.*, 2023a). If the green bridge is broken for even a short period after wheat harvest, minimal mite dispersal dramatically limits the potential for significant mite population establishment and buildup. Other management options in winter wheat contribute to reducing risk, including avoiding early planting and using varieties with increased tolerance. While these actions help, they are not adequate to avoid serious damage under epidemic conditions (Hein *et al.*, 2023a, b).

In the fall, mites disperse from green bridge hosts into newly planted winter wheat fields, resulting in virus spread. Early fall infection results in the most serious disease impact from this complex (Wosula *et al.*, 2018), but severe symptoms seldom are expressed until spring when warm temperatures return. Therefore, the spatial occurrence of fields with volunteer wheat serving as a mite–virus source will determine disease risk and virus spread in newly planted wheat fields.

Stilwell *et al.* (2013) determined that remote sensing via the reflectance in the red edge region of the spectrum works well at detecting virus symptom severity, and thus the extent of virus spread the following spring. Using this approach, Stilwell *et al.* (2019) mapped the spatial distribution of symptoms in the spring as a proxy for the spatial dispersal of mites from a volunteer wheat source area the previous fall. It was found that mites moved throughout the fall, spreading in all directions driven by varying wind direction occurring through the fall; however, greater movement occurred to the southeast correlating with the direction of the strongest winds (>9 m/s) coming from the northwest. In western Nebraska, this overall fall dispersal pattern was oval shaped with the longest axis extending from the northwest to southeast (Fig. 1.4). The spatial scale of this study was limited, but dispersal parameters were used to estimate a potential dispersal distance (i.e. significant virus spread) of up to 3 km from larger mite–virus source fields. They also found that the distance of mite dispersal and virus spread was related to the density of mites in the source population. This density dependence results in often dramatic dispersal gradients extending away from the source fields (Hein *et al.*, 2023a; Chen *et al.*, 2024) (Fig. 1.4). Further characterization and validation of the mites' dispersal potential are necessary to improve risk prediction.

In instances of optimally timed hailstorms (preharvest) with extensive spatial coverage, the risk of a widespread epidemic is exacerbated. In these extensive hail streaks, there are likely numerous fields with considerable preharvest volunteer wheat scattered through the landscape. If volunteer wheat in these fields is not adequately controlled, the increased density of mites dispersing can threaten wheat fields across the local landscape. Chen *et al.* (2024) demonstrated the extensive potential of community-wide

Continued

Case Study 1.2: Continued.

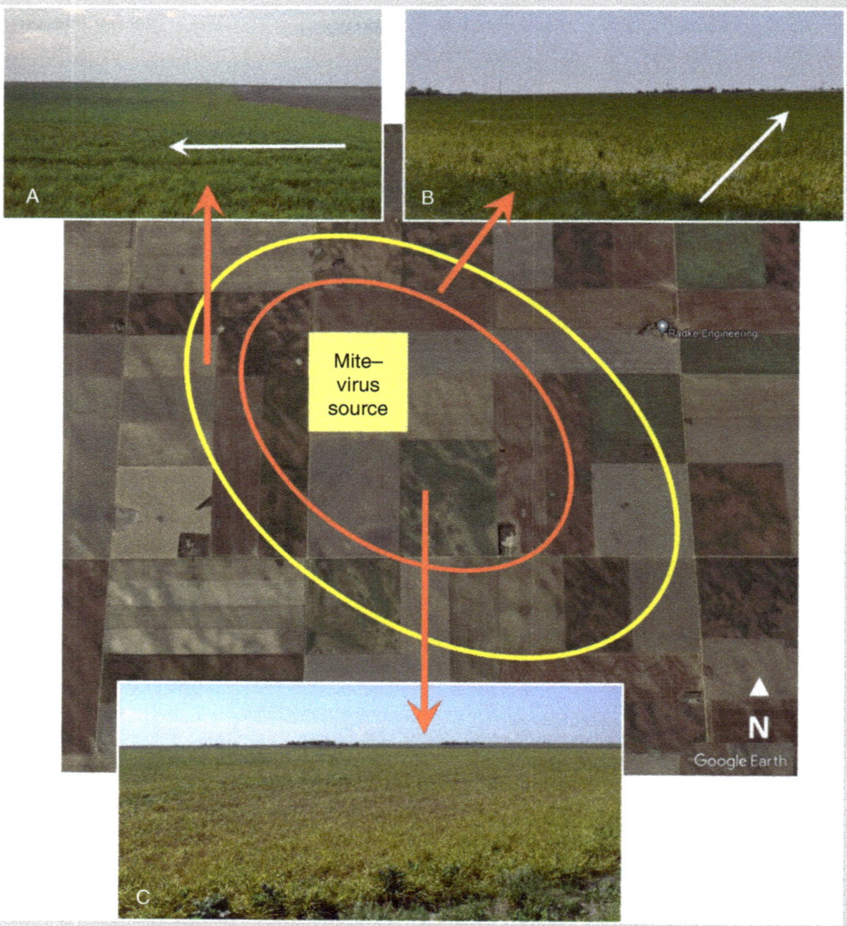

Fig. 1.4. View of local landscape surrounding an extensive mite–virus source field (yellow box) showing high-risk (red oval) and medium-risk (yellow oval) zones extending greater distances to the southeast due to prevailing strong northwest winds through the fall. Field photos show increasing disease severity likely at three sites: (A) mild severity gradient, (B) more severe gradient, and (C) complete field-wide loss. (Photo credits: G. Hein; landscape image: Google Earth.) (Original artwork by G. Hein.)

risk within the previous year's hailed area, but they also documented that serious risk also extended several miles outside of the immediate hailed area. This results from a concentration of inadequately controlled preharvest volunteer source fields and increased mite activity throughout the local landscape during the summer and fall following the hail. In these situations, serious epidemics will only be avoided with effective community-wide efforts aimed at targeting preharvest volunteer control.

In the Central Great Plains, the relationship of hail occurring within the 3-week window of risk just prior to harvest resulting in preharvest volunteer is by far the greatest risk of serious disease. Because of the extended oversummering period further south in the Great Plains, disease risk from postharvest volunteer wheat is likely to increase (Hein *et al.*, 2023a). Further research is needed to be able to model mite buildup in pre- and postharvest volunteer wheat to establish risk levels more precisely across the Great Plains.

Continued

Case Study 1.2: Continued.

Mite dispersal is the key to understanding the spread of this disease complex, and controlling volunteer wheat is central to management. However, in many instances optimal management is not within the farmer's control. This identifies the importance of "being a good neighbor" to reduce disease risk. Additionally, as the risk level increases with extended hail area coverage, the importance of community-wide management increases. Nguy-Robertson *et al.* (2016) used radar and model-based hail estimates along with satellite imagery to map these high-risk areas and establish a risk warning system. Chen *et al.* (2024) extended these ideas using similar methodology, but with more advanced and accurate hail estimation techniques, to connect hail events to subsequent virus occurrence. Advances in spatial modeling of these hail events along with mite population modeling in volunteer wheat and other potential green bridge hosts are needed to establish risk more effectively from this disease complex across the Great Plains.

that hatch, overwinter as larvae, and in some years produce dramatic larval densities in localized areas. The spatial distribution of fall immigrants is difficult to predict, but they are often concentrated in areas within the region, making the risk of these invasions an important consideration in management of vulnerable crops the following spring. Regional dispersal can impact risk assessment and pest management response at both local and regional levels. Boll weevil eradication efforts in the US discussed in Case Study 1.3 illustrates the value of understanding pest dispersal.

The boll weevil case study demonstrates the integral connection between local and regional landscapes and the need to understand a pest's ecological interactions at various spatial scales for effective management. Additionally, local and regional dispersal impacts gene flow, which in turn impacts development of pesticide resistance, making the scale of pest movement a critical consideration in developing resistance management strategies (see Chapters 9, 10, and 14, this volume). Eradication is not a practical goal for most pests, but the principle is the same: understanding a pest's population and movement dynamics (and preferably that of their natural enemies) is fundamental to effective long-term management. It is important to understand how these dynamics operating at larger scales will impact ecological processes across the various scales, affect local to regional farm economics, and inform management decisions.

Continental Scale

Long-range insect migration on a continental scale has long been recognized (Williams, 1957; Pedgley, 1993). Well-known examples include the Monarch butterfly, *Danaus plexippus* (L.) (Lepidoptera: Nymphalidae), and several species of migratory locusts, but migration contributes significantly to the dynamics of numerous other insects as well (Chapman *et al.*, 2015; Satterfield *et al.*, 2020). As defined earlier, migration is movement that is undistracted by normally acceptable resources (e.g. food), non-meandering, and persists to enable relocation well beyond the typical foraging range (Dingle and Drake, 2007). Migration can be classified into two categories reflecting different ecological outcomes. The first is non-directed migration, which is one-way because there is no spatial goal. Permanent range expansion may result if the migrating insect leaves its normal range and environmental conditions at the terminus are adequate. More often, however, host or climatic conditions encountered outside its normal range limit year-round survival. Instances of repeated long-distance one-way migration are usually found, on closer examination, to include a later return migration, but there may be exceptions. An example of this type of migration by a serious pest is seen in the northward movement of cereal aphids across the North American Great Plains (see Chapter 7, this volume).

The second category of long-range migration involves seasonal movement across latitudes and the subsequent return movement at a later season. Many insects, including several pest species, seasonally migrate long distances between overwintering and summer breeding regions, but short lifespans generally necessitate multiple generations to complete the geographic circuit. Predictions of location and timing of an influx of immigrant pests is highly desirable at the field level, but predictive accuracy becomes more challenging at that scale. Regional and local monitoring networks help to address these

Case Study 1.3: Boll weevil dispersal attributes critical to eradication efforts

The boll weevil, *Anthonomus grandis grandis* (Boheman) (Coleoptera: Curculionidae), historically has been one of the worst pests of cotton and row-crop agriculture in the US (Lange *et al.*, 2009). It entered South Texas from Mexico in 1892 through natural range expansion, spreading to the Atlantic Coast by 1916 (Burke *et al.*, 1986). A national boll weevil eradication program initiated in the late 1970s progressively purged it from east to west through the US Cotton Belt, and from cotton-growing areas in Arizona and California (Smith, 1998). This program was largely successful, with only a few areas of subtropical Texas still infested with very low numbers (Allen, 2008; Raszick, 2021).

Ideally, eradication would have proceeded spatially along a continuous front. However, eradication zones of multiple counties were established by referendum of cotton growers, and not all farmers were supportive due to the imposed assessment fees. Continued education efforts (and occasional reconfiguration of zone boundaries) were necessary to bring all areas into eventual participation. The result was a spatial mosaic of zones entering the eradication program in different years. This was a problem because as zones neared eradication, they were often at risk of reinfestation along multiple boundaries from neighboring infested zones. After eradication, a zone entered a maintenance phase of monitoring for boll weevil reintroductions using pheromone-baited traps. Reintroductions can occur by flight or by human-mediated transport (e.g. via farm equipment). Response to a reintroduction is swift and expensive, with the aim to prevent re-establishment (Allen, 2008; Raszick, 2021). Better understanding of boll weevil dispersal distances and spatial patterns thus became an urgent research imperative.

Most insect flight activity is appetitive, where an individual is seeking a resource, resulting in displacement by diffusion through the local landscape. Following the terminology of Dingle and Drake (2007), routine activities such as local (i.e. home range) foraging for food, a suitable oviposition site, a mate, or shelter is called *station keeping*. Appetitive flight that results in movement beyond the current home range is *ranging*. For example, a ranging insect may be searching for the source of a pheromone plume, a better food source, or overwintering habitat. Appetitive flight is generally meandering, takes place near ground level where low wind speeds allow directional control, and is terminated when the resource is encountered.

Some insect species also engage in *migratory* flight, which is non-appetitive, non-meandering, and undistracted by otherwise acceptable resources (Dingle and Drake, 2007). Migration often occurs at altitudes above the flight boundary layer where transport in fast-flowing winds results in spread by leaps over longer distances. Although not usually considered a migratory species, evidence strongly indicates boll weevil migration regularly occurs. Boll weevils were captured in airplane tow nets 100–600 m above ground level (Glick, 1939; Rummel *et al.*, 1977), indicative of migratory activity. Wind trajectory analyses showed migratory flight is wind-aided at high altitude (Kim *et al.*, 2010; Westbrook *et al.*, 2011). Long-distance movement is suggested by rates of range expansion averaging 95 km/year (maximum 235 km/year) (Culin *et al.*, 1990); captures in traps 160–240 km from the nearest likely source (Jones *et al.*, 1992; Lukefahr *et al.*, 1994; Spurgeon *et al.*, 1997); recapture of marked weevils 72–270 km from the release site (Johnson *et al.*, 1976; Guerra, 1988); historical gene flow between populations separated by more than 600 km (Kim and Sappington, 2013; Alvarado *et al.*, 2017; Raszick *et al.*, 2021); and detection of immigrants 640–740 km from the most likely source using genetic population assignment analyses (Kim and Sappington, 2006, 2013).

The spatial patchwork of zones at different stages of eradication made movement at all spatial scales a concern for program managers. Ranging and even station keeping flight could bring boll weevils from an infested zone a short distance across a boundary into an advanced eradication zone. Human-mediated transport over short distances was just as great a concern. Boll weevils can survive in harvested cotton lint packed into modules constructed for transport to the cotton gin (Sappington *et al.*, 2004a), and this was widely seen as a threat for weevil reintroductions. A series of studies showed this threat was minimal. Mark–release experiments demonstrated weevils infesting cotton at harvest could potentially disperse from the module surface only under limited environmental conditions (Sappington *et al.*, 2006). After transport, weevils could not survive the different ginning processes (Sappington *et al.*, 2004b).

Boll weevil reintroductions must be detected by pheromone traps as soon as possible to avoid successful establishment. Consequently, there has been considerable interest in maximizing trap

Continued

Case Study 1.3: Continued.

sensitivity by optimal placement in the landscape. Boll weevil movement can be influenced by several environmental and topographical features, thus affecting weevil capture. For example, boll weevils can control their flight direction only in winds up to 5–7 km/h (Hardee *et al.*, 1969; McKibben *et al.*, 1991), and placing traps on the lee side (relative to prevailing winds) of brush lines increased captures about 3-fold (Fig. 1.5) (Sappington and Spurgeon, 2000). This effect is greatest when traps are placed adjacent to prominent brush (Spurgeon and Raulston, 2006; Spurgeon, 2016). Using a GIS-based analysis strategy, Brewer *et al.* (2020) found that boll weevil captures in Texas' Lower Rio Grande Valley eradication zone increased with decreasing distance to landscape features such as water bodies, the Rio Grande River, and watermelon fields over a scale of 0–10 km.

Long-distance reintroductions of boll weevils to eradication zones are still occasionally detected (Allen, 2008; Raszick, 2021). Early-on the immigrants were often assumed to have been transported by contaminated equipment, and quarantines were considered the best solution. Although it was long known that major weather systems such as tropical storms and hurricanes could carry boll weevils long distances (Hinds, 1916; Culin *et al.*, 1990), the role of migratory flight in long-distance reintroductions to eradication zones was underappreciated. Molecular markers were developed and surveyed to characterize genetic variation and gene flow among 18 boll weevil populations across eight states and northeastern Mexico, revealing regular dispersal distances over 200 km (Kim and Sappington, 2006, 2013; Choi *et al.*, 2011) (Fig. 1.6). In addition, this DNA database allowed assignment of boll weevils captured in eradication zones in the US and Mexico to probable source regions (Kim *et al.*, 2006, 2008, 2010), thus enabling appropriate remedial responses. For example, demonstrating that natural migration commonly occurs from the south into the next eradication zones proceeding northward, Kim and Sappington (2006) persuaded the Texas Department of Agriculture not to impose stricter quarantine measures on transporting cotton equipment outside the Lower Rio Grande Valley, averting an economic hardship within this region. For example, a large influx of boll weevils was detected in the Southern Rolling Plains eradication zone in Texas in 2007. Insect samples were analyzed with a combination of genetic markers, profiles of pollen species on the weevils, and back-track wind trajectories, narrowing down the probable source of immigrants due to storm-aided movement from Tropical Storm Erin (Kim *et al.*, 2010). These tools and advances in understanding of boll weevil dispersal in local and regional landscapes were essential to successful eradication throughout US cotton-growing regions.

Fig. 1.5. Boll weevil pheromone traps (inset) placed along lee side of tree and brush lines to maximize potential for weevil detection. (Photo credits: T. Sappington)

Continued

Case Study 1.3: Continued.

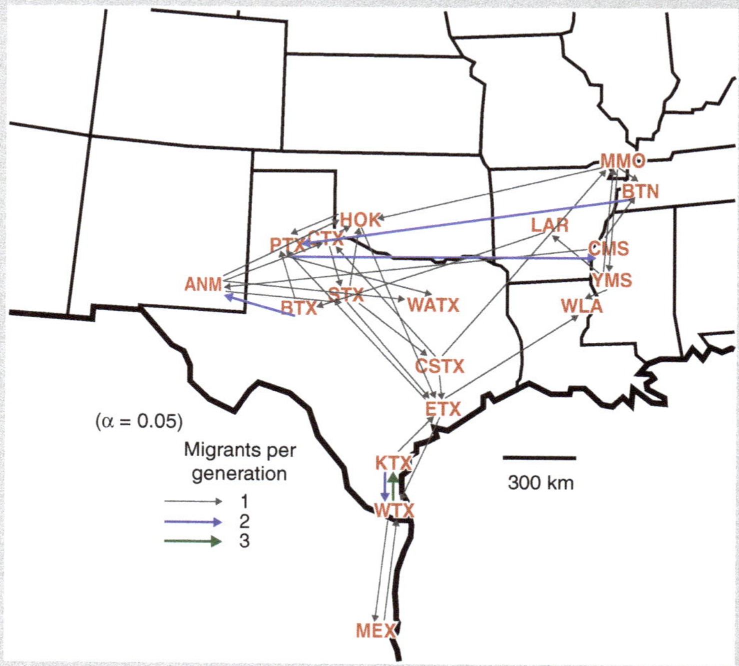

Fig. 1.6. Boll weevil dispersal in cotton-growing areas of the southern United States, visualized by genetic assignment analysis (Kim *et al.*, 2006), dramatically affected reintroductions to eradication zones throughout the region. (Original artwork by T. Sappington.)

challenges (see Chapters 10, and 14, this volume). In addition, seasonal migration impacts resistance management strategies at all scales by mixing populations over continental-scale distances, affecting gene flow and thus development and spread of insecticide resistance. Either category of long-range migration can be difficult to verify because of the difficulty of tracking long-range movement by these small organisms. However, the distinction between one-way and two-way migration is important because resistance management in distant areas may impact local management options, but if one-way migration frequently takes individuals out of their normal range, the alleles they carry will be removed from the gene pool.

Case Study 1.4 illustrates the importance of understanding continental-scale interactions for long-distance migrants. Risk assessment for these pests necessitates some connection with the insect's movement patterns. The greatest risk from black cutworm comes in the spring when the potential for stand loss is greatest. Effects at the continental scale may be difficult to identify at the individual field level; however, immigration of pests into a region does elevate the risk of pest impact at the field level. This elevated risk may spur further management action, such as sampling at the regional level or even the local landscape level, to establish risk more specifically. Understanding the factors that drive a pest's long-distance movement is critical to establishing an effective risk assessment process.

A very important factor in determining spatial and temporal aspects of insect migration is the interaction of insect movement with meteorological factors (Isard and Gage, 2001). Many insects are particularly good at taking advantage of wind to assist in long-range movement (Chapman *et al.*, 2015). Better understanding of

Case Study 1.4: Black cutworm migration triggering risk assessment

The black cutworm, *Agrotis ipsilon* (Lepidoptera: Noctuidae), is a serious, but sporadic pest of several crops in North America. As a seedling pest of corn, this species provides a particularly good example of the interaction of continental-scale migratory movement with local landscape-scale effects on risk assessment and management. This species cannot overwinter north of about 38°N latitude. Instead, the Corn Belt and Great Lakes states of the US are recolonized each spring by moths flying hundreds of kilometers northward from overwintering habitat in the Gulf Coast states and Mexico (Showers *et al.*, 1989). Migrating moths ascend at dusk into the low-level jet in advance of cold fronts for rapid transport northward (Showers, 1997; Chapman *et al.*, 2015) (Fig. 1.7).

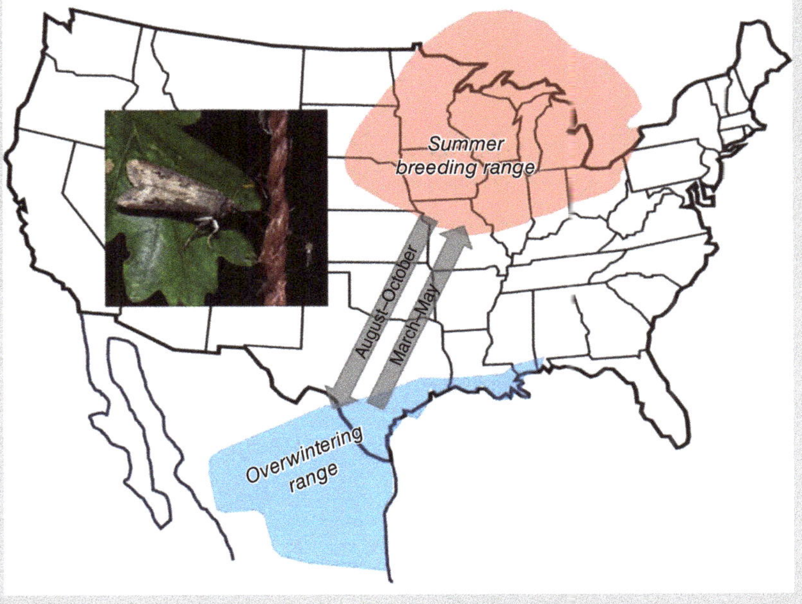

Fig. 1.7. Multi-generation, multi-night migration of black cutworm between its overwintering range near the Gulf Coast and its summer breeding range in the northcentral Corn Belt (Showers, 1997). Inset: migrating adult refueling at honey/beer-soaked bait station. (Original artwork by T. Sappington.)

Larvae are the damaging stage, growing through six or seven instars. Young larvae (instars 1–3) feed only on leaves and cause little injury to corn. Later instars attack the stalk near or below the soil surface, often cutting a young plant completely (Fig. 1.8). Corn is susceptible to cutting from seedling stage through the V7 growth stage (Abendroth *et al.*, 2011). A black cutworm larva can kill several plants before pupation depending on how long it is in a cutting stage of development while the corn plant is at V6 stage or less. Yield loss resulting from stand reduction may be great enough to warrant replanting part or all of a field, making infestations very costly to the grower (Sappington *et al.*, 2018).

The first influx of black cutworm moths into US Corn Belt states often occurs in early spring before corn planting. Most females arrive already mated with developing eggs (Sappington and Showers, 1992; Showers, 1997), allowing rapid oviposition in the new landscape. It is a polyphagous species; females prefer to oviposit on weed seedlings and even on plant debris in the absence of green vegetation (Busching and Turpin, 1976; Sherrod *et al.*, 1979). Black cutworm infestations are more common in low-lying areas, flood plains, cloddy fields, rotated corn following soybean, and reduced tillage fields, but such patterns may be related to weed control issues. Early season weed growth is more preferred for oviposition than corn, but larvae will feed on corn seedlings after weeds have been destroyed by herbicide or tillage (Busching and Turpin, 1976; Engelken *et al.*, 1990) and larvae can survive at least

Continued

Case Study 1.4: Continued.

Fig. 1.8. Two-leaf corn plant cut off by black cutworm larvae and showing prior leaf feeding damage. Insets: black cutworm moth (A) and larva (B). (Photo credits: J. Kalisch, University of Nebraska.)

10 days in plant debris (Showers *et al.*, 1985). Thus, by the time corn seedlings emerge most black cutworm larvae are already present and partly grown (Sherrod *et al.*, 1979).

Because of black cutworm reliance on frequent but transient weather transport systems to accomplish long-distance northward displacement during spring migration, the temporal and spatial risk of infestation varies considerably between and within years (e.g. Bailey, 2004). The size of populations arriving in an area after a favorable transport event also depends on population dynamics, demographics, and rate of emigration in the southern source areas, adding further variability to local threats in the north. Despite occasional widespread outbreaks, few cornfields experience an economic infestation of black cutworm in most years (Sappington *et al.*, 2018). But every year some fields in some areas inevitably suffer severe damage, leaving a lasting impression on the farmers affected. Low or moderate infestations may be suppressed by neonicotinoid seed treatments, at-planting soil insecticides, or certain transgenic *Bacillus thuringiensis* (Bt) hybrids; however, the best protection against losses from heavy infestations of black cutworm is scouting followed by a rescue treatment with a foliar insecticide if needed (Kullik *et al.*, 2011; Ostlie and Potter, 2021; Seiter, 2021).

Determining the need for a rescue insecticide treatment requires timely scouting and evaluation of several factors. Ideally, scouting seedling corn should begin in time to detect leaf feeding by early-instar cutworms and before plant cutting by fourth instars begins. Recommended thresholds for rescue treatment are based on per cent plants with leaf feeding by pre-cutting instars and/or per cent plants killed by cutting instars (e.g. 6–8% plants with leaf feeding or 2–4% cut plants and larvae under 2.4 cm long; Wright *et al.*, 2013). When damage is found, it is important to identify the larvae responsible and the range of larval instars. Multiple black cutworm immigration events can occur throughout the spring, so the range of instars present and their damage potential can be narrow or broad. Other factors that can influence a decision to use a rescue insecticide treatment include plant growth stage, plant population, soil moisture, weed status and control timing, insecticide cost and projected efficacy, and commodity price (Showers *et al.*, 1983; Levine *et al.*, 1983; Bailey, 2004; Ostlie and Potter, 2021).

Scouting of vulnerable fields can be expensive, so risk assessment methods prioritize the need to initiate scouting. Pheromone trapping of males is used to establish arrival time by a substantial influx of immigrants to an area (Troester *et al.*, 1983; Smelser *et al.*, 1991). The date a "significant"

Continued

Case Study 1.4: Continued.

threshold of moths per unit time (often 8–9 moths per 2-night period) is caught provides a biofix to initiate a simple degree-day model to predict dates of expected first and final cutting of corn (Bailey, 2004).

Farmers can deploy their own pheromone traps, but universities and government agencies in several states coordinate pheromone trapping networks to provide early warning of significant influxes of black cutworm adults (i.e. risk) and optimal timing to start scouting in their area (e.g. Ostlie and Potter, 2021; DATCP, 2023; Dean and Hodgson, 2023; Baute, 2024). Some provide a website where a user can enter a biofix date for their location to generate a forecast of predicted larval development times (e.g. Bailey, 2004). The Corn ipmPIPE (2019) website provides time-lapse heatmaps of black cutworm trap captures across four states and an interactive map of projected dates of first cutting of plants by larvae in counties with significant catches. The risk management strategy outlined for black cutworm is quite effective, but its foundation is maintenance of a robust volunteer trapping network during the migratory period.

Looking into the future, the changing climate may impact risk assessment. Showers (1997) hypothesized that the springtime cue to emigrate northward may be temperature rather than photoperiod. Although this question remains unresolved (Gemeno and Haynes, 2001), a detailed analysis of black cutworm phenology, meteorological data, and spring migration patterns from 1991 to 2017 in China shows a progressive acceleration of development of the overwintering generation in the south, but not a concomitant acceleration of arrival of immigrants in the north (Zeng *et al.*, 2020). This lack of immediate facultative tracking of ambient temperatures with time of northward migration is consistent with a photoperiodic cue for spring emigration. Earlier emigration under control of photoperiod may evolve, but on the slower timescale required for response to selection. Though speculative, the observed decoupling of accelerated timing of development of overwintering black cutworm with static timing of photoperiod-controlled emigration may help explain the steady decline in this species' overall abundance in China since 2003 (Guo *et al.*, 2020). As already documented in China (Zeng *et al.*, 2020), the geographic limits of overwintering in North America may be shifting northward with warming temperatures, potentially changing the dynamics of risk in those areas, as well as the geography of early springtime colonization in the north. This underscores the importance that spatial influences resulting from migration will continue to have on risk assessment for black cutworm management.

the interactions of insects and these meteorological drivers is essential to enhancing the ability to predict increased risk resulting from these movements (see Chapters 7, 10, and 14, this volume).

The immediate benefit of understanding the factors impacting a pest's migration is in predicting elevated risk of infestation in local situations, but migratory behavior must also be accounted for when modeling resistance evolution (see Chapters 10 and 14, this volume). Thus, there is need for improved predictive power for both spatial and temporal dimensions of movement and infestation risk. Numerous advancements in technology have expanded and refined the tools available for probing this scale of movement (Schellhorn *et al.*, 2014), including telemetry (Kissling *et al.*, 2014; Fisher *et al.*, 2021), radar (Dwivedi *et al.*, 2020; Noskov *et al.*, 2021), and isotope analysis (Quinby *et al.*, 2020; Hobson, 2023). Additionally, advancements in genetic techniques provide powerful means to address migration source and return issues (see Chapter 14, this volume).

IPM Across all Landscape Scales

IPM implementation at the farm level

IPM programs addressing insect pests important in the production system must always be focused at the field or farm level, while considering influences and impacts from across all landscape scales. Farm managers continually make decisions that by necessity incorporate IPM strategies into their integrated crop management plans. IPM decisions must be made intentionally, but they are often made by default because of a mismatch with crop management plans. Successful IPM programs should incorporate interactions across all landscape scales and consider habitat management tools where possible to enhance agroecosystem services. Development of technology enabling site-specific management has broadened IPM options for those pest species where it is appropriate and available (see Chapter 13, this volume). Farmers need to understand the benefits of site-specific management to local biological control, resistance management, and environmental protection. However, in many instances the details of technical application, ecological interactions, and economic benefits must be better elucidated for these to become viable options.

In their IPM programs, farmers also should consider their connection to the local landscape and the beneficial services that can accrue. A farmer's ability to manage across the local landscape may be limited by lack of ownership, but where they have control, they ought to factor in how management of these areas can provide beneficial services to support their IPM program. Local landscape research has often made a connection between habitat diversity and increased natural enemy or pollinator abundance (Schellhorn *et al.*, 2015; Lundin *et al.*, 2021). However, the connection to pest populations, their management, and plant productivity is often unclear or highly variable; therefore, it is particularly important to expand this work to provide a more accurate measure of the impact of landscape effects on pest damage and economics (Schellhorn *et al.*, 2015; Brewer and Elliott, 2023). Haan *et al.* (2020) suggest that research into landscape services identify direct and indirect factors affecting pest and natural enemy populations. To identify these, research must expand beyond observational studies to include experimental manipulation and modeling. Because of the scale involved, these efforts will be challenging. For landscape management efforts to be effectively implemented, variability in the extent and timing of benefits must be better understood to allow their prediction. Additionally, an honest appraisal of potential environmental disservices (e.g. increases in pest density) must be included (Karp *et al.*, 2018; see Chapter 2, this volume).

Farm managers may have little control over the regional- and continental-scale factors that impact their operations, but it is important they consider the contributions to their systems (e.g. ecosystem services, pest risk prediction and assessment) that proceed across these scales. It is also important they understand their potential contribution to these regional agroecological services (e.g. resistance management, residue management, crop rotation) and plan their management decisions accordingly.

IPM implementation at the broader spatial scales

Increased understanding of local and regional landscape factors that impact pest, natural enemy, and pollinator populations in large-scale agroecosystems has reinforced the need to develop regional pest management efforts to take advantage of these larger agroecosystem services (see Chapters 2, 10, and 12, this volume). Many of these benefits have been documented through areawide IPM programs (e.g. Koul *et al.*, 2008; Schellhorn *et al.*, 2015; Chaplin-Kramer *et al.*, 2019; Brewer and Elliott, 2023).

Because of the complexity of interactions that occur across larger landscapes, modeling will be an important tool for predicting environmental services. However, useful models will require an expansion of the knowledge base on landscape interactions with pests and natural enemies across all scales. It is also important that ecologists account for risk management factors considered by growers in evaluating the co-benefits of multi-functional landscapes (Chaplin-Kramer *et al.*, 2019). Studies on interactions at local and regional scales need to include direct damage and loss estimates to provide the grower with stronger incentives for positive management decisions (Schellhorn *et al.*, 2015; Karp *et al.*, 2018).

Regional pest management efforts present significant challenges in implementation. Encouraging positive involvement by farmers and other land managers in broader agroecosystem objectives will be central to mitigating Hardin's "tragedy of the commons" (Feeny *et al.*, 1990). In this scenario the overall effect of reasonable, often economic-driven decisions leads to resource overexploitation, or in this case, a lack of contribution to positive regional pest management services. A significant factor in successful implementation of practices promoting broader landscape benefits lies with individuals making decisions beneficial to the overall agroecosystem in addition to their immediate economic gain. Educational efforts to explain longer-term and broader agroecosystem benefits are certainly central to any advancement of these ideas. To maximize buy-in, short- and long-term benefits from management actions must be quantified and clearly demonstrated, along with establishing the factors that impact the variability of those benefits.

In some instances, social and economic incentives will be necessary to maximize implementation and benefits of broader agroecosystem efforts (Brewer and Goodell, 2012; Zhao *et al.*, 2016; see Chapters 2 and 12, this volume). Voluntary management decisions by the farmer may benefit the overall community, but these decisions are not assured. Economic policy incentives could establish payment for adherence to certain beneficial practices as seen in the US, European Union, Australia, and selected other countries (e.g. USDA conservation programs) (Brewer *et al.*, 2009; Brewer and Goodell, 2012). However, some practices that benefit the entire agroecosystem community could be made mandatory, as in the boll weevil eradication program in Case Study 1.3 (Allen, 2008). Any incentive program will require clear demonstration of significant economic, social, and environmental benefit to the individuals and communities involved.

Addressing the social challenges involved in this process will require coordinated community-wide decision making to build socio-economic and behavioral science understanding to enable/enhance implementation (Brewer and Goodell, 2012; Chaplin-Kramer *et al.*, 2019; see also Chapter 4, this volume, for examples of these issues applied to geographically widespread small plantings). Community-wide decisions will require sensitivity to community dynamics and the farmer's interactions with those dynamics. Effective decisions will require knowledgeable local leadership and community-wide involvement and buy-in. Community dynamics are likely to be variable and complex, so significant effort needs to go into understanding those dynamics and identifying potential incentives important within those communities. This will require a diverse team of IPM and agronomic experts. One component important to these collaborations that is often not present or may be excluded in the process is the consultant or "trusted advisor." This advisor, through establishing a long-term relationship and trust with the farmer, is best able to match the scientific knowledge of the system to the needs of the farmer (Briese, 2019). To maximize success, a final and critical component of this team will be significant involvement by social scientists and economists. A recent successful model of managing herbicide-resistant weeds illustrates the importance and value of participation by social scientists in helping to drive community-wide support for and implementation of best management practices (Dentzman, 2023).

Summary and Conclusions

The spatial scale of landscape research addressing IPM in large-scale agroecosystems is a primary consideration throughout this book. New knowledge from this research impacts implementation of IPM programs across four different spatial scales: (i) intrafield, (ii) local landscape, (iii) regional, and (iv) continental scales. Understanding the variability of pest and beneficial insect populations and their interactions across these agroecological landscape scales can expand management options to encompass wider interactions and impacts from mobile pests, natural enemies, and pollinators.

IPM programs are generally applied at the field level, and data from clear and predictable intrafield spatial distributions of pest insects have enabled development of site-specific management (i.e. precision agriculture) within fields. Local landscapes surrounding fields have a direct effect on the dynamics of field populations of pests and natural enemies. Thus, the makeup of these landscapes and their effects on pests will impact management decisions, and hopefully, expand considerations for managing the makeup and status of those local landscapes. Beyond the local landscape (i.e. regional landscape), longer range movement by pests and natural enemies and their interactions with crop and non-crop habitats, can dramatically influence pest risk potential, natural enemy prevalence, and management itself. Finally, long-distance movement and continental-scale migration of pests can contribute significantly to risk assessment that is important for IPM program development and implementation.

Much of the research highlighted in this book addresses larger spatial scales from the local landscape to the continental scales, for which more efficient and accurate spatial technologies have enabled extensive research. Findings over the last few decades have demonstrated the potential impact of landscape services at different spatial scales on pest management. Sustainability of large-scale crop production systems will be enhanced by farm-level IPM decisions that account for local, regional, and continental landscape factors affecting pest and natural enemy movement and interactions both spatially and temporally. Better understanding of population dynamics at larger scales will also enable development and implementation at a regional scale (e.g. areawide programs) of IPM practices with increased overall effectiveness of landscape services and environmental resilience. The effectiveness of such management efforts will rely on a comprehensive interdisciplinary team with involvement of social and economic scientists to identify strategies that maximize program participation and buy-in. The success of future IPM programs will depend on understanding their ecological effects and how the variability in agroecosystem services can be better predicted and leveraged, not only for improved farm-level management, but also more effective and efficient IPM implemented across entire agroecosystems.

References

Abendroth, L.J., Elmore, R.W., Boyer, M.J. and Marlay, S.K. (2011) *Corn Growth and Development.* PMR 1009. Ames, IA: Iowa State University, University Extension.

Adamchuk, V.I., Ferguson, R.B. and Hergert, G.W. (2010) Soil heterogeneity and crop growth. In: Oerke, E.C., Gerhards, R., Menz, G. and Sikora, R. (eds) *Precision Crop Protection – The Challenge and Use of Heterogeneity*. Dordrecht, Netherlands: Springer, pp. 3–16.

Allen, C.T. (2008) Boll weevil eradication: an areawide pest management effort. In: Koul, O., Cuperus, G. and Elliott, N. (eds) *Areawide Pest Management: Theory and Implementation*. Wallingford, UK: CAB International, pp. 467–559. DOI: 10.1079/9781845933722.0467

Alvarado, A., Jones, R.A., Pedraza-Lara, C., Villanueva, O. and Pfeiler, E. (2017) Reassessment of the phylogeography and intraspecific relationships of western and eastern populations of the boll weevil, *Anthonomus grandis* Boheman (Coleoptera: Curculionidae), in North America. *Biological Journal of the Linnean Society* 122, 29–45. DOI: 10.1093/biolinnean/blx049

Bailey, W. (2004) *Black Cutworm Monitoring and Forecasting Program*. Columbia, MO: University of Missouri-Columbia, Integrated Pest Management Program, Department of Entomology, Commercial Agriculture Program. Available at: http://agebb.missouri.edu/weather/reports/bcwforecast.htm (accessed July 9, 2023).

Baute, T. (2024) *Great Lakes and Maritimes Pest Monitoring Network*. Available at: https://experience.arcgis.com/experience/7164d23d488246d198dcf7a07d8c9021 (accessed May 9, 2024).

Brewer, M.J. and Elliott, N.C. (2023) Recent advances in agroecological research for increasing scope of areawide pest management of arthropods in cropping systems. *Current Opinion in Insect Science* 56, 101019. DOI: 10.1016/j.cois.2023.101019

Brewer, M.J. and Goodell, P.B. (2012) Approaches and incentives to implement integrated pest management that addresses regional and environmental issues. *Annual Review of Entomology* 57, 41–59. DOI: 10.1146/annurev-ento-120709-144748

Brewer, M.J., Rajotte, E.G., Kaplan, J.R., Goodell, P.B., Biddinger, D.J. *et al.* (2009) Opportunities, experiences, and strategies to connect integrated pest management to US Department of Agriculture conservation programs. *American Entomologist* 55, 140. DOI: 10.1093/ae/55.3.140

Brewer, M.J., Deleon, L. and Esquivel, I.L. (2020) Geographic Information System (GIS)-based mapping and spatial analyses applied to risk assessment and resource allocation for boll weevil (Coleoptera: Curculionidae) detection. *Annals of the Entomological Society of America* 113, 71–78. DOI: 10.1093/aesa/saz048

Brewer, M.J., Elliott, N.C., Esquivel, I.L., Jacobson, A.L., Faris, A.M. *et al.* (2022) Natural enemies, mediated by landscape and weather conditions, shape response of the sorghum agroecosystem of North America to the invasive aphid *Melanaphis sorghi*. *Frontiers in Insect Science* 2, 830997. DOI: 10.3389/finsc.2022.830997

Briese, L. (2019) Science communication in agriculture: the role of the trusted adviser. Doctor of Plant Health doctoral document. Lincoln, NE: University of Nebraska-Lincoln. Available at: https://digitalcommons.unl.edu/planthealthdoc/11/ (accessed July 9, 2023).

Burke, H.R., Clark, W.E., Cate, J.R. and Fryxell, P.A. (1986) Origin and dispersal of the boll weevil. *Bulletin of the Entomological Society of America* 32, 228–238. DOI: 10.1093/besa/32.4.228

Burrows, M., Franc, G., Rush, C., Blunt, T., Ito, D. *et al.* (2009) Occurrence of viruses in wheat in the Great Plains region, 2008. *Plant Health Progress* 10, 14. DOI: 10.1094/PHP-2009-0706-01-RS

Burton, R.L., Starks, J.K. and Peters, D.C. (1980) *The Army Cutworm*. Bulletin B-749. Stillwater, OK: Oklahoma State University Agricultural Experimental Station.

Busching, M.K. and Turpin, F.T. (1976) Oviposition preferences of black cutworm moths among various crop plants, weeds, and plant debris. *Journal of Economic Entomology* 69, 587–590. DOI: 10.1093/jee/69.5.587

Byamukama, E., Seifers, D., Hein, G., DeWolf, E., Tisserat, N. *et al.* (2013) Occurrence and distribution of Triticum mosaic virus in the central Great Plains. *Plant Disease* 97, 21–29. DOI: 10.1094/PDIS-06-12-0535-RE

Byamukama, E., Tatineni, S., Hein, G.L., McMechan, A.J. and Wegulo, S.N. (2015) Incidence of *Wheat streak mosaic virus*, *Triticum mosaic virus*, and *Wheat mosaic virus* in wheat curl mites recovered from maturing winter wheat spikes. *Plant Disease* 100, 318–323. DOI: 10.1094/PDIS-06-15-0692-RE

Carroll, M.W. (2005) Spatial distribution of the green peach aphid, *Myzus persicae* (Sulzer), and management applications in seed potato. PhD thesis. Twin Cities, MN: University of Minnesota.

Carroll, M.W., Radcliffe, E.B., MacRae, I.V., Ragsdale, D.W., Olson, K.D. *et al.* (2009) Border treatment to reduce insecticide use in seed potato production: biological, economic, and managerial analysis. *American Journal of Potato Research* 86, 31–37. DOI: 10.1007/s12230-008-9058-7

Chaplin-Kramer, R., O'Rourke, M., Schellhorn, N., Zhang, W., Robinson, B.E. *et al.* (2019) Measuring what matters: actionable information for conservation biocontrol in multifunctional landscapes. *Frontiers in Sustainable Food Systems* 3, 60. DOI: 10.3389/fsufs.2019.00060

Chapman, J.W., Reynolds, D.R. and Wilson, K. (2015) Long-range seasonal migration in insects: mechanisms, evolutionary drivers and ecological consequences. *Ecology Letters* 18, 287–302. DOI: 10.1111/ele.12407

Chen, D., Hein, G.L., Adams-Selin, R., Wang, L., Zhang, J. *et al.* (2024) Spatial relationship between pre-harvest hail and the impact from the wheat streak mosaic disease complex by using remote sensing data. *Crop Protection* 179, 106627. DOI: 10.1016/j.cropro.2024.106627

Choi, S.K., Kim, K.S., Lee, H., Adamczyk, J.J., Greenberg, S.M. *et al.* (2011) Temporal changes in genetic variation of boll weevil (Coleoptera: Curculionidae) populations, and implications for population assignment in eradication zones. *Annals of the Entomological Society of America* 104, 816–825. DOI: 10.1603/AN11012

Corn ipmPIPE (2019) *Black Cutworm*. Available at: https://corn.ipmpipe.org/insects/black-cutworm/ (accessed July 9, 2023).

Culin, J., Brown, S., Rogers, J., Scarborough, D., Swift, A. *et al.* (1990) A simulation model examining boll weevil dispersal: historical and current situations. *Environmental Entomology* 19, 195–208. DOI: 10.1093/ee/19.2.195

DATCP (2023) *Black Cutworm*. Madison, WI: State of Wisconsin: Department of Agriculture, Trade and Consumer Protection. Available at: https://datcp.wi.gov/Pages/Programs_Services/BlackCutworm.aspx (accessed July 9, 2023).

Davidson, S.A., Peairs, F.B. and Khosla, R. (2007) Insect pest densities across site-specific management zones of irrigated corn in Northeastern Colorado. *Journal of Economic Entomology* 100, 781–789. DOI: 10.1093/jee/100.3.781

Davis, J.A., Radcliffe, E.B. and Ragsdale, D.W. (2009) Planter skips and impaired stand favors potato virus Y spread in potato. *American Journal of Potato Research* 86, 203–208. DOI: 10.1007/s12230-009-9073-3

Davis, J.A., Radcliffe, E.B., Ragsdale, D.W. and MacRae, I. (2015) Increasing in-row spacing enhances Potato virus Y and Potato leafroll virus spread in potato. *American Journal of Potato Research* 92, 497–501. DOI: 10.1007/s12230-015-9462-8

Dean, A. and Hodgson, E. (2023) *Join the Iowa Moth Trapping Network in 2023*. Ames, IA: Iowa State University, Extension and Outreach, Integrated Crop Management. Blog Post, February 15. Available at: https://crops.extension.iastate.edu/blog/ashley-dean-erin-hodgson/join-iowa-moth-trapping-network-2023 (accessed July 9, 2023).

Deleon, L., Brewer, M.J., Esquivel, I.L. and Halcomb, J. (2017) Use of a geographic information system to produce pest monitoring maps for south Texas cotton and sorghum land managers. *Crop Protection* 101, 50–57. DOI: 10.1016/j.cropro.2017.07.016

Dentzman, K. (2023) Governance of emerging pests and pathogens in production landscapes: pesticides resistance and collaboration governance. *Current Opinions in Environmental Sustainability* 58, 101220. DOI: 10.1016/j.cosust.2022.101220

DiFonzo, C.D. (1996) Epidemiology and control of potato virus Y in the Red River Valley of Minnesota and North Dakota. PhD dissertation. Twin Cities, MN: University of Minnesota.

DiFonzo, C.D., Ragsdale, D.W., Radcliffe, E.B., Gudmestad, N.C. and Secor, G.A. (1996) Crop borders reduce potato virus Y incidence in seed potato. *Annals of Applied Biology* 129, 289–302. DOI: 10.1111/j.1744-7348.1996.tb05752.x

Dingle, H. and Drake, V.A. (2007) What is migration? *BioScience* 57, 113–121. DOI: 10.1641/B570206

Doerge, T.A. (1999) *Management Zone Concepts*. Site-Specific Management Guidelines (SSMG)-2. Potash & Phosphate Institute. Available at: http://www.ipni.net/publication/ssmg.nsf/0/C0D052F04A53E0B-F852579E500761AE3/$FILE/SSMG-02.pdf (accessed January 22, 2024).

Dwivedi, M., Shadab, M.H. and Santosh, V.R. (2020) Insect pest detection, migration and monitoring using radar and lidar systems. In: Chakravarthy, A. (ed.) *Innovative Pest Management Approaches for the 21st Century*. Singapore: Springer, pp. 61–76. DOI: 10.1007/978-981-15-0794-6_4

Engelken, L.K., Showers, W.B. and Taylor, S.E. (1990) Weed management to minimize black cutworm (Lepidoptera: Noctuidae) damage in no-till corn. *Journal of Economic Entomology* 83, 1059–1063. DOI: 10.1 093/jee/83.3.1058

Feeny, D., Berkes, F., McCay, B.J. and Acheson, J.M. (1990) The tragedy of the commons: twenty-two years later. *Human Ecology* 18, 1–19. DOI: 10.1007/BF00889070

Fisher, K.E., Dixon, P.M., Han, G., Adelman, J.S. and Bradbury, S.P. (2021) Locating large insects using automated VHF radio telemetry with a multi-antennae array. *Methods in Ecology and Evolution* 12, 494–506. DOI: 10.1111/2041-210X.13529

Fleischer, S.J., Blom, P.E. and Weisz, R. (1999) Sampling in precision IPM: when the objective is a map. *Phytopathology* 89, 1112–1118. DOI: 10.1094/PHYTO.1999.89.11.1112

Gemeno, C. and Haynes, K.F. (2001) Impact of photoperiod on the sexual behavior of the black cutworm moth (Lepidoptera: Noctuidae). *Environmental Entomology* 30, 189–195. DOI: 10.1603/0046-225X-30.2.189

Gerhards, R., Andujar Sanchez, D., Hamouz, P., Peteinatos, G.G., Christensen, S. *et al.* (2022) Advances in site-specific weed management in agriculture – a review. *Weed Research* 62, 123–133. DOI: 10.1111/wre.12526

Glick, P.A. (1939) *The Distribution of Insects, Spiders, and Mites in the Air*. Technical Bulletin 673. Washington, DC: US Department of Agriculture.

Griffel, L.M., Delparte, D. and Edwards, J. (2018) Using support vector machines classification to differentiate spectral signatures of potato plants infected with Potato Virus Y. *Computers and Electronics in Agriculture* 153, 318–324. DOI: 10.1016/j.compag.2018.08.027

Guerra, A.A. (1988) Seasonal boll weevil movement between northeastern Mexico and the Rio Grande Valley of Texas, USA. *Southwestern Entomologist* 13, 261–271.

Guo, J., Fu, X., Zhao, S., Shen, X., Wyckhuys, K.A.G. *et al.* (2020) Long-term shifts in abundance of (migratory) crop-feeding and beneficial insect species in northeastern Asia. *Asian Journal of Pest Science* 93, 583–594. DOI: 10.1007/s10340-019-01191-9

Haan, N.L., Zhang, Y. and Landis, D.A. (2020) Predicting landscape configuration effects on agricultural pest suppression. *Trends in Ecology & Evolution* 35, 175–186. DOI: 10.1016/j.tree.2019.10.003

Hardee, D.D., Cross, W.H., Mitchell, E.B., Huddleston, P.M., Mitchell, H.C. *et al.* (1969) Biological factors influencing responses of the female boll weevil to the male sex pheromone in field and large-cage tests. *Journal of Economic Entomology* 62, 161–165. DOI: 10.1093/jee/62.1.161

Hein, G., McKelvy, U., Wegulo, S., Hunger, R., Burrows, M. *et al.* (2023a) *Mite-transmitted Virus Disease Complex of Wheat in the Great Plains of the United States*. EC 3063. Lincoln, NE: University of Nebraska Extension. Available at: https://extensionpubs.unl.edu/publication/2115/html/view (accessed April 20, 2024).

Hein, G.L., McMechan, A.J. and Overmyer, L. (2023b) Wheat curl mite ecology and epidemiology of its associated wheat viruses. In: Eigenbrode, S.D. and Rashed, A. (eds) *Advances in Understanding Insect Pests Affecting Wheat and Other Cereals*. Cambridge, UK: Burleigh Dodds, pp. 271–310.

Hinds, W.E. (1916) *Boll weevil in Alabama*. Bulletin 188. Auburn, AL: Alabama Polytechnic Institute, Alabama Agricultural Experiment Station, pp. 23–64. Available at: https://aurora.auburn.edu/xmlui/bitstream/handle/11200/2149/1370BULL.pdf?sequence=1 (accessed July 9, 2023).

Hobson, K.A. (2023) Stable isotopes and a changing world. *Oecologia* 203, 233–250. DOI: 10.1007/s00442-023-05387-w

Hollandbeck, G.F., Onofre, K.A., DeWolf, E. and Todd, T. (2021) 2021 Kansas wheat disease loss estimates. *Kansas Cooperative Plant Disease Survey Report*, October 9.

Isard, S.A. and Gage, S.H. (2001) *Flow of Life in the Atmosphere*. East Lansing, MI: Michigan State University Press.

Johnson, W.L., Cross, W.H. and McGovern. W.L. (1976) Long-range dispersal of marked boll weevils in Mississippi during 1974. *Annals of the Entomological Society of America* 69, 421–422. DOI: 10.1093/aesa/69.3.421

Jones, R.W., Cate, J.R., Hernandez, E.M. and Navarro, R.T. (1992) Hosts and seasonal activity of the boll weevil (Coleoptera: Curculionidae) in tropical and subtropical habitats of northeastern Mexico. *Journal of Economic Entomology* 85, 74–82. DOI: 10.1093/jee/85.1.74

Karp, D.S., Chaplin-Kramer, R., Meehan, T.D., Martin, E.A., DeClerck, F. *et al.* (2018) Crop pests and predators exhibit inconsistent responses to surrounding landscape composition. *Proceedings of the National Academy of Sciences* 115, E7863–E7870. DOI: 10.1073/pnas.1800042115

Khosla, R., Fleming, K., Delgado, J.A., Shaver, T.M. and Westfall, D.G. (2002) Use of site-specific management zones to improve nitrogen management for precision agriculture. *Journal of Soil and Water Conservation* 57, 513–517.

Kim, K.S. and Sappington, T.W. (2006) Molecular genetic variation of boll weevil populations in North America estimated with microsatellites: implications for patterns of dispersal. *Genetica* 127, 143–161. DOI: 10.1007/s10709-005-2673-z

Kim, K.S. and Sappington, T.W. (2013) Population genetics strategies to characterize long-distance dispersal of insects. *Journal of Asia-Pacific Entomology* 16, 87–97. DOI: 10.1016/j.aspen.2012.11.004

Kim, K.S., Cano-Ríos, P. and Sappington, T.W. (2006) Using genetic markers and population assignment techniques to infer origin of boll weevils (Coleoptera: Curculionidae) unexpectedly captured near an eradication zone in Mexico. *Environmental Entomology* 35, 813–826. DOI: 10.1603/0046-225X-35.4.813

Kim, K.S., Sappington, T.W. and Allen, C.T. (2008) Genetic profiling to determine potential origins of boll weevils (Coleoptera: Curculionidae) captured in a Texas eradication zone: endemicity, immigration, or sabotage? *Journal of Economic Entomology* 101, 1729–1736. DOI: 10.1603/0022-0493-101.6.1729

Kim, K.S., Jones, G.D., Westbrook, J.K. and Sappington, T.W. (2010) Multidisciplinary fingerprints: forensic reconstruction of an insect reinvasion. *Journal of the Royal Society Interface* 7, 677–686. DOI: 10.1098/rsif.2009.0345

Kissling, W.D., Pattemore, D.E. and Hagen, M. (2014) Challenges and prospects in the telemetry of insects. *Biological Reviews* 89, 511–530. DOI: 10.1111/brv.12065

Koul, O., Cuperus, G.W. and Elliott, N. (eds) (2008) *Areawide Pest Management: Theory and Implementation*. Wallingford, UK: CAB International.

Kullik, S.A., Sears, M.K. and Schaafsma, A.W. (2011) Sublethal effects of Cry1F Bt corn and clothianidin on black cutworm (Lepidoptera: Noctuidae) larval development. *Journal of Economic Entomology* 98, 1594–1602. DOI: 10.1603/EC10360

Lange, F., Olmstead, A.L. and Rhode, P.W. (2009) The impact of the boll weevil, 1892–1932. *The Journal of Economic History* 69, 685–718. DOI: 10.1017/S0022050709001090

Levine, E., Clement, S.L., Rubink, W.L. and McCartney, D.A. (1983) Regrowth of corn seedlings following injury at different growth stages by black cutworm larvae. *Journal of Economic Entomology* 76, 389–391. DOI: 10.1093/jee/76.2.389

Liebhold, A.M., Rossi, R.E. and Kemp, W.P. (1993) Geostatistics and geographic information systems in applied insect ecology. *Annual Review of Entomology* 38, 303–327. DOI: 10.1146/annurev.en.38.010193.001511

Lowenberg-DeBoer, J. and Erickson, B. (2019) Setting the record straight on precision agriculture adoption. *Agronomy Journal* 111, 1552–1569. DOI: 10.2134/agronj2018.12.0779

Lukefahr, M.J., Barbosa, S. and Sabrinho, R.B. (1994) The introduction and spread of the boll weevil, *Anthonomus grandis* Boh. (Coleoptera) in Brazil. *Southwestern Entomologist* 19, 414–417.

Lundin, O., Rundlöf, M., Jonsson, M., Bommarco, R. and Williams, N.M. (2021) Integrated pest and pollinator management – expanding the concept. *Frontiers in Ecology and the Environment*, 19, 283–291. DOI: 10.1002/fee.2325

MacGillivray, M.E. (1981) Aphids. In: Hooker, W.J. (ed.) *Compendium of Potato Diseases*. St Paul, MN: APS Press, pp. 101–103.

MacRae, I., Baker, T. and Thompson, A. (2021) Notes on UAS and remote sensing of insect damage. *American Journal of Potato Research* 98, 16–23. DOI: 10.1007/s12230-020-09816-x

Majer, A., Laska, A., Hein, G.L., Kuczynski, L. and Skoracka, A. (2021) Propagule pressure rather than population growth determines colonization ability: a case study using two phytophagous mite species differing in their invasive potential. *Ecological Entomology* 46, 1136–1147. DOI: 10.1111/een.13058

Marston, Z.P., Cira, T.M., Hodgson, E.W., Knight, J.F., MacRae, I.V. and Koch, R.L. (2020) Detection of stress induced by soybean aphid (Hemiptera: Aphididae) using multispectral imagery from unmanned aerial vehicles. *Journal of Economic Entomology* 113, 779–786. DOI: 10.1093/jee/toz306

McKibben, G.H., Willers, J.L., Smith, J.W. and Wagner, T.L. (1991) Stochastic model for studying boll weevil dispersal. *Environmental Entomology* 20, 1327–1332. DOI: 10.1093/ee/20.5.1327

McMechan, A. (2016) Over-summering ecology of the wheat curl mite (*Aceria tosichella* Keifer). PhD dissertation. Lincoln, NE: University of Nebraska.

McMechan, A.J. and Hein, G.L. (2017) Population dynamics of the wheat curl mite (Acari: Eriophyidae) during the heading stages of winter wheat. *Journal of Economic Entomology* 10, 355–361. DOI: 10.1093/jee/tox028

Nguy-Robertson, A.L., Zygielbaum, A.I., McMechan, A.J., Hein, G.L., Wegulo, S.N. *et al.* (2016) Developing the framework for a risk map for mite vectored viruses in wheat resulting from pre-harvest hail damage. *Crop Protection* 89, 21–31. DOI: 10.1016/j.cropro.2016.06.014

Noskov, A., Bendix, J. and Friess, N. (2021) A review of insect monitoring approaches with special reference to radar techniques. *Sensors* 21, 1474. DOI: 10.3390/s21041474

Ostlie, K. and Potter, B. (2021) *Insect Pests of Corn – Stand Reducers: Black cutworm,* Agrotis ipsilon *Hüfnagel (Lepidoptera: Noctuidae).* Twin Cities, MN: University of Minnesota Extension. Available at: https://extension.umn.edu/corn-pest-management/black-cutworm (accessed July 9, 2023).

Park, Y.L., Krell, R.K. and Carroll, M. (2007) Theory, technology, and practice of site-specific insect pest management. *Journal of Asia-Pacific Entomology* 10, 89–101. DOI: 10.1016/S1226-8615(08)60337-4

Pedgley, D.E. (1993) Managing migratory insect pests – a review. *International Journal of Pest Management* 39, 3–12. DOI: 10.1080/09670879309371751

Pelletier, Y., Nie, X., Giguère, M.A., Nanayakkara, U., Maw, E. *et al.* (2012) A new approach for the identification of aphid vectors (Hemiptera: Aphididae) of potato virus Y. *Journal of Economic Entomology* 105, 1909–1914. DOI: 10.1603/EC12085

Polder, G., Blok, P.M., De Villiers, H.A., Van der Wolf, J.M. and Kamp, J. (2019) Potato virus Y detection in seed potatoes using deep learning on hyperspectral images. *Frontiers in Plant Science* 10, 209. DOI: 10.3389/fpls.2019.00209

Quinby, B.M., Creighton, J.C. and Flaherty, E.A. (2020) Stable isotope ecology in insects: a review. *Ecological Entomology* 45, 1231–1246. DOI: 10.1111/een.12934

Radcliffe, E.B. and Ragsdale, D.W. (2002) Aphid transmitted potato viruses: the importance of understanding vector biology. *American Journal of Potato Research* 79, 353–386. DOI: 10.1007/BF02870173

Radcliffe, E.B., Ragsdale, D.W., Suranyi, R.A., Difonzo, C.D. and Hladilek, E.E. (2008) Aphid alert: how it came to be, what it achieved and why it proved unsustainable. In: Koul, O., Cuperus, G.W. and Elliott, N. (eds) *Areawide Pest Management: Theory and Implementation.* Wallingford, UK: CAB International, pp. 244–260. DOI: 10.1079/9781845933722.0244

Ragsdale, D., Radcliffe, E.B. and DiFonzo, C.D. (2001) Epidemiology and field control of PVY and PLRV. In: Loebenstein, G., Berger, P.H., Brunt, A.A. and Lawson, R.H. (eds) *Virus and Virus-like Diseases of Potatoes and Production of Seed-potatoes*. Dordrecht, Netherlands: Springer, pp. 237–270. DOI: 10.1007/978-94-007-0842-6_22

Raszick, T.J. (2021) Boll weevil eradication: a success story of science in the service of policy and industry. *Annals of the Entomological Society of America* 114, 702–708. DOI: 10.1093/aesa/saab031

Raszick, T.J., Dickens, C.M., Perkin, L.C., Tessnow, A.E. and Suh, C.P-C. (2021) Population genomics and phylogeography of the boll weevil, *Anthonomus grandis* Boheman (Coleoptera: Curculionidae), in the United States, northern Mexico, and Argentina. *Evolutionary Applications* 14, 1778–1793. DOI: 10.1111/eva.13238

Ribeiro, A.V., Lacerda, L.N., Windmuller-Campione, M.A., Cira, T.M., Marston, Z.P.D. *et al.* (2024) Economic-threshold-based classification of soybean aphid, *Aphis glycines*, infestations in commercial soybean fields using Sentinel-2 satellite data. *Crop Protection* 177, 106557. DOI: 10.1016/j.cropro.2023.106557

Rummel, D.R., Jordan, L.B., White, J.R. and Wade, L.J. (1977) Seasonal variation in the height of boll weevil flight. *Environmental Entomology* 6, 674–678. DOI: 10.1093/ee/6.5.674

Sappington, T.W. and Showers, W.B. (1992) Reproductive maturity, mating status, and long-duration flight behavior of *Agrotis ipsilon* (Lepidoptera: Noctuidae), and the conceptual misuse of the oogenesis-flight syndrome by entomologists. *Environmental Entomology* 21, 677–688. DOI: 10.1093/ee/21.4.677

Sappington, T.W. and Spurgeon, D.W. (2000) Variation in boll weevil captures in pheromone traps (Coleoptera: Curculionidae) arising from wind speed moderation by brush lines. *Environmental Entomology* 29, 807–814. DOI: 10.1603/0046-225X-29.4.807

Sappington, T.W., Brashears, A.D., Parajulee, M.N., Carroll, S.C., Arnold, M.D. *et al.* (2004a) Potential for transport of boll weevils (Coleoptera: Curculionidae) to the cotton gin within cotton modules. *Journal of Economic Entomology* 97, 934–940. DOI: 10.1093/jee/97.3.934

Sappington, T.W., Brashears, A.D., Parajulee, M.N., Carroll, S.C., Arnold, M.D. *et al.* (2004b) Potential for escape of live boll weevils (Coleoptera: Curculionidae) into cottonseed, motes, and cleaned lint at the cotton gin. *Journal of Economic Entomology* 97, 1773–1781. DOI: 10.1093/jee/97.6.1773

Sappington, T.W., Arnold, M.D., Brashears, A.D., Parajulee, M.N., Carroll, S.C. *et al.* (2006) Dispersal of boll weevils (Coleoptera: Curculionidae) from cotton modules before ginning. *Journal of Economic Entomology* 99, 67–75. DOI: 10.1093/jee/99.1.67

Sappington, T.W., Hesler, L.S., Allen, K.C., Luttrell, R.G. and Papiernik, S.K. (2018) Prevalence of sporadic insect pests of seedling corn, and factors affecting risk of infestation. *Journal of Integrated Pest Management* 9, 16. DOI: 10.1093/jipm/pmx020

Satterfield, D.A., Sillett, T.S., Chapman, J.W., Altizer, S. and Marra, P.P. (2020) Seasonal insect migrations: massive, influential, and overlooked. *Frontiers in Ecology and the Environment* 18, 335–344. DOI: 10.1002/fee.2217

Schellhorn, N.A., Bianchi, F.J.J.A. and Hsu, C.L. (2014) Movement of entomophagous arthropods in agricultural landscapes: links to pest suppression. *Annual Review of Entomology* 59, 559–581. DOI: 10.1146/annurev-ento-011613-161952

Schellhorn, N.A., Parry, H.R., Macfadyen, S., Wang, Y. and Zalucki, M.P. (2015) Connecting scales: achieving in-field pest control from areawide and landscape ecology studies. *Insect Science* 22, 35–51. DOI: 10.1111/1744-7917.12161

Seiter, N. (2021) Managing insect pests. In: *Illinois Agronomy Handbook*. Urbana-Champaign, IL: University of Illinois, Illinois Extension, Chapter 13. Available at: https://extension.illinois.edu/sites/default/files/managing_insect_pests_2022web.pdf (accessed July 11, 2023).

Sherrod, D.W., Shaw, J.T. and Luckmann, W.H. (1979) Concepts on black cutworm field biology in Illinois. *Environmental Entomology* 8, 191–195. DOI: 10.1093/ee/8.2.191

Showers, W.B. (1997) Migratory ecology of the black cutworm. *Annual Review of Entomology* 42, 393–425. DOI: 10.1146/annurev.ento.42.1.393

Showers, W.B., Kaster, L.V. and Mulder. P.G. (1983) Corn seedling growth stage and black cutworm (Lepidoptera: Noctuidae) damage. *Environmental Entomology* 12, 241–244. DOI: 10.1093/ee/12.1.241

Showers, W.B., Kaster, L.V., Sappington, T.W., Mulder, P.G. and Whitford, F. (1985) Development and behavior of black cutworm (Lepidoptera: Noctuidae) populations before and after corn emergence. *Journal of Economic Entomology* 78, 588–594. DOI: 10.1093/jee/78.3.588

Showers, W.B., Whitford, F., Smelser, R.B., Keaster, A.J., Robinson, J.F. *et al.* (1989) Direct evidence for meteorologically driven long-range dispersal of an economically important moth. *Ecology* 70, 987–992. DOI: 10.2307/1941366

Sishodia, R.P., Ray, R.L. and Singh, S.K. (2020) Applications of remote sensing in precision agriculture: a review. *Remote Sensing* 12, 3136. DOI: 10.3390/rs12193136

Skoracka, A., Rector, B., Kuszynski, L., Szydło, W., Hein, G.L. and French, R. (2014) Global spread of wheat curl mite by its most polyphagous and pestiferous lineages. *Annals of Applied Biology* 165, 222–235. DOI: 10.1111/aab.12130

Smelser, R.B., Showers, W.B., Shaw, R.H. and Taylor. S.E. (1991) Atmospheric trajectory analysis to project long-range migration of black cutworm (Lepidoptera: Noctuidae) adults. *Journal of Economic Entomology* 84, 879–885. DOI: 10.1093/jee/84.3.879

Smith, J.W. (1998) Boll weevil eradication: area-wide pest management. *Annals of the Entomological Society of America* 91, 239–247. DOI: 10.1093/aesa/91.3.239

Spurgeon, D.W. (2016) Captures of boll weevils (Coleoptera: Curculionidae) in relation to trap orientation and distance from brush lines. *Journal of Economic Entomology* 109, 676–683. DOI: 10.1093/jee/tov392

Spurgeon, D.W. and Raulston, J.R. (2006) Captures of boll weevils (Coleoptera: Curculionidae) in traps associated with different habitats. *Journal of Economic Entomology* 99, 752–756. DOI: 10.1093/jee/99.3.752

Spurgeon, D.W., Raulston, J.R. Zamora, O.Z. and Loera, J. (1997) Spatial and temporal patterns of boll weevil trap captures in northeastern Mexico. In: *Proceedings of the Beltwide Cotton Conference 2*. Memphis, TN: National Cotton Council, pp. 984–986. Available at: https://www.cotton.org/beltwide/proceedings/getPDF.cfm?paper=K048.pdf&year=1997 (accessed August 8, 2023).

Staples, R. and Allington, W.B. (1956) *Streak Mosaic of Wheat in Nebraska and its Control*. Research Bulletin 178. Lincoln, NE: Nebraska Agricultural Experiment Station.

Stevenson, W.R., Loria, R., Franc, G.D. and Weingartner, D.P. (eds) (2001) *Compendium of Potato Diseases*. St Paul, MN: APS Press.

Stilwell, A., Hein, G., Zygielbaum, A., Rundquist, D. (2013) Proximal sensing to detect symptoms associated with wheat curl mite-vectored viruses. *International Journal Remote Sensing* 34, 4951–4966. DOI: 10.1080/01431161.2013.783945

Stilwell, A.R., Rundquist, D.C., Marx, D.B. and Hein, G.L. (2019) Differential spatial gradients of wheat streak mosaic virus into winter wheat from a central mite-virus source. *Plant Disease* 103, 338–344. DOI: 10.1094/PDIS-01-18-0025-RE

Sugawara, M., Kojima, M. and Murayama, D. (1974) Latent period and retention of potato leaf-roll virus in its vector, *Myzus persicae*. *Japanese Journal of Phytopathology* 40, 39–45.

Suranyi, R., Radcliffe, E., Lockhart, B. and MacRae, I.V. (1999) Aphid Alert: a research/outreach initiative addressing potato virus problems in the northern Midwest. In: *Radcliffe's IPM World Textbook – An Electronic IPM Textbook*. Twin Cities, MN: University of Minnesota. Available at: http://ipmworld.umn.edu/chapters/aphidalert.htm (accessed April 20, 2024).

Suranyi, R.A. (2000) Management of potato leafroll virus in Minnesota. PhD dissertation. Twin Cities, MN: University of Minnesota.

Temple, J.H. (2007) Evaluating precision agricultural technologies in a Louisiana cotton insect pest management system. MS thesis. Monroe, LA: Louisiana State University and Agricultural & Mechanical College, Monroe.

Thomas, J., Hein, G. and Lyon, D. (2004) Spread of wheat curl mite and wheat streak mosaic virus is influenced by volunteer wheat control methods. *Plant Health Progress* 5, 2. DOI: 10.1094/PHP-2004-1206-01-RS

Troester, S.J., Meyer, R.H., Steffey, K., Godfrey, G.L., Bremer, C.D. *et al.* (1983) Projecting the beginning of black cutworm (Lepidoptera: Noctuidae) damage to field corn. *Journal of Economic Entomology* 76, 363–367. DOI: 10.1093/jee/76.2.363

van Helden, M. (2010) Spatial and temporal dynamics of arthropods in arable fields. In: Oerke, E.C., Gerhards, R., Menz, G. and Sikora, R. (eds) *Precision Crop Protection – The Challenge and Use of Heterogeneity*. Dordrecht, Netherlands: Springer. DOI: 10.1007/978-90-481-9277-9_4

Verbeek, M., Piron, P.G.M., Dullemans, A.M., Cuperus, C. and Van Der Vlugt, R.A.A. (2010) Determination of aphid transmission efficiencies for N, NTN and Wilga strains of *Potato virus Y*. *Annals of Applied Biology* 156, 39–49. DOI: 10.1111/j.1744-7348.2009.00359.x

Westbrook, J.K., Eyster, R.S. and Allen, C.T. (2011) A model for long-distance dispersal of boll weevils. *International Journal of Biometeorology* 55, 585–593. DOI: 10.1007/s00484-010-0359-4

Williams, C.B. (1957) Insect migration. *Annual Review of Entomology* 2, 163–180.

Wilson, L.J., Whitehouse, M.E. and Herron, G.A. (2018) The management of insect pests in Australian cotton: an evolving story. *Annual Review of Entomology* 63, 215–237. DOI: 10.1146/annurev-ento-020117-043432

Wosula, E., McMechan, A. and Hein, G. (2015) The effect of temperature, relative humidity, and virus infection status on off-host survival of the wheat curl mite (Acari: Eriophyidae). *Journal of Economic Entomology* 108, 1545–1552. DOI: 10.1093/jee/tov185

Wosula, E.N., McMechan, A.J., Knoell, E., Tatineni, S., Wegulo, S.N. and Hein, G.L. (2018) Impact of timing and method of virus inoculation on the severity of wheat streak mosaic disease. *Plant Disease* 102, 645–650. DOI: 10.1094/PDIS-08-17-1227-RE

Wright, R.J., Hunt, T.E. and Jarvi. K.J. (2013) *Insects & Pests, Field Crop Insects: Corn Cutworms.* NebGuide G1153. Lincoln, NE: University of Nebraska-Lincoln Extension. Available at: https://extensionpublications.unl.edu/assets/pdf/g1153.pdf (accessed July 9, 2023).

Zeng, J., Liu, Y., Zhang, H., Liu, J., Jiang, Y. *et al.* (2020) Global warming modifies long-distance migration of an agricultural insect pest. *Journal of Pest Science* 93, 569–581. DOI: 10.1007/s10340-019-01187-5

Zhao, Z., Reddy, G.V.P., Hui, C. and Li, B. (2016) Approaches and mechanisms for ecologically based pest management across multiple scales. *Agriculture, Ecosystems & Environment* 230, 199–209. DOI: 10.1016/j.agee.2016.06.010

Zhu, M., Radcliffe, E.B., Ragsdale, D.W., MacRae, I.V. and Seeley, M.W. (2006) Low-level jet streams associated with spring aphid migration and current season spread of potato viruses in the US northern Great Plains. *Agricultural and Forest Meteorology* 138, 192–202. DOI: 10.1016/j.agrformet.2006.05.001

2 Pest Management for Spatially Variable Arthropod Pests in Large-scale Agroecosystems

Michael J. Brewer,[1,*] Paul A. Umina,[2] and Norman C. Elliott[3]

[1]Texas A&M AgriLife Research, Corpus Christi, TX, USA; [2]School of BioSciences, The University of Melbourne, Parkville, Victoria, Australia; [3]US Department of Agriculture, Agricultural Research Service, Peanut and Other Field Crops Research Unit, Stillwater, OK, USA

Abstract

In large-scale agroecosystems, a growing trend in managing spatially variable pests is to recognize pest suppression innate to the system and to supplement this suppression using regional-based pest management tactics in context of agricultural and environmental conditions that influence pests. This chapter presents the rationale of this agroecosystem-focused approach, its roots in integrated pest management (IPM), and its connections to areawide pest management and insecticide resistance management (IRM) more broadly. Understanding sensitivity of trophic interactions that drives pest suppression and affects regional pest risk are basic principles of this approach. Spatially variable conditions that influence pest ecology and crop sensitivity to pests are selective biotic and abiotic components of the agroecosystem, inclusive of landscape structure (e.g. amount and mix of crops and semi-natural vegetation), natural enemies and other suppression agents, and weather. Using regional IPM tactics may reduce the need for pest management actions needed at the farm level and reduce the overall impact of pests regionally. Case examples are taken from the pest management literature: (i) *Diuraphis noxia* (Kurdjumov) (Hemiptera: Aphididae) invading the North American wheat agroecosystem where resident aphid natural enemies provided spatially sensitive aphid suppression and regional use of aphid-resistant wheat supported aphid suppression where pest risk was high, (ii) regional IRM as a way to minimize insecticide resistance risk in *Myzus persicae* (Sulzer) (Hemiptera: Aphididae) in Australia, although not fully realized due to multi-crop complexity and allied operational challenges, and (iii) spatially and temporally variable suppression of aphids by coccinellids in large-scale field crops which in some cases provides biological control sufficient to prevent economic crop injury depending on landscape conditions. Although challenging to implement, these case examples illustrate that favorable pest management and crop health outcomes become more likely when using agroecosystem-focused IPM in large-scale agroecosystems.

Introduction

In large-scale agroecosystems, a growing trend in managing spatially variable pests is to recognize pest suppression innate to the system and to supplement this suppression using regional-based pest management tactics in context of variable agricultural and environmental conditions

*Corresponding author: brewermj@tamu.edu

 Arthropod Management and Landscape Considerations in Large-Scale Agroecosystems (eds M.J. Brewer and G.L. Hein)
DOI: 10.1079/9781800622777.0002

that influence pests. For this agroecosystem-focused integrated pest management (IPM) approach, understanding spatial sensitivity of pest, crop, and suppression agent (trophic) interactions is valuable in characterizing pest suppression innate to the system and sensitivity of the trophic interactions to agricultural and environmental conditions (Brewer and Elliott, 2023). Guided by this spatial information, preventative IPM tactics can target regions where innate pest suppression is overshadowed by environmental and agricultural conditions that favor pest outbreaks or increase crop sensitivity to pests. Regionally applied IPM tactics appropriate to the agroecosystem may also relieve selection for insecticide resistance. These tactics tend to be preventative in nature and designed to reduce region-level pest status and the extent of remedial pest management actions needed at the farm level. While reducing region-level pest status, acute threats to crop health may still occur that justify use of remedial pest management at the farm level, such as using an insecticide to respond to local pest increases due to weather or other favorable local conditions.

This chapter presents the rationale of an agroecosystem-focused approach to pest management, its roots in IPM, and its connections to areawide pest management and insecticide resistance management (IRM) more broadly. Understanding sensitivity of trophic interactions that drives pest suppression and affects regional pest risk are basic principles of this approach that can lead to improved guidance when applying IPM and IRM practices regionally. Compatibility of innate pest suppression and regional-based IPM tactics may also support other agroecosystem services and resilience to pest threats (Kogan, 1998; Brewer and Goodell, 2012). A review of the literature and three case examples demonstrate the value of agroecosystem-focused IPM in large-scale agroecosystems, as well as the challenges to estimate the benefits of innate pest suppression and justify use of regionally applied IPM tactics.

Rationale of Agroecosystem-focused IPM

The foundation of agroecosystem-focused IPM is the recognition of pest suppression innate to the system, supported with regional-based IPM (and IRM) tactics, in context of spatial variability of agricultural and environment conditions that influence pests. This agroecosystem orientation has benefitted from pest management research rooted in the development of the IPM concept over the last 75 years (Pedigo *et al.*, 1986; Kogan, 1998; Brewer and Goodell, 2012). An operational distinction is that crop protection using IPM tactics tends to be individual field-oriented, while an agroecosystem focus brings spatial variability of pest suppression to the forefront. A tenet in crop protection as commonly considered is to maintain pest populations below economic injury levels (EILs), with minimal disruptions to non-target organisms (Pedigo *et al.*, 1986). Non-crop areas may be considered as pest management decisions are made and pest management tactics are deployed (e.g. pest or natural enemy movement into a crop from semi-natural vegetation), but the individual crop field is the primary focus in pest management decision-making and application of insect management tactics (Brewer and Goodell, 2012).

The classic example is use of an economic threshold (ET)—a value below the EIL that considers the lag time in implementing and seeing results from an IPM tactic—for a specific pest, crop, and growing region that triggers the application of an insecticide once the ET is exceeded. The aim of this remedial action is to prevent pest populations from causing economic yield loss (Pedigo *et al.*, 1986). In principle, other remedial pest management tactics may be used, such as timing the release of natural enemies in an inundative biocontrol strategy (Van Driesche and Bellows, 1996). Sampling methods for estimating pest abundance and forecasting crop risk are a mainstay of IPM practiced at the individual field level (Pedigo and Buntin, 1994). Within the IPM concept, spatial context has been considered when modifying EILs and ETs for a variety of crop production regions and other environmental considerations (Pedigo and Higley, 1992). But the field-based decision-making process is typically maintained, given the emphasis on whether the field-level estimate of monetary benefits of crop protection offsets the cost of pest control.

Merging IPM and agroecological viewpoints, regional IPM actions likely complement and possibly reduce the extent of pest management actions needed at the farm level, but the value is difficult to monetize at the farm level. Ecologically, the benefits of a larger-scale agroecosystem orientation for IPM are reinforced when source/sink insect dynamics, insect movement, habitat affiliation, and crop sensitivity to the pest vary spatially. Spatial arrangement of different field crops and semi-natural vegetation may deter pest population growth and increase pest suppression by natural enemies, dependent on insect mobility across crops and semi-natural habitats that differ in resource availability for insects (and other invertebrates) and plant sensitivity to pest attacks (Carriére *et al.*, 2006; Schellhorn *et al.*, 2015; Downes *et al.*, 2017). The development of other preventative IPM tactics for more persistent and severe pests, such as host plant resistance and classical biological control, is well recognized and often takes a regional management perspective as well (Panda and Khush, 1995; Van Driesche and Bellows, 1996). Furthermore, the spatial variation in pest suppression occurring across large-scale systems often makes an individual field focus operationally impractical due to human resource constraints. An individual field focus may also miss opportunities to use pest management tactics deployed regionally to complement pest suppression (such as by resident natural enemies) that is occurring but overshadowed by environmental conditions that favor pest population increase.

But challenges to justify use of agroecosystem-focused IPM can be seen in light of the ecological and economic aspects of how IPM and IRM are deployed and valued. For example, effectively managing insecticide resistance evolution may be hindered by the deployment of individual management tactics at the field level. This is particularly the case for highly mobile species, such as lepidopterans, thrips, and aphids, where the chemical practices of one grower can impact the resistance status of pests in neighboring fields. In another example of addressing an invasive pest, adoption of aphid-resistant soybean cultivars is advised to manage *Aphis glycines* (Matsumura) (Hemiptera: Aphididae), soybean aphid (Tilmon *et al.*, 2021), but adoption has lagged in large-scale United States (US) soybean production. This lag time may be due in part to strategies taken by the seed industry to address perceived marketability challenges using resistant varieties in selected areas of high pest risk while retaining seed options in other growing regions. Despite these challenges, a large-scale agroecosystem orientation to IPM holds promise where spatially variable pest and natural enemy abundances are common. It is this premise, from concept to application, that is explored in this chapter.

Links to Areawide Management and IRM

Agroecosystem-focused IPM shares a regional management perspective with areawide pest management (Elliott *et al.*, 2008; Brewer and Elliott, 2023). The term areawide pest management is well embedded in pest management literature. Specific projects have been sponsored by selected government programs with grower consent, and with the goal to substantially reduce or eliminate significant pests in a large area (see chapters in Koul *et al.*, 2008). The agroecosystem focus of IPM is more rooted in the understanding of the ecosystem service of pest suppression in the agroecology literature (Jeanneret *et al.*, 2021). Areawide pest management places more emphasis on widespread application of both preventative and remedial IPM tactics to drive the pest population downward as much as possible, inclusive of eliminating the pest (i.e. termed eradication in the areawide pest management literature; Elliott *et al.*, 2008).

For agroecosystem-focused IPM, an assessment of pest risk guides the use of preventative pest management tactics deployed regionally. This approach is based on the expectation that innate pest suppression may be sufficient in some areas, but in other areas pest suppression may not function well (e.g. low abundance of natural enemies due to poor quality habitat for natural enemies) or simply cannot offset conditions that favors pest outbreaks (e.g. weather-driven insect movement, temperatures that favor pest population increase, growing one genetically uniform pest susceptible crop). Consistent with areawide pest management concepts, an agroecosystem-focused IPM approach may be more effective, and provide more pest management and environmental benefits, compared with use of remedial pest management tactics deployed field-by-field based on ETs (Brewer and Elliott, 2023).

A regional orientation of IPM also mirrors the principles and challenges of IRM, which is typically more effective when applied on a regional basis rather than at the individual field level. IRM strategies may be introduced in response to chemical control failures (curative [equivalent to remedial in IPM literature] IRM) or to avoid such failures from occurring in the first place (preventative IRM). Regardless of the approach, insecticide resistance among agricultural pests is more likely in polyphagous and highly mobile arthropods (Rane *et al.*, 2016; Umina *et al.*, 2019; Mota-Sanchez and Wise, 2023) and therefore almost always affects multiple farmers and often impacts multiple agricultural industries. One such example is *Helicoverpa armigera* (Hübner) (Lepidoptera: Noctuidae), cotton bollworm, and its management within Australia, where wide-scale pest management programs have been adopted over many decades to alleviate escalating insecticide resistance issues (Forrester, 1990). The successes achieved through regional IPM strategies implemented in the Australian cotton industry have been well documented (Zalucki *et al.*, 2009; Wilson *et al.*, 2018). Analogous approaches have also been attempted in other crops for *H. armigera* management, such as in sorghum and maize (Murray *et al.*, 1998).

Pest Suppression Influenced by Spatially Variable Conditions

A widely accepted tenet in the agroecological literature is that monotypic large-scale cropping systems are prone to pest outbreaks and lack habitat for pest suppression agents such as natural enemies (Schneider *et al.*, 2022). Which raises the question: so why bother investigating regionally applied IPM tactics to support innate pest suppression that may be lacking in large monotypic cropping systems? While the trend of higher pest risk may broadly hold in large-scale systems, meta-analysis of over 50 studies has indicated that pests and natural enemies are spatially variable across a variety of agroecosystems including large systems (Karp *et al.*, 2018). This variability actually may be consistent with the viewpoint that pest and natural enemy abundance (and natural enemy species diversity) are spatially variable and conditional on a range of agricultural (inclusive of crop genetics, crop field size and shape, and crop diversity) and environmental (inclusive of weather and climate, and local to regional semi-natural vegetative structure) conditions. From a landscape view, large-scale cropping systems may be more heterogeneous in structure than indicated by simple features such as the extent of a crop grown in a region and uniformity of crop genetics. The extent and form of heterogeneity "in-the-shadows" may support innate pest suppression in some areas but not others, resulting in a mix of pest-prone and pest-resilient regions of crop production. Regional IPM inputs, such as planting insect-resistant host plants in selected regions, often supplements innate pest suppression. These regional IPM inputs become vital when relevant heterogeneity that supports innate suppression is low and environmental conditions tend to favor pest population increase (e.g. Brewer *et al.*, 2019).

IPM research has demonstrated effects on pest biology and suppression as a result of changes in agricultural practices and shifts in weather patterns. These include effects on pest population dynamics by agronomic practices (i.e. irrigation access, agrochemicals application methods, cultivation techniques, planting date modifications such as host-free periods), resistance to insecticides over broad areas, and plant improvement and breeding efforts targeting pests (Kogan, 1998). Characterizing landscape influences on natural enemies and pests (as done by using one landscape metric; see Karp *et al.*, 2018) has been traditionally challenging in research, but spatial data and geographic information systems (GIS) technology in agriculture broadens research capabilities where they are available (Sun *et al.*, 2021; Tang, 2021). These include comprehensive high quality spatial data of land use inclusive of agricultural and semi-natural vegetation, weather, and records of insecticide usage and insect pest and beneficial insect monitoring data (Martin *et al.*, 2019; Brewer *et al.*, 2022; Courson *et al.*, 2022).

These spatial data and advances in GIS have enabled research to investigate the extent that agricultural and environmental conditions influence spatial variation of pest and natural enemy activity (Brewer and Elliott, 2023). Taken from the agroecological pest management literature, research has increased on characterizing pest and beneficial insect life history and trophic interactions under changing conditions (Martin *et al.*, 2019; Perrot

Case Study 2.1: Suppression of *Diuraphis noxia* supported with aphid-resistant wheat

The need of region-wide use of aphid-resistant wheat to manage *Diuraphis noxia* (Kurdjumov) (Hemiptera: Aphididae), Russian wheat aphid, was anticipated in response to its invasion into the North American small grains production system. In contrast, the recognition and value of regionally variable suppression of *D. noxia* by resident natural enemies was realized as the invasion matured.

Diuraphis noxia feeds on the upper surface of younger leaves of wheat, barley, and cool season grasses. Plant injury includes chlorotic tightly furled leaves, stunted plants, bleached spikes containing poorly formed grain, and awns trapped in the injured flag leaf. Yield response is correlated with the percentage infested tillers and the duration of infestation (Archer *et al.*, 1998). Based on a review of three cereal aphid species including *D. noxia* (Brewer *et al.*, 2019), aphid invasions of North American small grain crops generally have started with colonization of a cereal host, followed by geographic range expansion and aphid population outbreaks in part of the potentially exploitable area. The pace of geographic range expansion and locations where population establish are determined primarily by climate and host plant availability. Population outbreaks that cause economic crop injury in a region appear to be related to periods of favorable weather. *Diuraphis noxia* was first detected in North America in 1978 near Mexico City and was first detected in the US in Texas in 1986. The following range expansion was rapid across wheat and barley grain production and cool season grasses in rangeland of the North American Great Plains: by 1988, infestations occurred from the southern reaches of the Great Plains that extends into Mexico and progressing northward into 16 US western states and three Canadian prairie provinces (Brewer *et al.*, 2019) (Fig. 2.1).

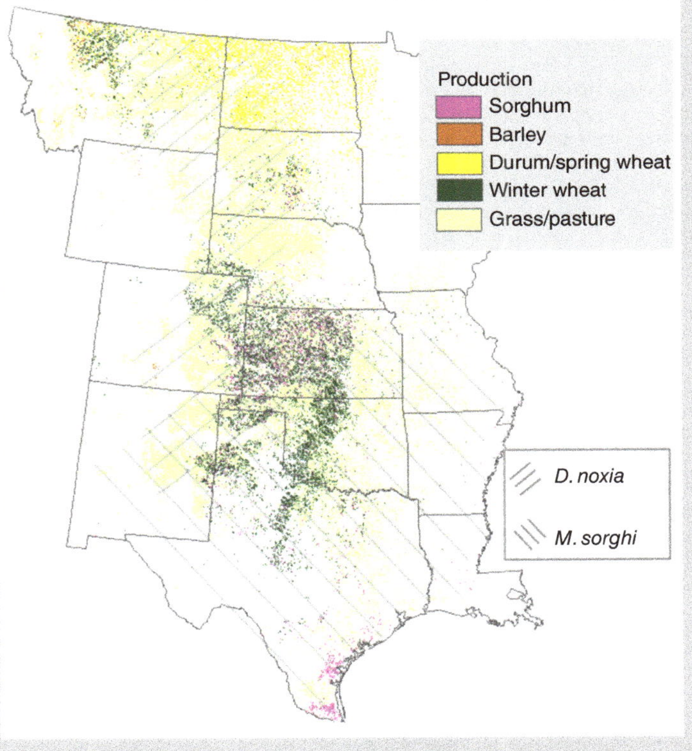

Fig. 2.1. A classified image of cropland of the North American Great Plains. Small grain production is widespread with substantial variation in field sizes and spatial arrangement across this large agroecosystem. This image is color-coded, with the five primary land uses provided (see legend) as derived from CropScape (USDA-NASS, 2023, public domain resource). The diagonal lines indicate the range of two cereal aphid pests, *Diuraphis noxia*, Russian wheat aphid, and *Melanaphis sorghi*, sorghum aphid (added by M.J. Brewer based of published reports; see citations in Brewer *et al.*, 2019).

Continued

Case Study 2.1: Continued.

Diuraphis noxia caused widespread loss to winter wheat and spring barley, with economic losses peaking in the late 1980s. Approximately 65% of losses attributed to *D. noxia* infestations occurred in areas of relatively low rainfall west of the 100th meridian (Morrison and Peairs, 1998), suggesting pest sensitivity to environmental conditions. Year-round anholocycly predominated in North America, and aphids survived temperatures as low as –25°C (Armstrong and Peairs, 1996). These attributes facilitated winter survival on green leaves of winter wheat that led to a heightened risk of outbreaks during warm, dry spring weather (Brewer *et al.*, 2019). As the *D. noxia* invasion matured, less frequent and more local outbreaks were the norm and were associated with an increased abundance of natural enemies in many regions (Brewer *et al.*, 2005, 2008; Giles *et al.*, 2008). A similar pattern occurred for *Melanaphis sorghi* (Theobald) (Hemiptera: Aphididae), sorghum aphid, that invaded sorghum production: rapid spread and severe outbreaks across the US Great Plains followed by lessening and more localized outbreaks associated with increases in natural enemy abundance (Brewer *et al.*, 2022; see Chapter 8, this volume). Range expansion of *M. sorghi* overlapped with *D. noxia* in the Southern and Southcentral Great Plains into Mexico (Fig. 2.1), with landscape associations detected for *M. sorghi* (Brewer *et al.*, 2022; Elkins *et al.*, 2022) and *D. noxia* (Noma *et al.*, 2010).

Crop diversification, conservation tillage and grasslands, and improved small grain cultivars have been broadly implemented in the Great Plains (Dhuyvetter *et al.*, 1996; Giles *et al.*, 2008), providing the opportunity for recognition, and manipulation, if possible, of landscape features that support aphid suppression. Considering the varied landscape across the Great Plains, local (e.g. strip cropping winter wheat with sunflower) and regional (e.g. crop and grassland) landscape diversity benefitted parasitism of *D. noxia* (Ahern and Brewer, 2002; Brewer *et al.*, 2008). In the northwest part of the US Great Plains (northeast Colorado, eastern Wyoming, western Nebraska), increased abundance of the parasitoid *Aphelinus albipodus* (Hayat and Fatima) (Hymenoptera: Aphelinidae) was associated with declines in *D. noxia* abundance over 5 years of data collection (Brewer *et al.*, 2005). Across much of the Great Plains, abundances of coccinellids, nabids, chrysopids, and selective aphid parasitoids were positively correlated with habitat diversity adjacent to cereal production (Elliott *et al.*, 1998b, 2018; Brewer *et al.*, 2008; Giles *et al.*, 2008). This spatial information on aphid abundance and outbreak risk partly guided the need for regional preventive IPM tactics in the form of using aphid-resistant host plants for *D. noxia*. The use of aphid-resistant plants was particularly attractive for wheat, which is extensively grown and a low profit margin crop (Dhuyvetter *et al.*, 1996). The main advantages of improved cultivars with aphid resistance were ease of adoption and potential compatibility with natural enemies innate to the system (Brewer *et al.*, 1999; Randolph *et al.*, 2005). Most deployed *D. noxia*-resistant winter wheat cultivars utilized the *Dn4* gene (Saidi and Quick, 1996). Use of these cultivars was high in southeastern Colorado where mild winters supported *D. noxia* overwintering conducive to spring and early summer *D. noxia* outbreaks. Resistant winter wheat cultivars were used for more than a decade (USDA-NASS, 2004). Slowing aphid population growth rates using resistant host plants can increase the effectiveness of natural enemies (Brewer and Elliott, 2004), which illustrates how a regional IPM tactic can supplement innate pest suppression. A working hypothesis of synergies in plant resistance and natural enemies was seen in laboratory settings using resistant barley (Brewer *et al.*, 1999), and observational field data supported the hypothesis in winter wheat (Brewer and Elliott, 2004). Opportunities to collect field evidence in originally *D. noxia*-prone regions became more limited as the invasion subsided over about 15 years to the point that planting of aphid-resistant wheat declined substantially.

Case Study 2.2: Management of *Myzus persicae* in the context of insecticide resistance

Within Australia, regional IRM has been seen as a way to minimize insecticide resistance risk in *Myzus persicae* (Sulzer) (Hemiptera: Aphididae), the peach-potato aphid. Although not realized in this large-scale system, the Australian experience illustrates the approaches, challenges, and potential outcomes of encouraging IRM at the field level and regionally. Strategies in China and the US are also presented to showcase examples where IPM successes can be achieved to manage *M. persicae* by using regionally focused approaches.

Myzus persicae is one of the most damaging aphid pests worldwide. It has a host range of more than 400 plant species across 40 different families, including many economically important crop plants (Blackman and Eastop, 2000). In addition to causing direct feeding damage, *M. persicae* is

Continued

Case Study 2.2: Continued.

responsible for the transmission of over 100 plant viruses (Van Emden *et al.*, 1969). *Myzus persicae* has an impressive ability to evolve insecticide resistance, with resistance to at least 80 different insecticide compounds reported worldwide (Mota-Sanchez and Wise, 2023). Further challenging the management of this pest is the ability of *M. persicae* to disperse readily between crop fields and over large distances between agricultural regions through the development of alate (winged) forms (e.g. Cocu *et al.*, 2005; Hemming *et al.*, 2022). The ability to effectively protect crop yields from *M. persicae* attack, while at the same time minimizing the selection pressure for insecticide resistance evolution, is greatly enhanced through control strategies applied at a regional level, but this regional approach has rarely been implemented in the field. Rather, individual farmers typically target *M. persicae* populations using chemical tools with little consideration for the pest and resistance management approaches being employed in neighboring fields. Increasing the research knowledge base and farmer awareness of the spatial implications of IRM could improve understanding and adoption of regional-based IRM strategies.

In Australia, *M. persicae* is a major threat to the grain, horticulture, and forage industries, and like other countries, this threat is exacerbated by widespread resistance to insecticides (Umina *et al.*, 2019). In response to growing concerns surrounding resistance in major crop pests, the Australian grains industry commissioned the development of several Insecticide Resistance Management Strategies (IRMSs) between 2014 and 2018. This included a national IRMS for *M. persicae* in canola, first launched in 2014. This IRMS was centered on management actions applicable at the field level, chiefly underpinned by the judicious use of insecticides: (i) only applying chemicals when the pest infestation warrants it, (ii) avoiding the application of broad-spectrum formulations as much as is practical, and (iii) rotation of formulations whereby the same insecticide mode of action is not applied across consecutive aphid generations. This IRMS also recommended numerous IPM tactics, such as encouraging natural enemies and minimizing the risk of aphid buildup on weeds prior to sowing (Department of Agriculture and Fisheries, 2016) (Fig. 2.2A).

Prior to the introduction of the IRMS in 2014, Australian *M. persicae* were already known to possess numerous resistance mechanisms, including widespread resistances to pyrethroid, organophosphate, and carbamate insecticides (Edwards *et al.*, 2008; Umina *et al.*, 2014) and localized resistance to neonicotinoids (de Little *et al.*, 2017). It was hoped the adoption of the IRMS by Australian farmers

Fig. 2.2. Agroecosystem-focused integrated pest management (IPM) has an important role to play in sustainably managing agricultural pests such as *Myzus persicae*, which is highly mobile and has evolved widespread insecticide resistances. A national Insecticide Resistance Management Strategy for *Myzus persicae* was developed in Australia in 2014 (A) and widely disseminated to farmers and farm advisors through multiple forums, including field walks and training workshops (B). (Graphic (A) produced from literature in the public domain and photograph (B) is the author's own: P.A. Umina.)

Continued

Case Study 2.2: Continued.

would prevent or delay new resistances evolving and help regain susceptibility in field populations in which resistance had already arisen. Despite considerable efforts to encourage the adoption of this IRMS by industry (Fig. 2.2B), resistance issues have continued to increase in Australia over the last decade. Pyrethroid, organophosphate, and carbamate resistances remain commonplace in *M. persicae*, while neonicotinoid resistance is also now widespread (P.A. Umina, unpublished data). Moreover, several new resistances have arisen, including resistance to sulfoximines (Pym *et al.*, 2022) and ketoenols (Umina *et al.*, 2022). This inability to curb insecticide resistance evolution in *M. persicae* has been observed in other countries (e.g. Bass *et al.*, 2014) and provides strong justification for agroecosystem-focused pest management approaches when dealing with highly mobile pests that possess insecticide resistance.

The spatial context of IRM and need for large-scale coordination are well recognized, but application is not straightforward for *M. persicae* due to the myriad of logistic, information, economic, and social challenges. Short-term economic priorities at the individual level continue to outweigh chemical stewardship goals, and there is often a conflict between local management objectives (e.g. "insurance" insecticide sprays) and those based on regional priorities (e.g. reducing chemical applications to minimize resistance risks) (Umina *et al.*, 2019). None the less, challenges in applying regional IPM have been successfully overcome in Australia for other pest species impacting large-scale field crop systems. Perhaps the most well-known example comes from the Australian cotton industry, where IPM has been widely adopted, principally in response to widespread issues with insecticide resistance in *H. armigera* (cotton bollworm) and other key pests in the mid-late 1990s (Wilson *et al.*, 2013, 2018). Importantly, an areawide approach enabled these IPM strategies to be coordinated across large areas and different industries, including the management of native vegetation to support the buildup of natural enemy populations over large scales (Kelly *et al.*, 2004; Wilson *et al.*, 2018). Cotton as a large-scale crop and the persistent regional impact of insecticide resistance were valuable to gaining support for regional coordination.

Outside of Australia, other forms of regional application of management tactics have been applied for *M. persicae*. One such case comes from the US, where in the 1970s an IPM program was initiated in response to potato leafroll virus vectored by *M. persicae*. Although little publicized, this program was reported to be quite successful. This program entailed the identification and treatment of peach trees across large spatial areas in central and eastern Idaho (Byrne, 2008). Peach trees were the primary host of *M. persicae* and were identified to be responsible for harboring aphid populations overwintering in the region, allowing immigration of aphid populations into potato fields in the spring period. The implementation of this program across large areas reportedly led to a reduction in the number of spring migrant *M. persicae* that moved into potato crops (Byrne, 2008). Clearly this success could not have been realized without a well-coordinated regional approach formulated with consideration of vector competence and movement of *M. persicae* across the landscape.

A more recent case comes from China, where an impressive areawide approach has been undertaken to help manage insecticide-resistant *M. persicae* in the tobacco industry by using biological control. This program is centered around the mass rearing and augmentative release of the parasitoid *Aphidius gifuensis* (Ashmead) (Hymenoptera: Braconidae) (Yu *et al.*, 2021). First established in 2010, this program has grown to encompass more than 500 mass rearing facilities across 16 provinces. A key element of this program has been the development of a training system, which has been delivered to thousands of technical experts, technicians, extension workers, and farmers in China. Importantly, from a resistance evolution perspective, this areawide program has directly resulted in a decrease in chemical usage by almost 2000 tons (Yu *et al.*, 2021). This ambitious project, covering 3 million ha between 2010 and 2015, has reportedly achieved large-scale suppression of *M. persicae* in tobacco and other crops, whereby aphid management has been transformed from a predominantly passive, reactive and insecticide-based approach to an active, preventive, and ecologically founded program. Further efficiencies in the mass release of *A. gifuensis* could be achieved by mapping crops, aphid movement and abundance, and influencing landscape features using GIS combined with modeling and big data analysis. Managing landscapes surrounding crop fields to support the persistence and buildup of natural enemies may further increase aphid suppression (Schellhorn *et al.*, 2015). Although unlikely to be replicated in other countries, at least not to the same scale, this example illustrates the potential of coordinated regional use of IPM and IRM tactics in minimizing resistance issues.

Case Study 2.3: Biological control by Coccinellidae

Suppression of aphids by coccinellids in large-scale field crops is spatially and temporally variable. Despite this variability, coccinellids appear to be key natural enemies that suppress aphids in field crops and at times achieve biological control that prevents economic crop loss.

Aphids can thrive in agricultural landscapes and several species are serious pests of large-scale field crops in the US and worldwide (Andow, 1983; Way, 1988; Brewer *et al.*, 2019). Coccinellids are often the most abundant among aphid predators in agricultural crops. Coccinellid populations fluctuate widely from year to year. Fluctuations of populations of coccinellid species are often independent (Elliott and Kieckhefer, 1990; Obrycki *et al.*, 2009)., indicating that species, perhaps unsurprisingly, have varying life histories, habitat and prey affinities, and ecological requirements (Hodek and Honěk, 1996; Dixon, 2000; Obrycki *et al.*, 2009). Aphidophagous Coccinellidae can suppress aphid populations in crops although effective control can be inconsistent and sometimes insufficient to curtail economic losses (Hodek and Honěk, 1996; Dixon, 2000; Obrycki *et al.*, 2009; Michaud, 2012). In the US, suppression of aphids by coccinellids is common in large-scale field crops, although achieving effective biological control is spatially and temporally variable (Kring *et al.*, 1985; Michels *et al.*, 2001; Michels and Natis, 2008; Gardiner *et al.*, 2009; Obrycki *et al.*, 2009).

At a local scale, simple explanations of population fluctuations and subsequent aphid suppression, such as variation in prey density and winter survival among years, are inadequate for understanding the population dynamics of coccinellid in agricultural crops. Several factors occurring in the crop, nearby crops, and semi-natural non-crop areas are involved in coccinellid population dynamics (Obrycki *et al.*, 2009). Ranking the importance of these factors and certainly estimating suppression in particular circumstances is often not possible due to limited understanding of the intricacies of coccinellid biology and ecology. However, understanding of factors involved in coccinellid population and community dynamics has progressed in recent years. Strategies in vegetation management to enhance biological control using coccinellids have advanced (Obrycki *et al.*, 2009). There has been widespread acceptance of this approach in the conservation biological control literature (Barbosa, 1998), and some governments provide growers with financial incentives to implement such strategies (Kleijn *et al.*, 2006; Brewer and Goodell, 2012).

Ecologically, an issue challenging development of best strategies is that the processes governing and factors influencing aphid suppression by coccinellids may be best understood from a landscape and temporal perspective. Coccinellid population and community dynamics are determined by factors operating at multiple spatial and temporal scales (Gardiner *et al.*, 2009; Chaplin-Kramer *et al.*, 2011). Factors that appear to be important at a particular scale may not be relevant to the population dynamics of a species. For example, if agricultural crops are not preferred habitats, population fluctuations in agricultural fields would provide little insight into landscape-scale population fluctuations, especially if the species did not reproduce in the fields. In this case, a species that reproduces in semi-natural habitat may visit a crop to forage for abundant aphid prey, thereby contributing to biocontrol. For other species, an annual crop may have sufficient prey and other resources to support natural enemy reproduction in the crop, further increasing the likelihood of aphid biocontrol. This knowledge is relevant to developing best approaches to manage semi-natural vegetation to support coccinellids, such as inserting flowering strips along individual field edges or broader scale conservation of semi-natural areas in a growing region (see Chapter 5, this volume).

Population fluctuations of some coccinellid species are correlated over great distances. For example, from 1988 to 1992, coccinellid populations in several alfalfa fields in eastern South Dakota and central Iowa were sampled by sweep nets during the growing season. These two locations were separated by over 300 km. Spearman rank correlations of abundance at the two locations were significant for three of six species, indicating that populations of these species probably responded similarly to environmental factors, such as macro-weather events, that were often correlated over large geographic areas (Obrycki *et al.*, 2000). Dispersal over broad geographic areas also may occur among populations of some coccinellid species as indicated by similarity in gene frequencies over great distances (Krafsur *et al.*, 1992, 1996).

At the regional scale, factors such as topography, climate, and spatial heterogeneity (e.g. the ratio and spatial mixing of cropland and semi-natural vegetation) determine the size and diversity of the pool of potential colonists of a particular habitat (e.g. the number and diversity of coccinellids moving into a focal crop threatened by aphids) (Honěk, 1985; Colunga-Garcia *et al.*, 1997; Elliott *et al.*, 1998b). Landscape

Continued

Case Study 2.3: Continued.

structure influences the availability and spatial distribution of food and basic requirements such as overwintering habitat, but its effects are intertwined with the life history of a species. For example, dispersal behaviors differ within the seasonal life cycle of a species and among species (Hagen, 1962; Ewert and Chiang, 1966; Hodek *et al.*, 1993; Elliott *et al.*, 1998a). The successful exploitation of resources by a species in a mosaic of land use types is largely dependent on characteristics of movement in relation to landscape structure (Dunning *et al.*, 1992). At a local scale, plant community composition (e.g. the ratio of the length of field edges with flowering strips to the crop size) and physical structure (due to its influence on microclimate), prey abundance (e.g. aphids in a crop, field edge, or nearby semi-natural vegetation), and interspecies interactions are key determinants of coccinellid abundance in agricultural crops (Hodek and Honěk, 1996). But from a regional perspective, the local processes and ability to support coccinellids by vegetation management in or near the crop are contingent on the size and diversity of the species pool as affected by larger-scale influences. Understanding the pests, natural enemies, and landscape influences locally and regionally is a key challenge that spills over into applied challenges in recognizing and supporting biological control of coccinellids in the following examples.

Aphis glycines, soybean aphid, was first discovered in the northcentral US in the summer of 2000. Intensive research effort ensued to identify the existing natural enemies that fed on *A. glycines* and their roles in biological control of the aphid. Over 40 species of arthropod predators from several taxa were associated with *A. glycines* in the field (Rutledge *et al.*, 2004). Several species of parasitoids were also present (Noma *et al.*, 2010). The species complex of aphid predators in soybean fields varied geographically and over time, but overall coccinellid species were among the most abundant predators and exerted greatest predation on *A. glycines* (Mignault *et al.*, 2006). In a multi-state study of *A. glycines* suppression by natural enemies, it was found that suppression was sufficient to prevent economic injury when initial aphid colonization of the crop was low but not when it was high (Landis *et al.*, 2008). Initial colonization fluctuated among years and geographically in relation to factors such as episodic extreme weather events and variation in cropping systems. For example, Landis *et al.* (2008) noted that an approximately 20% increase in corn production in four midwestern US states decreased the supply of natural enemies to soybean fields, thereby reducing effectiveness of biological control of *A. glycines* in soybean. Predicting the successful suppression of *A. glycines* in fields is difficult because of the highly variable abiotic and biotic environment, but *A. glycines* suppression by resident predators has a significant economic impact valued in millions of dollars annually (Landis *et al.*, 2008). From an IPM viewpoint, the recognition of varying aphid suppression that is linked to the mix of soybean and corn production may help assess the region-scaled aphid risk to soybean.

In another example, *Schizaphis graminum* (Rondani) (Hemiptera: Aphididae), greenbug, and *D. noxia* are severe aphid pests of wheat and grain sorghum in the Southwestern Great Plains. Insect predators, predominantly coccinellids, cause 50–60% reductions of infestations of *S. graminum* and *D. noxia* in winter wheat in this large area of grain production (Michels *et al.*, 2001), although the degree of aphid suppression varies among years (Michels *et al.*, 2001; Lee *et al.*, 2005). Adult coccinellid foraging in wheat also affects aphids early in the growing season by disrupting aphids on plants and increasing aphid falling rates, and hence aphid mortality (Kerzicnik *et al.*, 2010). The economic impact of biological control services provided by coccinellids has not been determined in winter wheat in the southwestern US, but field observations support a working hypothesis that aphid suppression by coccinellids was sufficient to prevent economic injury to wheat in some circumstances (Michels and Matis, 2008). From a farm-level view, assessment of aphid risk by monitoring individual fields for pests remains vital given this uncertainty, but periodic biological control of aphids is welcomed.

et al., 2021). The composition, configuration, diversity, and scale of agricultural and environmental features have been shown to affect pest management services innate to agroecosystems (Perović *et al.*, 2018; Martin *et al.*, 2019; Brewer *et al.*, 2022; Dolezal *et al.*, 2022). Spatial data management and analysis can also help tackle the escalating issues with insecticide resistance in intensively cropped agroecosystems (Umina *et al.*, 2019; Onstad and Knolhoff, 2022). Such advancements set the stage to unmask pest and pest suppression associations with landscape (agricultural and environmental) conditions and help guide the appropriate application of

IPM and IRM tactics regionally. These managed regional tactics may be chosen to complement innate pest suppression and minimize the risks associated with insecticide resistance. They may be supported by underappreciated positive influences of landscape structure.

Future Needs and Directions

Advances in landscape ecology and GIS have helped reveal pest and natural enemy associations with landscape (agricultural and environmental) conditions. Good understanding of these relationships coupled with on-the-ground information on regional pest status set the stage to increase use and reap the benefits of agroecosystem-focused IPM. Indeed, past accomplishments in using regionally applied IPM tactics have demonstrated substantial benefits, such as use of host plant resistance to manage *D. noxia* in large-scale cereal production in North America. Present day information and technical resources (e.g. agricultural outreach and GIS technologies where available; see Chapters 3 and 4, this volume) are largely sufficient for planning regional use of pest resistance crops and insecticide strategies to manage resistance risks (e.g. *M. persicae*). But recommendations for other forms of preventive IPM tactics require a finer grain of understanding of trophic interactions. An example presented in this chapter is the use of semi-natural (non-crop) vegetation management from local to regional scales to bolster suppression of crop aphids by coccinellids in context of landscape influences. This example demonstrates the need for experimental evaluation to properly understand the influence of specific types, structures, and diversity of semi-natural vegetation on pest and natural enemy populations.

Such an expansion of agroecological research is critical to bridge concepts into the practice of applying regional IPM tactics (see Chapters 3, 5, 6, and 8, this volume). Currently, agroecology studies in pest management tend to report landscape influences as coarse-grain (positive or negative) effects on natural enemy activity and more rarely pest suppression. Improved estimation of these landscape influences to justify predictions of pest suppression under a range of agroecosystem conditions should be pursued when candidate regional IPM tactics and good indicator landscape metrics have been identified (Brewer and Elliott, 2023). To further increase adoption among farmers, fine-grain data is likely needed to achieve high predictability of pest management and crop production outcomes on a regional basis. Although more difficult to achieve, these resources could also be used to monitor pest management and crop health outcomes as regional-based IPM tactics are deployed to supplement innate pest suppression (e.g. conservation of natural enemies, plant resistance, augmentative biological control) (Gurr *et al.*, 2004; Carrière *et al.*, 2006; Segre *et al.*, 2020; Albrecht *et al.*, 2021). Additionally, future research should consider the sensitivity of pests and natural enemies to major agricultural and environmental changes such as climate change, adaptation of invasive species to a new geographic range, and policies that affect agricultural land use. Ecology, management, policy, and socio-economic conditions (see Chapters 3 and 4, this volume) must come together as research and demonstration projects are proposed at scales relevant to large-scale agroecosystems. Selected large-scale systems of production, some introduced in this chapter, are well positioned to justify such detailed research and efforts to increase use of agroecosystem-focused IPM.

Regarding implementation, the level of adoption of regionally applied IPM tactics has been mixed. The buy-in by land managers and public agencies and the social and technical aspects of sharing information on pest status and use of IPM and IRM tactics remain an important hurdle to increase use of regional IPM tactics. Implementing and coordinating preventative IPM strategies (such as in the form of vegetative management in- and off-crop) have in some cases been financially supported by "green" or "conservation" financial incentives in the US, European Union (EU), and other countries with significant governmental agricultural programs tied to environmental conservation goals (Kleijn *et al.*, 2006; Brewer and Goodell, 2012). In other cases, there has been a concerted effort coupled with considerable investment by industry to address escalating pest issues, as illustrated by the mass rearing and release of *A. gifuensis* to control *M. persicae* in China (Yu *et al.*, 2021). Programs in the US, Australia, and many other countries tend to be farm-level focused (Brewer and Goodell, 2012), while EU programs are inclusive of regional efforts in implementing strategies such as hedgerows for joint bird, pollinator, and natural enemy conservation (Kleijn *et al.*, 2006). Availability and regional focus of such programs to coordinate and financially support use of IPM tactics to meet pest suppression and other conservation goals would help overcome the economic and social aspects of buy-in of agroecosystem-focused IPM that includes regional action.

Summary and Conclusions

Agroecosystem-focused IPM considers landscape-level processes and influences affecting pest ecology and management. A region-level assessment of pest abundance, severity, and innate suppression is placed in the spatial context of agricultural and environmental conditions. This regional pest risk assessment guides selection and regional use of primarily preventative IPM tactics to supplement innate pest suppression, decrease pest risk, and in some cases reduce insecticide resistance risk. This approach is becoming more feasible as availability of spatial data and advances in spatial analysis tools have increased research capacity to characterize pest suppression regionally and help guide use of preventative pest management tactics. The case studies presented here illustrated that agroecosystem-focused IPM provides a pathway to increase the likelihood of favorable pest management and crop health outcomes and the overall resiliency to pests in large-scale agroecosystems.

To improve adoption in large-scale production agriculture, the approach needs improved regional coordination to implement and measure agricultural outcomes as well as environmental value of regionally applied IPM tactics. Feedback loops of pest assessment and management-oriented information (e.g. natural enemy activity, extent of use of insecticides, plant resistance, vegetation management) are also needed between the regional and local levels. And importantly, an agroecosystem orientation of IPM must be balanced with the recognition of the continued need to respond to localized pest threats associated with local weather and other conditions that favor pest outbreaks at the farm level. Local and regional IPM and IRM tactics working together may reveal best approaches to reduce the overall impact of pests on a regional basis while maintaining flexibility to respond to localized pest threats in large-scale agroecosystems.

Acknowledgments

The concepts and practices put forward here have been developed through discussions with colleagues and study of the rich literature of IPM and IRM, areawide pest management, and agroecology. This research was conducted partly under the umbrella of the USDA National Institute of Food and Agriculture, Hatch project TEXO-2-9394 assigned to M.J.B. at Texas A&M AgriLife Research. There is no intent to recommend a specific course of action or use specific commercial products in specific crop production systems.

References

Ahern, R.G. and Brewer, M.J. (2002) Effect of different wheat production systems on the presence of two parasitoids (Hymenoptera: Aphelinidae, Braconidae) of the Russian wheat aphid in the North American Great Plains. *Agriculture Ecosystems & Environment* 92, 201–210. DOI: 10.1016/S0167-8809(01)00296-1

Albrecht, M., Kleijn, D., Williams, N.M., Tschumi, M., Blaauw, B.R. *et al.* (2021) The effectiveness of flower strips and hedgerows on pest control, pollination services and crop yield: a quantitative synthesis. *Ecology Letters* 10, 1488–1498. DOI: 10.1111/ele.13576

Andow, D. (1983) The extent of monoculture and its effects on insect pest populations with particular reference to wheat and cotton. *Agriculture, Ecosystems & Environment* 9, 25–35. DOI: 10.1016/0167-8809(83)90003-8

Archer, T.L., Peairs, F.B., Pike, K.S., Johnson, G.D. and Kroening, M. (1998) Economic injury levels for the Russian wheat aphid (Homoptera: Aphididae) on winter wheat in several climate zones. *Journal of Economic Entomology* 91, 741–747. DOI: 10.1093/jee/91.3.741

Armstrong, J.S. and Peairs, F. (1996) Environmental parameters related to winter mortality of the Russian wheat aphid (Homoptera: Aphididae): basis for predicting mortality. *Journal of Economic Entomology* 89, 1281–1287. DOI: 10.1093/jee/89.5.1281

Barbosa, P. (1998) *Conservation Biological Control*. San Diego, CA: Academic Press.

Bass, C., Puinean, A.M., Zimmer, C.T., Denholm, I., Field, L.M. *et al.* (2014) The evolution of insecticide resistance in the peach potato aphid, *Myzus persicae*. *Insect Biochemistry and Molecular Biology* 51, 41–51. DOI: 10.1016/j.ibmb.2014.05.003

Blackman, R.L. and Eastop, V.F. (2000) *Aphids on the World's Crops: An Identification and Information Guide*, 2nd edn. Chichester, UK: Wiley.

Brewer, M.J. and Elliott, N.C. (2004) Biological control of cereal aphids in North America and mediating effects of host plant and habitat manipulations. *Annual Review of Entomology* 49, 219–242. DOI: 10.1146/annurev.ento.49.061802.123149

Brewer, M.J. and Elliott, N.C. (2023) Recent advances in agroecological research for increasing scope of areawide pest management of arthropods in cropping systems. *Current Opinion in Insect Science* 56, 101019. DOI: 10.1016/j.cois.2023.101019

Brewer, M.J. and Goodell, P.G. (2012) Approaches and incentives to implement integrated pest management that addresses regional and environmental issues. *Annual Review of Entomology* 57, 41–59. DOI: 10.1146/annurev-ento-120709-144748

Brewer, M.J., Mornhinweg, D.W. and Huzurbazar, S. (1999) Compatibility of insect management strategies: *Diuraphis noxia* (Hom: Aphididae) abundance on susceptible and resistant barley in the presence of parasitoids. *BioControl* 43, 479–491. DOI: 10.1023/A:1009929301953

Brewer, M.J., Noma, T. and Elliott, N.C. (2005) Hymenopteran parasitoids and dipteran predators of the invasive aphid *Diuraphis noxia* after enemy introductions: temporal variation and implication for future aphid invasions. *Biological Control* 33, 315–323. DOI: 10.1016/j.biocontrol.2005.03.013

Brewer, M.J., Noma, T., Elliott, N.C., Kravchenko, A.N. and Hild, A.L. (2008) A landscape view of cereal aphid parasitoid dynamics reveals sensitivity to farm- and region-scale vegetation structure. *European Journal of Entomology* 105, 503–511. DOI: 10.14411/eje.2008.066

Brewer, M.J., Peairs, F.B. and Elliott, N.C. (2019) Invasive cereal aphids of North America: ecology and pest management. *Annual Review of Entomology* 64, 73–93. DOI: 10.1146/annurev-ento-011118-111838

Brewer, M.J., Elliott, N.C., Esquivel, I.L., Jacobson, A.L., Faris, A.M. *et al.* (2022) Natural enemies, mediated by landscape and weather conditions, shape response of the sorghum agroecosystem of North America to the invasive aphid *Melanaphis sorghi*. *Frontiers in Insect Science* 2, 830997. DOI: 10.3389/finsc.2022.830997

Byrne, D.N. (2008) Dispersal and migration of insects and their importance in pest management. In: Koul, O., Cuperus, G.W. and Elliott, N.C. (eds) *Areawide Pest Management: Theory to Implementation*. Wallingford, UK: CAB International, pp. 60–80.

Carriére, Y., Ellsworth, P., Dutilleul, P., Kirk, C., Barkley, V. and Antilla, L. (2006) A GIS-based approach for areawide pest management: the scales of *Lygus hesperus* movements to cotton from alfalfa, weeds, and cotton. *Entomologia Experimentalis et Applicata* 118, 203–210. DOI: 10.1111/j.1570-7458.2006.00384.x

Chaplin-Kramer, R., O'Rourke, M.E., Blitzer, E.J. and Kremen, C. (2011) A meta-analysis of crop pest and natural enemy response to landscape complexity. *Ecology Letters* 14, 922–932. DOI: 10.1111/j.1461-0248.2011.01642.x

Cocu, N., Harrington, R., Hullé, M. and Rounsevell, M.D.A. (2005) Spatial autocorrelation as a tool for identifying the geographical patterns of aphid annual abundance. *Agricultural and Forest Entomology* 7, 31–43. DOI: 10.1111/j.1461-9555.2005.00245.x

Colunga-Garcia, M., Gage, S.H. and Landis, D.A. (1997) Response of all assemblage of Coccinellidae (Coleoptera) to a diverse agricultural landscape. *Environmental Entomology* 26, 797–804. DOI: 10.1093/ee/26.4.797

Courson, E., Petit, S., Poggi, S. and Ricci, B. (2022) Weather and landscape drivers of the regional level of pest occurrence in arable agriculture: a multi-pest analysis at the French national scale. *Agriculture, Ecosystems & Environment* 338, 108105. DOI: 10.1016/j.agee.2022.108105

de Little, S.C., Edwards, O., van Rooyen, A., Weeks, A. and Umina, P. (2017) Discovery of metabolic resistance to neonicotinoids in green peach aphids (*Myzus persicae*) in Australia. *Pest Management Science* 73, 1611–1617. DOI: 10.1002/ps.4495

Department of Agriculture and Fisheries. (2016) *IPM Guidelines for Grains*. National Invertebrate Pest Initiative, Queensland Government, Australia. Available at: https://ipmguidelinesforgrains.com.au/ (last accessed July 2023).

Dhuyvetter, K.C., Thompson, C.R., Norwood, C.A. and Halvorson, A.D. (1996) Economics of dryland cropping systems in the Great Plains: a review. *Journal of Production Agriculture* 9, 216–222. DOI: 10.2134/jpa1996.0216

Dixon, A.F.G. (2000) *Insect Predator–Prey Dynamics: Ladybird Beetles and Biological Control*. Cambridge, UK: Cambridge University Press.

Dolezal, A.J., Esch, E.H. and MacDougall, A.S. (2022) Restored marginal farmland benefits arthropod diversity at multiple scale. *Restoration Ecology* 30, e13485. DOI: 10.1111/rec.13485

Downes, S., Kriticos, D., Parry, H., Paull, C., Schellhorn, N. and Zalucki, M.P. (2017) A perspective on management of *Helicoverpa armigera*: transgenic Bt cotton, IPM, and landscapes. *Pest Management Science* 73, 485–492. DOI: 10.1002/ps.4461

Dunning, J.B., Danielson, B.J. and Pulliam, H.R. (1992) Ecological processes that affect populations in complex landscapes. *Oikos* 65, 169–175. DOI: 10.2307/3544901

Edwards, O.R., Franzmann, B., Thackray, D. and Micic, S. (2008) Insecticide resistance and implications for future aphid management in Australian grains and pastures: a review. *Australian Journal of Experimental Agriculture* 48, 1523–1530. DOI: 10.1071/EA07426

Elkins, B.H., Eubanks, M.D., Faris, A.M., Wang, H.H. and Brewer, M.J. (2022) Landscape complexity has mixed effects on an invasive aphid and its natural enemies in sorghum agroecosystems. *Environmental Entomology* 51, 660–669. DOI: 10.1093/ee/nvac036

Elliott, N.C. and Kieckhefer, R.W. (1990) Dynamics of aphidophagous coccinellid assemblages in small grain fields in eastern South Dakota. *Environmental Entomology* 19, 1320–1329. DOI: 10.1093/ee/19.5.1320

Elliott, N.C., Kieckhefer, R.W., French, B.W. and Lee, J.H. (1998a) Variation in size, sex ratio, and dispersal in three coccinellid (Coleoptera) species. *Southwestern Entomologist* 23, 137–145.

Elliott, N.C., Kieckhefer, R.W., Lee, J.H. and French, B.W. (1998b) Influence of within-field and landscape factors on aphid predator populations in wheat. *Landscape Ecology* 14, 239–252. DOI: 10.1023/A:1008002528345

Elliott, N.C., Onstad, D.W. and Brewer, M.J. (2008) History and ecological basis for areawide pest management. In: Koul, O., Cuperus, G.W. and Elliott, N.C. (eds) *Areawide Pest Management: Theory to Implementation*. Wallingford, UK: CAB International, pp. 15–33.

Elliott, N.C., Brewer, M.J. and Giles, K.L. (2018) Landscape context affects aphid parasitism by *Lysiphlebus testaceipes* in wheat fields. *Environmental Entomology* 47, 803–811. DOI: 10.1093/ee/nvy035

Ewert, M.A. and Chiang, H.C. (1966) Dispersal of three species of Coccinellids in corn fields. *The Canadian Entomologist* 98, 999–1003. DOI: 10.4039/Ent98999-9

Forrester, N.W. (1990) Designing, implementing, and servicing an insecticide resistance management strategy. *Pesticide Science* 28, 167–179. DOI: 10.1002/ps.2780280205

Gardiner, M.M., Landis, D.A., Gratton, C., DiFonzo, C.D., O'Neal, M. *et al.* (2009) Landscape diversity enhances biological control of an introduced crop pest in the north-central USA. *Ecological Applications* 19, 143–154. DOI: 10.1890/07-1265.1

Giles, K., Hein, G. and Peairs, F. (2008) Areawide pest management of cereal aphids in dryland wheat systems of the Great Plains, USA. In: Koul, O., Cuperus, G.W. and Elliott, N.C. (eds) *Areawide Pest Management: Theory to Implementation*. Wallingford, UK: CAB International, pp. 441–466.

Gurr, G., Scarratt, S., Wratten, S., Brendt, L. and Irvin, N. (2004) Ecological engineering, habitat manipulation and pest management. In: Gurr, G.M., Wratten, S.D. and Altieri, M.A. (eds) *Ecological Engineering for Pest Management: Advances in Habitat Manipulation for Arthropods*. Collingwood, Australia: CSIRO, pp. 1–12.

Hagen, K.S. (1962) Biology and ecology of predaceous Coccinellidae. *Annual Review of Entomology* 7, 289–326. DOI: 10.1146/annurev.en.07.010162.001445

Hemming, D., Bell, J., Collier, R., Dunbar, T., Dunstone, N. *et al.* (2022) Likelihood of extreme early flight of *Myzus persicae* (Hemiptera: Aphididae) across the UK. *Journal of Economic Entomology* 115, 1342–1349. DOI: 10.1093/jee/toac012

Hodek, I. and Honěk, A. (1996) *Ecology of Coccinellidae*, Vol. 54. Dordrecht, Netherlands: Springer.

Hodek, I., Iperti, G. and Hodkova, M. (1993) Long-distance flights in Coccinellidae (Coleoptera). *European Journal of Entomology* 90, 403–414.

Honěk, A. (1985) Habitat preferences of aphidophagous coccinellids [Coleoptera]. *Entomophaga* 30, 253–264. DOI: 10.1007/BF02372226

Jeanneret, P., Aviron, S., Alignier, A., Lavigne, C., Helfenstein, J. *et al.* (2021) Agroecology landscapes. *Landscape Ecology* 36, 2235–2257. DOI: 10.1007/s10980-021-01248-0

Karp, D.S., Chaplin-Kramer, R., Meehan, T.D., Martin, E.A., DeClerck, F. *et al.* (2018) Crop pests and predators exhibit inconsistent responses to surrounding landscape composition. *Proceedings of the National Academy of Sciences* 115, E7863–E7870. DOI: 10.1073/pnas.1800042115

Kelly, D., Grundy, P. and Sequeira, R. (2004) The silverleaf whitefly management challenge: a new pest in Central Queensland. Presented at: *12th Australian Cotton Conference, August 10–12, Broadbeach, Australia*.

Kerzicnik, L.M., Peairs, F.B. and Harwood, J.D. (2010) Implications of Russian wheat aphid, *Diuraphis noxia*, falling rates for biological control in resistant and susceptible winter wheat. *Arthropod-Plant Interactions* 4, 129–138. DOI: 10.1007/s11829-010-9085-4

Kleijn, D., Baquero, R.A., Clough, Y., Diaz, M., De Esteban, J. *et al.* (2006) Mixed biodiversity benefits of agri-environment schemes in five European countries. *Ecology Letters* 9, 243–254. DOI: 10.1111/j.1461-0248.2005.00869.x

Kogan, M. (1998) Integrated pest management: historical perspectives and contemporary developments. *Annual Review of Entomology* 43, 243–270. DOI: 10.1146/annurev.ento.43.1.243

Koul, O., Cuperus, G.W. and Elliott, N.C. (eds) (2008) *Areawide Pest Management: Theory to Implementation*. Wallingford, UK: CAB International.

Krafsur, E.S., Obrycki, J.J. and Flanders, R.V. (1992) Gene flow in populations of the seven-spotted lady beetle, *Coccinella septempunctata*. *Journal of Heredity* 83, 440–444. DOI: 10.1093/oxfordjournals.jhered.a111248

Krafsur, E.S., Nariboli, P. and Obrycki, J.J. (1996) Gene flow and diversity at allozyme loci in the two spotted lady beetle (Coleoptera: Coccinellidae). *Annals of the Entomological Society of America* 89, 410–419. DOI: 10.1093/aesa/89.3.410

Kring, T.J., Gilstrap, F.E. and Michels, G.J., Jr (1985) Role of indigenous coccinellids in regulating greenbugs (Homoptera: Aphididae) on Texas grain sorghum. *Journal of Economic Entomology* 78, 269–273. DOI: 10.1093/jee/78.1.269

Landis, D.A., Gardiner, M.M., van der Werf, W. and Swinton, S.M. (2008) Increasing corn for biofuel production reduces biocontrol services in agricultural landscapes. *Proceedings of the National Academy of Sciences* 105, 20552–20557. DOI: 10.1073/pnas.0804951106

Lee, J.H., Elliott, N.C., Kindler, S.D., French, B.W., Walker, C.B. and Eikenbary, R.D. (2005) Natural enemy impact on the Russian wheat aphid in southeastern Colorado. *Environmental Entomology* 34, 115–123. DOI: 10.1603/0046-225X-34.1.115

Martin, E.A., Dainese, M., Clough, Y., Báldi, A., Bommarco, R. *et al.* (2019) The interplay of landscape composition and configuration: new pathways to manage functional biodiversity and agroecosystem services across Europe. *Ecology Letters* 22, 1083–1094. DOI: 10.1111/ele.13265

Michaud, J.P. (2012) Coccinellids in biological control. In: Hodek, A., van Emden, H.F. and Honěk, A. (eds) *Ecology and Behavior of the Ladybird Beetles (Coccinellidae)*. New York: Wiley, pp. 488–519.

Michels, G.J., Jr and Matis, J.H. (2008) Corn leaf aphid, *Rhopalosiphum maidis* (Hemiptera: Aphididae), is a key to greenbug, *Schizaphis graminum* (Hemiptera: Aphididae), biological control in grain sorghum, *Sorghum bicolor*. *European Journal of Entomology* 105, 513–520. DOI: 10.14411/eje.2008.068

Michels, G.J., Elliott, N.C., Romero, R.A., Owings, D.A. and Bible, J.B. (2001) Impact of indigenous coccinellids on Russian wheat aphids and greenbugs (Homoptera: Aphididae) infesting winter wheat in the Texas Panhandle. *Southwestern Entomologist* 26, 97–114.

Mignault, M.P., Roy, M. and Brodeur, J. (2006) Soybean aphid predators in Quebec and the suitability of *Aphis glycines* as prey for three Coccinellidae. *BioControl* 51, 89–106. DOI: 10.1007/s10526-005-1517-1

Morrison, W.P. and Peairs, F.B. (1998) Response model concept and economic impact. In: Quisenberry, S.S. and Peairs, F.B. (eds) *Response Model for an Introduced Pest – The Russian Wheat Aphid*. Lanham, MD: Entomological Society of America, pp. 1–11.

Mota-Sanchez, D. and Wise, J.C. (2023) *The Arthropod Pesticide Resistance Database*. East Lansing, MI: Michigan State University. Available at: http://www.pesticideresistance.org (accessed June 20, 2023).

Murray, D.A.H., Boddington, J., Lloyd, R., Rogers, J., Zalucki, M.P. and Ward, A. (1998) Regional management of *Heliothis* on the Darling Downs. In: *Proceedings 9th Australian Cotton Conference, Broadbeach, Queensland*. Wee Waa, Australia: Australian Cotton Grower's Research Organization, pp. 351–355.

Noma, T., Gratton, C., Colunga-Garcia, M., Brewer, M.J., Mueller, E.E. *et al.* (2010) Relationship of soybean aphid (Hemiptera: Aphididae) to soybean plant nutrients, landscape structure, and natural enemies. *Environmental Entomology* 39, 31–41. DOI: 10.1603/EN09073

Obrycki, J.J., Elliott, N.C. and Giles, K.L. (2000) Coccinellid introductions: potential for and evaluation of nontarget effects. In: Follett, P.A. and Duan, J.J. (eds) *Nontarget Effects of Biological Control*. New York: Springer, pp. 127–145.

Obrycki, J.J., Harwood, J.D., Kring, T.J. and O'Neil, R.J. (2009) Aphidophagy by Coccinellidae: application of biological control in agroecosystems. *Biological Control* 51, 244–254. DOI: 10.1016/j.biocontrol.2009.05.009

Onstad, D.W. and Knolhoff, L.M. (2022) Major issues in insect resistance management. In: Onstad, D.W. and Knolhoff, L.M. (eds) *Insect Resistance Management*, 3rd edn. San Diego, CA: Academic Press, pp. 1–29.

Panda, N. and Khush, G.S. (1995) *Host Plant Resistance to Insects*. Wallingford, UK: CAB International.

Pedigo, L.P. and Buntin, G.D. (1994) *Handbook of Sampling Methods for Arthropods in Agriculture*. Boca Raton, FL: CRC Press.

Pedigo, L.P. and Higley, L.G. (1992) A new perspective of the economic injury level concept and environmental quality. *American Entomologist* 38, 12–21. DOI: 10.1093/ae/38.1.12

Pedigo, L.P., Hutchins, S.H. and Higley, L.G. (1986) Economic injury levels in theory and practice. *Annual Review of Entomology* 31, 341–368. DOI: 10.1146/annurev.en.31.010186.002013

Perović, D.J., Gámez-Virués, S., Landis, D.A., Wäckers, F., Gurr, G.M. *et al.* (2018) Managing biological control services through multi-trophic trait interactions: review and guidelines for implementation at local and landscape scales. *Biological Reviews* 93, 306–321. DOI: 10.1111/brv.12346

Perrot, T., Rusch, A., Coux, C., Gaba, S. and Bretagnolle, V. (2021) Proportion of grassland at landscape scale drives natural pest control services in agricultural landscapes. *Frontiers in Ecology and Evolution* 9, 607023. DOI: 10.3389/fevo.2021.607023

Pym, A., Umina, P.A., Reidy-Crofts, J., Troczka, B.J., Matthews, A. *et al.* (2022) Overexpression of UDP-glucuronosyltransferase and cytochrome P450 enzymes confers resistance to sulfoxaflor in field populations of the aphid, *Myzus persicae*. *Insect Biochemistry and Molecular Biology* 143, 103743. DOI: 10.1016/j.ibmb.2022.103743

Randolph, T.L., Peairs, F.B., Koch, M., Walker, C.B., Stubbs, J.R. *et al.* (2005) Yield response and categories of resistance to Russian wheat aphid in four *Dn*4 hard red winter wheat cultivars. *Journal of Economic Entomology* 98, 588–594. DOI: 10.1093/jee/98.2.588

Rane, R.V., Walsh, T.K., Pearce, S.L., Jermiin, L.S., Gordon, K.H., Richards, S. *et al.* (2016) Are feeding preferences and insecticide resistance associated with the size of detoxifying enzyme families in insect herbivores? *Current Opinion in Insect Science* 13, 70–76. DOI: 10.1016/j.cois.2015.12.001

Rutledge, C.E., O'Neil, R.J., Fox, T.B. and Landis, D.A. (2004) Soybean aphid predators and their use in integrated pest management. *Annals of the Entomological Society of America*, 97, 240–248.

Saidi, A. and Quick, J.S. (1996) Inheritance and allelic relationships among Russian wheat aphid resistance genes in winter wheat. *Crop Science* 36, 256–258. DOI: 10.2135/cropsci1996.0011183X003600020007x

Schellhorn, N.A., Parry, H.R., Macfadyen, S., Wang, Y. and Zalucki, M.P. (2015) Connecting scales: achieving in-field pest control from areawide and landscape ecology studies. *Insect Science* 22, 35–51. DOI: 10.1111/1744-7917.12161

Segre, H., Segoli, M., Carmel, Y. and Shwartz, A. (2020) Experimental evidence of multiple ecosystem services and disservices provided by ecological intensification in Mediterranean agro-ecosystem. *Journal of Applied Ecology* 10, 1873–2096. DOI: 10.1111/1365-2664.13713

Schneider, L., Rebetez, M. and Rasmann, S. (2022) The effect of climate change on invasive crop pests across biomes. *Current Opinion in Insect Science* 50, 100895. DOI: 10.1016/j.cois.2022.100895

Sun, Z., Di, L., Fang, H., Guo, L., Tan, X. *et al.* (2021) Agro-geoinformatics data sources and sourcing. In: Di, L. and Üstündağ, B. (eds) *Agro-geoinformatics*. Cham, Switzerland: Springer, pp. 41–66. DOI: 10.1007/978-3-030-66387-2_4

Tang, J. (2021) GIS fundamentals for agriculture. In: Di, L. and Üstündağ, B. (eds) *Agro-geoinformatics*. Cham, Switzerland: Springer, pp. 25–39. DOI: 10.1007/978-3-030-66387-2_3

Tilmon, K.Y., Michel, A. and O'Neal, M.E. (2021) Aphid resistance is the future for soybean production and has been since 2004: efforts towards a wider use of host plant resistance in soybean. *Current Opinion in Insect Science* 45, 53–58. DOI: 10.1016/j.cois.2021.01.003

Umina, P.A., Edwards, O., Carson, P., van Rooyen, A. and Anderson, A. (2014) High levels of resistance to carbamate and pyrethroid chemicals widespread in Australian *Myzus persicae* (Hemiptera: Aphididae) populations. *Journal of Economic Entomology* 107, 1626–1638. DOI: 10.1603/EC14063

Umina, P.A., McDonald, G., Maino, J., Edwards, O. and Hoffmann, A.A. (2019) Escalating insecticide resistance in Australian grain pests: contributing factors, industry trends and management opportunities. *Pest Management Science* 75, 1494–1506. DOI: 10.1002/ps.5285

Umina, P.A., Bass, C., van Rooyen, A., Chirgwin, A., Arthur, A. *et al.* (2022) Spirotetramat resistance in *Myzus persicae* (Sulzer) (Hemiptera: Aphididae) and its association with the presence of the A2666V mutation. *Pest Management Science* 78, 4822–4831. DOI: 10.1002/ps.7103

USDA-NASS (2004) *Colorado Winter Wheat Varieties – 2004 Crop*. Washington, DC: US Department of Agriculture, National Agricultural Statistics Service Special Interest Report. Available at:

https://www.nass.usda.gov/Statistics_by_State/Colorado/Publications/Special_Interest_Reports/CO-Winter-Wheat-Varieties-01202004.pdf (accessed July 20, 2023).

USDA-NASS (2023) *CropScape – Cropland Data Layer*. Washington, DC: US Department of Agriculture, National Agricultural Statistics Service. Available at: https://nassgeodata.gmu.edu/CropScape (accessed July 20, 2023).

Van Driesche, R.G. and Bellows, T.S., Jr (1996) *Biological Control*. New York: Chapman & Hall.

Van Emden, H.F., Eastop, V.F., Hughes, R.D. and Way, M.J. (1969) The ecology of *Myzus persicae*. *Annual Review of Entomology* 14, 197–270. DOI: 10.1146/annurev.en.14.010169.001213

Way, M.J. (1988) Entomology of wheat. In: Harris, M.K. and Rogers, C.E. (eds) *The Entomology of Indigenous and Naturalized Systems in Agriculture*. Boulder, CO: Westview Press, pp. 183–206.

Wilson, L., Downes, S., Khan, M., Whitehouse, M., Baker, G. *et al*. (2013) IPM in the transgenic era: a review of the challenges from emerging pests in Australian cotton systems. *Crop and Pasture Science* 64, 737–749. DOI: 10.1071/CP13070

Wilson, L.J., Whitehouse, M.E. and Herron, G.A. (2018) The management of insect pests in Australian cotton: an evolving story. *Annual Review of Entomology* 63, 215–237. DOI: 10.1146/annurev-ento-020117-043432

Yu, Y.B., Yang, H.L., Lin, Z., Yang, S.Y., Zhang, L.M. *et al*. (2021) Development and area-wide application of biological control using the parasitoid *Aphidius gifuensis* against *Myzus persicae* in China. In: Yu, Y.B., Yang, H.L., Lin, Z., Yang, S.Y., Zhang, L.M. *et al*. (eds) *Area-wide Integrated Pest Management*. Boca Raton, FL: CRC Press, pp. 3–16.

Zalucki, M.P., Adamson, D. and Furlong, M.J. (2009) The future of IPM: whither or wither? *Australian Journal of Entomology* 48, 85–96. DOI: 10.1111/j.1440-6055.2009.00690.x

3 Landscape Considerations in Pest Management: Case Study of the Arizona Cotton IPM System

Steven E. Naranjo[1,*] and Peter C. Ellsworth[2]
[1]*US Department of Agriculture, Agricultural Research Service, Arid-Land Agricultural Research Center, Maricopa, AZ, USA (Emeritus);* [2]*Department of Entomology, University of Arizona, Maricopa Agricultural Center, Maricopa, AZ, USA*

Abstract
Integrated pest management (IPM) for cotton in Arizona has been under continual evolution for many decades. For over 30 years its foundation has rested on the careful integration of avoidance tactics, thresholds, sampling, and the use of selective insecticides and technologies to foster conservation biological control along with a crucial community-based approach to enable broad-scale education of and adoption by cotton growers and their consultants. The success of this strategy and its evolution involves both implicit and explicit consideration of the landscape context because all the key insect pests of this system are mobile, and several are highly polyphagous. Most tactics such as sampling, thresholds, effective insecticide use, conservation biological control, and crop management are practiced on individual farms. But because many growers use these tactics, they have profound regional impact on pest and natural enemy populations. Cotton growers explicitly engage in cooperative practices such as pest eradication, insecticide resistance management, and crop placement that contribute further to sustainable IPM on a landscape scale. Many of these gains were enabled through the development of new "hard" technologies such as Bt cotton and selective insecticides. However, the Arizona cotton IPM system's success is rooted in the research, development, and deployment of the requisite set of "soft" technologies or use instructions needed to collectively implement individual practices and harness the power of landscape-level cooperative management. The overall result has been a savings of over US$600 million in control costs and yield loss savings for growers since 1996, along with an associated 85–90% reduction in insecticide use, and an increasing focus on natural control elements as the engine driving further innovations. Here we discuss the development, implementation, and evolution of the cotton IPM program for Arizona, with emphasis on the critical importance of a landscape perspective for its current and future success.

Introduction

There is a long history of interest in developing pest management systems that take a broader and more inclusive view of the ecological matrix in which our crops are embedded (Stern *et al.*, 1959; Levins and Wilson, 1980; Kogan, 1998). Many of our most significant insect pests are highly polyphagous and mobile, moving among a mosaic of crop and non-crop host plants throughout the year. Describing, understanding, and managing temporal population changes

*Corresponding author: senaranjo@cox.net and steve.naranjo@usda.gov

 Arthropod Management and Landscape Considerations in Large-Scale Agroecosystems (eds M.J. Brewer and G.L. Hein)
DOI: 10.1079/9781800622777.0003

of such pests in a single location is inexorably tied to their dynamics over a larger landscape. More than 40 years ago, Rabb (1978) pointed to the critical need for comprehensive areawide approaches for more effective and sustainable integrated pest management (IPM). Here, we provide an example of the power of a landscape perspective to effect positive change through a case study of the Arizona cotton IPM program.

Cotton (*Gossypium* spp.) comprises about 25% of the world demand for natural fiber and is grown in nearly 70 countries (National Cotton Council, 2023; USDA-ERS, 2023). Lint is the primary product with a global production of 25.3 billion kg in 2022 (National Cotton Council, 2023), and the seed and oil are used in food products and animal feed. The United States (US) is the third largest producer of cotton after China and India with a value of US$21 billion and an associated 125,000 jobs in the farming and industrial sectors (USDA-ERS, 2023). Within the US, there are four major production regions comprised of 17 states with total 2022 production of 3.14 billion kg of lint. The western production region encompasses irrigated farming systems of Arizona and California. Arizona has the highest levels of lint production per area in the nation and among the highest lint yields globally. The western region also has unique pest issues compared with other US production regions.

In Arizona, cotton has been grown for at least the last 18 centuries, first by native peoples of the region (and still today), then with modern, large-scale production over the last 100 years (Shapiro, 1989). Cotton is cultivated in the low desert region of the state, and production as well as the types of cotton grown have fluctuated greatly since the early 1900s with peak production of 281,263 ha in 1953 (264,306 ha of upland cotton, *G. hirsutum* L., and 16,957 ha of long-staple Pima cotton, *G. barbadense* L. Over the past decade, upland production has averaged over 56,000 ha (National Cotton Council, 2023). Pima cotton, which was historically almost exclusive to Arizona and the first cotton species to be commercially grown in the state, has largely relocated to the central valley of California over the past two decades. There has been renewed interest in growing this cotton species in Arizona in recent years with about 8000 ha planted in 2022.

Key arthropod pests

As a long-season crop grown in generally lower latitudes of the world, cotton harbors a plethora of arthropods. Rainwater (1952) lamented that "cotton is a plant that nature seems to have designed specifically to attract insects," and Hargreaves (1948) estimated that more than 1300 herbivorous insect species can be found in cotton globally. Thankfully, only a small proportion of those species are of economic importance. In the US, about 100 species of arthropods are recognized as pests of some note, but only about 20% would be considered routinely important and capable of inflicting economic damage if unmanaged (Leigh *et al.*, 1996). The remainder are sporadic or occasional secondary pests that might become important in certain years depending on environmental conditions and through disruption of natural control forces precipitated by the use of broader spectrum insecticides. Many major pest species in the US are polyphagous and highly mobile, and these characteristics point to the need for a landscape-level perspective for effective management. Advances in insecticide development, crop biotechnology, and overall IPM tactics and strategies have lessened opportunities for primary and secondary pests, and as we will show, this is the hallmark of the cotton IPM system in Arizona that has enabled successful and sustainable arthropod management.

In Arizona, there are three primary pests of cotton: the pink bollworm (*Pectinophora gossypiella* [Saunders] [Lepidoptera: Gelechiidae]), the western tarnished plant bug or Lygus bug (*Lygus hesperus* [Knight] [Hemiptera: Miridae]), and the silverleaf whitefly (*Bemisia argentifolii* [Bellows and Perring] = *B. tabaci* MEAM1 [Hemiptera: Aleyrodidae]). The latter two pest species are highly polyphagous, and all three are moderately to highly mobile.

The pink bollworm is an invasive species that has colonized most cotton growing regions of the world. It was first detected in Texas in 1917 and became the key pest of cotton in Arizona and the low desert valleys of southern California around the mid 1960s (Spears, 1968; Noble, 1969). The insect is functionally monophagous on cotton even though its known host range includes around 70 plant species globally. Okra, *Abelmoschus esculentus* (L.) Moench, is the only other cultivated crop of importance in the

US (Noble, 1969; Ingram, 1994; Henneberry and Naranjo, 1998). The pink bollworm attacks reproductive structures, completing an initial small first generation in flower buds and then moving to seeds within the fruit (boll) for up to four additional generations. This feeding disrupts lint development and introduces pathways for contamination (e.g. fungi that produce aflatoxins), leading to significant economic damage. Areawide approaches have long been considered and used for this pest. Repeated efforts focused on eradication of the pest in Arizona and elsewhere in the early to mid 1900s (Henneberry and Naranjo, 1998), but not until the advent of transgenic cotton producing the insecticidal proteins of *Bacillus thuringiensis* Berliner (Bt) was a binational, multi-tactic control program successful in eradicating the pest from the US and northern Mexico (Staten and Walters, 2021; Tabashnik *et al.*, 2021) (see "Areawide elements" below). Prior to eradication, millions of sterile males were released annually for nearly 40 years into the central valley of California as a pest exclusion tactic for the once vast acreage of cotton grown in that region (Henneberry, 1994).

Lygus hesperus is native to western North America and is a major pest of several agronomic and horticultural crops throughout this region (Scott, 1977; Jackson *et al.*, 1995; Wheeler, 2001). In Arizona, *L. hesperus* has been considered the most significant pest of cotton yield for over two decades due to its direct feeding on reproductive structures (Diehl *et al.*, 1998; Ellsworth and Jones, 2001). Their feeding activity on cotton causes the abscission of flower buds (squares). The loss of potential bolls results in a direct impact on yield, and their feeding on developing buds and bolls can disrupt pollination, lint formation, and quality (Strong, 1970; Mauney and Henneberry, 1984; Leigh *et al.*, 1988). Lygus bugs are highly mobile, have an overwinter reproductive diapause, and infest several crops at the same time as well as sequentially over the season. Taking advantage of the source–sink relationship of Lygus bugs between alfalfa and cotton, strip-cropping alfalfa was an early landscape-level innovation for pest management (Stern *et al.*, 1964). By leaving undisturbed strips of alfalfa after harvest, movement of Lygus bugs into cotton could be delayed or mitigated on an areawide scale. New systems were established in the 1990s to refine this approach, and they still represent a recommended management approach (Diehl *et al.*, 1998; Goodell, 2009).

The final key pest is silverleaf whitefly, an invasive species that colonized the southern tier of the US beginning in the late 1980s (Brown *et al.*, 1995). In the southwest, the putatively native *B. tabaci* (Gennadius) was quickly displaced by the invasive *B. argentifolii* in the late 1980s, turning an occasional pest into a key pest of many agronomic and horticultural crops.

It is a highly polyphagous pest of global significance. It can cause direct feeding damage, is a vector of more than 100 plant viruses, induces plant disorders, and causes quality issues such as sticky cotton through honeydew excretion and associated growth of sooty molds (Naranjo and Ellsworth, 2001). The lint quality issue is paramount because it could, and did, lead to market discounts and even outright rejection of cotton from the region (Ellsworth *et al.*, 1999). The insect lacks a quiescent stage, enabling year-round population growth on an expansive number of cultivated and non-cultivated host plants in the landscape. Silverleaf whitefly also has a propensity to develop resistance to a large array of insecticide classes, adding to management challenges (Palumbo *et al.*, 2001; Horowitz *et al.*, 2020). Despite its small size, the insect is a capable disperser (Blackmer *et al.*, 1995; Isaacs and Byrne, 1998; Byrne, 1999), and its broad host range enables it to effectively exploit the agricultural system in the desert southwest. An understanding of these landscape relationships has allowed exploitation of the insect's life system for effective and sustainable IPM (Naranjo *et al.*, 2022).

Case study of the Arizona cotton IPM system

We will use the Arizona cotton IPM system as a case study for how landscape-level considerations can inform and strengthen the development and implementation of IPM strategies. The current system is mature, validated, highly effective, and economically efficient in managing all arthropod pests of the system (Ellsworth and Martinez-Carrillo, 2001; Naranjo, 2001; Ellsworth *et al.*, 2006; Naranjo and Ellsworth, 2009b). The development and continual evolution of this IPM system represents a large collaborative and cooperative effort involving multiple entities, including federal and university scientists, Cooperative Extension, the agrochemical and seed industries, and most importantly, cotton growers and their consultants who have implemented the program on the ground. In this chapter, we will describe the component tactics of the system and how they are integrated into an overall strategy. Specific focus will be placed, both implicitly and explicitly, on the broader landscape perspectives that have shaped the elements of the current system and how the program's evolution continues to engage in an inclusive agroecosystem approach.

The Arizona Cotton IPM System

As with many insect management programs there are always iterations, modifications, technical advances, and wholesale restructuring of approaches. This is no different for the cotton IPM story in Arizona. The genesis of the most recent version of cotton IPM begins with the invasion of the silverleaf whitefly into Arizona in the late 1980s to early 1990s. The putatively native whitefly species, *B. tabaci* (=biotype A) (Brown *et al.*, 1995), present in this region since at least the 1920s (Russell, 1975), had been an occasional problem in cotton and other crops for many decades (Gerling, 1967; Duffus and Flocks, 1982; Coudriet *et al.*, 1985; Brown and Nelson, 1987). However, the incursion of *B. argentifolii* set the stage for transformative changes in insect control tactics and ultimately IPM strategies for the entire pest complex. The whitefly, like the pink bollworm before (Henneberry, 1986), had the potential to completely derail the long history of cotton production in the state. Sweeping changes were urgently needed in pest management to preserve the industry.

IPM can be envisioned as sets of multiple, hierarchical, and interacting tactics that effectively control pests while reducing risk for producers and the environment and resulting in positive outcomes for agriculture and society. A convenient organizing metaphor is a pyramid. From the top down, the three main tiers are: (i) sampling and monitoring to quantify pest populations and plant health; (ii) effective chemical tools and tactics for optimizing their use; and (iii) a large assortment of natural and human-mediated elements that contribute to maintaining or lowering pest populations below economic levels, i.e. things that help growers avoid pest issues before they require action from higher levels in the pyramid (Fig. 3.1). Not every tactic fits neatly nor exclusively into a single tier or section within a tier. For example, sampling impacts all levels of the pyramid, and tactics like biological control and resistance management interact with multiple tiers. While the metaphor is imperfect, it provides a useful framework to describe and especially teach the Arizona cotton IPM program to others. Each facet of the pyramid describes the specific information needed to manage each of the three key pests. In the discussion below we will start with the upper layers of the pyramid and work down for several reasons. First, the two upper layers represent the prescriptive elements of IPM with which most are familiar (Ellsworth and Martinez-Carrillo, 2001). Second, similar to many other systems, the Arizona cotton IPM program was born in the upper levels of the pyramid and then grew to encompass the multiple foundational elements that have ultimately contributed to its effectiveness and sustainability. Throughout, we will emphasize elements that connect directly or indirectly to broader landscape perspectives.

Sampling and monitoring

Sampling to determine pest abundance and plant health and using this information to make informed management decisions are fundamental elements of any IPM program (Stern *et al.*, 1959). Efficient and precise sampling plans are crucial for understanding pest dynamics, both locally and regionally, and for conducting the research necessary to develop IPM systems. While the typical focus of scouting and decision making for pest management is a single field or farm rather than the entire landscape, their independent but broad-scale use over large regions is an example where even unintentional cooperation and coordination can result in areawide suppression of pests. In Arizona, over 95% of cotton fields are professionally scouted by pest control advisors (Naranjo and Luttrell, 2009). These advisors often service numerous growers

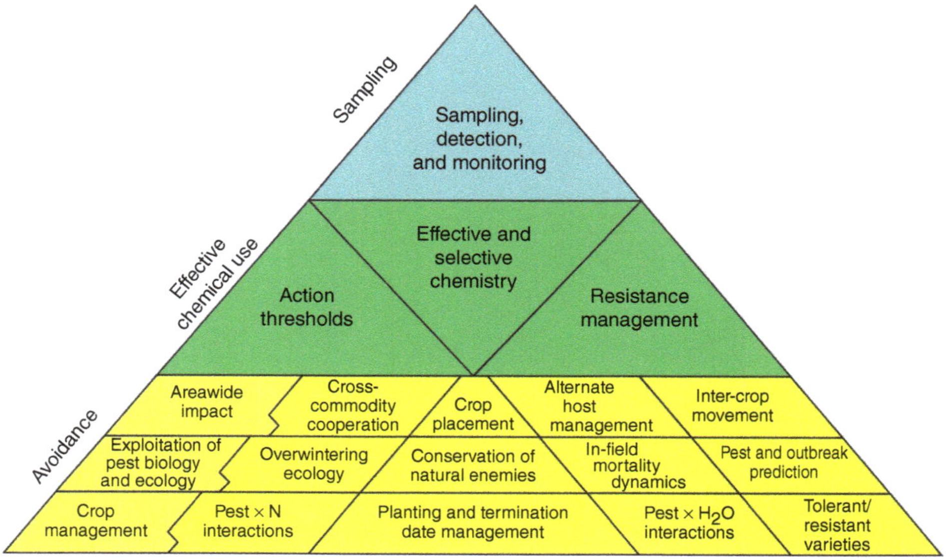

Fig. 3.1. A conceptual, multi-tiered, and multi-tactic model of integrated pest management (IPM) emphasizing three main elements: sampling, effective insecticide use, and a suite of avoidance components that can act to maintain pest populations below economic levels. Adapted from Ellsworth and Martinez-Carrillo (2001). (Used with permission from Elsevier.)

over areas of multi-crop production that encompass vast areas of the agroecosystem. For example, on average 97.3% of upland cotton was professionally scouted over the last 5 years with just 25 pest control advisors scouting more than half of the state's acreage in 2020 (total 27,209 ha; mean 1088 ha per advisor; range 32–4000 ha) (P.C. Ellsworth, unpublished data). As such, pest control advisors intuitively understand pest and natural enemy dynamics over a broad scale, and they use this knowledge to advise their collective grower clientele. For each of the three major pests, there are also examples of intentionally collective areawide monitoring and management programs (Moore *et al.*, 1992; Diehl *et al.*, 1994; Thacker *et al.*, 1994a, b; Ellsworth *et al.*, 1996, 2000; Ellsworth and Jones, 2001).

Sampling systems have been developed for all three key pests. Pheromone traps (adult males) and boll sampling to assess larval infestation are used for pink bollworm (Henneberry and Naranjo, 1998). Pheromone traps continue to be important for this pest as a large-scale detection tool for possible reinvasion following eradication (National Cotton Council, 2009; Staten and Walters, 2021). Sweep nets are used to sample Lygus bugs (Ellsworth and Barkley, 2001; Ellsworth *et al.*, 2023), secondary pests like stink bugs, caterpillars, and cotton fleahoppers, as well as predatory arthropods that play key roles in the Arizona IPM system (see "Pest biology/ecology" below). Silverleaf whitefly sampling plans have been developed for both management and research purposes (Naranjo and Flint, 1994, 1995; Naranjo *et al.*, 1996b; Diehl *et al.*, 1997; Ellsworth *et al.*, 2006). Most of these tools have been validated through commercial scale assessments (Naranjo *et al.*, 1997; Ellsworth and Barkley, 2001; Ellsworth and Jones, 2001).

Effective insecticide use and optimization

Decision thresholds

Closely allied with sampling is the decision framework that governs when control actions are necessary to prevent economic damage. This concept is among the most basic in any IPM program and yet, surprisingly, development and implementation of such tools has lagged need (Pedigo *et al.*, 1986; Castle and Naranjo, 2009). Stern *et al.* (1959) defined these as the economic injury level (EIL), the level of pest abundance for which the cost of control equals the value of crop damage prevented, and the economic threshold (ET), the pest level at which action should be taken to prevent populations from exceeding the EIL.

In the Arizona system, action thresholds are available for all key pests. These are accurate and effective thresholds that have been developed and implemented primarily through empirical means based on experimental field study. Nominal thresholds for pink bollworm were developed decades ago based on both adult pheromone traps and larval infestation levels in cotton bolls (Henneberry and Naranjo, 1998). These decision aids played an important role in the eradication program (Staten and Walters, 2021), but are no longer a consideration in day-to-day pest management. The highly coordinated deployment of monitoring and decision-based control along with areawide use of sterile insect release, host-free periods, and Bt cotton ultimately resulted in eradication of pink bollworm. In this case, coordinated monitoring and control actions on a landscape scale were explicit and highly effective.

Lygus bug action thresholds have been developed through comparative experimental approaches in the field (Ellsworth and Barkley, 2001). The threshold is based on both the abundance of adults and nymphs, reflecting the more voracious feeding activity and plant damage posed by nymphs. This threshold effectively mitigates economic yield loss in balance with the cost of control and has been robust to prevailing lint value and control costs over time.

Thresholds for silverleaf whitefly were developed from a combination of approaches including multi-state, comparative experimental field studies (Naranjo *et al.*, 1998), estimation of EILs (Naranjo *et al.*, 1996a), and experiences of commercial pest control advisors. Initially, when available control tactics focused on broad-spectrum adulticidal materials, thresholds were based solely on counts of adults. The advent of new selective insecticides in 1996 with primary activity against immature stages, necessitated evolution to a dual threshold based on both adults and nymphs that accounted for the interacting dynamics of these different life stages (Diehl *et al.*, 1997). The most recent refinement has been the inclusion of key predatory arthropods along with pest abundance in the decision matrix (Vandervoet *et al.*, 2018, 2022; Ellsworth *et al.*, 2019, 2022). This matrix reflects the critical role that conservation biological control plays in suppression of key pests (see "Pest biology/ecology" below). Instead of assessment of only whitefly abundance, predator to prey ratios derived from counts of six key predators from sweep net samples and binomial samples of whitefly adult and nymphs define decision zones based on biological control capacity. When ratios are favorable, sprays are deferred or eliminated, and inaction is supported with confidence. But when ratios do not support inaction, spraying might be advanced relative to the traditional pest-centric threshold (Fig. 3.2). Pest control advisors and growers in Arizona and northern Mexico were taught these monitoring and threshold advances through extensive, statewide education and outreach meetings and workshops.

Sampling and decision making in general is up to each individual farm. But common understanding and collective practice ultimately led to areawide impacts on pest dynamics. The impact of this community-based approach was greatly amplified by the evolving set of selective chemistry options that started to become available by the mid 1990s. These selective products helped enable conservation biological control and continue to be introduced today.

Effective and selective insecticides

There has been a complete transformation in the types and amounts of insecticides used in the Arizona cotton IPM system. Prior to 1996, the primary pest suppression tools were those that would be common for cotton and many other production systems globally, including typical organophosphates, carbamates, pyrethroids, and cyclodienes (Ellsworth *et al.*, 1994; Dennehy *et al.*, 1995). These materials were generally effective, but also highly disruptive to the diverse natural enemy communities that inhabit cotton in this region (Wene and Sheets, 1962; Van den Bosch and Hagen, 1966). Arizona maintains careful tracking of statewide insecticide use and other pest management behaviors (Fournier *et al.*, 2007; Ellsworth and Fournier, 2024). Between 1990 and 1995 cotton growers averaged nine sprays for all arthropod pests with

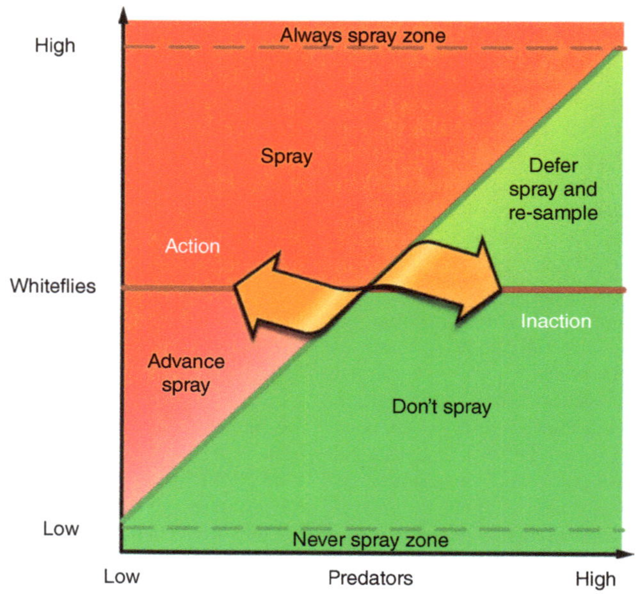

Fig. 3.2. Conceptual diagram of whitefly action thresholds showing the pest-based standard threshold (red horizontal line) and the predator threshold that supports inaction decisions based on specific predator–prey relationships (green diagonal line). These predator thresholds lead to better and more confident decision making by growers. Adapted from Ellsworth *et al.* (2019). (Creative Commons Attribution 4.0 International License.)

broadly toxic or broad-spectrum materials that were indiscriminate in their effects on target and non-target arthropods (Ellsworth *et al.*, 2015). In 1995, a key inflection point for Arizona pest management, cotton growers averaged 13 sprays with individual growers spraying as much as 25 times, and there was clear evidence that whiteflies were resistant to commonly used pyrethroid plus organophosphate insecticide mixtures (Dennehy and Antilla, 1996).

In 1996, US Environmental Protection Agency (EPA) emergency use exemptions were given for two new insect growth regulators (IGRs), buprofezin and pyriproxyfen, that were specific to and highly effective on whitefly and also highly selective, and thus, non-disruptive to the natural enemy community (Ellsworth and Martinez-Carrillo, 2001; Naranjo, 2001; Naranjo *et al.*, 2003, 2004; Naranjo and Ellsworth, 2009a). Extensive grower and pest control advisor training was conducted on these insecticides, and new sampling and decision aids resulted in the new chemistries being readily adopted (Ellsworth *et al.*, 1997; Ellsworth and Jones, 2001). Coincidentally, transgenic Bt cotton was first introduced for pink bollworm management the same year, and it also was quickly adopted (see "Crop management" below). Statewide use of insecticides for cotton pest management dropped to an average of about four sprays per hectare between 1996 and 2005. A selective feeding inhibitor, flonicamid, for Lygus bugs was introduced in 2006 (Ellsworth and Barkley, 2005), largely replacing acephate, and contributed to another dramatic drop in insecticide use in cotton to fewer than two sprays per hectare for all pests. That same year, the pink bollworm eradication program got under way in Arizona, and by 2008 grower sprays for this pest fell to zero.

The introduction and use of selective insecticide options by growers regionally in a community-based paradigm has transformed the agricultural landscape into one in which natural biological control can thrive. At least five new modes of action for whitefly control and two new modes of action for Lygus bug control have been introduced over the last 20 years. As each is under development for the Arizona market, field-based research tests both efficacy and selectivity of these materials, preparing growers for rapid and informed adoption (Bordini *et al.*, 2021). Starting in 2021, Bt cotton with activity against Lygus bugs and thrips (see "Crop management" below) was introduced and is now grown on more than 18% of Arizona's upland cotton acreage. With areawide adoption and multiple selective options for pest control, growers averaged just 1.6 sprays per hectare over the last 5 years. More than 85% of what is sprayed are fully selective materials (Fig. 3.3). This implicit spatial outcome has created a positive feedback loop where natural pest mortality increases while insecticide use declines, akin to a remarkable reversal of the classic pesticide treadmill.

Insecticide resistance management

The evolution of insecticide resistance is an inevitable outcome for many arthropod pest species globally (Onstad and Knolhoff, 2022). Of the three key pests in the Arizona cotton agroecosystem, the whitefly was immediately recognized as a species that would require special attention in this regard. Over 700 cases of resistance have been documented for *Bemisia* spp. worldwide, including 74 insecticides representing ten modes of action (IRAC, 2022). The rapid evolution of resistance to synergized pyrethroids (mixtures of pyrethroids usually with organophosphates) by 1995 in Arizona punctuates this history. Thus, from the very beginning it was understood that any management system involving insecticides also had to include a proactive strategy to try and minimize resistance and in turn reduce the risk of losing valuable control agents. This includes approaches that limit insecticide use through adherence to sound IPM practices (e.g. sampling, thresholds, maximizing biological control), and tactics to diversify and partition use when insecticides are needed. All three of these insecticide resistance management (IRM) elements have grown in sophistication and scope since the early 1990s.

Prior to the availability of selective materials, the Arizona IRM plan consisted of a two-stage system that first recommended non-pyrethroid insecticides either singly or in mixtures, followed later by synergized pyrethroids (Dennehy *et al.*, 1995). Arguably, there were four modes of action available, though only the cyclodiene endosulfan provided any commercial-level control when used alone. All others were used in mixtures, thus limiting the application of IRM goals to rotate and diversify modes of action. Despite this early IRM attempt, strong resistance developed to synergized pyrethroids by 1995 (Dennehy and Antilla, 1996).

Through an industry–government–academic collaboration, limited use of two IGRs was proposed to the EPA as a Section 18 Emergency Use Exemption (Ellsworth and Diehl, 1996). Section 18s are rarely granted for multiple active ingredients to resolve a specific pest condition and usually only because of supply limitations. In this case, however, this coalition was acutely concerned about the resistance potential of this insect and therefore argued that single uses were needed of each to prevent resistance evolution. The justification for this Section 18 to the EPA was that both products were needed because neither agent was believed to be sufficient to alleviate the emergency condition. With the granting of this historic Section 18 in 1996, the IRM plan grew to three stages of chemical use. Stage 1 comprised the two selective IGRs used individually and in succession, if necessary (Ellsworth and Watson, 1996). Stages 2 and 3 included non-pyrethroids followed by synergized pyrethroids as before. The hope was that growers would never reach Stage 3, and insecticide use data support that outcome. In fact, the Stage 1 materials were so effective that it was rarely necessary to go beyond Stage 1 for the entire season if growers chose to use those materials first (Ellsworth and Martinez-Carrillo, 2001; Naranjo, 2001; Naranjo *et al.*, 2004) (Fig. 3.4). As more selective and partially selective options reached the market, they were tested through research and placed accordingly into the three-stage system, with the modification that Stage 2 now emphasizes partially selective materials (Ellsworth *et al.*, 2006). Today the system encourages a more comprehensive assessment of risks associated with insecticide use, especially impacts on natural enemies, a reflection of the major advances made to avoid resistance and expand access to numerous effective and selective modes of action (Bordini *et al.*, 2024). The current program also adds the second key pest, Lygus bugs, and classifies insecticides for their efficacy, safety to natural enemies, safety to other non-target organisms including humans, in addition to their resistance potential. Today, the system includes nine insecticides with seven to

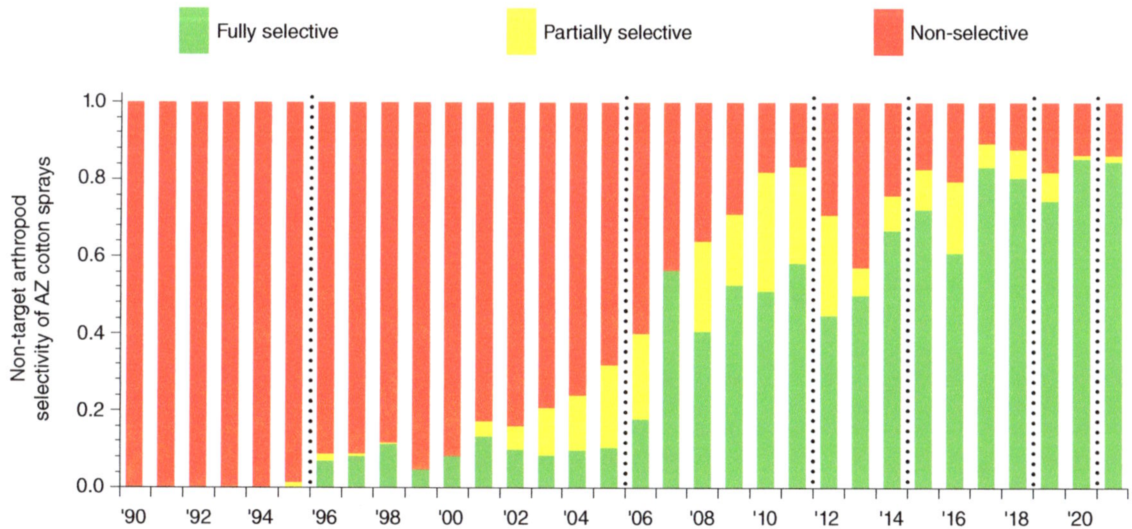

Fig. 3.3. Estimates of proportions of selective insecticide use in Arizona cotton. In 2021, over 85% of hectare-sprays made are with fully selective insecticides that are safe to key predators and other natural enemies in the cotton system. Dotted lines denote initiation of significant pest management events (see Fig. 3.10). Data drawn from the Arizona Pest Management Center Pesticide Use Database (Fournier *et al.*, 2007). (Original artwork by P. Ellsworth.)

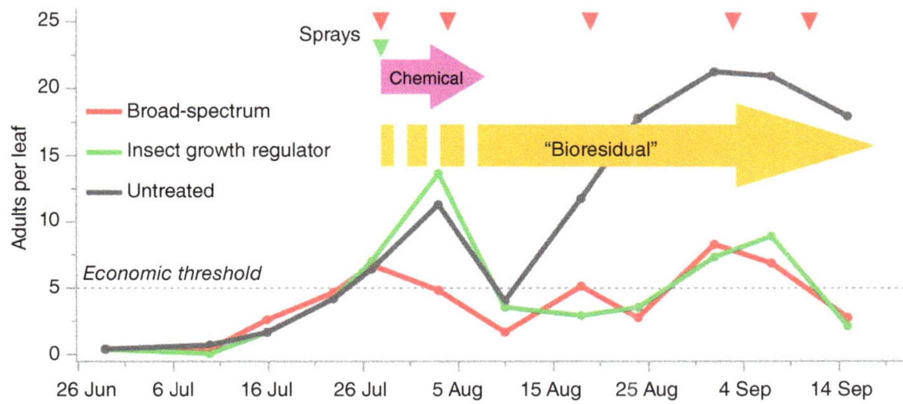

Fig. 3.4. Representative dynamics of whitefly populations in cotton relative to different management scenarios. Both broad-spectrum and selective (insect growth regulator) insecticide options result in pest control but do so in different ways. Repeated broad-spectrum sprays are required to achieve the same control as a single spray of a selective insecticide (triangles). Chemical residual is relatively short in both scenarios, but in the selective case, the conserved natural enemies and other natural forces can provide season-long control (the "bioresidual"). Adapted from Naranjo (2001). (Used with permission from Elsevier.)

nine modes of action in Stage 1, five insecticides with two modes of action in Stage 2, and more than six insecticides with two modes of action in Stage 3 (Fig. 3.5), giving growers multiple options for sustainable IRM through diversification of use. As emphasized before, the regional practices enabled by this knowledge have had broad negative impacts on pest abundance and positive impacts on the natural enemies that provide critical ecosystem services in pest control.

Long-term, statewide monitoring of silverleaf whitefly resistance to multiple insecticides (Dennehy *et al.*, 2008) occasionally identified issues (Li *et al.*, 2001). For instance, laboratory bioassays of field-collected insects showed that there was reduced susceptibility to pyriproxyfen for whiteflies as early as 2001 (Crowder *et al.*, 2007, 2008a) despite relatively limited use of this once-per-season material. Further modeling work with spatial/temporal components suggested possible genetic underpinnings and also environmental and operational elements that might speed or mitigate this resistance (Crowder *et al.*, 2006, 2008b). Empirical field research tied with laboratory bioassays demonstrated that this reduced efficacy did not result in control failure disruptive of the overall IPM program (Ellsworth *et al.*, 2007).

As the theme of this book emphasizes, many components of an IPM program would be more effective and sustainable if practiced in a coordinated fashion over the entire agroecosystem rather than a single field or farm (see Chapter 2, this volume). IRM represents a special case because while the evolution of resistance may develop in a specific location, its ramifications will ultimately be felt by growers within a region or even larger geographic scopes depending on the extent of pest movement and gene flow. The loss of an effective insecticide could eventually harm the entire agricultural community when polyphagous pests are involved. Further for insecticides, the tenets of resistance management are: (i) limit insecticide use to the lowest practical level; (ii) diversify insecticide use patterns, for example through rotation practices; and (iii) partition modes of action through space and time such that refugia are maintained for all chemistries in a region (Ellsworth and Diehl, 1998; Ellsworth and Fournier, 2024). While an individual grower can limit insecticide use and can diversify and even partition modes of action to avoid repeated use on their own farms, the effective scale for the ecology and evolution of polyphagous mobile pests usually exceeds the land holdings of any single grower. Spatial and temporal partitioning of modes of action can only be achieved through cooperation at a landscape scale. That is, what decisions a grower makes can be informed by

Cotton Insecticide Use Guide. Insecticides have been screened for efficacy against target pests, *Lygus hesperus*, *Bemisia argentifolii* (MEAM1; silverleaf whitefly, SWF), and *Euschistus servus* (brown stink bug); as well as for their impact on non-target beneficial arthropods including more than 20 predators common in Arizona cotton. Those insecticides with full selectivity or safety towards these beneficial predators are in green; those that are partially selective or safe are in yellow; non-selective insecticides are in red. Some insecticides pose environmental and human health risks that require mitigations such as buffer zones and additional personal protective equipment (PPE). IRAC group numbers are provided to facilitate rotation of chemistry and SWF resistance risks are presented.

Product Name	Common Name	IRAC No.[1]	Chemical Group	Lygus Bug	Silverleaf Whitefly	Brown Stink Bug	Risk to Aquatic Life	Risk to Wildlife	Risk to Pollinators	Inhalation Risk	SWF, Risk of Resistance
Carbine	flonicamid	29	Feeding inhibitor	****							
Courier	buprofezin	16	Chitin inhibitor		**** (N)						under investigation
Exirel / Benevia	cyantraniliprole	28	Diamide		****						
Knack / Farewell	pyriproxyfen	7C	Juvenoid		**** (E,N)						mild-moderate
Oberon[2]	spiromesifen	23	Lipid synthesis inhibitor		**** (N)						under investigation
PQZ	pyrifluquinazon	9B	Pyridine azomethine		****						
Sefina Inscalis	afidopyropen	9D	Pyropene		***						
Sivanto prime	flupyradifurone	4D	Butenolide		****						
Transform	sulfoxaflor	4C	Sulfoxamine	****	*						
Assail / Intruder[3]	acetamiprid	4A	Neonicotinoid		****		Yes				moderate-severe
Belay	clothianidin[4]	4A	Neonicotinoid	**	**		Yes		Yes		
Centric	thiamethoxam[4]	4A	Neonicotinoid		**		Yes		Yes		
Plinazolin	*isocycloseram*	*30*	*Isoxazoline*	****			*not yet registered in the U.S.*				
Venom	dinotefuran	4A	Neonicotinoid		***		Yes		Yes		
Acephate	acephate	1B	Organophosphate (OP)	***		*		Yes	Yes		
Bidrin	dicrotophos[5]	1B	Organophosphate	*		*	Yes	Yes	Yes	Yes	
Cormoran	novaluron + acetamiprid	15 + 4A	Chitin inhibitor	** (N)	**	* (N)	Yes				
Diamond / Mayhem	novaluron	15	Chitin inhibitor	* (N)	*	* (N)	Yes				
Synergized pyrethroids	various[6]	3A + 1B	Pyrethroid + OP		**		Yes	Yes	Yes		moderate-severe
Vydate C-LV	oxamyl[5]	1A	Carbamate	****			Yes	Yes	Yes	Yes	

Background color: **Green** = Fully selective and safe to beneficials; **Yellow** = Partially selective or safe to beneficials; **Red** = broad spectrum, not safe to beneficials; *Italics* = not registered in U.S.
Risks as calculated from ipmPRIME (Jepson et al. 2014) and/or the Pesticide Risk Tool at , where available; 'Yes' indicates moderate to high risk for the given category.
****, Excellent control; ***, Good control; **, Fair control; *, Suppression only; E, N = Efficacy against eggs or nymphs only, respectively.

Rev. 02/02/24

The label is law! Please follow all instructions on the pesticide label.

A PDF of this publication is available on-line at:

[1] The (IRAC) assigns numbers for each unique mode of action or class of chemistry. Many appear on U.S. insecticide labels and are helpful for resistance management.
[2] At 0.125–0.156 lbs ai / A only; higher rates are more destructive of natural enemies.
[3] The State of Arizona has approved a Special Local Needs (SLN) increase in acetamiprid use rates by up to +50% against difficult-to-control whiteflies. Impact to beneficials is more severe at these higher rates.
[4] This active ingredient can significantly affect bee populations, other pollinators and birds, can persist for years in soils, and can leach into waterways and groundwater.
[5] This active ingredient is considered highly hazardous by the Word Health Organization (WHO Ib), a restricted use pesticide with signal words DANGER and POISON, requiring posting, additional PPE, and closed systems. Avoid if possible.
[6] Beta-cyfluthrin[ab], bifenthrin[b] and lambda-cyhalothrin[b] are considered highly hazardous by the [a]WHO (Ib) or in the [b]Globally Harmonized System of Classification and Labelling of Chemicals (GHS Category 2). Avoid if possible.

Fig. 3.5. Current insecticide recommendations for cotton pest control, including the broad set of risks associated with use including efficacy, safety to beneficials (selectivity), safety to other non-target organisms, and resistance. Reproduced from Bordini *et al.* (2024). (Creative Commons Attribution 4.0 International License.)

actions of other growers in their region. In Arizona, three related efforts have explicitly considered landscape-level IRM for cotton and other crops, primarily those impacted by silverleaf whitefly.

The first approach was developed and implemented for multiple crops affected by silverleaf whitefly and a single class of insecticides (neonicotinoids) of great value to the vegetable industry in Arizona (Palumbo *et al.*, 2001, 2003). The voluntary community plan, a coordinated effort among researchers, Cooperative Extension, industry and growers, was an explicit strategy to limit the exposure of sequential generations of the pest to the same mode of action over space and time. It was prompted by the impending registration of effective neonicotinoids for foliar use in cotton, potentially drastically expanding the geographic footprint for use of this class of insecticide. Three cropping communities were identified, posing three different temporal and spatial risks for resistance development and with differing dependencies on the key practice of soil-applied imidacloprid in vegetables and melons. All communities had some amount of cotton production in them. The goal was to limit each community to no more than two non-sequential uses of neonicotinoids (seed, soil, or foliar applied) per year among all crops through sharing agreements. Each community was defined by what crops were grown annually within a 3-km radius of each other. This was the presumed resistance management unit, rooted in movement ecology and other research of the time (Blackmer and Byrne, 1993, 1999a; Byrne *et al.*, 1996; Byrne, 1999). Cotton-dominated communities were characterized by areas where cotton was the only whitefly-sensitive crop in cultivation along with non-host crops. Thus, two non-consecutive uses of neonicotinoids, usually foliar acetamiprid, were permitted here, albeit with emphasis given to first making use of the Stage 1 products like the IGRs wherever possible. In communities with cotton and melons (a favored host of whitefly), the two uses of neonicotinoids were shared between the two crops, one use in melons, usually as soil-applied imidacloprid to fall melons, and one use in cotton. Finally, in the most complex communities made up of cotton, melons, and vegetables, the two uses were partitioned between the melon and vegetable growers, one each. Perhaps the largest concession made in this voluntary system, growers of cotton in these complex systems agreed to forgo any use of the neonicotinoid class of chemistry. They recognized that there were large advantages not only to melon and vegetable growers who required access to neonicotinoids but also to their own cotton production due to areawide lowering of whitefly pressures coming out of nearby vegetable and melon crops. The goal was to preserve the use of valuable neonicotinoids, primarily for all growers statewide but based on ecologically relevant scales (i.e. 3 km radius or ~28 km^2 communities) that they could perceive, understand, and implement (Palumbo *et al.*, 2001, 2003; Ellsworth *et al.*, 2008a, b).

Ultimately cotton growers within the more complex cotton-melon-vegetable landscapes altered their behavior by reducing or eliminating their uses of neonicotinoids to support areawide resistance management (Ellsworth *et al.*, 2008a, b, 2009). Through analyses of pesticide use records, we could show that between 2003 and 2005 in complex systems, growers adopted and followed these voluntary guidelines 90% of the time in cotton, i.e. they elected not to use the newly available foliar neonicotinoids in cotton despite fourfold increases of their use in cotton grown in cotton-only communities. Furthermore, we revisited adoption of the guidelines ten seasons later specifically in areas where guidelines suggested constraints on neonicotinoid usage in cotton and found still very high subscription rates (Ellsworth *et al.*, 2013). Based on section level (~2.6 km^2) pesticide use reports of acetamiprid use in cotton in Yuma County, where these more complex multi-cropping systems exist, there was extremely limited usage of this insecticide. Despite evidence for a large need to chemically control whiteflies in cotton, growers there were still constraining use of this product as set forth in the voluntary guidelines and instead used available alternatives such as pyriproxyfen (Ellsworth *et al.*, 2013).

Later research confirmed the important role cotton plays in the landscape-level relationships among refuges and the evolution of resistance to a key whitefly control compound, pyriproxyfen (Carrière *et al.*, 2012a). This research effort mapped the locations of focal cotton fields, and surrounding melon, alfalfa, and other crop and fallow fields in large, replicated landscapes out to 3 km from focal fields. Whiteflies from the focal fields were bio-assayed in the laboratory using a discriminating dose of pyriproxyfen, and spatial ring analyses estimated the distance over which resistance effects could be measured. Research showed that areas of untreated cotton (i.e. refuges) were negatively related to resistance out to 3 km and areas of pyriproxyfen-treated cotton were positively related to resistance out to 2.75 km. These findings were validated and predicted with additional replicated landscapes in subsequent years. This clearly showed the association between actions within the landscape and the potential for resistance evolution and the spatial scale over which it can occur. While this retrospective understanding is generally useful, it does not provide operational and actionable information directly to growers or pest control advisors making field-specific decisions on where and what to spray. Further advances are needed to enable proactive resistance management at the agroecosystem level of practice.

Our current efforts in IRM target all whitefly chemistries and host crops across entire cropping landscapes contained within ~23 km^2 of a focal field of management interest (Dayoob, 2021; P.C. Ellsworth *et al.*, unpublished data). The goal is to identify the temporal and spatial abundance of refuges relevant to the field of interest over varying historical timescales. Rooted in first principles where more use of a compound results in larger risks for the evolution of resistance, we hypothesize that pest managers equipped with knowledge about the availability of refuges for each mode of action in their community will opt to choose a product with more existing refuges than one without significant refuge availability. Both individually and collectively, growers should benefit from efficacies that are optimally sustained by ongoing choices of products with larger refuges on a field-to-field basis.

The chosen spatial scale is based on prior understanding of the extent of silverleaf whitefly movement and resistance development (Byrne *et al.*, 1996; Carrière *et al.*, 2012a) as well as operational constraints. Pesticide use reporting in Arizona is granular only to the section (~2.6 km^2) level. A 3 × 3 grid of sections is equivalent to ~23.3 km^2, which is very close to the 2.75–3.0 km radii (~23.8–28.3 km^2) found to be predictive of pyriproxyfen resistances. The chosen temporal scale or look-back periods for examination of pesticide usage remains an open research question. To tackle this, Dayoob (2021) constructed temporal periods measured in 90-day increments up to 540 days prior to the moment when a pest manager is facing a choice of products to spray on the subject field (Dayoob, 2021). Ongoing work is also examining temporal refuge periods as physiological time or whitefly generations to improve model predictive performance. Partition models have shown that the specific look-back period used to predict resistances from refuge abundance is variable depending on the chemistry used and major region of the state, likely reflecting variation in resistance potentials and broad-scale regional differences in cropping and resistant patterns. But in general, where resistances already exist, increased availability of temporal refuges, particularly during one or more 90-day periods of the previous calendar year, resulted in bioassayed whitefly populations with lower levels of resistance (Dayoob, 2021). A future grower digital tool would permit pest managers to access contemporaneous information about refuge availability for major modes of action for their specific field of interest (Fig. 3.6). This information might be broadly useful and applicable to insect pests beyond whiteflies, again based in the presumption that access to larger refuges helps lower resistance risks. Limiting use of insecticides to the lowest practical level is bedrock IPM and well serves IRM. Diversifying chemical use patterns through rotation practices is also bedrock to IRM. These two actions can be practiced by individual growers to some beneficial effect. However, partitioning modes of action through space and time at a scale that produces resistance management benefits has heretofore not been feasible for the individual grower. We believe these innovations will for the first time enable pest managers to engage this key practice of IRM and support landscape-level lowering of risks to multiple modes of action.

Avoidance as the foundation of IPM

To this point the discussion has focused on the upper layers of the conceptual IPM pyramid that encompasses those tactics brought into play when pest abundance warrants remedial action. Underlying these largely prescriptive elements is a substantial suite of tactics and natural agroecosystem assets, that if properly harnessed, can help producers avoid or reduce the risk of economic pest densities in the first place. As such, they are the fundamental building blocks of an effective and sustainable IPM strategy. Below, we will discuss these components and their landscape connections under three broad categories, crop management, pest biology and ecology, and explicit areawide elements (see Fig. 3.1).

Crop management

The connections among crop health, crop production practices, and pest dynamics have been long recognized as central to effective IPM (Stern *et al.*, 1959; Kogan, 1998). Essentially all crops, including cotton, produced in low desert regions of Arizona are irrigated, and research has shown this to be an important consideration in pest avoidance. For example, whiteflies perform better on water-stressed cotton (Flint *et al.*, 1994, 1995, 1996) likely due to higher leaf temperatures and other nutritional components (Blackmer and Byrne, 1999b). In contrast, Lygus bugs tend to perform better on lush plant growth (Leigh *et al.*, 1974; Munk and Goodell, 2002). Further, irrigation management to terminate crop growth and enable shorter production seasons were important for whitefly management in the early years of the invasion (Nuessly *et al.*, 1994). Early cotton termination and plow-down (to disrupt and bury larvae initiating diapause in the soil) also were part of a successful campaign to reduce pink bollworm overwintering prior to eradication. This mandated program in southern California was effective in greatly reducing pest infestations regionally (Chu *et al.*, 1996).

Crop fertilization is another important input that can potentially affect pest performance. There is some evidence to suggest that excess nitrogen in the cotton system can be favorable to whitefly population development (Blua and Toscano, 1994; Watson *et al.*, 1994; Bi *et al.*,

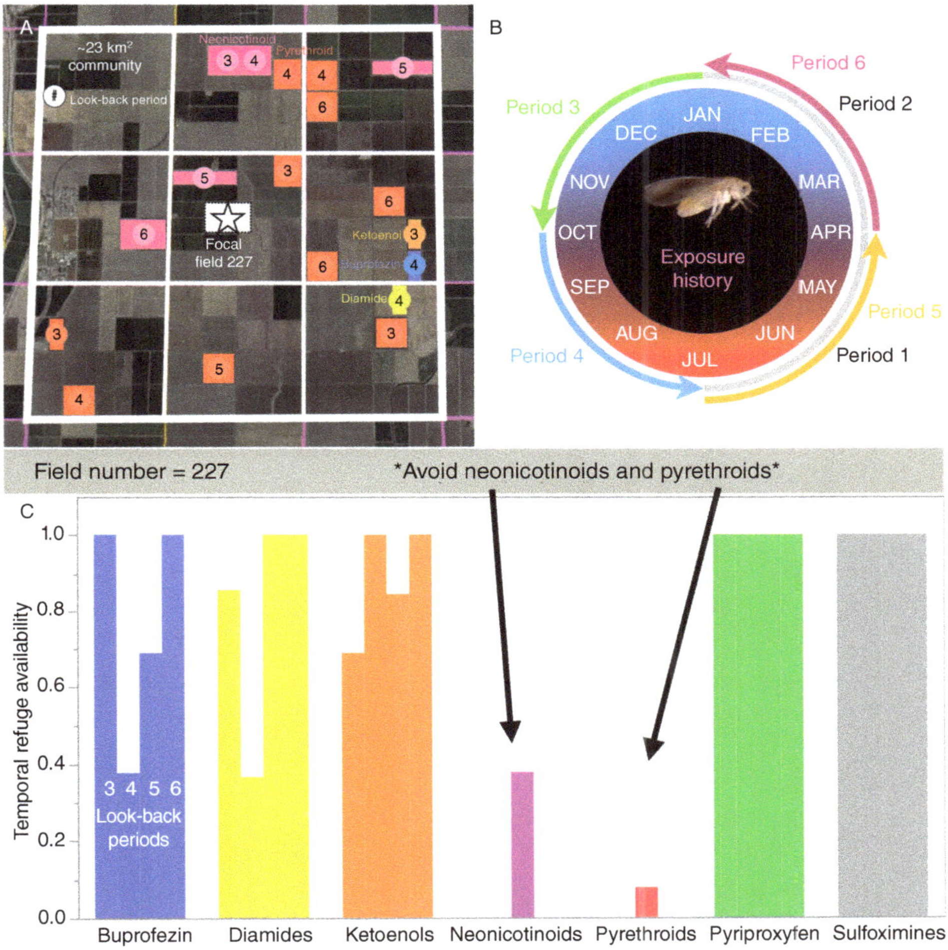

Fig. 3.6. Proactive resistance management system under development to guide grower product choices for whitefly control using landscape-level information about refuge availability for seven different modes of action in a 23 km² community. (A) Example map depicting field locations and chemistries (color-coded in C) sprayed during different 90-day look-back periods (periods 3, 4, 5, or 6 reflect exposure histories from the previous calendar year) that are used in determining refuge availability for the focal field when making a decision on what product to spray. (B) Whitefly exposure history look-back calendar depicting the six periods analyzed. For most chemistries, the exposure periods in the previous year were most predictive of resistance risk (periods 3, 4, 5, and/or 6). (C) Mock-up of digital tool that analyzes and presents temporal refuge availability based on exposure histories of whiteflies during the previous year for each of seven modes of action. Pyriproxyfen or sulfoxamines were not used in the prior year. Ample refuge was available for three other chemistries. Scant refuge was available for neonicotinoids or pyrethroids. The grower should avoid their use in the spray made to the focal field (starred field in map). Based on Dayoob (2021). (Original artwork by P. Ellsworth.)

2001). The impact of crop nitrogen fertility on Lygus bug populations has been equivocal (Leigh *et al.*, 1969). However, work with the related tarnished plant bug, *L. lineolaris* (Palisot de Beauvois), suggests that excessive nitrogen may be beneficial to this pest (Samples *et al.*, 2019).

Other crop management practices such as planting date can impact the pest complex.

Timing of planting along with knowledge of the cessation of overwintering diapause of pink bollworm can act to maximize "suicidal" emergence (pest emergence when no host fruit are available) and reduce the initial generation of this pest while still allowing planting during an optimal production window (Brown *et al.*, 1992; Moore *et al.*, 1992, 1993; Thacker *et al.*, 1994a, b). Optimal planting windows also can reduce the potential for early whitefly population growth (Ellsworth *et al.*, 2006). Overall, growers that effectively manage irrigation, nitrogen fertility, and planting dates optimally (not too much/early, not too little/late) are likely to reduce pest pressures and reap positive economic outcomes relative to yield and management inputs, once again with positive implicit landscape effects when practiced widely.

Host plant resistance is another critical crop management element and one that has long formed a key cornerstone in the IPM foundation (Smith, 2005). There has been considerable effort to develop whitefly-resistant crops (Belloti and Arias, 2001; Nombela and Muñiz, 2009; Shelby *et al.*, 2020) and crops resistant to the viruses they transmit (Morales, 2001). In cotton, hairy leaf varieties tend to be more attractive to whiteflies than glaborous varieties (Chu *et al.*, 1995), and this also is true for other whitefly host crops (Heinz and Zalom, 1995; Lambert *et al.*, 1997). Soon after whiteflies invaded Arizona, glabrous cotton varieties were recommended to growers (Ellsworth, 1993), and fortuitously these were the dominant varieties available at the time. This trend continues today, albeit not at this initial level due to changing variety availability and more effective IPM programs.

By far, the largest contribution of host plant resistance to the Arizona cotton IPM program has been the introduction in 1996 of Bt cotton targeting lepidopteran pests, primarily pink bollworm. While considerable effort had been placed on developing cotton varieties resistant to pink bollworm in the preceding decades (Wilson, 1992), nothing could compare with the nearly 100% efficacy of Bt cotton (Naranjo and Ellsworth, 2010). The adoption of Bt cotton in Arizona was rapid and widespread, going from 11% in the initial year to 86% by 2006. This was indicative of a global trend in Bt cotton adoption for management of several lepidopteran species (Naranjo *et al.*, 2020). In 2007, the EPA permitted Arizona growers to plant 100% Bt cotton as part of the eradication program under the condition that ongoing pink bollworm sterile insect releases would substitute for the mandated refuge requirements (Tabashnik *et al.*, 2010). Impacts on pink bollworm populations were rapid and dramatic. By 1999, regional reductions in pink bollworm populations could be measured (Carrière *et al.*, 2003). Correspondingly, insecticide application for this pest declined rapidly and went to zero within 2 years after the initiation of the eradication program in Arizona in 2006. Even with the 100% exemption, the adoption and use of Bt cotton was driven by individual grower decisions, once again displaying the power of collective, but uncoordinated actions over a broad landscape. These extraordinary reductions in regional target pest populations have been noted for Bt cotton in China (Wu *et al.*, 2008) and Bt maize in the US (Hutchison *et al.*, 2010; Dively *et al.*, 2018). Despite the severe selection pressure from Bt cotton to the functionally monophagous pink bollworm, field-evolved resistance never developed in the US, a result that cannot be claimed for Bt cotton targeting other lepidopteran pest species and pink bollworm in India and Bt maize targeting both lepidopteran and coleopteran pests in the US and other parts of the world (Tabashnik *et al.*, 2023). This success was likely achieved because of adherence of the industry to the refuge requirements both before and during pink bollworm eradication efforts that incorporated use of Bt cotton.

In addition to its high efficacy against pests, Bt cotton and other Bt crops have been extensively evaluated for non-target effects, particularly for arthropod natural enemies, and been found highly selective (Naranjo, 2005a, b, 2009; Meissle *et al.*, 2022). As noted above, pink bollworm was traditionally managed with broad-spectrum materials, often first applied early in the growing season (Moore *et al.*, 1993). Thus, an emergent outcome of the use of selective Bt cotton in Arizona was the elimination of these insecticides and the corresponding cascading and positive feedback loop that conserves arthropod natural enemies. This increased suppression of the remaining key pests and other minor pests on a regional scale, leading to even more reductions in insecticide use statewide. Although the main lepidopteran threat to cotton has been eliminated, growers still plant a high percentage of Bt cotton for several reasons. The main one is that genetic gains in cotton varieties for yield and quality have been greatest among the Bt varieties as seed companies invested and concentrated more of their breeding efforts into these germplasm lines. Another large factor is the bundling of other useful traits, especially herbicide tolerance, that growers want and need in these Bt varieties. Other reasons include guarding against pink bollworm reinvasion and protection from other sporadic lepidopteran pests such as *Helicoverpa zea*, *Chloridea virescens*, and *Spodoptera exigua* (Lepidoptera: Noctuidae) and *Estigmene acrea* (Lepidoptera: Erebidae), *Bucculatrix thurberiella* (Lepidoptera: Bucculatricidae), and others. Arizona is also a major contract producer of Bt cotton planted seed for the industry nationally.

The arsenal of Bt traits in cotton was expanded in 2021 with the soft launch of ThryvOn™, a new Bt trait that confers some resistance to Lygus bugs and thrips (*Frankliniella* spp.) (Thysanoptera: Thripidae). In Arizona, about 5600 ha were grown in 2023, the first year of full commercial release without export restrictions. Early surveys suggest that it can reduce insecticide sprays for Lygus bugs (Ellsworth *et al.*, 2021a, b). There is no efficacy against whitefly. Thrips are not generally a pest of concern in Arizona compared with other parts of the cotton belt, and in fact function as important predators of mites, whiteflies, and other small insects (Gonzalez and Wilson, 1982; Trichilo and Leigh, 1986; Ellsworth *et al.*, 2021a; P.C. Ellsworth and S.E. Naranjo, personal observation). Some early non-target assessments suggest this new Bt trait is selective (Bachman *et al.*, 2017; Asiimwe *et al.*, 2023; P.C. Ellsworth and S.E. Naranjo, unpublished data) and will contribute to conservation biological control as do current lepidopteran-targeted Bt cottons.

Pest biology/ecology

The biology and ecology of pests represent important facets that can be exploited in the development of IPM systems (Stern *et al.*, 1959; Clark *et al.*, 1967). The local population dynamics of an insect, or any organism, depends on additive (growth and immigration) and subtractive (mortality and emigration) processes. We touch here on dispersal processes for the key cotton pests. Primary focus will be placed on population dynamics and in-field mortality dynamics of whiteflies, which can be found on a wide range of wild and cultivated hosts in the landscape over the course of the year (Fig. 3.7). Thus, their metapopulation dynamics are driven by host plant availability and suitability, natural mortality forces, environmental conditions, movement, and pest management activities.

Extensive life table studies conducted in the field have provided a detailed understanding of the natural forces that impact whitefly mortality. Immature whiteflies are subject to very high levels of mortality in cotton, even when insecticides are not used (Naranjo and Ellsworth, 2005), and in most major crops and wild hosts that they inhabit (Naranjo *et al.*, 2022) (Fig. 3.8). Generational mortality is greater than 95% on all hosts

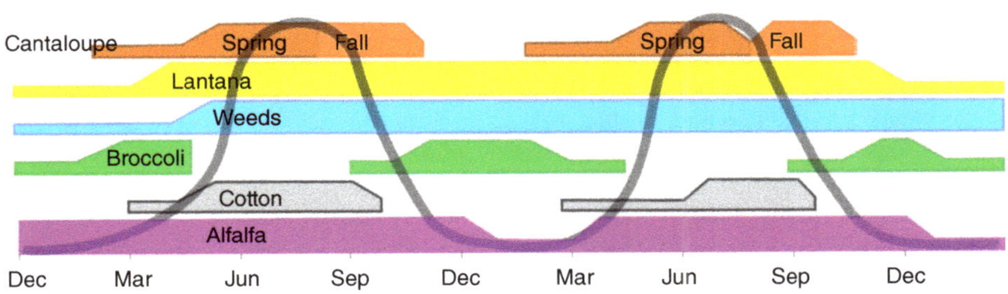

Fig. 3.7. Seasonality of the host crops over time from planting to harvest and the population trajectory of the silverleaf whitefly (gray line). The shallow heights indicate very low populations. Lantana represents the influence of ornamentals on the system. Adapted from Naranjo *et al.* (2022). (Open access—Wiley, Creative Commons.)

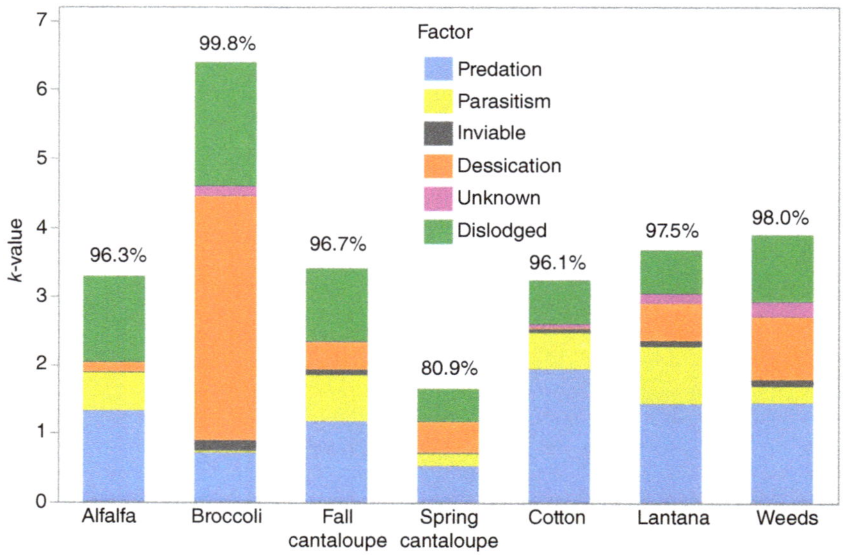

Fig. 3.8. Mortality forces impacting silverleaf whitefly immature stages on various crop and non-crop hosts over the season. Mortality is depicted as *k*-values (–ln[1 – mortality rate]). Percentages above the bars denote the total generational mortality. Adapted from Naranjo *et al.* (2022). (Open access—Wiley, Creative Commons.)

but spring cantaloupes (80%). This sets the stage for large increases in regional populations in the spring on cantaloupe that cascade into the summer and fall crops before lower temperatures and high mortality drive population decline. Predation by sucking predators naturally occurring in the system is the largest source of mortality and represents the key factor on most host plants followed by dislodgement that results from a combination of chewing predation and weather events such as rain, blowing dust, and leaf senescence. Parasitism by a combination of native and exotic parasitoids is generally low on all host plants. Long-term assessments in cotton have shown that classical introductions of exotic species have contributed nothing to changes in mortality rates of silverleaf whitefly on this crop in Arizona (Naranjo, 2018).

The vital agents are a suite of indigenous arthropod predators, and it is this complex that has been a fundamental focus of conservation biological control in the Arizona IPM system. We have identified six key arthropod predators useful in advancing decision making for pest control. These include *Misumenops celer* (Hentz) (Araneae: Thomisidae), *Drapetis* nr *divergens* (Diptera: Empididae), *Geocoris* spp. (Hemiptera: Geocoridae), *Orius tristicolor* (White) (Hemiptera: Anthocoridae), *Chrysoperla carnea sensu lato* (Neuroptera: Chrysopidae), and *Collops* spp. (Coleoptera: Melyridae). The predator thresholds (inaction thresholds) we developed help growers understand when and where natural enemies are abundant enough to forgo insecticide applications and represent a rare example of operationalizing conservation biological control measures in open-field cropping systems and potentially fostering even further reductions in insecticide use (Vandervoet *et al.*, 2018, 2022; Ellsworth *et al.*, 2019, 2022).

Extensive field life table research also examined the interactions between chemical and biological controls in our system (Naranjo and Ellsworth, 2009a). We found that insecticides, broad-spectrum or selective, were the key mortality factors immediately following applications, and this replaced the mortality provided by natural enemies. However, 2–6 weeks later, well after chemical residuals had degraded, mortality from predation rebounded when selective insecticides were used and resumed the role as key mortality factor. In contrast, mortality from natural enemies continued to be low when broad-spectrum insecticides were deployed. Generally, a single, well-timed selective insecticide application was sufficient for season-long pest suppression, while multiple applications of broad-spectrum materials were needed to provide the same result (Fig. 3.4). We coined the term "bioresidual" to define this extended pest suppression in selective insecticide regimes after the chemical residual had dissipated. Bioresidual represents the mortality imposed by conserved natural enemies and other natural environmental forces such as weather.

Comparatively less is known about natural enemy-induced mortality in Lygus bugs and pink bollworm. The pink bollworm was the subject of classical biological control efforts in the 1970s and 1980s, but none of the releases of introduced parasitoids resulted in establishment and pest suppression (Henneberry and Naranjo, 1998). Several naturally occurring arthropod predators are capable of attacking multiple stages of pink bollworm, but most studies were conducted under controlled conditions in the laboratory and field. Gut content research demonstrated that several predator species can attack eggs and possibly adult female moths (Hagler and Naranjo, 1994a, b), but the level of potential pest suppression is low and not well understood (Naranjo and Hagler, 1998). Likewise, little is known about natural enemy-induced mortality on Lygus bugs. Several exotic parasitoids have been introduced, but establishments have been rare, and impacts are largely unknown (Jackson *et al.*, 1995). Several predators have been shown to attack eggs, nymphs, and adult females in the field using gut content analyses (Hagler *et al.*, 1992; Hagler, 2006, 2011), but again, little is known about the levels of suppression these predators impose on Lygus bug populations (Zink and Rosenheim, 2008; Hagler *et al.*, 2020).

None the less, there is a diverse and abundant (largely indigenous) natural enemy community in Arizona agroecosystems. These communities, along with an IPM system that enables their conservation through the use of selective pest control technologies in cotton (e.g. Bt cottons and fully selective insecticides), have enabled biological control of all key and other cotton pests on a landscape scale. While it has not been measured explicitly, enhanced conservation of natural enemies in cotton is likely providing for increased natural control of pests in other crops in the agroecosystem. Our life table studies with whitefly showing high levels of predation throughout multi-crop/host systems provide some evidence (Naranjo *et al.*, 2022). Indeed, this is the integrated control concept first espoused by Stern *et al.* (1959), which is or should be core to modern-day IPM at both the local and landscape scale (Naranjo and Ellsworth, 2009b).

Areawide elements

All key pests in our system are moderately to highly mobile and seek hosts throughout the landscape. As such, all our key pests and many others are "shared" agroecosystem pests and not just the concern of an individual field, crop, or farm. Given this shared context, what one grower does or does not do can affect neighbors and even others further away, including urban areas. The most striking example of this latter concern was the huge waves of whitefly clouds inundating urban landscapes from agricultural areas many kilometers away in the early years after first invasion (Ellsworth, 2014; S.E. Naranjo and P.C. Ellsworth, personal observation).

Pink bollworm adults have been shown to disperse locally within and between fields and over 100s of kilometers in search of hosts, with these latter phenomena often aided by weather patterns (Glick and Noble, 1961; Bariola *et al.*, 1973; Stern, 1979; Flint and Merkle, 1981). Since the late 1960s, large numbers of sterile pink bollworm males were routinely released in the central valley of California where they successfully excluded migrating pests from the south from establishing in the once vast acreages of cotton in that region (Henneberry, 1994). Understanding dispersal behavior is critical to optimum implementation of resistance management strategies for pests targeted by Bt cotton. Trapping networks were used to better define the distance over which there would be sufficient interbreeding between Bt susceptible and putatively resistant moths arising from fields planted to Bt cotton. The work suggested that refuges of non-Bt cotton should be within 1.6 km of Bt cotton for effective cross mating and dilution of resistance genes (Tabashnik *et al.*, 1999).

Although there is abundant anecdotal knowledge that Lygus bugs move among wild and cultivated hosts in the landscape (Goodell *et al.*,

1998; Goodell, 2009), there are relatively few studies that have attempted to quantify the scale and rate of this movement. Bancroft (2005) estimated diffusion coefficients for *L. hesperus* in a small mixed landscape of cotton, alfalfa, and beans, and Sivakoff *et al.* (2012) examined the relative scale of movement in alfalfa. Larger landscape studies examined the source–sink relationships for this species among a broad range of affected crops and wild hosts (Carrière *et al.*, 2006, 2012b). Using ring analyses with replicated focal fields of cotton, Lygus bug densities in focal fields could be predicted by both positive and negative relationships with crop and non-crop resources. Relationships included predictable source crops like seed alfalfa (at a scale of 1.5 and 2.5 km for Arizona and California, respectively), sugar beets (1.25 km), and weeds (0.5 km). Sinks also were identified, such as cotton (0.75 and 2.5 km for Arizona and California, respectively), uncultivated fields (1 km), and a new crop, guayule (0.5–1.0 km). However, two crops demonstrated both landscape effects, sink and source for Lygus bugs, depending on whether they were managed or unmanaged. Untreated safflower serves as a long-range source to 2.5 km, but that role reverses over the same spatial scale if it is sprayed with one timely application for Lygus bugs. Similarly, forage alfalfa functions as a source up to 1.5 km, but as a sink over the same distance if it is specifically managed for Lygus bug retention via strip- or block-cutting. These spatial relationships were used to develop grower-friendly gaming models that could be used to explore the development of crop configurations in landscapes to minimize Lygus bug damage and maximize cotton revenues (Ellsworth, 2013).

A multi-user gaming platform was built where up to 16 grower-players could simulate Lygus bug distributions across a 4000-ha valley or community of growers (Ellsworth, 2013). Growers were presented with a planting palette of eight different crops that they could "plant" into their own virtual farmscape including for example cotton, safflower, guayule, seed alfalfa, alfalfa, wheat, sorghum, and fallow with the option to designate their safflower and/or alfalfa as managed or not for Lygus bugs. The simulations were run iteratively with growers given an opportunity to replant their farm as they wished. The only constraint on play was that every grower had to plant at least one cotton field. At the conclusion of each run, only then was a grower able to examine the pattern of Lygus bug distributions over the entire 4000-ha community and see a tally of source and sink areas under cultivation. Generally, growers grasped the landscape concepts very quickly and in as few as three to five runs arrived at a solution optimized for minimizing Lygus bug losses in cotton both for the individual and community of growers.

The most explicit exploitation of the movement dynamics for Lygus bugs is the strip harvesting of alfalfa pioneered in California (Stern *et al.*, 1964; Summers *et al.*, 2004; Goodell, 2009). This cultural tactic takes advantage of the preference of Lygus bugs for alfalfa and helps to manage the abundance of this pest in nearby cotton fields. Although it can be challenging to implement in modern production systems and requires some coordination among growers, it remains a recommended practice for the management of this insect in cotton/alfalfa systems (Goodell, 2009). When alfalfa production is locally very large, harvest operations cannot clear-cut entire areas. Under these conditions, block-harvesting can be practiced where alternate blocks or fields of alfalfa are harvested and re-grown, such that half-grown hay is always available to Lygus bugs in a nearby block. In Arizona where the ratio of alfalfa to cotton area in any given location can be 3–5 (alfalfa) to 1 (cotton), block-harvesting is a de facto, even passive, outcome of harvest and production capacities in alfalfa cultivation. Thus, as local demands and prices go up for forage to supply regional dairies as they have over the last decade, fewer issues have developed where forage alfalfa is sourcing Lygus bugs to cotton because of block-cutting.

As with Lygus bugs, much of our understanding of the dispersal of silverleaf whitefly is anecdotal, but some experimental studies have shown that dispersal appears to be bi-modal, with a combination of local and longer range movement up to about 3 km during short intervals (Byrne *et al.*, 1996). In cotton, life table data and modeling were used to define temporal patterns of movement related to crop development (Naranjo and Ellsworth, 2005) (Fig. 3.9). In the early part of the summer growing season, adults immigrate into cotton fields from sources such as spring-planted melons and then emigrate towards the end of the growing cycle as cotton

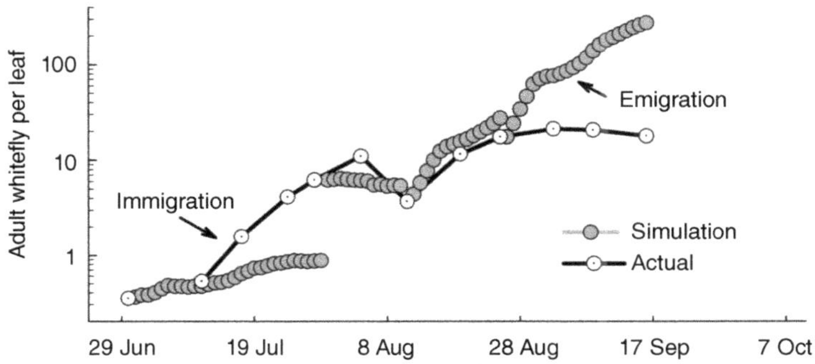

Fig. 3.9. Simulated and observed population dynamics for whiteflies in cotton. Simulations are based on a temperature-dependent, stage-structured model initiated with observed insect densities and observed rates of egg and nymph mortality from life tables. Departures from the actual whitefly adult densities are very likely the result of immigration and emigration processes. Adapted from Naranjo and Ellsworth (2005). (Used with permission from Wiley Publishing.)

plants mature and near harvest. The dynamics during most of the cotton season can be predicted solely by in-field demographics of reproduction, development, and mortality. Such a pattern likely occurs in all crop and wild hosts that whiteflies inhabit over the course of the year. As noted above in the IRM section, gene flow relative to resistance development in whiteflies occurs at a scale of roughly 3 km. Further landscape investigations using ring analyses, showed that the spatial scale of whitefly abundance and the incidence of an important virus they transmit to melons is on the order of 1.75–3 km (Carrière *et al.*, 2014). Later validation of this work suggested that disease transmission in melons can be predicted by the abundance of whitefly in cotton fields 1.5 km away (Carrière *et al.*, 2017). In addition to the value of this knowledge in configuring crop landscape, it also validates some prior work on the movement potential of this pest.

Ever since the invasion of pink bollworm into the western US, there have been numerous attempts at its eradication and coordinated management plans over various spatial scales (Henneberry and Naranjo, 1998). These ranged from small area, cooperative, grower-sponsored scouting programs that enabled uniform sampling of pink bollworm infestations with data provided to growers for decision making (Carruth and Moore, 1973), and coordinated multi-tactic programs involving uniform planting dates, trap crops, and insecticides (Moore *et al.*, 1992, 1993; Thacker *et al.*, 1994a, b), to larger scale efforts to manage the pest with pheromone disruption (El-Lissy *et al.*, 1993). While these programs had short-term success in reducing regional insecticide use and pest damage, they were ultimately not sustainable without continued investment from public institutions. The short season production system in southern California has already been mentioned and was another example of successful areawide management (Chu *et al.*, 1996) that was ultimately a victim of declining cotton production in the region and the advent of Bt cotton. As also mentioned, the sterile insect release program in the central valley of California, funded by growers, was a long standing and successful program that excluded pink bollworm in that region for many decades.

However, the pink bollworm eradication program that was initiated in 2001 in all infested areas of the US and northern Mexico was by far the most successful and, so far, lasting program for this pest. The grower-approved program involved a consortium of growers, industry, and federal and state institutions in two countries (Staten and Walters, 2021). It was multi-tactic and included transgenic Bt cotton, sterile insect release, pheromones, cultural practices, and insecticides as needed. Growers planted Bt cotton at their own expense but were charged by the acre for other control tactics if they planted non-Bt cotton at roughly the equivalent cost of the Bt trait. This created large incentives to plant Bt cotton and further incentive by regulatory relief from the EPA on the planting of refuges during eradication. Federal support funded sterile insect production, and state personnel mapped all cotton fields, scouted, and implemented other program elements. The program was phased in over a 7-year period starting in south Texas, mainly due to constraints on sterile insect production capacity and other operational issues. In most of Arizona the program was initiated in 2006, with final regions in western and southwest Arizona phased in by 2007–2008. The success in Arizona was rapid and, as already noted, grower sprays for pink bollworm ceased in 2008, aided greatly by the huge regional reductions in pest abundance resulting from the widespread adoption and use of Bt cotton before and during the program. Post-eradication analyses of the program in Arizona suggested that it was the combined action of Bt cotton and sterile insect release that enabled eradication (Tabashnik *et al.*, 2021), neither tactic alone would have achieved the desired outcome. These authors also estimate that cotton growers in the US saved US$192 million between 2014 and 2019. In 2018, the pest was officially declared eradicated from continental US and the northern portions of three Mexican states (Perdue, 2018). Post-eradication activities, mainly trap monitoring, continue to this day to prevent reintroduction of pink bollworm or boll weevil to Arizona.

Soon after the invasion of the silverleaf whitefly, there were attempts to establish coordinated, large-scale pest management programs. Arizona Cooperative Extension organized and implemented a community-wide whitefly management program in a 5000-ha community west of Phoenix from 1993 to 1995 (Ellsworth and Watson, 1996). The central feature of this program was grower communication and education and the introduction of newly developed whitefly adult sampling techniques. This included regular meetings and sharing of data from program scouting of all community fields and observance of action thresholds that were being actively researched at the time. The program was successful in introducing and demonstrating pest management techniques to growers and pest control advisors and represented a model of community-based education, validation and implementation (Ellsworth and Watson, 1996). A quasi-state-based organization, the Arizona Cotton Research and Protection Council, and USDA-APHIS tested an areawide pest management program on an area of cotton production in Gila Bend, Arizona, in 1994 and 1995. They used an ad hoc but uniform sampling method, nominal thresholds significantly lower than was being researched at the time and an aggressive border spraying program to manage the pest. The effort had some success in lowering whitefly populations and damage in program fields but was ultimately not sustained due to the need for state and federal personnel longer-term and the acceleration of resistance to synergized pyrethroids. With significantly lower thresholds in use, spray frequencies were very high and consisted of repeated usages of synergized pyrethroids areawide. A follow-on voluntary program was implemented in the same production area in 1995 to 1997 by Cooperative Extension at the University of Arizona. This program scouted fields using newly developed adult and large nymph binomial sampling and thresholds discussed above along with newly available IGRs and provided information to growers for ultimate decision making (Husman and Jech, 1998). The program was successful in organizing growers areawide and deriving benefits from collective action and informed decision making. The approach used today is informed both by the challenges and successes of the past. However, the success of the Arizona cotton IPM strategy is rooted in education, stakeholder engagement, a professional workforce dedicated to pest management (i.e. licensed pest control advisors), the advent of selective technologies accompanied by sophisticated "soft" technology (see next section), and a very tight integration of research and Extension (Ellsworth and Naranjo, 2017; Reisig *et al.*, 2019), each functioning even better when practiced over a wide area.

The counterfactual case: grower education and IPM implementation

As a complex system of tactics combined variably to achieve producer and societal goals, IPM is a difficult strategy to evaluate, absent proper controls as in a pre-planned experiment. The landscape-level processes covered thus far make this proposition even more intractable.

Replication over landscapes the scale of "Arizona" is virtually impossible. Even case-control studies are inadequate to capture the scales over which both education and introduced technologies are deployed. However, one thing is clear. The broad-scale progressive improvements in the Arizona IPM strategy were in part enabled, if not fueled, by the introduction of selective technologies like Bt cotton, IGRs, and other selective insecticides. Gains, even landscape-level ones, might be possible via passive deployment and diffusion processes; however, it can be argued that education of growers, pest managers, and supporting industries are critical to the success of the Arizona IPM strategy. It is these sets of "soft" technologies (Reisig *et al.*, 2019)—the human-mediated sampling, threshold, resistance management, conservation biological control, crop placement, and other cultural plans and innovations accompanied by actively taught "use instructions"—that drive the success of the Arizona cotton IPM program.

Rarely is there a counterfactual case where one can see and measure the behavior of complex systems over a large comparable landscape. While Arizona's program success today was largely put into practice by 2006, in 2010 nearby agricultural communities in northern Mexico were still operating as if it were 1991. This was despite all the same hard technologies being available to them, Bt seeds, selective insecticides, and even the benefits of boll weevil and pink bollworm eradication. Through an EPA "Border 2012" grant, we engaged Mexican cotton stakeholders in a 17-month intensive program of validation research, grower demonstration trials and education, totaling over 1200 contact hours and 100 site visits. In that time, we estimate growers saved over US$1.6 million in 2012 alone with a 31–40% decrease in sprays, a 34% decrease in spray costs, and a 23–86% reduction in the use of broad-spectrum insecticides (Ellsworth *et al.*, 2018; Ellsworth and Naranjo, 2017). This had the added benefit of demonstrating to us that indeed the benefits of any pest management tactic cannot be realized without the full integration of hard and soft technologies. Further, the landscape-level benefits of decreased pest pressures, reduced risks of resistance development, and conservation biological control were all enhanced by the individual and collective actions of these Mexican growers.

Ecological and economics impacts

The ecological and economic outputs of the Arizona cotton IPM program have been transformative (Fig. 3.10). From 1990 to 1995 cotton growers sprayed an average of nine times per hectare to control their arthropod pest complex, peaking at 13 sprays in 1995 when whiteflies evolved resistance to synergized pyrethroids and pink bollworm and Lygus bugs were problematic. With the introduction of Bt cotton and two new selective insecticides for whitefly suppression, along with the roll out of new decision-making tools for whitefly, insecticide use dropped dramatically and averaged 3.9 sprays from 1996 to 2005. In 2006, a selective insecticide for Lygus bugs became available and the pink bollworm eradication commenced in much of Arizona, driving insecticide use to historically low levels of around two total sprays per hectare for all pests. A slight uptick occurred in 2012–2014 with the emergence of the stink bug, *Euschistus servus* (Hemiptera: Pentatomidae). A rapid research response showed growers that ineffectively controlling a modest density of *E. servus* with broad-spectrum insecticides was uneconomical and disruptive to the natural enemy community resulting in resurgences of whiteflies and outbreaks of other secondary pests (Brown, 2017; P.C. Ellsworth *et al.*, unpublished data). Still today, almost all growers continue to refrain from using these broadly toxic chemistries for stink bug control.

In 2019, a new decision aid was introduced that included consideration of adult and nymphal whiteflies along with the abundance of six key arthropod predators (Vandervoet *et al.*, 2018, 2022; Ellsworth *et al.*, 2019, 2022). Efforts to refine and teach this new system are ongoing. Lygus Bt cotton saw a soft introduction in 2021. Adoption has so far been small (almost 19% of upland cotton acreage in 2023) due to both seed supply and to uncertainty by growers about additional costs of the technology and how it might fit into their production systems and be beneficial. During the last five seasons, spraying for arthropods has reached historic lows (Fig. 3.10). Whiteflies and Lygus bugs were the target of 79% of the total sprays made against arthropods in Arizona cotton (2019–2023). Broadly toxic insecticide use in cotton is down more than 96.9% with pyrethroids, organophosphate, and

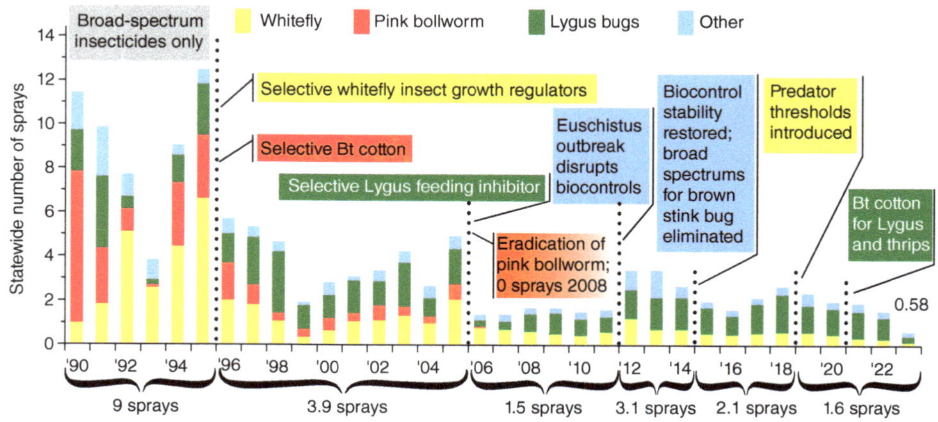

Fig. 3.10. A 34-year summary of insecticide use in the Arizona cotton system along with key milestones that have driven usage to historic lows and saved growers over US$600 million in the last 28 years. Vertical dotted lines depict dates of initiation of major events in the evolution of the Arizona IPM program (see Fig. 3.3). Sprays to control silverleaf whitefly are at their lowest levels since invasion of this species and insecticide sprays overall are at their lowest point in history. Most sprays made in the last 4 years are fully selective and safe to natural enemies. More than 67% of cotton went unsprayed for arthropod pests in 2023. Derived from the Arizona Pest Management Center Crop Pest Losses Database (Fournier *et al.*, 2007; Ellsworth and Fournier, 2024). (Original artwork by P. Ellsworth.)

carbamate insecticides specifically down more than 99.2% since the early 1990s; all insecticide use is down more than 87% from 1995. In 2023, a statewide average of only 0.58 sprays were required to control arthropods with over 67% of the state's cotton production not receiving a foliar spray for any arthropod pest. Overall, we estimate that well over 88,000 kg of insecticide active ingredient have been eliminated in the past 28 years (Ellsworth *et al.*, 2018; Ellsworth and Fournier, 2022). Combine low insecticide use with the fact that 85–90% of the sprays involve fully selective materials (Fig. 3.3) and it is easy to see why biological control provided by natural enemies has assumed such a fundamental role in IPM over the broad Arizona landscape.

Economic gains are commensurately impressive (Ellsworth *et al.*, 2018). We estimate that Arizona cotton growers have saved an average of US$25 million annually in control costs and yield losses due to arthropod pests since 1996. This equates to more than US$600 million in revenue gains since the first introduction of selective insecticides and Bt cotton in 1996. Furthermore, we estimate that about 42% of these overall economic gains have been directly due to the conservation of natural enemies that are residents in our system (Ellsworth *et al.*, 2018). While few agricultural systems have valuated biological control (Naranjo *et al.*, 2019), it represents a key foundational element in IPM (Stern *et al.*, 1959) that has rarely been exploited to its fullest potential. In Arizona, while growers and their pest control advisors may not all use the new decision aid with predators, most are very aware of their natural enemies and of the important pest suppression they can provide. At a recent field day, 82% of respondents indicated they "always" consider predators when timing whitefly sprays in cotton. In a program survey, 100% of respondents strongly (75%) or somewhat (25%) agreed that reliance on broadly toxic pesticides has been reduced, and 91% agreed that economic returns have improved because of better cotton IPM practices. The new landscape that has been created by the IPM program is largely the result of a positive feedback loop based on natural enemy conservation as a central consideration.

Some new opportunities have emerged from this enhanced landscape with the potential to augment positive environmental gains and turn them into value-added supply chains. One of these is the Better Cotton Initiative (Better Cotton,

2023), a worldwide organization with the goal of fostering sustainable cotton production and providing an assured conduit for customers that value and want to market sustainable agricultural production. The US Cotton Trust (US Cotton Trust Protocol, 2023) is an organization with similar goals to Better Cotton, but it is focused on developing verifiable sustainability guidelines and protocols for US growers. One key element of the Better Cotton Initiative is reduction in use of certain broadly toxic insecticides. Arizona's cotton is far less dependent on highly hazardous insecticides than any other state or region of the US. Our recent analyses show that less than 1% of Arizona's cotton acres make use of any of the seven highly hazardous insecticides identified by Better Cotton (Ellsworth and Fournier, 2022). In contrast, the rest of the cotton belt uses these same insecticides sometimes multiple times, on average (averaging 64–281% application-acres). Due to the success of cotton IPM in Arizona, growers are poised to be able to take advantage of such initiatives and potentially grow markets and value from their cotton.

Summary and Conclusions

The remarkable success of the Arizona cotton IPM program is, perhaps, unparalleled by any other cotton system in the world. However, there remain excellent opportunities for improvement, including those that can be realized on broader social, economic, and agroecosystem scales. We have shown how a community-based approach to developing and implementing IPM for cotton in Arizona contains both implicit and explicit connections to the landscape in which agricultural systems are embedded. While cotton IPM is not practiced as a single strategy and the industry does not move monolithically, providing individual growers with the tools needed, including critical "soft" technology or use instructions, to manage their pests more effectively and sustainably can reap rewards that go well beyond their fields and farms. These implicit landscape outcomes have paid huge dividends on many fronts, from environmental to economic to social. More formal "areawide" programs have been attempted for all the key pests, but for one exception (pink bollworm eradication) they have been limited in scope, longevity, and impact. In part this is due to the difficult task of organizing, funding, and sustaining such programs as more formal collectives. The community-based approach circumvents many of these limitations while still providing landscape-level benefits. One example is the massive reduction in insecticide use, particularly broad-spectrum materials, that has enabled true integrated control—*sensu* Stern *et al.* (1959)—that permits natural pest suppression on a regional scale. Other elements, like IRM, can profit from explicit consideration of the landscape and enable partitioning of insecticides to stem the evolution of resistance. This potentially benefits all cotton growers and even growers of other crops battling the same pests.

Sustainability of production is a private and public good. There is broad agreement to this basic concept, which supports farmers and their economy and the needs of society for food and fiber. And, there has been a steady shift in the marketplace mainly by buyers to identify, benchmark, and market sustainable products. Researching, validating, teaching, and ultimately developing tools that expressly harness and quantify landscape-level processes should be part of the scientific and sustainability dialog. In Arizona, simple tools for measuring and monitoring pest populations have been advanced to a point of collective common understanding. We have advanced the private and public good of natural enemy conservation by quantifying predator to prey relationships that can be used by individual growers and in individual fields but with far reaching and additive benefits that span the landscape. Resistance management has gone from retrospective science—the veritable coroner on the scene pronouncing the patient dead with great accuracy—to advanced understanding that guides a successful pest eradication campaign and produces the hope for proactive digital tools for understanding individual risks for resistance with additive landscape common pool benefits. Computing power, artificial intelligence, common pool imagery resources, other "big" data, and other major digital advances to understanding complex systems should be fertile ground for future gains in landscape-level IPM (see Chapter 13, this volume). The Arizona cotton IPM plan has been advanced via a plethora of smaller incremental changes and improvements of individual tactics like sampling, action thresholds, and cultural controls, and via major breakthroughs in the development of hard technologies like Bt cottons, selective insecticides, and sterile moth technique. However, the connective tissue that binds all these elements together is the very tight integration of research with Extension and the heavy investment in the human-mediated "soft" technologies, the use instructions for each of IPM's component tactics, if you will, that have made this IPM system the success that it is. It is a reminder that no matter how advanced hard technologies become, IPM is still rooted in the people and cultures in which it is implemented. Arizona has a progressive agricultural industry with a highly skilled and professional pest management workforce that amplify the use instructions to a landscape scale through their advisory services to growers. As data becomes richer and more complex, these pest managers will need to be taught how best to understand, evaluate, and implement these new resources for the benefit of all agriculture and society over the next generation. Just in the last generation there has been a shift from one of the major concerns for environmental pollution and human health and safety of users of cotton pesticides to one largely focused on safety concerns for invertebrates. Beyond just food security, agricultural production *because* of its large footprint in the landscape should become the eco-efficient broad-scale solution to some of society's grand challenges for the future, including biodiversity and habitat preservation, natural resource management, carbon sequestration, greenhouse gas emissions, and climate change.

Acknowledgements

The work that gave rise to the successful IPM system present in Arizona cotton was fueled by many grants, gifts, and contracts that supported the considerable research and Extension efforts conducted. The authors were supported by their respective institutions, USDA-ARS and the University of Arizona. P.C.E. would like to acknowledge that material presented here was enabled by current support from the National Institute of Food and Agriculture, USDA, under award number 2021-70006-35385 and grants from the Arizona Cotton Growers Association and Cotton Incorporated. Any recommendations, products, services, or organizations that are mentioned, shown, or indirectly implied in this publication do not imply endorsement by the USDA or the University of Arizona.

References

Asiimwe, P., Brown, C.R., Ellsworth, P.C., Reisig, D.D., Bertho, L. *et al.* (2023) Transgenic cotton expressing Mpp51Aa2 does not adversely impact beneficial non-target Hemiptera in the field. *Crop Protection* 173, 106384. DOI: 10.1016/j.cropro.2023.106384

Bachman, P.M., Ahmad, A., Ahrens, J.E., Akbar, W., Baum, J.A. *et al.* (2017) Characterization of the activity spectrum of MON 88702 and the plant-incorporated protectant Cry51Aa2.834_16. *PLoS ONE* 12, e0169409. DOI: 10.1371/journal.pone.0169409

Bancroft, J.S. (2005) Dispersal and abundance of *Lygus hesperus* in field crops. *Environmental Entomology* 34, 1517–1523. DOI: 10.1603/0046-225X-34.6.1517

Bariola, L.A., Keller, J.C., Turley, D.L. and Farris, J.R. (1973) Migration and population studies of the pink bollworm in the arid west. *Environmental Entomology* 2, 205–208. DOI: 10.1093/ee/2.2.205

Belloti, A.C. and Arias, B. (2001) Host plant resistance to whiteflies with emphasis on cassava as a case study. *Crop Protection* 20, 813–823. DOI: 10.1016/S0261-2194(01)00113-2

Better Cotton (2023) *Better Cotton*. London, UK: Better Cotton. Available at: https://bettercotton.org (accessed August 9, 2023).

Bi, J.L., Ballmer, G.R., Hendrix, D.L., Henneberry, T.J. and Toscano, N.C. (2001) Effect of cotton nitrogen fertilization on *Bemisia argentifolii* populations and honeydew production. *Entomologia Experimentalis et Applicata* 99, 25–36. DOI: 10.1046/j.1570-7458.2001.00798.x

Blackmer, J.L. and Byrne, D.N. (1993) Flight behaviour of *Bemisia tabaci* in a vertical flight chamber: effect of time of day, sex, age and host quality. *Physiological Entomology* 18, 223–232. DOI: 10.1111/j.1365-3032.1993.tb00592.x

Blackmer, J.L. and Byrne, D.N. (1999a) Changes in amino acids in *Cucumis melo* in relation to life-history traits and flight propensity of *Bemisia tabaci*. *Entomologia Experimentalis et Applicata* 93, 29–40. DOI: 10.1046/j.1570-7458.1999.00559.x

Blackmer, J.L. and Byrne, D.N. (1999b) The effect of *Bemisia tabaci* on amino acid balance in *Cucumis melo*. *Entomologia Experimentalis et Applicata* 93, 315–319. DOI: 10.1046/j.1570-7458.1999.00592.x

Blackmer, J.L., Byrne, D.N. and Tu, Z. (1995) Behavioral, morphological, and physiological traits associated with migratory *Bemisia tabaci* (Homoptera: Aleyrodidae). *Journal of Insect Behavior* 8, 251–267. DOI: 10.1007/BF01988909

Blua, M.J. and Toscano, N.C. (1994) *Bemisia argentifolii* (Homoptera: Aleyrodidae) development and honeydew production as a function of cotton nitrogen status. *Environmental Entomology* 23, 316–321. DOI: 10.1093/ee/23.2.316

Bordini, I., Ellsworth, P.C., Naranjo, S.E. and Fournier, A. (2021) Novel insecticides and generalist predators support conservation biological control in cotton. *Biological Control* 154, 4502–4502. DOI: 10.1016/j.biocontrol.2020.104502

Bordini, I.C., Fournier, A.J., Naranjo, S.E., Pier, N.M. and Ellsworth, P.C. (2024) *Cotton Insecticide Use Guide: Knowing and Balancing Risks*. Arizona Pest Management Center. IPM Short. Tucson, AZ: Cooperative Extension, College of Agriculture and Life Sciences, University of Arizona. Available at: http://hdl.handle.net/10150/670916 (accessed February 14, 2024).

Brown, J.K. and Nelson, M.R. (1987) Host range and vector relationships of cotton leaf crumple virus. *Plant Disease* 71, 522–524. DOI: 10.1094/PD-71-0522

Brown, J.K., Frohlich, D.R. and Rosell, R.C. (1995) The sweetpotato or silverleaf whiteflies: biotypes of *Bemisia tabaci* or a species complex? *Annual Review of Entomology* 40, 511–534. DOI: 10.1146/annurev.en.40.010195.002455

Brown, L.M. (2017) Biology, ecology, and economics of brown stink bug, *Euschistus servus* (Heteroptera: Pentatomidae), in desert cotton agroecosystems. MS thesis. Tucson, AZ: University of Arizona. Available at: http://hdl.handle.net/10150/625340 (accessed January 3, 2024).

Brown, P., Silvertooth, J., Moore, L. and Watson, T. (1992) Revised planting window for full season cotton varieties. In: *Cotton: A College of Agriculture Report*. P-91. Tucson, AZ: College of Agriculture, University of Arizona, pp. 240–251. Available at: https://repository.arizona.edu/handle/10150/208660 (accessed August 9, 2023).

Byrne, D., Rathman, R., Orum, T. and Palumbo, J. (1996) Localized migration and dispersal by the sweet potato whitefly *Bemisia tabaci*. *Oecologia* 105, 320–328. DOI: 10.1007/BF00328734

Byrne, D.N. (1999) Migration and dispersal by the sweet potato whitefly, *Bemisia tabaci*. *Agricultural and Forest Meteorology* 97, 309–316. DOI: 10.1016/S0168-1923(99)00074-X

Carrière, Y., Ellers-Kirk, C., Sisterson, M., Antilla, L., Whitlow, M. *et al.* (2003) Long-term regional suppression of pink bollworm by *Bacillus thuringiensis* cotton. *Proceedings of the National Academy of Sciences* 100, 1519–1523. DOI: 10.1073/pnas.0436708100

Carrière, Y., Ellsworth, P.C., Dutilleul, P., Ellers-Kirk, C. and Antilla, L. (2006) A GIS-based approach for area-wide pest management: the scales of *Lygus hesperus* movements to cotton from alfalfa, weeds and cotton. *Entomologia Experimentalis et Applicata* 118, 203–210. DOI: 10.1111/j.1570-7458.2006.00384.x

Carrière, Y., Ellers-Kirk, C., Hartfield, K., Larocque, G., Degain, B. *et al.* (2012a) Large-scale, spatially-explicit test of the refuge strategy for delaying insecticide resistance. *Proceedings of the National Academy of Sciences* 109, 775–780. DOI: 10.1073/pnas.1117851109

Carrière, Y., Goodell, P.B., Ellers-Kirk, C., Larocque, G., Dutilleul, P. *et al.* (2012b) Effects of local and landscape factors on population dynamics of a cotton pest. *PLoS ONE* 7, e39862. DOI: 10.1371/journal.pone.0039862

Carrière, Y., Degain, B., Hartfield, K.A., Nolte, K.D., Marsh, S.E. *et al.* (2014) Assessing transmission of crop diseases by insect vectors in a landscape context. *Journal of Economic Entomology* 107, 1–10. DOI: 10.1603/EC13362

Carrière, Y., Degain, B., Liesner, L., Dutilleul, P. and Palumbo, J.C. (2017) Validation of a landscape-based model for whitefly spread of the cucurbit yellow stunting disorder virus to fall melons. *Journal of Economic Entomology* 110, 2002–2009. DOI: 10.1093/jee/tox213

Carruth, L.A. and Moore, L. (1973) Cotton scouting and pesticide use in eastern Arizona. *Journal of Economic Entomology* 66, 187–190. DOI: 10.1093/jee/66.1.187

Castle, S. and Naranjo, S.E. (2009) Sampling plans, selective insecticides and sustainability: the case for IPM as 'informed pest management'. *Pest Management Science* 65, 1321–1328. DOI: 10.1002/ps.1857

Chu, C.C., Henneberry, T.J. and Cohen, A.C. (1995) *Bemisia argentifolii* (Homoptera: Aleyrodidae): host preference and factors affecting oviposition and feeding site preference. *Environmental Entomology* 24, 354–360. DOI: 10.1093/ee/24.2.354

Chu, C.C., Henneberry, T.J., Weddle, R.C., Natwick, E.T., Carson, J.R. *et al.* (1996) Reduction of pink bollworm (Lepidoptera: Gelechiidae) populations in the Imperial Valley, California, following mandatory short-season cotton management systems. *Journal of Economic Entomology* 89, 175–182. DOI: 10.1093/jee/89.1.175

Clark, L.R., Geier, P.W., Hughes, R.D. and Morris, R.F. (1967) *The Ecology of Insect Populations: in Theory and Practice*. London, UK: Methuen.

Coudriet, D.L., Prabhaker, N., Kishaba, A.N. and Meyerdirk, D.E. (1985) Variation in developmental rate on different hosts and overwintering of the sweetpotato whitefly, *Bemisia tabaci* (Homoptera: Aleyrodidae). *Environmental Entomology* 14, 516–519. DOI: 10.1093/ee/14.4.516

Crowder, D.W., Carrière, Y., Tabashnik, B.E., Ellsworth, P.C. and Dennehy, T.J. (2006) Modeling evolution of resistance to pyriproxyfen by the sweetpotato whitefly (Homoptera: Aleyrodidae). *Journal of Economic Entomology* 99, 1396–1406. DOI: 10.1603/0022-0493-99.4.1396

Crowder, D.W., Dennehy, T.J., Ellerskirk, C., Yafuso, C.M., Ellsworth, P.C. *et al.* (2007) Field evaluation of resistance to pyriproxyfen in *Bemisia tabaci* (B biotype). *Journal of Economic Entomology* 100, 1650–1656. DOI: 10.1603/0022-0493(2007)100[1650:feortp]2.0.co;2

Crowder, D.W., Ellerskirk, C., Yafuso, C.M., Dennehy, T.J., Degain, B.A. *et al.* (2008a) Inheritance of resistance to pyriproxyfen in *Bemisia tabaci* (Hemiptera: Aleyrodidae) males and females (B biotype). *Journal of Economic Entomology* 101, 927–932. DOI: 10.1603/0022-0493(2008)101[927:IORTPI]2.0.CO;2

Crowder, D.W., Ellsworth, P.C., Tabashnik, B.E. and Carrière, Y. (2008b) Effects of operational and environmental factors on evolution of resistance to pyriproxyfen in the sweetpotato whitefly (Hemiptera: Aleyrodidae). *Environmental Entomology* 37, 1514–1524. DOI: 10.1603/0046-225X-37.6.1514

Dayoob, N. (2021) Whitefly resistance management: time and space refugia in cross-commodity systems of Arizona and California. MS thesis. Tucson, AZ: University of Arizona. Available at: https://repository.arizona.edu/handle/10150/663218 (accessed January 4, 2024).

Dennehy, T.J. and Antilla, L. (1996) Whitefly resistance to insecticides in Arizona. In: Dugger, P. and Richter, D.A. (eds) *Proceedings Beltwide Cotton Conferences.* Memphis, TN: National Cotton Council, pp. 144–145.

Dennehy, T.J., Ellsworth, P.C. and Watson, T.F. (1995) *Whiteflies in Arizona: Pocket Guide*. Publication 195009. Tucson, AZ: Cooperative Extension, College of Agriculture, University of Arizona. Available at: https://cales.arizona.edu/crops/cotton/insects/wf/wfly795.pdf (accessed July 23, 2023).

Dennehy, T.J., Degain, B., Harpold, G., Li, X., Crowder, D.W. *et al.* (2008) Management of *Bemisia* resistance: cotton in the southwestern USA. *Journal of Insect Science (Tucson)* 8, 17. DOI: 10.1673/031.008.0401

Diehl, J.W., Ellsworth, P.C. and Husman, S. (1994) A community-wide approach to whitefly management. In: *Cotton: A College of Agriculture Report.* P-96. Tucson, AZ: College of Agriculture, University of Arizona, pp. 299–303. Available at: http://hdl.handle.net/10150/209642 (accessed January 3, 2024).

Diehl, J.W., Naranjo, S.E. and Ellsworth, P.C. (1997) Whitefly growth regulators: a field sampling protocol for nymphs. In: *Cotton: A College of Agriculture Report*. P-108. College of Agriculture, University of Arizona, Tucson, AZ, pp. 265–271. Available at: https://repository.arizona.edu/handle/10150/211111 (accessed January 3, 2024).

Diehl, J.W., Ellsworth, P.C. and Moore, L. (1998) *Lygus in Cotton: Identification, Biology and Management*. No. 1. Tucson, AZ: Cooperative Extension. University of Arizona. Available at: https://cales.arizona.edu/crops/cotton/insects/lygus/lygus1.pdf (accessed January 3, 2024).

Dively, G.P., Venugopal, P.D., Bean, D., Whalen, J., Holmstrom, K. *et al.* (2018) Regional pest suppression associated with widespread Bt maize adoption benefits vegetable growers. *Proceedings of the National Academy of Sciences* 115, 320–3325. DOI: 10.1073/pnas.1720692115

Duffus, J.E. and Flocks, R.A. (1982) Whitefly-transmitted disease complex of the desert southwest. *California Agriculture*, November–December, 4–6. Available at: https://calag.ucanr.edu/download_pdf.cfm?article=ca.v036n11p4 (accessed January 3, 2024).

El-Lissy, O., Staten, R.T. and Antilla, L. (1993) Control of pink bollworm, *Pectinophora gossypiella* (Saunders) (Lepidoptera: Gelechiidae) in Parker Valley, Arizona by mating disruption using commercial formulations of gossyplure. In: *Proceedings of the International Cotton Pest Work Committee.* Sacramento, CA: California Department of Food and Agriculture, pp. 114–117.

Ellsworth, P.C. (1993) Sweetpotato whitefly in Arizona. In: *Cotton: A College of Agriculture Report*. P-94. Tucson, AZ: College of Agriculture, University of Arizona, pp. 283–293. Available at: http://hdl.handle.net/10150/209578 (accessed July 23, 2023).

Ellsworth, P.C. (2013) Landscape influences on Lygus bugs in western cotton. *Grow: Plant Health Exchange*. St Paul, MN: American Phytopathological Society. DOI: 10.1094/GROW-COT-08-13-031 (accessed January 3, 2024).

Ellsworth, P.C. (2014) Whitefly management and prevention of excess sugars in cotton. Webinar. *Plant Management Network*. DOI: 10.1094/GROW-COT-08-14-049 (accessed February 14, 2024).

Ellsworth, P.C. and Barkley, V. (2001) Cost-effective *Lygus* management in Arizona cotton. In: *Cotton: A College of Agriculture Report*. P-125. Tucson, AZ: University of Arizona, pp. 299–307. Available at: https://repository.arizona.edu/handle/10150/211330 (accessed July 23, 2023).

Ellsworth, P.C. and Barkley, V. (2005) Transitioning Lygus chemical controls to more selective options for Arizona Cotton. In: *Cotton: A College of Agriculture Report*. P-142. Tucson, AZ: University of Arizona, pp. 165–178. Available at: https://repository.arizona.edu/handle/10150/198176 (accessed July 23, 2023).

Ellsworth, P.C. and Diehl, J.W. (1996) *Whiteflies in Arizona No. 5: Insect Growth Regulators*. Tucson, AZ: Cooperative Extension, University of Arizona. Available at: http://ag.arizona.edu/crops/cotton/insects/wf/wfly5.pdf (accessed February 14, 2024).

Ellsworth, P.C. and Diehl, J.W. (1998) *Lygus in Cotton 2: An Integrated Management Plan for Arizona*. Tucson, AZ: Cooperative Extension, University of Arizona. Available at: https://cales.arizona.edu/crops/cotton/insects/lygus/lygus2.pdf (accessed January 3, 2024).

Ellsworth, P.C. and Fournier, A. (2022) *Highly Hazardous Pesticide Phase-out for US Cotton Growers: Alternatives, Risks, and Opportunities: Better Cotton initiative*. Tucson, AZ: University of Arizona. Available at: http://hdl.handle.net/10150/667318 (accessed January 3, 2024).

Ellsworth, P.C. and Fournier, A. (2024) Theory versus practice: are insecticide mixtures in Arizona cotton used for resistance management? *Pest Management Science* (in press). DOI: 10.1002/ps.8241

Ellsworth, P.C. and Jones, J.S. (2001) Cotton IPM in Arizona: a decade of research, implementation & education. In: *Cotton: A College of Agriculture Report*. P-125. Tucson, AZ: College of Agriculture, University of Arizona, pp. 195–215. Available at: http://hdl.handle.net/10150/211323 (Last accessed January 3, 2024).

Ellsworth, P.C. and Martinez-Carrillo, J.L. (2001) IPM for *Bemisia tabaci*: a case study from North America. *Crop Protection* 20, 853–869. DOI: 10.1016/S0261-2194(01)00116-8

Ellsworth, P.C. and Naranjo, S.E. (2017) Implementing IPM in cotton in Arizona & Mexico. In: *14th International Symposium on the Biosafety of GMOs, Guadalajara, Mexico*. Available at: https://cals.arizona.edu/crops/presentations/2017/17ISBGMOGuadalajaravF1uplo.pdf (accessed January 3, 2024).

Ellsworth, P.C. and Watson, T.F. (1996) *Whiteflies in Arizona No. 7: Pocket Guide '96*. Publication 196005. Tucson, AZ: Cooperative Extension, University of Arizona. http://ag.arizona.edu/crops/cotton/insects/wf/wfly7.pdf (accessed January 3, 2024).

Ellsworth, P., Moore, L., Watson, T.F. and Dennehy, T. (1994) *Insect Pest Management for Cotton*. Publication 194022. Tucson, AZ: Cooperative Extension, University of Arizona.

Ellsworth, P.C., Dennehy, T.J., Diehl, J.W., Ansolabehere, M., Antilla, L. *et al*. (1997) Statewide training sessions for use of insect growth regulators in Arizona. In: Henneberry, T.J. (ed.) *Silverleaf Whitefly: 1997 Supplement to the Five-Year National Research & Action Plan*. Washington, DC: US Department of Agriculture, p. 195.

Ellsworth, P.C., Tronstad, R., Leser, J., Goodell, P.B., Godfrey, L.D. *et al*. (1999) *Sticky Cotton Sources and Solutions*. IPM Series 13. Tucson, AZ: Cooperative Extension, University of Arizona. Available at: https://ag.arizona.edu/crops/cotton/insects/wf/stickycss.pdf (accessed July 9, 2023).

Ellsworth, P.C., Husman, S. and Langston, D. (2000) *The Communitywide (Central AZ) Lygus Action Plan Newsletter*. Cooperative Extension Series 1-8. Tucson, AZ: College of Agriculture and Life Sciences, University of Arizona. Available at: https://cales.arizona.edu/crops/cotton/insects/lygus/cwlapjuly700.pdf (accessed January 4, 2024).

Ellsworth, P.C., Palumbo, J.C., Naranjo, S.E., Dennehy, T.J. and Nichols, R.L. (2006) *Whitefly management in Arizona Cotton 2006*. IPM Series 18. Tucson, AZ: Cooperative Extension, University of Arizona. Available at: https://extension.arizona.edu/sites/extension.arizona.edu/files/pubs/az1404.pdf (accessed January 3, 2024).

Ellsworth, P.C., Barkley, V., Dennehy, T., Degain, B., Ellingson, B. *et al*. (2007) Assessment of Knack field performance through precision field and laboratory biosassays in cotton. In: *Cotton: A College of Agriculture and Life Sciences Report*. P-151. Tucson, AZ: University of Arizona, pp. 167–182. Available at: http://hdl.handle.net/10150/198208 (accessed July 23, 2023).

Ellsworth, P., Palumbo, J., Fournier, A. and Carrière, Y. (2008a) Beyond field borders: cross-commodity resistance management of *Bemisia tabaci* – spatial evaluation of group adoption of neonicotinoid guidelines. In: *Plant–Insect Interactions Symposium*, Reno, NV. Available at: https://ag.arizona.edu/crops/presentations/Reno_x-com_adoption.pdf (accessed January 3, 2024).

Ellsworth, P., Palumbo, J., Fournier, A. and Carrière, Y. (2008b) Cross-commodity insecticide usage: spatial analysis of management practices, control & risks. In: *WERA-069 Western IPM Coordinators Meeting,*

Chena Hot Springs, AK. Available at: https://cales.arizona.edu/crops/presentations/SpatialX-commodity-Flo.pdf (accessed January 3, 2024).

Ellsworth, P., Palumbo, J., Fournier, A. and Carrière, Y. (2009) Measuring adoption of cross-commodity IPM guidelines for whitefly control. In: *Fifth International Bemisia Workshop, Guangzhou, China*. Available at: https://cales.arizona.edu/crops/presentations/5thBemisia_x-com_adoption.pdf (accessed January 3, 2024).

Ellsworth, P.C., Li, X., Dennehy, T., Palumbo, J., Castle, S. *et al*. (2013) Is monitoring susceptibility of *Bemisia tabaci* to insecticides useful to management? In: *First International Whitefly Symposium, Kolymbari, Crete, Greece*. Available at: https://ag.arizona.edu/crops/presentations/2013/13WhiteflyResistanceMgtCretevF4.pdf (accessed January 17, 2024).

Ellsworth, P.C., Fournier, A. and Smith, T.D. (2015) *Arizona Cotton Insect Losses*. Publication AZ1183. Tucson, AZ: College of Agriculture and Life Sciences, University of Arizona. Available at: http://cals.arizona.edu/crops/cotton/insects/cil/cil.html (accessed January 3, 2024).

Ellsworth, P.C., Fournier, A., Frisvold, G. and Naranjo, S.E. (2018) Chronicling the socio-economic impact of integrating biological control, technology, and knowledge over 25 years of IPM in Arizona. In: Mason, P.G., Gillespie, D.R. and Vincent, C. (eds) *Proceedings of the 5th International Symposium on Biological Control of Arthropods*. Wallingford, UK: CAB International, pp. 214–216.

Ellsworth, P.C., Pier, N., Fournier, A., Naranjo, S.E. and Vandervoet, T. (2019) *Whitefly Predator "Thresholds" in Cotton*. Arizona Pest Management Center. IPM Short. Tucson, AZ: University of Arizona. Available at: http://hdl.handle.net/10150/665534 (accessed January 3, 2024).

Ellsworth, P.C., Bordini, I. and Pier, N. (2021a) *ThryvOn™ Cotton, Frequently Asked Questions*. Arizona Pest Management Center. IPM Short. Tucson, AZ: University of Arizona. Available at: http://hdl.handle.net/10150/659845 (accessed January 3, 2024).

Ellsworth, P.C., Bordini, I. and Pier, N. (2021b) *Tips on How to Manage Lygus Efficiently in ThryvOn™ Cotton*. Arizona Pest Management Center. Arizona Pest Management Center. IPM Short. Tucson, AZ: University of Arizona. Available at: http://hdl.handle.net/10150/668029 (accessed January 3, 2024).

Ellsworth, P.C., Bordini, I., Pier, N. and Naranjo, S.E. (2022) *Integrating Chemical & Biological Control in Cotton*. Arizona Pest Management Center. IPM Short. Tucson, AZ: University of Arizona. http://hdl.handle.net/10150/665531 (accessed January 3, 2024).

Ellsworth, P.C., Brown, L. and Pier, N. (2023) *Anatomy of a Cotton Sweep for Pests and Predators*. Arizona Pest Management Center. IPM Short. Tucson, AZ: University of Arizona. Available at: http://hdl.handle.net/10150/668520 (accessed January 3, 2024).

Flint, H.M. and Merkle, J.R. (1981) Early-season movements of pink bollworm male moths between selected habitats. *Journal of Economic Entomology* 74, 366–371. DOI: 10.1093/jee/74.3.366

Flint, H.M., Wilson, F.D., Hendrix, D., Leggett, J., Naranjo, S.E. *et al*. (1994) The effect of plant water stress on beneficial and pest insects including the pink bollworm and the sweetpotato whitefly in 2 short-season cultivars of cotton. *Southwestern Entomologist* 19, 11–22.

Flint, H.M., Radin, J.W., Parks, N.J. and Reaves, L.L. (1995) The effects of drip or furrow irrigation of cotton on *Bemisia argentifolii* (Homoptera: Aleyrodidae). *Journal of Agricultural Entomology* 12, 25–32.

Flint, H.M., Naranjo, S.E., Leggett, J.E. and Henneberry, T.J. (1996) Cotton water stress, arthropod dynamics, and management of *Bemisia tabaci* (Homoptera: Aleyrodidae). *Journal of Economic Entomology* 89, 1288–1300. DOI: 10.1093/jee/89.5.1288

Fournier, A., Ellsworth, P. and Barkley, V. (2007) Economic impact of Lygus in Arizona cotton: a comparative approach. In: *Cotton: A College of Agriculture and Life Sciences Report*. Tucson, AZ: College of Agriculture, University of Arizona, pp. 155–166. Available at: http://hdl.handle.net/10150/198220 (accessed January 3, 2024).

Gerling, D. (1967) Bionomics of the whitefly-parasite complex associated with cotton in southern California. *Annals of the Entomological Society of America* 60, 1306–1321. DOI: 10.1093/aesa/60.6.1306

Glick, P.A. and Noble, L.W. (1961) *Airborne Movement of the Pink Bollworm and Other Arthropods*. Technical Bulletin 1255. Washington, DC: US Department of Agriculture. Available at: https://ageconsearch.umn.edu/record/170872/files/tb1255.pdf (accessed January 3, 2024).

Gonzalez, D. and Wilson, L.T. (1982) A food web approach to economic thresholds: a sequence of pests/predaceous arthropods on California cotton. *Entomophaga* 27, 31–43. DOI: 10.1007/BF02371853

Goodell, P.B. (2009) Fifty years of the integrated control concept: the role of landscape ecology in IPM in San Joaquin valley cotton. *Pest Management Science* 65, 1293–1297. DOI: 10.1002/ps.1859

Goodell, P.B., Brazzle, J.R., Godfrey, L.D. and Grafton-Cardwell, B. (1998) Guidelines for *Lygus* management. *California-Arizona Cotton* 34, 13–15.

Hagler, J.R. (2006) Development of an immunological technique for identifying multiple predator-prey interactions in a complex arthropod assemblage. *Annals of Applied Biology* 149, 153–165. DOI: 10.1111/j.1744-7348.2006.00076.x

Hagler, J.R. (2011) An immunological approach to quantify consumption of protein-tagged *Lygus hesperus* by the entire cotton predator assemblage. *Biological Control* 58, 337–345. DOI: 10.1016/j.biocontrol.2011.05.015

Hagler, J.R. and Naranjo, S.E. (1994a) Determining the frequency of heteropteran predation on sweetpotato whitefly and pink bollworm using multiple ELISAs. *Entomologia Experimentalis et Applicata* 72, 63–70. DOI: 10.1111/j.1570-7458.1994.tb01802.x

Hagler, J.R. and Naranjo, S.E. (1994b) Qualitative survey of two Coleopteran predators of *Bemisia tabaci* (Homoptera: Aleyrodidae) and *Pectinophora gossypiella* (Lepidoptera, Gelechiidae) using a multiple prey gut content ELISA. *Environmental Entomology* 23, 193–197. DOI: 10.1093/ee/23.1.193

Hagler, J.R., Cohen, A.C., Bradley-Dunlop, D. and Enriquez, F.J. (1992) Field evaluation of predation on *Lygus hesperus* (Hemiptera: Miridae) using a species-specific and stage-specific monoclonal antibody. *Environmental Entomology* 21, 896–900. DOI: 10.1093/ee/21.4.896

Hagler, J.R., Casey, M.T., Hull, A.M. and Machtley, S.A. (2020) A molecular approach for detecting stage-specific predation on *Lygus hesperus* (Hemiptera: Miridae). *Journal of Insect Science* 20, 35–35. DOI: 10.1093/jisesa/ieaa136

Hargreaves, H. (1948) *List of the Recorded Cotton Insects of the World*. London, UK: Commonwealth Institute of Entomology.

Heinz, K.M. and Zalom, F.G. (1995) Variation in trichome-based resistance to *Bemisia argentifolii* (Homoptera: Aleyrodidae) oviposition on tomato. *Journal of Economic Entomology* 88, 1494–1502. DOI: 10.1093/jee/88.5.1494

Henneberry, T.J. (1986) *Pink Bollworm Management in Cotton in the Southwestern United States*. ARS-51. Washington, DC: US Department of Agriculture, Agricultural Research Service.

Henneberry, T.J. (1994) Pink bollworm sterile moth releases: suppression of established infestations and exclusion from noninfested areas. In: Calkins, C.O., Klassen, W. and Liedo, P. (eds) *Fruit Flies and the Sterile Insect Techniques*. Boca Raton, FL: CRC Press, pp. 181–207.

Henneberry, T.J. and Naranjo, S.E. (1998) Integrated management approaches for pink bollworm in the southwestern United States. *Integrated Pest Management Reviews* 3, 31–52. DOI: 10.1023/A:1009673622862

Horowitz, A.R., Ghanim, M., Roditakis, E., Nauen, R. and Ishaaya, I. (2020) Insecticide resistance and its management in *Bemisia tabaci* species. *Journal of Pest Science* 93, 893–910. DOI: 10.1007/s10340-020-01210-0

Husman, S.H. and Jech, L.E. (1998) Voluntary area-wide whitefly monitoring project implementation 1995–1997 Gila Bend, Arizona. In: Dugger, P. and Richter, D. (eds) *Proceedings of the Cotton Beltwide Conferences*. Memphis, TN: National Cotton Council, pp. 1084–1088.

Hutchison, W.D., Burkness, E.C., Mitchell, P.D., Moon, R.D., Leslie, T.W. *et al*. (2010) Areawide suppression of European corn borer with Bt maize reaps savings to non-Bt maize growers. *Science* 330, 222–225. DOI: 10.1126/science.1190242

Ingram, W.R. (1994) *Pectinophora* (Lepidoptera: Gelechiidae). In: Matthews, G.A. and Tunstall, J.P. (eds) *Insect Pests of Cotton*. Wallingford, UK: CAB International, pp. 107–149.

IRAC (2022) *Arthropod Pesticide Resistance Database*. East Lansing, MI: Michigan Sate University, Insecticide Resistance Action Committee. Available at: https://www.pesticideresistance.org/ (accessed April 30, 2024).

Isaacs, R. and Byrne, D.N. (1998) Aerial distribution, flight behaviour and eggload: their inter-relationship during dispersal by the sweetpotato whitefly. *Journal of Animal Ecology* 67, 741–750. DOI: 10.1046/j.1365-2656.1998.00236.x

Jackson, C.G., Debolt, J.W. and Ellington, J. (1995) Lygus bugs. In: Nechols, J.R., Andres, L.A., Beardsley, J.W., Goeden, R.D. and Jackson, C.G. (eds) *Biological Control in the Western United States*. Oakland, CA: Division of Agriculture and Natural Resources, University of California, pp. 87–90.

Kogan, M. (1998) Integrated pest management: historical perspectives and contemporary developments. *Annual Review of Entomology* 43, 243–270. DOI: 10.1146/annurev.ento.43.1.243

Lambert, A.L., McPherson, R.M. and Herzog, G.A. (1997) Field evaluation of fourteen soybean genotypes for resistance to whitefly (Homoptera: Aleyrodidae) infestations. *Journal of Economic Entomology* 90, 658–662. DOI: 10.1093/jee/90.2.658

Leigh, T.F., Grimes, D.W., Yamada, H., Stockton, J.R. and Basset, D. (1969) Arthropod abundance in cotton in relation to some cultural management variables. *Proceeding of the Tall Timbers Conference on*

Ecological Animal Control and Habitat Management 1. Tallahassee, FL: Tall Timbers Research Station, pp. 71–83.

Leigh, T.F., Grimes, D.W., Dickens, W.L. and Jackson, C.E. (1974) Planting pattern, plant population, irrigation, and insect interactions in cotton. *Environmental Entomology* 3, 492–496. DOI: 10.1093/ee/3.3.492

Leigh, T.F., Kirby, T.A. and Wynholds, P.F. (1988) Cotton square damage by the plant bug, *Lygus hesperus* (Hemiptera: Heteroptera: Miridae), and abscission rates. *Journal of Economic Entomology* 81, 1328–1337. DOI: 10.1093/jee/81.5.1328

Leigh, T.F., Roach, S.H. and Watson, T.F. (1996) Biology and ecology of important insect and mites pests of cotton. In: King, E.G., Phillips, J.R. and Coleman, R.J. (eds) *Cotton Insects and Mites: Characterization and Management*. Memphis, TN: Cotton Foundation, pp. 17–85. Available at: https://www.cotton.org/foundation/reference-books/insects/upload/CI-M_Chapter2.pdf (accessed January 3, 2024).

Levins, R. and Wilson, M. (1980) Ecological theory and pest management. *Annual Review of Entomology* 25, 287–308. DOI: 10.1146/annurev.en.25.010180.001443

Li, A.Y.-S., Dennehy, T.J., Li, S.X.-H., Wigert, M.E., Zaborac, M. *et al.* (2001) Sustaining Arizona's fragile success in whitefly resistance management. In: Dugger, P. and Richter, D. (eds) *Proceedings Beltwide Cotton Conferences*. Memphis, TN: National Cotton Council, pp. 1108–1114.

Mauney, J.R. and Henneberry, T.J. (1984) Causes of square abscission in cotton in Arizona. *Crop Science* 24, 1027–1030. DOI: 10.2135/cropsci1984.0011183X002400060006x

Meissle, M., Naranjo, S.E. and Romeis, J. (2022) Does the growing of Bt maize change abundance or ecological function of non-target animals compared to the growing of non-GM maize? A systematic review. *Environmental Evidence* 11, 21. DOI: 10.1186/s13750-022-00272-0

Moore, L., Thacker, G., Watson, T., Ellsworth, P. and Combs, J. (1992) Community-wide insect management program in Pima County, 1991. In: *Cotton: A College of Agriculture Report*. Tucson, AZ: College of Agriculture, University of Arizona, pp. 99–104. Available at: http://hdl.handle.net/10150/208625 (accessed January 3, 2024).

Moore, L., Watson, T.F., Natwick, E., Allen, C., Henneberry, T. *et al.* (1993) Season-long pink bollworm management. *Pink Bollworm Management Newsletter* Vol. 1(1). Memphis, TN: Cotton Foundation. Available at: https://ag.arizona.edu/crops/cotton/insects/pbw/NCCPBWnewsNo1.pdf (accessed January 3, 2024).

Morales, F.J. (2001) Conventional breeding for resistance to *Bemisia tabaci*-transmitted geminiviruses. *Crop Protection* 20, 825–834. DOI: 10.1016/S0261-2194(01)00114-4

Munk, D.S. and Goodell, P.B. (2002) Irrigation management's influence on arthropod populations. In: Dugger, P. and Richter, D. (eds) *Proceedings of the Beltwide Cotton Conference*. Memphis, TN: National Cotton Council.

Naranjo, S.E. (2001) Conservation and evaluation of natural enemies in IPM systems for *Bemisia tabaci*. *Crop Protection* 20, 835–852. DOI: 10.1016/S0261-2194(01)00115-6

Naranjo, S.E. (2005a) Long-term assessment of the effects of transgenic Bt cotton on the abundance of nontarget arthropod natural enemies. *Environmental Entomology* 34, 1193–1210. DOI: 10.1093/ee/34.5.1193

Naranjo, S.E. (2005b) Long-term assessment of the effects of transgenic Bt cotton on the function of the natural enemy community. *Environmental Entomology* 34, 1211–1223. DOI: 10.1093/ee/34.5.1211

Naranjo, S.E. (2009) Impacts of *Bt* crops on non-target organisms and insecticide use patterns. *CAB Reviews: Perspectives in Agriculture, Veterinary Science, Nutrition and Natural Resources* 4, 011. DOI: 10.1079/PAVSNNR20094011

Naranjo, S.E. (2018) Retrospective analysis of a classical biological control programme. *Journal of Applied Ecology* 55, 2439–2450. DOI: 10.1111/1365-2664.13163

Naranjo, S.E. and Ellsworth, P.C. (2001) Introduction: challenges and opportunities for pest management of *Bemisia tabaci* in the new century. *Crop Protection* 20, 707. DOI: 10.1016/S0261-2194(01)00107-7

Naranjo, S.E. and Ellsworth, P.C. (2005) Mortality dynamics and population regulation in *Bemisia tabaci*. *Entomologia Experimentalis et Applicata* 116, 93–108. DOI: 10.1111/j.1570-7458.2005.00297.x

Naranjo, S.E. and Ellsworth, P.C. (2009a) The contribution of conservation biological control to integrated control of *Bemisia tabaci* in cotton. *Biological Control* 51, 458–470. DOI: 10.1016/j.biocontrol.2009.08.006

Naranjo, S.E. and Ellsworth, P.C. (2009b) Fifty years of the integrated control concept: moving the model and implementation forward in Arizona. *Pest Management Science* 65, 1267–1286. DOI: 10.1002/ps.1861

Naranjo, S.E. and Ellsworth, P.C. (2010) Fourteen years of Bt cotton advances IPM in Arizona. *Southwestern Entomologist* 35, 437–444. DOI: 10.3958/059.035.0329

Naranjo, S.E. and Flint, H.M. (1994) Spatial distribution of preimaginal *Bemisia tabaci* (Homoptera: Aleyrodidae) in cotton and development of fixed-precision sequential sampling plans. *Environmental Entomology* 23, 254–266. DOI: 10.1093/ee/23.2.254

Naranjo, S.E. and Flint, H.M. (1995) Spatial distribution of adult *Bemisia tabaci* (Homoptera: Aleyrodidae) in cotton and development and validation of fixed-precision sampling plans for estimating population density. *Environmental Entomology* 24, 261–270. DOI: 10.1093/ee/24.2.261

Naranjo, S.E. and Hagler, J.R. (1998) Characterizing and estimating the impact of heteropteran predation. In: Coll, M. and Ruberson, J. (eds) *Predatory Heteroptera: Their Ecology and Use in Biological Control. Thomas Say Symposium Proceedings.* Lanham, MD: Entomological Society of America, pp. 170–197. DOI: 10.4182/YGQF2785.1998.171

Naranjo, S.E. and Luttrell, R.G. (2009) Cotton arthropod IPM. In: Radcliff, E.B., Hutchison, W.D. and Cancelado, R.E. (eds) *Integrated Pest Management.* Cambridge, UK: Cambridge University Press, pp. 324–340. DOI: 10.1017/CBO9780511626463.026

Naranjo, S.E., Chu, C.C. and Henneberry, T.J. (1996a) Economic injury levels for *Bemisia tabaci* (Homoptera: Aleyrodidae) in cotton: impact of crop price, control costs, and efficacy of control. *Crop Protection* 15, 779–788. DOI: 10.1016/S0261-2194(96)00061-0

Naranjo, S.E., Flint, H.M. and Henneberry, T.J. (1996b) Binomial sampling plans for estimating and classifying population density of adult *Bemisia tabaci* in cotton. *Entomologia Experimentalis et Applicata* 80, 343–353. DOI: 10.1111/j.1570-7458.1996.tb00947.x

Naranjo, S.E., Diehl, J.W. and Ellsworth, P.C. (1997) Sampling whiteflies in cotton: validation and analysis of enumerative and binomial plans. *Environmental Entomology* 26, 777–788. DOI: 10.1093/ee/26.4.777

Naranjo, S.E., Ellsworth, P.C., Chu, C.C., Henneberry, T.J., Riley, D.G. *et al.* (1998) Action thresholds for the management of *Bemisia tabaci* (Homoptera: Aleyrodidae) in cotton. *Journal of Economic Entomology* 91, 1415–1426. DOI: 10.1093/jee/91.6.1415

Naranjo, S.E., Hagler, J.R. and Ellsworth, P.C. (2003) Improved conservation of natural enemies with selective management systems for *Bemisia tabaci* (Homoptera: Aleyrodidae) in cotton. *Biocontrol Science and Technology* 13, 571–587. DOI: 10.1080/0958315031000151828

Naranjo, S.E., Ellsworth, P.C. and Hagler, J.R. (2004) Conservation of natural enemies in cotton: role of insect growth regulators in management of *Bemisia tabaci*. *Biological Control* 30, 52–72. DOI: 10.1016/j.biocontrol.2003.09.010

Naranjo, S.E., Frisvold, G.B. and Ellsworth, P.C. (2019) Economic value of arthropod biological control. In: Onstad, D.W. and Crain, P.R. (eds) *The Economics of Integrated Pest Management of Insects.* Wallingford, UK: CAB International, pp. 49–85. DOI: 10.1079/9781786393678.0049

Naranjo, S.E., Hellmich, R.L., Romeis, J., Shelton, A.M. and Vélez, A.M. (2020) The role and use of genetically engineered insect-resistant crops in integrated pest management systems. In: Kogan, M. and Heinrichs, E.A. (eds) *Integrated Management of Insect Pests: Current and Future Developments.* Cambridge, UK: Burleigh Dodds, pp. 283–340. DOI: 10.19103/as.2019.0047.10

Naranjo, S.E., Cañas, L. and Ellsworth, P.C. (2022) Mortality dynamics of a polyphagous invasive herbivore reveal clues in its agroecosystem success. *Pest Management Science* 78, 3988–4005. DOI: 10.1002/ps.7018

National Cotton Council (2009) *Minimum Standards for Pink Bollworm Post-eradication*. Washington, DC: National Cotton Council of America. Available at: https://www.aphis.usda.gov/plant_health/plant_pest_info/cotton_pests/downloads/PBW-post-eradication-minimum-standards.pdf (accessed July 20, 2023).

National Cotton Council (2023) *Crop Information*. Washington, DC: National Cotton Council of America. Available at: https://www.cotton.org/econ/cropinfo/ (accessed July 14, 2023).

Noble, L.W. (1969) *Fifty Years of Research on the Pink Bollworm in the United States*, Washington, DC: US Department of Agriculture, Agricultural Research Service. Available at: https://www.gcvinfo.gov/content/pkg/GOVPUB-A-PURL-gpo23717/pdf/GOVPUB-A-PURL-gpo23717.pdf (accessed January3, 2024).

Nombela, G. and Muñiz, M. (2009) Host plant resistance for the management of *Bemisia tabaci*: a multi-crop survey with emphasis on tomato. In: Stansly, P.A. and Naranjo, S.E. (eds) *Bemisia: Bionomics and Management of a Global Pest.* Dordrecht, Netherlands: Springer, pp. 357–383. DOI: 10.1007/978-90-481-2460-2_14

Nuessly, G.S., Meyerdirk, D.E., Coudriet, D.L. and Henneberry, T.J. (1994) The effect of short season cotton production schedules on *Bemisia tabaci* (Gennadius). *Southwestern Entomologist* 19, 209–217.

Onstad, D.W. and Knolhoff, L.M. (2022) *Insect Resistance Management: Biology, Economics, and Prediction*, 3rd edn. New York: Academic Press.

Palumbo, J.C., Horowitz, A.R. and Prabhaker, N. (2001) Insecticidal control and resistance management for *Bemisia tabaci*. *Crop Protection* 20, 739–765. DOI: 10.1016/S0261-2194(01)00117-X

Palumbo, J.C., Ellsworth, P.C., Dennehy, T.J. and Nichols, R.L. (2003) *Cross-commodity Guidelines for Neonicotinoid Insecticides in Arizona*. IPM Series 17. Tucson, AZ: Cooperative Extension, University of Arizona. Available at: https://extension.arizona.edu/sites/extension.arizona.edu/files/pubs/az1319.pdf (accessed July 15, 2023).

Pedigo, L.P., Hutchins, S.H. and Higley, L.G. (1986) Economic injury levels in theory and practice. *Annual Review of Entomology* 31, 341–368. DOI: 10.1146/annurev.en.31.010186.002013

Perdue, S. (2018) *Eradication of the Pink Bollworm: A Proclamation*. Washington, DC: US Department of Agriculture. Available at: https://www.usda.gov/sites/default/files/documents/usda-pink-bollworm-proclamation.pdf (accessed August 23, 2023).

Rabb, R.L. (1978) A sharp focus on insect populations and pest management from a wide-area view. *Bulletin of the Entomological Society of America* 24, 55–61. DOI: 10.1093/besa/24.1.55

Rainwater, C.F. (1952) Progress in research on cotton insects. In: *Insects: The Yearbook of Agriculture*. Washington, DC: US Government Printing Office, pp. 497–500.

Reisig, D., Ellsworth, P.C. and Hodgson, E. (2019) The roles of soft technologies and Cooperative Extension in solving wicked IPM problems. In: Onstad, D.W. and Crain, P.R. (eds) *The Economics of Integrated Pest Management of Insects*. Wallingford, UK: CAB International, pp. 155–178. DOI: 10.1079/9781786393678.0155

Russell, L.M. (1975) *Collection Records of* Bemisia tabaci (Gennadius) *in the United States*. *Cooperative Economic Insect Report*. Washington, DC: US Department of Agriculture, pp. 229–230.

Samples, C.A., Dodds, D.M., Gore, J., Catchot, A.L., Golden, B.R. *et al.* (2019) Impact of nitrogen application rate on tarnished plant bug (Heteroptera: Miridae) populations and management in cotton. *Journal of Cotton Science* 23, 28–37. DOI: 10.56454/husz7224

Scott, D.R. (1977) An annotated listing of host plants of *Lygus hesperus* Knight. *Bulletin of the Entomological Society of America* 23, 19–22. DOI: 10.1093/besa/23.1.19

Shapiro, E.A. (1989) Cotton in Arizona: a historical geography. MS thesis. Tucson, AZ: University of Arizona. Available at: https://repository.arizona.edu/handle/10150/291975 (accessed August 9, 2023).

Shelby, E.A., Moss, J.B., Andreason, S.A., Simmons, A.M., Moore, A.J. *et al.* (2020) Debugging: strategies and considerations for efficient RNAi-mediated control of the whitefly *Bemisia tabaci*. *Insects* 11, 723. DOI: 10.3390/insects11110723

Sivakoff, F., Rosenheim, J. and Hagler, J. (2012) Relative dispersal ability of a key agricultural pest and its predators in an annual agroecosystem. *Biological Control* 63, 296–303. DOI: 10.1016/j.biocontrol.2012.09.008

Smith, C.M. (2005) *Plant Resistance to Arthropods: Molecular and Conventional Approaches*. Dordrecht, Netherlands: Springer.

Spears, J.F. (1968) The westward movement of the pink bollworm. *Bulletin of the Entomological Society of America* 14, 118–119. DOI: 10.1093/besa/14.2.118

Staten, R.T. and Walters, M.L. (2021) Technology used by field managers for pink bollworm eradication with its successful outcome in the United States and Mexico. In: Hendrichs, J., Pereira, R. and Vreysen, M.J.B. (eds) *Area-wide Integrated Pest Management: Development and Field Application*. Boca Raton, FL: CRC Press, pp. 51–92. DOI: 10.1201/9781003169239-5

Stern, V.M. (1979) Long and short range dispersal of the pink bollworm *Pectinophora gossypiella* over southern California. *Environmental Entomology* 8, 524–527. DOI: 10.1093/ee/8.3.524

Stern, V.M., Smith, R.F., Van den Bosch, R. and Hagen, K.S. (1959) The integrated control concept. *Hilgardia* 29, 81–101. DOI: 10.3733/hilg.v29n02p081

Stern, V.M., Van den Bosch, R. and Leigh, T.F. (1964) Strip cutting alfalfa for lygus bug control. *California Agriculture* 18, 4–6.

Strong, F.E. (1970) Physiology of injury caused by *Lygus hesperus*. *Journal of Economic Entomology* 63, 808–814. DOI: 10.1093/jee/63.3.808

Summers, C.G., Goodell, P.B. and Mueller, S.C. (2004) Lygus bug management by alfalfa harvest manipulation. In: Pimentel, D. (ed.) *Encyclopedia of Pest Management*, Vol. 2. Boca Raton, FL: CRC Press, pp. 322–325.

Tabashnik, B.E., Patin, A.L., Dennehy, T.J., Liu, Y.B., Miller, E. *et al.* (1999) Dispersal of pink bollworm (Lepidoptera: Gelechiidae) males in transgenic cotton that produces a *Bacillus thuringiensis* toxin. *Journal of Economic Entomology* 92, 772–780. DOI: 10.1093/jee/92.4.772

Tabashnik, B.E., Sisterson, M.S., Ellsworth, P.C., Dennehy, T.J., Antilla, L. *et al.* (2010) Suppressing resistance to Bt cotton with sterile insect releases. *Nature Biotechnology* 28, 1304–1309. DOI: 10.1038/nbt.1704

Tabashnik, B.E., Liesner, L.R., Ellsworth, P.C., Unnithan, G.C., Fabrick, J.A. *et al.* (2021) Transgenic cotton and sterile insect releases synergize eradication of pink bollworm a century after it invaded the United States. *Proceedings of the National Academy of Sciences* 118, e2019115118. DOI: 10.1073/pnas.2019115118

Tabashnik, B.E., Carrière, Y., Wu, Y. and Fabrick, J.A. (2023) Global perspectives on field-evolved resistance to transgenic Bt crops: a special collection. *Journal of Economic Entomology* 116, 269–274. DOI: 10.1093/jee/toad054

Thacker, G., Moore, L., Ellsworth, P.C. and Combs, J. (1994a) Evaluation of trap crops as a component of a community-wide pink bollworm control program. In: *Cotton: A College of Agriculture Report.* P-96. Tucson, AZ: College of Agriculture, University of Arizona, pp. 259–266. Available at: http://hdl.handle.net/10150/209637 (accessed January 3, 2024).

Thacker, G.W., Ellsworth, P.C., Moore, L. and Combs, J. (1994b) Cotton producers working in unison: the multi-component IPM program in Marana, AZ. In: *Cotton: A College of Agriculture Report.* P-96. Tucson, AZ: College of Agriculture, University of Arizona, pp. 251–258. Available at: http://hdl.handle.net/10150/209636 (accessed January 3, 2024).

Trichilo, P.J. and Leigh, T.F. (1986) Predation on spider mite eggs by the western flower thrips, *Frankliniella occidentalis* (Thysanoptera: Thripidae), an opportunist in a cotton agroecosystem. *Environmental Entomology* 15, 821–825. DOI: 10.1093/ee/15.4.821

US Cotton Trust Protocol (2023) *Trust in a Smarter Cotton Future*. Cordova, TN: US Cotton Trust Protocol. Available at: https://trustuscotton.org/ (accessed August 9, 2023).

USDA-ERS (2023) *Cotton and Wool*. Washington, DC: US Department of Agriculture, Economic Research Service. Available at: https://www.ers.usda.gov/topics/crops/cotton-and-wool/ (accessed July 14, 2023).

Van den Bosch, R. and Hagen, K.S. (1966) *Predaceous and Parasitic Arthropods in California Cotton Fields*. No. 820. Berkeley, CA: California Agricultural Experiment Station, University of California.

Vandervoet, T.F., Ellsworth, P.C., Carrière, Y. and Naranjo, S.E. (2018) Quantifying conservation biological control for management of *Bemisia tabaci* (Hemiptera: Aleyrodidae) in cotton. *Journal of Economic Entomology* 111, 1056–1068. DOI: 10.1093/jee/toy049

Vandervoet, T., Ellsworth, P.C., Brown, L., Fournier, A. and Naranjo, S.E. (2022) *Making Whitefly and Predator Counts*. Arizona Pest Management Center. IPM Short. Tucson, AZ: University of Arizona. Available at: http://hdl.handle.net/10150/665535 (accessed January 3, 2024).

Watson, T.F., Silvertooth, J.C. and Tellez, A. (1994) Varietal and nitrogen-level effects on sweetpotato whitefly populations in cotton. In: Herber, D.J. and Richter, D.A. (eds) *Proceedings Beltwide Cotton Conferences.* Memphis, TN: National Cotton Council, pp. 868–869.

Wene, G.P. and Sheets, L.W. (1962) Relationship of predatory and injurious insects in cotton fields in the Salt River Valley area of Arizona. *Journal of Economic Entomology* 55, 395–398. DOI: 10.1093/jee/55.3.395

Wheeler, A.G., Jr (2001) *Biology of the Plant Bugs*. Ithaca, NY: Cornell University Press.

Wilson, F.D. (1992) Registration of 8 cotton germplasm lines with pink bollworm resistance traits. *Crop Science* 32, 288–289. DOI: 10.2135/cropsci1992.0011183X003200010075x

Wu, K.M., Lu, Y.H., Feng, H.Q., Jiang, Y.Y. and Zhao, J.Z. (2008) Suppression of cotton bollworm in multiple crops in China in areas with Bt toxin-containing cotton. *Science* 321, 1676–1678. DOI: 10.1126/science.1160550

Zink, A.G. and Rosenheim, J.A. (2008) Stage-specific predation on *Lygus hesperus* affects its population stage structure. *Entomologia Experimentalis et Applicata* 126, 61–66. DOI: 10.1111/j.1570-7458.2007.00634.x

4 Managing the Invasive Fall Armyworm in the Large- and Small-scale Maize Farming Landscapes of East and Southern Africa

Johnnie van den Berg* and Hannalene du Plessis
IPM Program, Unit for Environmental Sciences and Management, North-West University, Potchefstroom, South Africa

Abstract

The invasion of Africa by the fall armyworm, *Spodoptera frugiperda* (J.E. Smith) (Lepidoptera: Noctuidae), in 2016 changed the way in which maize pests are managed on the continent. Ninety-eight per cent of farms in Africa are small scale (<2 ha), and South Africa is the only country where large-scale systems dominate. Pest management approaches depend on the scale of these maize farming systems and although both systems employ integrated pest management (IPM) principles, the major differences are the amount and types of agricultural inputs used. Due to fall armyworm being a "new" pest on the continent, IPM programs are in the early stages of development. In small-scale (low-input) systems, yields are inherently low and agroecological methods are recommended for control. These include curative use of biopesticides, handpicking, and egg crushing, as well as long-term approaches (e.g. habitat management, host plant resistance), and biological control. In this chapter we discuss common approaches recommended for fall armyworm control and assess their relevance and adoption potential. While biological and cultural control approaches provide long-term solutions, the use of pesticides and other inputs to improve plant vigor and yield also contribute to pest management, especially in low-yielding environments. The success observed with genetically modified Bt maize, which is only cultivated in large-scale systems in South Africa, may influence future decisions regarding its approval for cultivation in other African countries.

Introduction

The African agricultural landscape is a complex environment in which maize is produced in small-scale mixed farming systems for local consumption, or large-scale systems for industrial use. Maize yields attained in small-scale farming systems are inherently low due to erratic rainfall, poor soils, low fertilizer use, poor seed, weeds, pests, and diseases. In East and parts of Southern Africa, where small-scale farming systems dominate, the risk of fall armyworm damage is high. The opposite is true for Southern Africa where large-scale production systems dominate, the use of pesticides and Bt maize is common, and environmental conditions are less favorable for this pest.

At the landscape level, small-scale systems are characterized by diverse agroecological zones and mosaics of small (<2 ha) and diverse

*Corresponding author: johnnie.vandenberg@nwu.ac.za

 Arthropod Management and Landscape Considerations in Large-Scale Agroecosystems (eds M.J. Brewer and G.L. Hein)
DOI: 10.1079/9781800622777.0004

crop fields. This diversity produces a heterogenous landscape in which the biophysical environment (agroecological zone) and socio-economic factors not only strongly influence farming and pest management practices, but also the vulnerability of systems to externalities such as drought or pest attack. Although not a dominant feature, large-scale farms are also part of the African maize landscape. The effect of scale of farming on the types of technologies used and which are relevant or not is clearly illustrated in the comparison between these systems in this chapter.

In this chapter, small- and large-scale farming are contrasted in terms of their diversity, farming practices and relevant pest management strategies. The role that socio-economic factors play in the adoption of certain pest management methods in the context of small- and large-scale systems is also discussed.

Maize Cultivation in Africa

Maize, a staple crop in Africa, is cultivated on approximately 36.9 million ha (Fig. 4.1A). Between 2007 and 2017, the area under maize cultivation experienced a notable 60% increase (Santpoort, 2020), with no increase in yield per surface area. Contrary to the largest maize-producing countries in the world (United States and China), which produce 90% yellow maize, African farmers cultivate largely white maize for human consumption. Maize landscapes exhibit significant variation across diverse agroecological zones on the continent (Fig. 4.2). While large-scale commercial maize farming systems are largely confined to South Africa, positioning maize as a commodity crop, small-scale systems produce maize as a staple crop. Ninety-eight per cent of the farms in Africa are small scale, with crops grown on plots of <2 ha. However, an increase in medium-sized maize farms (5–100 ha) in East and Southern Africa over the last decade has been reported (Jayne *et al.*, 2010; Lay *et al.*, 2018). In South Africa, there are approximately 9000 large-scale and 1 million small-scale farmers, accounting respectively for 98% and 2% of total maize production (Lacambra *et al.*, 2020). Consequently, maize production epitomizes a dual agrarian structure: large-scale commercial farming drives maize export to global markets, while small-scale farming supports local markets and subsistence needs.

The strategies for managing fall armyworm are shaped by the specific context in which maize is grown. Factors such as regional agroecological conditions, farmers' socio-economic status, access to resources and existing agricultural practices play a pivotal role in determining the type and effectiveness of pest management approaches. In East Africa, where the landscape is dominated by small-scale maize farming systems and pest pressure is high due to persistent fall armyworm populations (Fig. 4.1B), intercropping is practiced, and farmers have little access to pest management technologies such as Bt maize and pesticides. In contrast, the Southern African maize landscape is characterized by a lower and more variable risk of pest damage, due to climatic conditions that are less favorable for this pest. South Africa, where large-scale maize production dominates, is the region with the lowest risk of pest damage, but also the one where farmers have access to Bt maize and pesticides. It is within these differing contexts that management of fall armyworm within the larger East and Southern African regions is discussed below.

Maize cultivation in East and Southern Africa

The East African Community comprises Tanzania, Kenya, Uganda, Rwanda, Burundi, Democratic Republic of Congo, and South Sudan. Southern Africa encompasses Angola, Botswana, Eswatini, Lesotho, Madagascar, Malawi, Mauritius, Mozambique, Namibia, South Africa, Zambia, Zimbabwe, and several Indian Ocean islands (Fig. 4.1B).

In East Africa, maize occupied an area of 16.7 million ha in 2017, with around 36% dedicated to hybrids, 21% to improved open-pollinated varieties, and 43% to "local cultivars" that include landraces, recycled hybrids, and open-pollinated varieties recycled for over 3 years (Abate *et al.*, 2017). In Southern Africa, excluding South Africa, maize cultivation covered 336,700 ha in 2017 (Santpoort, 2020). South Africa, ranking as the ninth-largest global maize producer, predominantly hosts large-scale

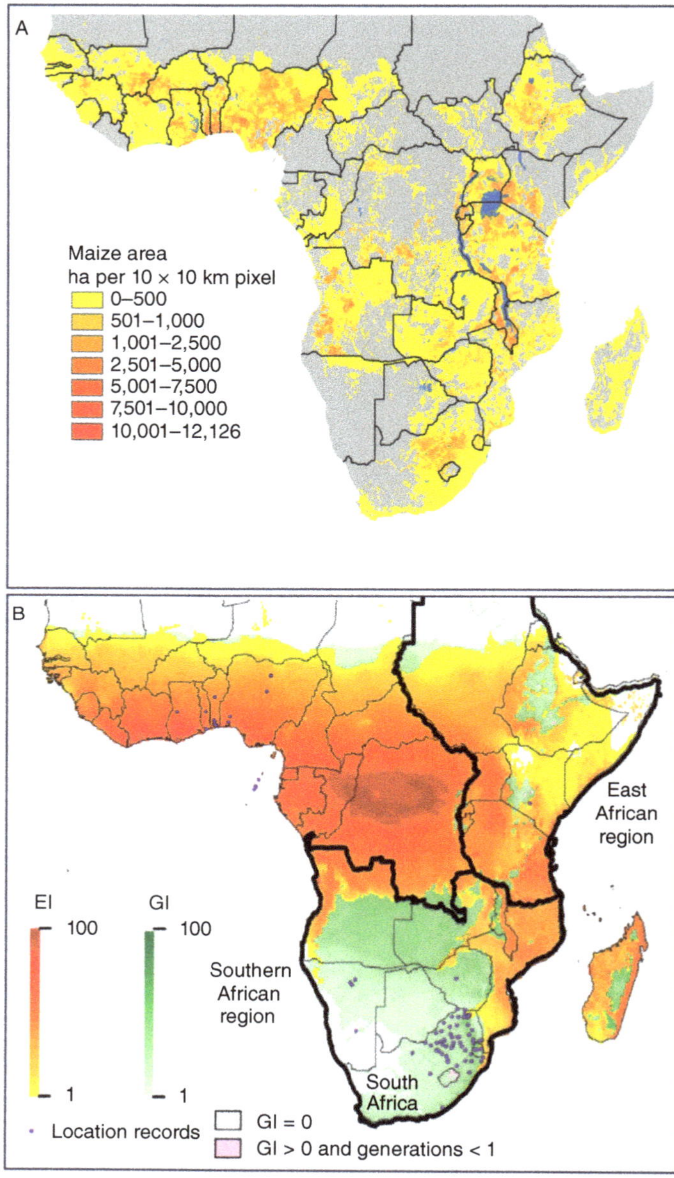

Fig. 4.1. (A) Maize cultivation in sub-Saharan Africa (hectares maize per 10 × 10 km). Source: Kai Sonder, CIMMYT (based on Spatial Production Allocation Model [SPAM] 2017 data from the International Food Policy Research Institute [IFPRI] and Global Administration Unit Layers [GAUL], Food and Agriculture Organization of the United Nations [FAO]). (B) Climate suitability for fall armyworm in Africa modelled using CLIMEX, including the spatially explicit effects of irrigation. The Ecoclimatic Index (EI) describes the potential suitability for persistence, while the Growth Index (GI) describes suitability for population growth. (Diagram (A) used with permission: B. Prasanna, international Maize and Wheat Improvement Center, Nairobi, Kenya. Diagram (B) used with permission: D. Kriticos, Cervantes Agritech, CLIMEX software developer.)

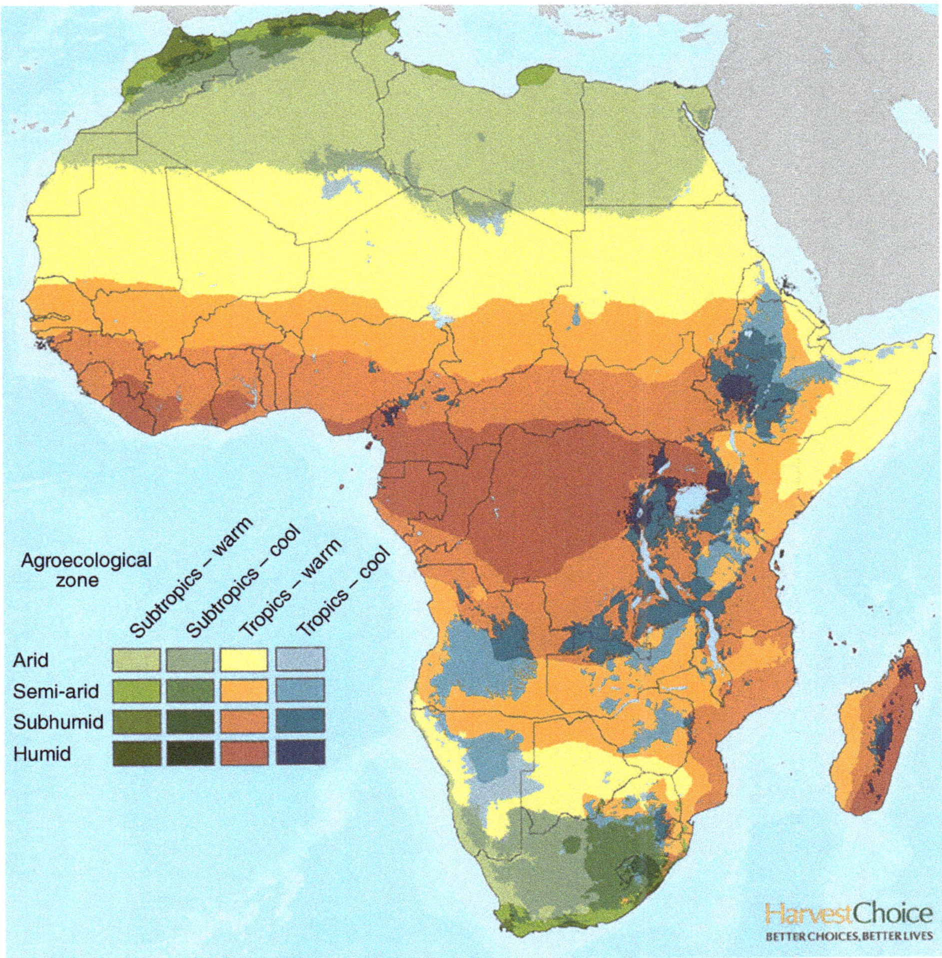

Fig. 4.2. Agroecological zones in Africa, categorized by temperature (subtropics and tropics, warm and cool) and by humidity (arid, semi-arid, subhumid, and humid) (Sebastian, 2009). (Diagram is open access online.)

commercial maize farming systems across 2–3 million ha. Complementary grain crops like sorghum, soybean, and rice are commonplace in maize-based farming systems across East and Southern Africa.

The impact of fall armyworm in Africa is exacerbated by the inherent low yield of maize in small-scale farming systems. Across sub-Saharan Africa, country-average yields range from 1.7 to 2.0 tonnes/ha (Van Ittersum *et al.*, 2016; Ten Berge *et al.*, 2019). These low yields are ascribed to factors such as poor soil fertility, cultivation practices, droughts, and restricted access to improved seeds and fertilizers (Baudron *et al.*, 2012a, b; Badu-Apraku and Fakorede, 2017; Tadele, 2017; Sikora *et al.*, 2019). Improved crop management through appropriate planting time, higher seeding rates, improved seed, fertilizer inputs, and weed control could more than double current yields (Silva *et al.*, 2023). There are, however, also views such as those of Rapsomanikis (2015) who stated that there is no room for high-input solutions in small-scale farming systems.

South Africa produces between 14 and 17 million tonnes of maize per year at an average 5.9 tonnes/ha (Statista, 2023). Yields of up to 18.1 tonnes/ha under dry land conditions and approximately 20 tonnes/ha under irrigation are achieved. In Kenya, where the maize landscape is dominated by small-scale systems, large-scale farms produce yields of up to 9 tonnes/ha under rainfed conditions. These comparatively high yields in commercial production systems are ascribed to the use of all possible available inputs such as hybrid maize, fertilizer, pesticides, and, in the case of South Africa, genetically modified hybrids with insecticidal and herbicide tolerance traits.

The Lepidopteran Pest Complex in Maize

The lepidopteran pest complex of maize in East and Southern Africa shares similarities, encompassing several species. The fall armyworm, *Spodoptera frugiperda* (J.E. Smith) (Lepidoptera: Noctuidae), has emerged as the predominant maize pest in the region, posing a severe threat to the food security of countless smallholder farmers (Diagne *et al.*, 2021). The economic ramifications of fall armyworm cannot be viewed in isolation, given its coexistence with other lepidopteran species in the same fields (Sokame *et al.*, 2020; Visser and Van den Berg, 2020). Before the invasion by fall armyworm, the most important pests were the indigenous African maize stem borer, *Busseola fusca* (Fuller) (Lepidoptera: Noctuidae), and an invasive species, *Chilo partellus* (Swinhoe) (Lepidoptera: Crambidae) (Van den Berg *et al.*, 2014; Ntiri *et al.*, 2016). The African bollworm, *Helicoverpa armigera* (Hübner) (Lepidoptera: Noctuidae), is a common pest in the region but with low pest status on maize (Van Wyk *et al.*, 2008; Van den Berg *et al.*, 2022). Additionally, *Chilo orichalcociliellus* (Strand) (Lepidoptera: Crambidae), another stem borer species, assumes significance in the low-altitude regions of East Africa (Ong'amo *et al.*, 2013; Moolman *et al.*, 2014). These lepidopteran species can coexist in mixed populations (Krüger *et al.*, 2008; Sokame *et al.*, 2020; Ntiri *et al.*, 2019; Mwalusepo *et al.*, 2018) and can inflict substantial yield losses depending on the agroecological zone, pest population density, and crop stage at infestation. The potential for confusion between these stem borer species and fall armyworm is high, complicating management and contributing to the adoption of ineffective management approaches.

Yield losses from maize stem borers range widely, spanning 10–100%. In Mozambique, losses have ranged between 28% and 34% (Cugala, 2007), and in Kenya, the average national crop loss due to *C. partellus* was reported as 13.5% (De Groote *et al.*, 2011). Yield loss estimates range between 2% and 33% in Ethiopia, Malawi, and Tanzania. Prior to the deployment of genetically modified Bt maize in South Africa, average annual yield losses from stem borer damage were estimated at 10% (Bate and Van Rensburg, 1992).

Very few on-farm studies have been conducted to assess yield losses caused by fall armyworm in Africa. Yield loss assessments done on experimental farms where pest pressure is usually very high, reported losses reaching 90% (Overton *et al.*, 2021; Van den Berg *et al.*, 2021). Farmer perceptions of yield loss across various African countries, however, tend to be lower, ranging from 11% to 54% (Rwomushana *et al.*, 2018; Kansiime *et al.*, 2019; Babendreier *et al.*, 2020; De Groote *et al.*, 2020). Yield loss assessments conducted under natural pest infestation, however, show that losses may be low and overestimated (Baudron *et al.*, 2019; Kassie *et al.*, 2020; Harrison *et al.*, 2022; Osae *et al.*, 2022). Day *et al.* (2017) projected the potential annual cost of yield losses due to fall armyworm across 12 maize-producing African countries to range from US$2.5 to 6.3 billion. In other studies, farmers' perceptions of yield losses caused by fall armyworm were 28% in Zambia (Kansiime *et al.*, 2019), 33–54% in Kenya (Kumela *et al.*, 2019, De Groote *et al.*, 2020), and 47% in Ethiopia (Kumela *et al.*, 2019). Recently, Eschen *et al.* (2021) estimated overall yield losses in Africa to be between US$7.7 and 12.1 billion.

Together with information on field-level impact in some regions and reports of a wide range of natural enemies of fall armyworm, the perception of the pest status is changing. Emergency measures have evolved into endorsing sustainable interventions, fostering biological control and habitat management initiatives (Abang *et al.*, 2021). It also seems that the indiscriminate spraying of pesticides has decreased, followed by identification of many beneficial insect species, particularly parasitoids, which could associate with this invasive pest on the continent. It is important to note that the predominant fall armyworm population in Africa consists of the corn (C) strain, typically infesting maize and sorghum, rather than the rice (R) strain, which typically targets rice, alfalfa, pasture grasses, and millet (Nagoshi, 2019). While the C-strain and hybrid strains dominate, the R-strain remains uncommon or absent from Africa (Nagoshi, 2019; Nagoshi *et al.*, 2022). This distinction is advantageous for fall armyworm management, as the presence of the R-strain could have led to a broader range of crop hosts, increasing the burden on farmers.

Spatial and temporal infestation patterns of fall armyworm in Africa

While the seasonal migration patterns of fall armyworm in North America is well known (Westbrook *et al.*, 2016, 2019), these have not yet been established in Africa. The pest has established throughout the tropical regions of the continent and persistent populations occur between similar latitudes as in the Americas (Fig. 4.3) (Du Plessis *et al.*, 2019). Variable rainfall patterns, seasonality of maize, and variation in planting dates within African countries and between regions strongly suggest that seasonal migrations occur as cropping seasons unfold (Niassy *et al.*, 2021). Limitations in natural migration between West and East Africa was, however, suggested by Nagoshi *et al.* (2022).

In East Africa, most countries experience two rainy seasons (September to January and February to March). Most Southern African countries have unimodal rainfall patterns and maize planting begins in November to December, with harvesting in March and April. These variations in rainfall patterns and cropping seasons imply potential fall armyworm migratory routes within East Africa and between East and Southern Africa considering the long off-season in the latter. Similarly, there is possibility of migration from Ethiopia, northwards to Sudan and Egypt following the Nile River or southwards during onset of the maize-cropping season in Kenya. In South Africa, which is the most southern distribution of fall armyworm on the continent, planting occurs from September to December and harvesting from May to August. This long period is due to maize having to dry on the field before harvest.

Since fall armyworm cannot survive the cold winter in large parts of Southern Africa, infestations are ascribed to seasonal invasions. The large-scale maize production region in South Africa is outside of the geographical region which is suitable for year-long pest presence (Du Plessis *et al.*, 2019), and invasion patterns are similar to those observed in the southern parts of Brazil and northern Argentina (Andrews, 1980). Eco-climatic maps developed by Adan *et al.* (2023) show a low establishment risk index for parts of Angola, Namibia, South Africa, and Zambia. Nearly all of Mozambique, regions in northern Angola and countries north of Zambia are highly suitable for pest development and are most likely the sources of invasions moving into other countries. Similarly, infestations in some parts of Zimbabwe (Chimweta *et al.*, 2019; Tambo *et al.*, 2020) and Zambia (Kasoma *et al.*, 2021) might therefore be due to immigrant populations.

Although fall armyworm may occur in areas where maize is produced under irrigation, this contribution in Southern Africa is very small. In Central and East Africa, there are areas with continuous farming under irrigation. These areas can serve as permanent hotspots from where fall armyworm might spread to other regions when weather is favorable, and host crops are available (Niassy *et al.*, 2021). Countries such as Angola, Botswana, and Namibia, and some areas along the eastern coast of South Africa, can also support the year-round survival of fall armyworm, provided host plants are cultivated under irrigation (Senay *et al.*, 2022).

The farming systems, seasonal variability and scale of farming described above ultimately influence the type of strategies that can be deployed to manage this pest. For example, the agroecological zone in which maize is planted influences regional planting patterns. Larval development rate and the role that specific geographical regions could play influence sources and extent of migrating populations. In most small-scale farming areas, farmers' socio-economic status is poor, pest pressure is high, and traditional farming practices are used. These factors all influence the types of technologies that are

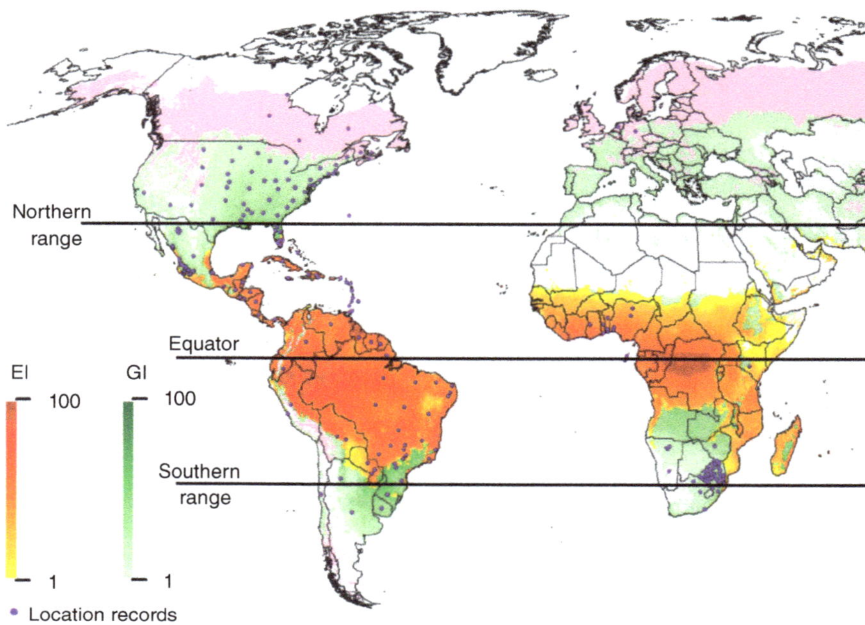

Fig. 4.3. Climate suitability for *Spodoptera frugiperda* globally modelled using CLIMEX, including the effects of irrigation. Black lines other than the equator indicate the northern and southern ranges where persistent population of fall armyworm may occur. The Ecoclimatic Index (EI) describes the potential suitability for fall armyworm persistence, while the Growth Index (GI) describes suitability for population growth. Adapted from Du Plessis *et al.* (2018). (Diagram used with permission: B. Prasanna, International Maize and Wheat Improvement Center, Nairobi, Kenya.)

applicable on farms of different scales, as well as the regional approaches to develop pest management solutions in the many different landscapes within East and Southern Africa.

Managing Fall Armyworm in the Context of Economic and Societal Viability

A noticeable divergence exists in the approaches needed to manage fall armyworm between small- and large-scale maize farming systems (Hruska, 2019). Both systems adhere to integrated pest management (IPM) principles, but the distinction lies in the quantity and types of agricultural inputs employed (Tadele, 2017). For small-scale systems agroecological approaches are recommended (Harrison *et al.*, 2019; CIFOR-ICRAF, 2023), while large-scale systems take advantage of technology such as genetically modified crops, e.g. Bt maize, and pesticides.

While research is ongoing to develop long-term solutions for fall armyworm control, the number of tools currently available for farmers to address these challenges remain limited. Insecticide application is one of the very few tools available to address the immediate problem at the field level. There are significant developments in the field of biopesticides, but its adoption is constrained by a lack of upscaling opportunities, availability, variable efficacy, and the cost of some of these products.

Due to fall armyworm being a "new" pest on the continent, IPM programs are in the early stages of development. Seen against the background that IPM uptake is very low in many developing countries (Deguine *et al.*, 2021; Tambo *et al.*, 2023), any changes in these traditional systems are highly unlikely over the short term.

In this chapter we focus on strategies and methods that are used by farmers in the region and information available on their use. We mention some of the promising technologies that could in future be used in IPM strategies. Although not discussed in any detail, they are placed in context of likely benefits and affordability for small- and large-scale maize production systems.

Managing fall armyworm in small-scale farming systems

In the realm of small-scale farming systems, research endeavors are focused on habitat management strategies, biological control, and the use of biopesticides (Tepa-Yotto *et al.*, 2022). However, context-specific IPM strategies that consider the existing natural enemy complex and environmental influences on pests (landscape considerations), social acceptance, available infrastructure, and labor intensiveness (social and economic viability) at the field level in different agroecological zones still need to be developed (Mendesil *et al.*, 2023).

Aside from insecticide application and physical control measures, there is virtually no other curative or remedial control measure to use once the pest is observed in fields. Currently available methods are largely limited to biopesticides, handpicking, trapping of moths, egg crushing, parasitoids and natural enemies, push–pull, and intercropping (Niassy *et al.*, 2021). Most of these best-bet technologies are long-term solutions. For example, at the planting and seedling stages when adult and larval counts are low, an initial release of parasitoids combined with egg crushing could suppress the buildup of populations. At the vegetative stage, the release of parasitoids and application of egg-killing biopesticides could be appropriate, and at the reproductive and later stages, predatory bugs should be recommended (Niassy *et al.*, 2021). Although some of these approaches are labor intensive and establishment of biocontrol facilities require significant capital investment, proper training and availability of products could in the long-term mitigate fall armyworm problems in small-scale maize systems.

Chemical control remains the predominant approach, and instances of high insecticide usage, regardless of it being warranted or not, have been documented. This is attributed to challenges in assessing pest status at the field level and the initial reactions of governments to subsidize pesticides as a response to the threat posed to food security (Kumela *et al.*, 2019; Makale *et al.*, 2021; Njuguna *et al.*, 2021). However, this scenario is projected to evolve over the coming years with the potential implementation of classical biological control programs and the adoption of habitat management and conservation biocontrol strategies. Successful initiatives in the field of host plant resistance and cultural control have also emerged. An areawide approach to manage fall armyworm is the most promising to mitigate the risk of pest damage in small-scale farming systems. For this reason, pest-resistant varieties, biopesticides, and biological control, due to their areawide applications, are highly promising technologies. Njuguna *et al.* (2021) emphasized the need to exercise caution, as several of these strategies might not be cost-effective or practical from a farmer's viewpoint of managing individual fields.

It is worth noting that historically, African farmers have not extensively relied on insecticides for low-value crops such as grains, as opposed to cash crops. For crops of higher value (e.g. tomato, coffee, potato), many smallholders turn to pesticide application for pest management (Constantine *et al.*, 2020). Only when maize is the primary crop, do farmers apply insecticides for control of fall armyworm (Constantine *et al.*, 2020). Kenyan farmers who only produce maize for home consumption are more likely to use cultural practices for pest management while market-orientated farmers typically make use of pesticides, either on their own, or in combination with cultural and biological practices (Kansiime *et al.*, 2021).

Chemical control

The dynamics of pesticide application in small-scale maize farming systems have undergone significant shifts in response to the introduction of fall armyworm (Van den Berg and Nur, 1998; Van den Berg *et al.*, 1998; Osae *et al.*, 2022; Chinwada *et al.*, 2023). In East and Southern Africa, a noticeable surge in insecticide use was observed. For instance, over two-thirds of a sample of 2356 maize-growing households in Rwanda and Uganda resorted to synthetic pesticides for fall armyworm control. Similar trends emerged with 60% of maize-growing households in Zambia and Zimbabwe reporting insecticide use (Tambo *et al.*, 2020). In Ethiopia and Kenya, Kumela *et al.* (2019) documented that 48% of farmers employed chemical sprays. In 2016, the governments of Zambia and Zimbabwe, in collaboration with the Food and Agriculture Organization of the United Nations, distributed free pesticides and knapsack sprayers to affected farmers (Rwomushana *et al.*, 2018). In Botswana, 27% of farmers applied insecticides three times during the crop cycle (Makale *et al.*, 2021). In Rwanda, Uzayisenga *et al.* (2020) reported that many farmers (28.4%) applied pesticides three times per season, while 2.8% applied them seven times and more. Unfortunately, reports on insecticide use patterns lack data on application frequency per cropping cycle, pesticide volume, and similar metrics, thereby complicating the interpretation of data from the region.

Pesticide use intensity is strongly influenced by farm size with market-orientated farmers being more likely to invest in pesticides (Kansiime *et al.*, 2021). The application of nearly all pesticides in small-scale maize farming is carried out using knapsack sprayers. However, these applications often yield poor results even after repeated applications (Kumela *et al.*, 2019). Factors contributing to this inefficiency include incorrect timing of application, application during larval stages that are too big, inappropriate insecticide selection, and incorrect calibration of spray equipment (Al-Sarar *et al.*, 2006; Suguiyama *et al.*, 2020; Makale *et al.*, 2021; Van den Berg and du Plessis, 2022).

Reports from various countries indicated variation in pesticide application rates, the mixing of chemicals in single sprays, and incorrect dosage rates (Kansiime *et al.*, 2019; Kassie *et al.*, 2020; Tambo *et al.*, 2020). Pesticides used include methomyl, methyl parathion, endosulfan, and lindane, classified as highly hazardous pesticides (Jepson *et al.*, 2018). The pesticides that are used differ between countries within the region with the most common ones in East Africa being alpha-cypermethrin, chlorpyriphos, malathion, and lufenuron (Niassy *et al.*, 2021). It seems that there is a changing tendency in the types of insecticides used, with spinosad being one of the most common products during the initial period, to products such as emamectin benzoate which is now the most common active ingredient used for control in Zambia and Zimbabwe (Nyamutukwa *et al.*, 2022). In Uganda, where pyrethroid use is common, emamectin benzoate use is also increasing (Kalyebi *et al.*, 2023).

Systemic seed treatments utilizing diamides offer potential as part of IPM strategies. The value of these treatments lies in their early protection against fall armyworm attacks and is particularly valuable given that many smallholder farmers do not recognize the need for scouting shortly after crop emergence (Chinwada *et al.*, 2023). In Zambia and Zimbabwe, a seed treatment combination employing cyantraniliprole for fall armyworm control and thiamethoxam for soil insect pest control has been introduced (Nyamutukwa *et al.*, 2022). This treatment could potentially provide safeguarding against fall armyworm during the initial 2–4 weeks after seedling emergence. The adoption of such technology will, however, depend on factors like cost, availability, and whether the existing seed supply systems in the region are effective.

A common scenario in many small-scale farming systems, particularly those with access to pesticides was detailed in a study conducted in Botswana (Makale *et al.*, 2021). Approximately 50% of the 1513 interviewed farmers refrained from taking any action after detecting fall armyworm infestations in their fields. This trend persisted even though 40–90% of the cultivated areas for sorghum and maize exhibited signs of infestation. The drivers behind this limited use of pesticides included their perceived high cost and inconvenient proximity of agrochemical dealerships to farms.

Although the figures on insecticide use in the region may be overestimated, generalization of data could imply that between 40% and 60% of farmers in the region apply pesticides. These farmers cannot be ignored in our efforts to develop effective pest management strategies. Instead of continuously criticizing the use of synthetic pesticides, the approach should rather be to develop spray service provider systems, and to train these providers and farmers in the safe and effective use of pesticides.

In any IPM program, insecticide application should always be reserved as a last resort. If insecticide applications are deemed necessary, products with low toxicity should be used. While no insecticide use, or the use of only low-toxicity products remains the ideal, it does not address the immediate problem faced by farmers once serious infestations are observed in the field.

Action threshold levels have been proposed for fall armyworm control in Africa (McGrath *et al.*, 2018), but evidence suggests that these are not used in decision-making processes. Implementing threshold levels on small fields may be impractical (see discussion below).

Cultural control practices

Cultural control strategies that create an unfavorable environment for pests encompass a range of tactics, including intercropping, implementing crop rotation, early planting, employing specific tillage methods, and enhancing soil fertility. Due to their cost-effectiveness and safety, these strategies hold promise for managing fall armyworm in small-scale systems (Ratnadass *et al.*, 2020). Options for non-chemical curative control of fall armyworm (eggs and larvae) in small-scale systems are limited to methods such as hand picking and application of plant extracts or wood ash and soil into whorls (Hruska, 2019; Yigezu and Wakgari, 2020; Makale *et al.*, 2021; Sanou *et al.*, 2023).

A typical example of a cultural control measure that is used as a curative measure, albeit not extensively, is the direct application of sand or soil into the whorls of plants (Kansiime *et al.*, 2019; Tambo *et al.*, 2020; Ahissou *et al.*, 2022; Chawanda *et al.*, 2023). Ash is commonly sourced from household kitchens or burnt wood (Oben *et al.*, 2015). Chimweta *et al.* (2019) described how Zimbabwean farmers applied ash (10 g per plant) into the maize plant whorls at weekly intervals, starting from 5 weeks after planting until tasseling. However, the efficacy of these alternative methods varies significantly, and in many cases efficacy is doubtful (Babendreier *et al.*, 2020; Maphumulo *et al.*, 2023). For example, the efficacy of different ash and soil treatments applied into maize whorls to control first instar larvae ranged between 19% and 28% and did not differ from the control treatment (Maphumulo *et al.*, 2023). Babendreier *et al.* (2020) also reported that wood ash, soil, and a combination of ash and soil provided poor control of fall armyworm after repeated applications.

Harrison *et al.* (2019) suggested various agroecological options for fall armyworm management. Several of these agroecological options, which largely contribute to improved soil and plant health, also enhance the role of natural enemies and have been shown to suppress fall armyworm infestations (Harrison *et al.*, 2019) in its native areas. The practices include minimum soil disturbance, crop residue mulching, planting leguminous intercrops, crop rotation, and maintaining diverse field margins. These agroecological approaches are founded on three interrelated strategies: sustainable soil fertility management that enhances crop health and pest resistance, promoting biodiversity at different scales from the field to the landscape, and undertaking targeted management activities to prevent outbreaks or minimize their impact.

Control methods based on agronomic management represent an interesting alternative, more affordable to resource-constrained smallholders and with lower risk for health and the environment (Thierfelder *et al.*, 2018b). Data regarding adoption rates of cultural control practices tailored for fall armyworm management remains scarce and only a fraction of smallholder farmers in tropical Africa practice IPM (Tamò *et al.*, 2022). While some cultural practices have been part of traditional farming methods, they are often reported in the literature as current fall armyworm management tools. However, there is little empirical data to guide recommendations for effective control of fall armyworm through agronomic management in Africa, as most of this knowledge is based on data from the Americas as well as local (sometimes anecdotal) observations from the region (Harrison *et al.*, 2019).

While agroecological options based on changes in agronomic management practices offer long-term strategies, their feasibility might be limited within certain agroecological zones and rainfed systems. Some strategies may also require extended periods for adoption (Thierfelder *et al.*, 2018a). Landscape-level environmental management holds the potential to bolster ecosystem health and curtail fall armyworm in the long term. For example, a study in Zimbabwe showed that the incidence of maize plants with fall armyworm damage increased significantly with increasing distances from forest patches (Clarkson *et al.*, 2022). Based on their review of predation by birds and bats on fall armyworm and other lepidopteran pests, Clarkson *et al.* (2022) concluded that predation by natural enemies like birds, bats and generalist predators could potentially lower pest infestations. However, the effectiveness of these natural enemies remains poorly understood.

Small-scale cropping systems are characterized by a high diversity of crop species (Fig. 4.4). Cereal–legume intercropping is common in Southern and East Africa, where maize is traditionally intercropped with common bean, pigeon pea, cowpea, or soybean (Kuyah *et al.*, 2021). Intercropping has been reported to provide beneficial ecosystem services such as pest suppression (Daryanto *et al.*, 2020) and could potentially contribute to managing fall armyworm. However, there is no single system that is well documented as to effectiveness in pest status reduction and affordability in costs and labor. Farmers do not typically employ these practices specifically for fall armyworm control. A study conducted on 791 smallholder maize plots across different cropping systems in Zimbabwe showed that frequent weeding operations and minimum or zero tillage reduced fall armyworm damage (Baudron *et al.*, 2019). Interestingly, the study found that not all intercropping methods were equally effective and that intercropping maize with pumpkin, for instance, increased fall armyworm damage. The recommendation of intercropping as a management tool is therefore not appropriate as a general recommendation and research reports often yield conflicting results. Baudron *et al.* (2019) and Kassie *et al.* (2020) reported that maize–legume intercropping was ineffective in reducing fall armyworm risk, contradicting findings by Hailu *et al.* (2018).

A notable development in terms of successful habitat management for control of fall armyworm in small-scale farming systems is the push–pull system. Initially developed for stem borer control in maize, this system employs trap plants and repellent intercrops, and is proving effective for fall armyworm management (Midega *et al.*, 2005, 2018; Ntiri *et al.*, 2019). In this companion cropping system, maize is intercropped with leguminous species (*Desmodium uncinatum* or *D. intortum*), while forage grasses (*Pennisetum purpureum* or *Brachiaria* cultivar "Mulato II") are planted around fields. Adaptation and extension of the push–pull system to drier agroecological zones and inclusion of drought and spider mite tolerant *Brachiaria* varieties (Cheruiyot *et al.*, 2018a, b, 2021) over the last decade gave rise to "climate-smart push–pull systems" and expanded its pest management functionality. Push–pull systems in Kenya led to an average reduction of 82.7% in fall armyworm larvae per plant and 86.7% reduction in plant damage per field (Midega *et al.*, 2018). This technology has been adopted by approximately 350,000 farmers in East Africa (icipe, 2023) with 20,000 farmers planting *Brachiaria* as the trap crop (Maass *et al.*, 2015). Despite the

Fig. 4.4. Small, irrigated plots with intercrops and staggered planting dates of maize. (Photo credit: J. van den Berg.)

success of the push–pull system, its adoption specifically for fall armyworm control in the larger East African region is reported to be quite low (<1%) in surveyed countries (Ethiopia, Kenya, Tanzania, Uganda, Rwanda, and Burundi) (Niassy *et al.*, 2022). This may further indicate the slow adoption rate of pest management strategies that involve significant changes in farming practices.

While the continued adoption of push–pull strategies may change the future small-scale farming landscapes in East and Southern Africa, it is not foreseen that this will happen in large-scale systems. Biodiversity-based management interventions such as the push–pull strategy are less effective in simple landscapes compared to landscapes of intermediate and high complexity (Tscharntke *et al.*, 2005; Kebede *et al.*, 2018). This is due to these landscapes hosting too little habitat to support effective natural enemy densities. Furthermore, cultivation of perennial crops such as *Desmodium* limits the adoption of push–pull technology in large-scale systems where cultivation is mechanized, and free grazing of animals in the off-season is common (Van den Berg *et al.*, 2001). Mechanized soil cultivation, even in small-scale systems (<1 ha), not only prevented adoption of perennial intercrops in South Africa, but also the spatial arrangement that farmers were willing to adopt for planting of Napier grass (i.e. only on contours between fields) (Van den Berg, 2013; Van den Berg and Van Hamburg, 2015).

Biological control

The development of biological control programs currently takes precedence across Africa. Classical biological control, involving the introduction of natural enemies from the pest's region of origin, is anticipated to be a long-term solution to this pest problem in the region. Initially, it was assumed that fall armyworm had been introduced without its natural predators, and that East and Southern Africa lacked indigenous natural enemies of this pest. However, numerous parasitoid species have already been documented, and parasitism rates are unusually high for a recently invasive species in a new distribution range (Kenis, 2023 and references therein). Surveys are ongoing throughout the continent (e.g. Sisay *et al.*, 2018), with findings indicating that several native parasitoid species target fall armyworm in East Africa (Durocher-Granger *et al.*, 2020). An exotic parasitoid species, *Telenomus remus* (Nixon) (Hymenoptera: Platygastridae), known to parasitize fall armyworm eggs in South America, has also been recorded throughout the region (Kenis *et al.*, 2019). *Cotesia icipe* (Fernández-Triana & Fiaboe) (Hymenoptera: Braconidae), a larval parasitoid, is notably abundant in East Africa (Sisay *et al.*, 2019) and common in Southern Africa (Durocher-Granger *et al.*, 2020). Various institutions are engaged in identifying suitable biocontrol agents for sustainable fall armyworm control (Kenis *et al.*, 2019; Kenis, 2023).

The International Centre of Insect Physiology and Ecology (icipe) in Kenya has initiated mass releases of indigenous natural enemies of fall armyworm. Towards the end of 2020, 140,000 wasps of each of the egg parasitoid species *T. remus* and *Trichogramma chilonis* (Ishii) (Hymenoptera: Trichogrammatidae), as well as 5000 wasps of *C. icipe* were released (icipe, 2023). Successful implementation of classical biocontrol could offer a lasting solution to the fall armyworm scourge in the region, like the highly effective biological control program with *Cotesia flavipes* (Cameron) (Hymenoptera: Braconidae) against the spotted stem borer, *C. partellus*, in East and Southern Africa (Midingoyi *et al.*, 2016).

Conservation biocontrol, potentially stemming from the agroecological approaches mentioned above, also shows promise in reducing pest populations and mitigating its impact. Unfortunately, this is a long-term strategy significantly influenced by farming practices and cropping patterns. A substantial challenge to adoption of such practices is the lack of awareness regarding the presence and value of natural enemies at the farm level. For example, Tambo *et al.* (2020) reported minimal adoption of biological control strategies involving predators and parasitoids for fall armyworm management in households across Ghana, Rwanda, Uganda, Zambia, and Zimbabwe. Nevertheless, Kansiime *et al.* (2021) noted that some Kenyan farmers were conscious of natural enemies and attempted to safeguard them through reduced insecticide use, intercropping, crop diversification, agroforestry, and the cultivation of flowering hedges. Additionally, farmers cited small plot sizes as a hindrance to adopting conservation agricultural practices for preserving natural enemies and pollinators on their farms (Kansiime *et al.*, 2021).

Biopesticides

Due to their generally lower risk profile for pest management, biopesticides, including microbial products and extracts (e.g. neem), emerged as promising avenues for exploration and utilization in small-scale farming systems (Babendreier *et al.*, 2020; Constantine *et al.*, 2020; Bateman *et al.*, 2021; Guo *et al.*, 2020; Tepa-Yotto *et al.*, 2022). The development and utilization of biopesticides in sub-Saharan Africa are, however, still in their early stages (Ndolo *et al.*, 2019). Biopesticides used for fall armyworm control in maize include nuclear polyhedrosis virus (NPV) formulations, *Bacillus thuringiensis* (Bt), *Metarhizium*, *Beauveria*, *Trichoderma*, and spinosad. These biopesticides offer a variety of options and are particularly abundant in the region, notably in Kenya (Akutse *et al.*, 2019; Constantine *et al.*, 2020; Bateman *et al.*, 2021). Scientists at icipe in Nairobi, Kenya, successfully developed and registered biopesticides for fall armyworm control within a remarkably short time frame, and reports indicate a growing adoption of this technology among farmers in the region (Akutse *et al.*, 2020).

Although these products hold promise, their use is markedly low and constitute only 2.9% of the insecticide volume used in East African country surveys (Niassy *et al.*, 2021). A study conducted in Botswana by Makale *et al.* (2021) also revealed minimal use of biopesticides. In Kenya, Constantine *et al.* (2020) showed that there was limited farmer awareness of specific biopesticide products, and that their low adoption could be attributed to farmers perceiving them as less effective (slower action), complex in application, and of having too specific a target range. Unfortunately, accurate information on the use of these products remains scarce. While some biopesticides and plant extracts can be effective (Hruska, 2019), their uptake in the region remains modest (Constantine *et al.*, 2020). Constantine *et al.* (2020) and references cited therein, have highlighted that even when biopesticides are included in national pest management decision guides, they are seldom recommended by advisors. Knowledge, availability, and price of biopesticides have been identified as the main factors leading to this low recommendation rate (Dougoud *et al.*, 2018).

Host plant resistance

Host plant resistance stands as an important component of IPM strategies. Fortunately, institutions such as the International Maize and Wheat Improvement Center (CIMMYT) swiftly launched host plant resistance breeding programs following the incursion of fall armyworm into Africa. Various international and national research centers are presently engaged in developing tolerant cultivars (Prasanna *et al.*, 2022).

One of the readily attainable tools to mitigate yield losses is the cultivation of improved maize varieties and hybrids, including Bt maize, conventional non-genetically modified hybrids, and improved open-pollinated varieties. Given the low yields obtained by smallholders in the region and the relatively low levels of fall armyworm-induced leaf damage recorded in most areas, IPM strategies should also include interventions aimed at improving plant vigor (e.g. through integrated soil fertility management) (Chisonga *et al.*, 2023). Improved cultivars and hybrids cultivated under favorable agronomic conditions may provide higher grain yields despite foliar and ear damage by fall armyworm (Matova *et al.*, 2020).

While recent progress has been marked by the release of maize varieties with native resistance (Prasanna *et al.*, 2022), these benefits may only materialize for farmers in the medium to long term. The successful adoption of such varieties hinges on factors such as seed availability and the state of seed systems within the region. Obstacles to the adoption of hybrids and improved open-pollinated varieties include their inability to meet farmers' agronomic and cultural needs, insufficient farmer awareness about new cultivars, elevated seed costs, and limited access to seed (Langyintuo *et al.*, 2010; Abate *et al.*, 2017; Badu-Apraku and Fakorede, 2017; Ekpa *et al.*, 2019).

Genetically modified Bt maize holds potential for fall armyworm management within small-scale farming systems (Kotey *et al.*, 2021; Van den Berg *et al.*, 2021; Prasanna *et al.*, 2022). There are, however, many intricacies associated with introducing Bt maize hybrids into small-scale farming systems (Assefa and Van den Berg, 2015; Fisher *et al.*, 2015; Mukarumbwa and Taruvinga, 2023), and it is not foreseen that Bt maize will gain a significant share of the African maize seed market in the foreseeable future. The National Agricultural Research Organizations of Kenya, Ethiopia, Nigeria, Tanzania, Uganda, and Mozambique are testing the performance of Bt maize technologies introgressed into locally adapted African maize hybrids under the TELA® Maize project (AATF, 2021). While Bt maize gained approval for cultivation in Kenya in 2022 and Nigeria in 2023, it has not been cultivated in these countries yet.

Studies done in South Africa before the arrival of fall armyworm, showed a mean yield gain of 12% over six growing seasons in small-scale farming systems where Bt maize was planted for stem borer control (Gouse, 2012). Despite the availability of seed, the current adoption rate of Bt maize by South African small-scale farmers is low (Van den Berg *et al.*, 2021). This will most likely also be the case in other African countries where this technology has not been approved yet, and where adoption is faced by many challenges. For example, regulatory approvals typically take many years and represent a significant barrier to the adoption of Bt maize (Prasanna *et al.*, 2022). Additionally, the MON810 Bt maize trait, which is earmarked for the African market is a dated, off-patent trait (Erasmus *et al.*, 2019) which was effective against stem borers, but not fall armyworm. The MON810 event (Cry1Ab protein) in single-gene Bt maize hybrids provides only partial control of fall armyworm and has never been registered as a single-gene event against fall armyworm in South Africa or in the Americas. The MON89034 event, which expressed Cry1A.105 + Cry2Ab toxins, is registered for fall armyworm control in South Africa (Bayer, 2022). Fall armyworm rapidly evolved resistance to Bt maize elsewhere (Botha *et al.*, 2019; Van den Berg and du Plessis, 2022; Tabashnik *et al.*, 2023). Challenges with insect resistance management in small-scale maize farming systems include recycling of seed, non-compliance to insecticide resistance management (IRM) requirements, and a poorly developed seed industry within the region (Kotey *et al.*, 2016; Van den Berg *et al.*, 2022; Prasanna *et al.*, 2022).

Managing fall armyworm in large-scale farming systems

In large-scale farming systems, farmers plant hybrid maize, make large-scale use of fertilizers, and rely on insecticides to control pests if needed. Because of mechanization of soil cultivation and planting processes, planting dates are similar over large geographical areas, which is contrary to those in small-scale farming systems. This mechanization limits the use of multi-cropping and intercropping such as in small-scale systems.

It was initially feared that fall armyworm would become a major pest of maize in large-scale maize farming systems in South Africa. In 2017, when the pest was first recorded in the country, large numbers of maize and sorghum fields were infested in the warmer northern parts of the country as well as occasional infestations on farms in the Highveld region which is considered the corn belt of South Africa. Over the last 5 years, the threat has declined, and it is considered a highly sporadic pest of maize, except on sweet corn which is produced under irrigation in the warm north-eastern parts of the country. In cases where infestation of fall armyworm does occur in large-scale systems, the impact is very low.

Chemical control

Because of the initial perception that fall armyworm threatened maize production in the country, and because of a lack of knowledge on its ecology, infestations were dealt with by means of immediate insecticide applications, irrespective of infestation levels. The strong agrochemical and seed industries in South Africa contribute to successful management of maize pests, through provision of hybrid seed, pesticides, and other inputs.

In 2017, within months after the first report of fall armyworm infestation of maize in large-scale production systems in South Africa, 46 insecticides received emergency registration for this pest (Agri-Intel, 2017). These consisted of 20 active ingredients and three biopesticides. Because agrochemical companies did not renew some of the emergency registrations, these numbers decreased over time and by 2023, there were 36 different active ingredients registered and only one biopesticide. Despite easy access to these pesticides, spray applications against fall armyworm soon declined and from 2018 onwards, was only done occasionally. This decline is ascribed to increased local knowledge of the pest and its ecology, and research that showed that fall armyworm was not capable of survival (Du Plessis *et al.*, 2019, 2020) in the ecoclimatic zones where large-scale maize farming is done in South Africa. Research projects funded by the large-scale farming community also showed that the Bt maize that farmers were planting could suppress fall armyworm damage, like it did for the maize stem borers against which insecticides are also not sprayed often anymore.

Bt maize

Bt maize dominates the large-scale maize landscape in South Africa and is the first line of defense against fall armyworm and the stem borers mentioned above (Van den Berg *et al.*, 2021). Large-scale farmers benefit from the use of Bt maize against lepidopteran pests (Kruger *et al.*, 2012). In 2019, farmers in South Africa cultivated Bt maize on 1.7 million ha, accounting for 74% of the nation's maize (ISAAA, 2020). In 2011, cultivation of single-gene Cry1Ab maize was followed by Bt maize producing two toxins (Cry1A.105 + Cry2Ab). Results of field trials conducted at 36 sites per year over an 18-year period show a mean yield gain of 7.9% (range 3.8–14.3%) for Bt maize hybrids (Ala-Kokko *et al.*, 2021; Shew *et al.*, 2021). The mean relative profit margin with Bt maize hybrids compared to their conventional counterparts over this period was 24.6%. An earlier study showed that Cry1Ab maize reduced damage from stem borers and provided a yield gain of 11% for commercial farmers from 2000 to 2002 (Gouse *et al.*, 2005).

The single-gene Cry1Ab maize event which is still planted on a large part of the main production region in South Africa is, however, not registered against fall armyworm in the country or other parts of the world. Cry1Ab maize hybrids do, however, suffer less leaf damage compared to non-Bt maize when infested by fall armyworm. Near complete control of fall armyworm in large-scale systems is provided by Bt maize hybrids that produce two toxins (Cry1A.105 + Cry2Ab) (Botha *et al.*, 2019), and these varieties are most likely responsible for the very low pest status of fall armyworm in these systems.

The evolution of resistance to Cry proteins is, however, a threat to the sustained use of Bt maize in South Africa. The role of concurrent planting of single-gene Bt maize and pyramid maize in the advancement of resistance evolution has been described (Tabashnik *et al.*, 2009). *Busseola fusca* evolved resistance to Cry1Ab maize and shifts in susceptibility were recorded for *B. fusca* populations against Cry1A.105 + Cry2Ab maize (Strydom *et al.*, 2019).

Resistance evolution by fall armyworm is less of a threat than for the maize stem borer species which go into diapause in South Africa. Since fall armyworm larvae cannot survive the very cold winters in the maize production region of the country and there is virtually no maize present in Southern Africa during the winter months, selection for resistance is not strong. This is because annual invasions originate from areas where Bt maize is not planted (countries to the north of South Africa) and small-scale farming systems in the northern parts of South Africa. It is doubtful that moths that complete their life cycle in large-scale farming systems in South Africa move north and eastward towards Mozambique during winter months since maize is also largely absent from these countries during the winter months.

Early warning, monitoring, and action threshold levels

Early detection and pheromone trapping

Early detection of moths is considered the key to successful fall armyworm control. Pheromone traps can therefore play an important role in the management of this pest. The application of this technology differs between regions where pest populations persist and those where annual invasions may occur. Early detection can alert farmers to the onset of seasonal immigration, while in regions where populations persist, monitoring of flight patterns can inform farmers of increased local activity.

The timing of insecticide applications based on pheromone trap catches is, however, not good practice (Abang *et al.*, 2022). For example, contrary to recommendations by Cruz *et al.* (2012), trap captures did not accurately guide pesticide intervention decisions in an African agroecological context (Abang *et al.*, 2022). Pheromone trap data may, therefore, have limited application value in small-scale systems if there is no supporting organization that takes care of the purchasing and maintenance of traps, as well as the collection, interpretation, and distribution of data to farmers. Furthermore, before any pest management decision is taken, field-level scouting should be done. Large-scale farmers in South Africa initially made use of pheromone traps to monitor for the presence of moths, but the practice was soon abandoned, due to the low pest status in the country.

Several web- and mobile-based applications (apps) for early detection, identification, and monitoring of fall armyworm in sub-Saharan Africa have been developed. The FAMEWS global platform maps data collected by the FAMEWS mobile app (FAO, 2023) and provides information on the risk, spread, and management of this pest. Unfortunately, the application of this technology is only useful to farmers who can read, have smartphones, possess a basic understanding of the software's functionality, and are able to interpret the data (Njuguna *et al.*, 2021) (see also Chapter 13, this volume). This issue of accessibility of information technologies reveals one of agricultural contrasts in the shadows between small- and large-scale maize production systems in East and Southern Africa.

Action thresholds in small- and large-scale farming systems

The action threshold concept is critical in IPM and should guide all decisions on insecticide application. The use and relevancy of action thresholds are strongly influenced by the scale of the farming operation. Since the window of opportunity for effective fall armyworm control, once infestations are detected, is short, the use of thresholds necessitates quick access to resources. If insecticides need to be applied, it should be done when larvae are still small (less than fourth instar) (Van den Berg and du Plessis, 2022).

The existing action thresholds for fall armyworm control in Africa were developed through expert opinion and distinguish between different plant growth stages (McGrath *et al.*, 2018). These action thresholds are as follows: for seedling stage maize, control should be applied when 20% (range 10–30%) of plants show signs of damage and for vegetative stages when 30% (range 20–40%) of plants show damage. For maize in the flowering stage, insecticide application is warranted when 20% (range 20–40%) of plants show symptoms of damage. These recommended action thresholds are only applicable to non-Bt maize, irrespective of the scale of farming. Since fall armyworm pest status in Bt-maize-dominated landscapes is very low, fields are not scouted for fall armyworm damage. However, in small-scale farming systems in South Africa, where Bt hybrids are less available due to infrastructure and marketing issues, farmers do inspect their fields for damage and apply pesticides when needed. Unfortunately, neither farmers nor extension services in small-scale systems make use of the action threshold concept when decisions are made on whether to apply insecticides (J. Van den Berg, personal observations).

These thresholds may need revision (Abang *et al.*, 2022; Chinwada *et al.*, 2023; Chisonga *et al.*, 2023), particularly for small-scale systems. Action thresholds have not been evaluated under conditions where yield potential is very low, field sizes are small, intercropping is common, and the crop growth stages of small adjacent fields differ largely due to staggered planting dates. Since intercropping is common in small-scale systems, maize plant stand per hectare may be low. For example, plant stands on a 1-ha field may range between 6000 and 10,000 in some intercropped fields. Application of action thresholds in the classical way may therefore not be relevant in such systems.

Timeliness of application of pesticides based on action thresholds is further challenged by the absence of a well-organized agrochemical industry in most of the regions where small-scale production is common. For example, most farmers do not own knapsack sprayers, and there are no spray service providers.

Future Needs and Prospects

There is no single solution to the management of fall armyworm in the region. While biological control and habitat management provide long-term solutions in terms of pest suppression, there is also a need to empower farmers with tools to control this pest when they are faced with infestations at the field level (e.g. chemical control).

Information provided in this chapter illustrates the differences in terms of farming systems, pest status, access to technology, and the resources available to manage pests at different production scales. The pest status of fall armyworm is much higher in agroecological zones where the pest persists. These are also the regions where small-scale farming dominates, where there is a lack of knowledge of pest ecology and IPM, and where socio-economic status influences access to information and other resources.

Pest management challenges can only be addressed through the development of IPM strategies that also consider the socio-economic environment within which production constraints are experienced. The long-term solution to the threat of fall armyworm in small-scale production systems therefore lies in the development of strategies that have areawide impact, and which are not heavily reliant on farmer interventions to be sustained. These are, for example, biological control programs and habitat management systems that are appropriate for different agroecological zones, and at different scales. For biocontrol programs to be successful, significant long-term investments by national and international institutions are required. Building on the proven success of the push–pull strategy, with possible inclusion of Bt maize (Midega *et al.*, 2006, 2018) and development of other climate-smart cropping systems, could also contribute to more diverse and sustainable small-scale farming landscapes across the region.

The information provided above also highlighted the value and role that improved open-pollinated varieties, hybrid seed, and pesticides could play in managing fall armyworm. To improve access to these types of technologies, the need exists for a strong agrochemical and seed industry. The socio-economic landscape in small-scale systems does not allow for many farmers to own pesticide spray equipment, and there is a lack of knowledge regarding action thresholds and the safe use of pesticides. These needs can be met by the establishment of service providers which are supported by the agrochemical industry. For example, local spray service providers that received training on the appropriate use of pesticides could provide much needed services when fall armyworm infestation levels necessitates control. Decisions to apply pesticides should be guided by action threshold levels. Unfortunately, no threshold levels in context of agroecological zones have been developed for fall armyworm control in Africa. The development of action thresholds that consider cropping system, variety, yield potential, and socio-economic environment should receive high priority.

Technologies such as Bt maize and seed treatments, where they are available, can provide solutions to fall armyworm infestations in both small- and large-scale systems. However, these technologies also have their unique constraints and challenges, and their use is not only influenced by the scale of the farming system, but also the socio-economic status of farmers, and especially the agroecological zone and pest complex where maize is cultivated. In agroecological zones where the pest is only of sporadic nature due to its migratory characteristics, "high preventative" input costs (costs of seed treatment and Bt maize) may be unwarranted if the pest does not annually invade the region (Fisher *et al.*, 2015). For chemical control to be effective, the correct pesticides need to be applied timeously, and only when action thresholds are exceeded. Due to uncertainty regarding pest risk, early-warning systems are needed, which, together with local pheromone trapping and regular scouting of large fields, inform management decisions. Early-warning systems, however, do not exist and without the labor-intensive scouting of large maize fields, the risk of unwarranted pesticide applications is high.

Uncertainty regarding the risk of fall armyworm in a particular season therefore influences the type of technology that is applicable, manageable, and affordable. A thorough understanding of the farming landscape is critical to develop and provide pest management solutions. In this context, the farming landscape includes factors such as the socio-economic status of farmers that work the land, affordability of crop protection technologies, access to information and knowledge of pest ecology, and the availability of pest management tools. For example, the additional cost of Bt maize seed may not be warranted in an environment where fall armyworm pest status is uncertain, and it is the only lepidopteran pest species that attack maize. This cost may, however, be offset in agroecological zones where maize yield is high and where mixed populations of fall armyworm and several stem borer pest species infest maize.

Fall armyworm is not the only reason for the low maize yields recorded in most small-scale farming systems. Agronomical aspects such as poor varieties, drought and poor nutrient status of soils play a very important role. Improved cultivars and hybrids cultivated under favorable agronomic conditions are generally more tolerant to pest damage and may provide higher yields despite damage. General improvement in farming practices which ensures higher yields therefore also contributes to mitigation of the fall armyworm threat.

Pest management, irrespective of the scale of farming, should consider landscape-level factors and not only involve single pests and crop species at field level. Long-term success lies in sustainable agricultural intensification which implies the use of all practices that are relevant to a particular farming system to increase yield without adverse environmental impact or conversion of additional non-agricultural land (Pretty *et al.*, 2011). In the context of fall armyworm, this means the use of methods that are applicable, effective, manageable, and affordable in a particular system. Traditional and modern techniques to improve crop genetics, as well as the agroecological methods, are pivotal components of sustainable intensification. Most importantly, sustainable agricultural intensification also considers socio-economic aspects and invests in development of socio-economic capital to empower farmers with knowledge, develop human capacity, and create conducive markets (Pretty *et al.*, 2011; Tadele, 2017; Kansiime *et al.*, 2021). IPM programs, in particular, benefit from investment in social and human capital development, and lead to increased technology adaptation and adoption (Pretty *et al.*, 2011). The concept of agricultural intensification applies to all farming systems, irrespective of scale, and does not exclude the use of improved technologies such as pesticides and hybrids (Pretty *et al.*, 2011) or Bt maize.

Summary and Conclusions

There is no single solution that can provide sustainable control of fall armyworm across East and Southern Africa that differs so much in terms of climate, pest status, the scale of farming, and access to resources. A wide range of technologies and management practices have been developed for other pests and some are available for control of fall armyworm, including host plant resistance, cultural control, biological control, biopesticides, pesticides, and agroecological management. These technologies are, however, not equally accessible to farming communities among countries in East and Southern Africa.

While future efforts should be directed towards developing technologies that have area-wide impact, the need for investment into, and training of farmers, extension staff and agro-dealers on fall armyworm management and responsible use of pesticides should not be overlooked. Fundamental shifts in pest management practices are needed, not only in terms of the use of environmentally safe methods such as biological and cultural controls where scale appropriate, but also with respect to the responsible use of pesticides and Bt maize where the fall armyworm threat is high and socio-economic conditions and governmental support allow affordable access and training to assure best use practices.

References

AATF (2021) *Tela Maize Project*. Nairobi, Kenya: African Agricultural Technology Foundation. Available at: https://www.aatf-africa.org/tela-maize-project/ (accessed October 27, 2023).

Abang, A.F., Nanga, S.N., Fotso Kuate, A., Kouebou, C., Suh, C. *et al.* (2021) Natural enemies of fall armyworm *Spodoptera frugiperda* (Lepidoptera: Noctuidae) in different agro-ecologies. *Insects* 12, 509. DOI: 10.3390/insects12060509

Abang, A.F., Nanga, S.N., Esi Ndanda, R.M., Doumtsop Fotio, A.R., Gonder, M.K. *et al.* (2022) Reliability of pheromone trap catches and maize plant damage as criteria for timing fall armyworm control interventions in humid forest agroecology of central Africa. *Journal of Economic Entomology* 115, 1806–1816. DOI: 10.1093/jee/toac087

Abate, T., Fisher, M., Abdoulaye, T., Kassie, G.T., Lunduka, R. *et al.* (2017) Characteristics of maize cultivars in Africa: how modern are they and how many do smallholder farmers grow? *Agriculture & Food Security* 6, 30. DOI: 10.1186/s40066-017-0108-6

Adan, M., Tonnang, H.E., Greve, K., Borgemeister, C. and Goergen, G. (2023) Use of time series normalized difference vegetation index (NDVI) to monitor fall armyworm (*Spodoptera frugiperda*) damage on maize production systems in Africa. *Geocarto International* 38, 2186492. DOI: 10.1080/10106049.2023.2186492

Agri-Intel (2017) *Agri-Intel: The Agrochemical Information Source of Preference*. CropLife, South Africa. Available at: https://www.agri-intel.com/ (accessed January 1, 2024).

Ahissou, B.R., Sawadogo, W.M., Bokonon-Gant, A.H., Somda, I. and Verheggen, F. (2022) Farmers' knowledge and management practices of the fall armyworm (*Spodoptera frugiperda* Smith) in Burkina Faso. *Biotechnology,Agronomy and Society and Environment* 26, 191–197. DOI: 10.25518/1780-4507.19909

Akutse, K.S., Kimemia, J.W., Ekesi, S., Khamis, F.M., Ombura, O.L. *et al.* (2019) Ovicidal effects of entomopathogenic fungal isolates on the invasive fall armyworm *Spodoptera frugiperda* (Lepidoptera: Noctuidae). *Journal of Applied Entomology* 143, 626–634. DOI: 10.1111/jen.12634

Akutse, K.S., Subramanian, S., Maniania, N.K., Dubois, T. and Ekesi, S. (2020) Biopesticide research and product development in Africa for sustainable agriculture and food security – experiences from the International Centre of Insect Physiology and Ecology (icipe). *Frontiers in Sustainable Food Systems* 4, 563016. DOI: 10.3389/fsufs.2020.563016

Ala-Kokko, K., Nalley, L.L., Shew, A.M., Tack, J.B., Chaminuka, P. *et al.* (2021) Economic and ecosystem impacts of GM maize in South Africa. *Global Food Security* 29, 100544. DOI: 10.1016/j.gfs.2021.100544

Al-Sarar, A., Hall, F.R. and Downer, R.A. (2006) Impact of spray application methodology on the development of resistance to cypermethrin and spinosad by fall armyworm *Spodoptera frugiperda* (JE Smith). *Pest Management Science* 62, 1023–1031. DOI: 10.1002/ps.1241

Andrews, K.L. (1980) The whorlworm, *Spodoptera frugiperda*, in Central America and neighboring areas. *The Florida Entomologist* 63, 456–467. DOI: 10.2307/3494530

Assefa, Y. and Van den Berg, J. (2015) Genetically modified maize: adoption practices of small-scale farmers in South Africa and implications for resource poor farmers on the continent. *Aspects of Applied Biology* 96, 215–224.

Babendreier, D., Koku Agboyi, L., Beseh, P., Osae, M., Nboyine, J. *et al.* (2020) The efficacy of alternative, environmentally friendly plant protection measures for control of fall armyworm, *Spodoptera frugiperda*, in maize. *Insects* 11, 240. DOI: 10.3390/insects11040240

Badu-Apraku, B. and Fakorede, M.A.B. (2017) Maize in Sub-Saharan Africa: importance and production constraints. In: Badu-Apraku, B. and Fakorede, M.A.B. (eds) *Advances in Genetic Enhancement of Early and Extra-early Maize for Sub-Saharan Africa*. Cham, Switzerland: Springer, pp. 3–10. DOI: 10.1007/978-3-319-64852-1

Bate, R. and Van Rensburg, J.B.J. (1992) Predictive estimation of maize yield loss caused by *Chilo partellus* (Swinhoe) (Lepidoptera: Pyralidae) in maize. *South African Journal of Plant and Soil* 9, 150–154. DOI: 10.1080/02571862.1992.10634619

Bateman, M.L., Day, R.K., Rwomushana, I., Subramanian, S., Wilson, K. *et al.* (2021) Updated assessment of potential biopesticide options for managing fall armyworm (*Spodoptera frugiperda*) in Africa. *Journal of Applied Entomology* 145, 384–393. DOI: 10.1111/jen.12856

Baudron, F., Andersson, J.A., Corbeels, M. and Giller, K.E. (2012a) Failing to yield? Ploughs, conservation agriculture and the problem of agricultural intensification: an example from the Zambezi Valley, Zimbabwe. *Journal of Development Studies* 48, 393–412. DOI: 10.1080/00220388.2011.587509

Baudron, F., Tittonell, P., Corbeels, M., Letourmy, P. and Giller, K.E. (2012b) Comparative performance of conservation agriculture and current smallholder farming practices in semi-arid Zimbabwe. *Field Crops Research* 132, 117–128. DOI: 10.1016/j.fcr.2011.09.008

Baudron, F., Zaman-Allah, M.A., Chaipa, I., Chari, N. and Chinwada, P. (2019) Understanding the factors conditioning fall armyworm (*Spodoptera frugiperda* J.E. Smith) infestation in African smallholder maize fields and quantifying its impact on yield: a case study in Eastern Zimbabwe. *Crop Protection* 120, 141–150. DOI: 10.1016/j.cropro.2019.01.028

Bayer (2022) *South Africa – Technology User Guide (TUG)*. Isando, South Africa: Bayer. Available at: https://www.cropscience.bayer.africa/za/en-za/resources/resource-guides.html (accessed November 1, 2023).

Botha, A.S., Erasmus, A., du Plessis, H. and Van den Berg, J. (2019) Efficacy of Bt maize for control of *Spodoptera frugiperda* (Lepidoptera: Noctuidae) in South Africa. *Journal of Economic Entomology* 112, 1260–1266. DOI: 10.1093/jee/toz048

Chawanda, G., Tembo, Y.L.B., Donga, T.K., Kabambe, V.H., Stevenson, P.C. *et al.* (2023) Agroecological management of fall armyworm using soil and botanical treatments reduces crop damage and increases maize yield. *Frontiers in Agronomy* 5, 1114496. DOI: 10.3389/fagro.2023.1114496

Cheruiyot, D., Midega, C.A., Van den Berg, J., Pickett, J.A. and Khan, Z.R. (2018a) Suitability of brachiaria grass as a trap crop for management of *Chilo partellus*. *Entomologia Experimentalis et Applicata* 166, 139–148. DOI: 10.1111/eea.12651

Cheruiyot, D., Midega, C.A.O., Van den Berg, J., Pickett, J.A. and Khan, Z.R. (2018b) Genotypic responses of Brachiaria grass (*Brachiaria* spp.) accessions to drought stress. *Journal of Agronomy* 17, 136–146. DOI: 10.3923/ja.2018.136.146

Cheruiyot, D., Chidawanyika, F., Midega, C.A., Pittchar, J.O., Pickett, J.A. *et al.* (2021) Field evaluation of a new third generation push-pull technology for control of striga weed, stemborers, and fall armyworm in western Kenya. *Experimental Agriculture* 57, 301–315. DOI: 10.1017/S0014479721000260

Chimweta, M., Nyakudya, I.W., Jimu, L. and Mashingaidze, A.B. (2019) Fall armyworm [*Spodoptera frugiperda* (J.E. Smith)] damage in maize: management options for flood-recession cropping smallholder farmers. *International Journal of Pest Management* 66, 142–154. DOI: 10.1080/09670874.2019.1577514

Chinwada, P., Fiaboe, K.K.M., Akem, C., Dixon, A. and Chikoyea, D. (2023) Assessment of effectiveness of cyantraniliprole and thiamethoxam maize seed treatments for fall armyworm, *Spodoptera frugiperda* (J.E. Smith) management. *Crop Protection* 174, 106418. DOI: 10.1016/j.cropro.2023.106418

Chisonga, C., Chipabika, G., Sohati, P.H. and Harrison, R.D. (2023) Understanding the impact of fall armyworm (*Spodoptera frugiperda* J. E. Smith) leaf damage on maize yields. *PLoS ONE* 18, e0279138. DOI: 10.1371/journal.pone.0279138

CIFOR-ICRAF (2023) *Managing Fall Armyworm: A Guide to Low Cost Pest Management Approaches.* Bogor, Indonesia: CIFOR-ICRAF Communications. Available at: https://www.cifor-icraf.org/publications/pdf/guide/Fall-Armyworm-guide.pdf (accessed April 30, 2024).

Clarkson, J., Borah, J.R., Baudron, F. and Sunderland, T.C.H. (2022) Forest proximity positively affects natural enemy mediated control of fall armyworm in Southern Africa. *Frontiers in Forests and Global Change* 5, 781574. DOI: 10.3389/ffgc.2022.781574

Constantine, K.L., Kansiime, M.K., Mugambi, I., Nunda, W., Chacha, D. *et al.* (2020) Why don't smallholder farmers in Kenya use more biopesticides? *Pest Management Science* 76, 3615–3625. DOI: 10.1002/ps.5896

Cruz, I., de Lourdes Corrêa Figueiredo, M., da Silva, R.B., da Silva, I.F., de Souza Paula, C. *et al.* (2012) Using sex pheromone traps in the decision-making process for pesticide application against fall armyworm (*Spodoptera frugiperda* [Smith] [Lepidoptera: Noctuidae]) larvae in maize. *International Journal of Pest Management* 58, 83–90. DOI: 10.1080/09670874.2012.655702

Cugala, D.R. (2007) Impact assessment of natural enemies on stem borer populations and maize yield in three agroecological zones in Mozambique. PhD dissertation. Kenyata University, Nairobi, Kenya. Available at: https://ir-library.ku.ac.ke/handle/123456789/14141 (accessed May 3, 2024).

Daryanto, S., Fu, B., Zhao, W., Wang, S., Jacinthe, P.A. and Wang, L. (2020) Ecosystem service provision of grain legume and cereal intercropping in Africa. *Agricultural Systems* 178, 102761. DOI: 10.1016/j.agsy.2019.102761

Day, R., Abrahams, P., Bateman, M., Beale, T., Clottey, V. *et al.* (2017) Fall armyworm: impacts and implications for Africa. *Outlooks on Pest Management* 28, 196–201. DOI: 10.1564/v28_oct_02

De Groote, H., Overholt, W.A., Ouma, J.O. and Wanyama, J. (2011) Assessing the potential economic impact of *Bacillus thuringiensis* (Bt) maize in Kenya. *African Journal of Biotechnology* 10, 4741–4751.

De Groote, H., Kimenju, S.C., Munyua, B., Palmas, S., Kassie, M. *et al.* (2020) Spread and impact of fall armyworm (*Spodoptera frugiperda* J.E. Smith) in maize production areas of Kenya. *Agriculture, Ecosystems & Environment* 292, 106804. DOI: 10.1016/j.agee.2019.106804

Deguine, J.P., Aubertot, J.N., Flor, R.J., Lescourret, F., Wyckhuys, K.A. *et al.* (2021) Integrated pest management: good intentions, hard realities. A review. *Agronomy for Sustainable Development* 41, 38. DOI: 10.1007/s13593-021-00689-w

Diagne, C., Turbelin, A.J., Moodley, D., Novoa, A., Leroy, B. *et al.* (2021) The economic costs of biological invasions in Africa: a growing but neglected threat? *NeoBiota* 67, 11–51. DOI: 10.3897/neobiota.67.59132

Dougoud, J., Cock, M.J., Edgington, S. and Kuhlmann, U. (2018) A baseline study using Plantwise information to assess the contribution of extension services to the uptake of augmentative biological control in selected low- to lower-middle-income countries. *BioControl* 63, 117–132. DOI: 10.1007/s10526-017-9823-y

Du Plessis, H., Van den Berg, J., Ota, N. and Kriticos, D.J. (2019) *Spodoptera frugiperda. CSIRO-InSTePP Pest Geography*. Canberra, Australia: CSIRO. Available at: https://pra.eppo.int/getfile/a180bbf9-abe1-40dd-a2db-f5c9ebe05195 (accessed May 10, 2024).

Du Plessis, H., Schlemmer, M. and Van den Berg, J. (2020) The effect of temperature on the development of *Spodoptera frugiperda* (Lepidoptera: Noctuidae). *Insects* 11, 228. DOI: 10.3390/insects11040228

Durocher-Granger, L., Mfune, T., Musesha, M., Lowry, A., Reynolds, K. *et al.* (2020) Factors influencing the occurrence of fall armyworm parasitoids in Zambia. *Journal of Pest Science* 94, 1133–1146. DOI: 10.1007/s10340-020-01320-9

Ekpa, O., Palacios-Rojas, N., Kruseman, G., Fogliano, V. and Linnemann, A.R. (2019) Sub-Saharan African maize-based foods-processing practices, challenges and opportunities. *Food Reviews International* 35, 609–639. DOI: 10.1080/87559129.2019.1588290

Erasmus, R., Pieters, R., Du Plessis, H., Hilbeck, A., Tritikova, M. *et al.* (2019) Introgression of a *cry1Ab* transgene into open pollinated maize and its effect on Cry protein concentration and target pest survival. *PLoS ONE* 14, e0226476. DOI: 10.1371/journal.pone.0226476

Eschen, R., Beale, T., Bonnin, J.M., Constantine, K.L., Duah, S. *et al.* (2021) Towards estimating the economic cost of invasive alien species to African crop and livestock production. *CABI Agriculture and Bioscience* 2, 18. DOI: 10.1186/s43170-021-00038-7

FAO (2023) *Global Action for Fall Armyworm Control*. Rome, Italy: Food and Agriculture Organization of the United Nations. Available at: https://www.fao.org/fall-armyworm/monitoring-tools/famews-global-platform/en/ (accessed January 1, 2024).

Fisher, K., Van den Berg, J. and Mutengwa, C. (2015) Is Bt maize effective in improving South African small holder agriculture? *South African Journal of Science* 111, 1–2. DOI: 10.17159/sajs.2015/a0092

Gouse, M. (2012) GM maize as subsistence crop: the South African smallholder experience. *AgBioForum* 15, 163–174.

Gouse, M., Pray, C.E., Kirsten, J. and Schimmelpfennig, D. (2005) A GM subsistence crop in Africa: the case of Bt white maize in South Africa. *International Journal of Biotechnology* 7, 84–94. DOI: 10.1504/IJBT.2005.006447

Guo, J., Wu, S., Zhang, F., Huang, C., He, K. *et al.* (2020) Prospects for microbial control of the fall armyworm, *Spodoptera frugiperda*: a review. *BioControl* 65, 647–662. DOI: 10.1007/s10526-020-10031-0

Hailu, G., Niassy, S., Zeyaur, K.R., Ochatum, N. and Subramanian, S. (2018) Maize–legume intercropping and push–pull for management of fall armyworm, stemborers, and striga in Uganda. *Agronomy Journal* 110, 2513–2522. DOI: 10.2134/agronj2018.02.0110

Harrison, R.D., Thierfelder, C., Baudron, F., Chinwada, P., Midega, C. *et al.* (2019) Agro-ecological options for fall armyworm (*Spodoptera frugiperda* JE Smith) management: providing low-cost, smallholder friendly solutions to an invasive pest. *Journal of Environmental Management* 243, 318–330. DOI: 10.1016/j.jenvman.2019.05.011

Harrison, R., Banda, J., Chipabika, G., Chisonga, C., Katema, C. *et al.* (2022) Low impact of fall armyworm (*Spodoptera frugiperda* Smith) (Lepidoptera: Noctuidae) across smallholder fields in Malawi and Zambia. *Journal of Economic Entomology* 115, 1783–1789. DOI: 10.1093/jee/toac113

Hruska, A.J. (2019) Fall armyworm (*Spodoptera frugiperda*) management by smallholders. *CAB Reviews* 14, 043. DOI: 10.1079/PAVSNNR201914043

icipe (2023) icipe *Launches Mass Release of Indigenous Natural Enemies to Control Fall Armyworm*. Nairobi, Kenya: International Centre of Insect Physiology and Ecology. Available at: http://www.icipe.org/news/icipe-launches-mass-release-indigenous-natural-enemies-control-fall-armyworm (accessed August 4, 2023).

ISAAA (2020) *2020 Accomplishments Report*. Nairobi, Kenya: International Service for the Acquisition of Agri-biotech Applications. Available at: https://www.isaaa.org/resources/publications/annualreport/2020/pdf/ISAAA-2020-Accomplishment-Report.pdf (accessed December 8, 2023).

Jayne, T.S., Mather, D. and Mghenyi, E. (2010) Principal challenges confronting smallholder agriculture in sub-Saharan Africa. *World Development* 38, 1384–1398. DOI: 10.1016/j.worlddev.2010.06.002

Jepson, P.C., Murray, K., Bach, O., Kachigamba, D., Ndeithi, F. *et al.* (2018) Pesticide hazard and risk management, and compatibility with IPM. In: Prasanna, B.M., Huesing, J.E., Eddy, R. and Peschke, V.M. (eds) *Fall Armyworm in Africa: A Guide for Integrated Pest Management*, 1st edn. Mexico, CDMX: CIMMYT, pp. 29–44. Available at: http://hdl.handle.net/10883/19204 (accessed January 1, 2024).

Kalyebi, A., Otim, M.H., Walsh, T. and Tay, W.T. (2023) Farmer perception of impacts of fall armyworm (*Spodoptera frugiperda* J.E. Smith) and transferability of its management practices in Uganda. *CABI Agriculture and Bioscience* 4, 9. DOI: 10.1186/s43170-023-00150-w

Kansiime, M.K., Mugambi, I., Rwomushana, I., Nunda, W., Lamontagne-Godwin, J. *et al.* (2019) Farmer perception of fall armyworm (*Spodoptera frugiperda* J.E. Smith) and farm-level management practices in Zambia. *Pest Management Science* 75, 2840–2850. DOI: 10.1002/ps.5504

Kansiime, M.K., Girling, R.D., Mugambi, I., Mulema, J., Oduor, G. *et al.* (2021) Rural livelihood diversity and its influence on the ecological intensification potential of smallholder farms in Kenya. *Food and Energy Security* 10, e254. DOI: 10.1002/fes3.254

Kasoma, C., Shimelis, H. and Laing, M.D. (2021) Fall armyworm invasion in Africa: implications for maize production and breeding. *Journal of Crop Improvement* 35, 111–146. DOI: 10.1080/15427528.2020.1802800

Kassie, M., Wossen, T., De Groote, H., Tefera, T., Sevgan, S. *et al.* (2020) Economic impacts of fall armyworm and its management strategies: evidence from southern Ethiopia. *European Review of Agricultural Economics* 47, 1473–1501. DOI: 10.1093/erae/jbz048

Kebede, Y., Baudron, F., Bianchi, F. and Tittonell, P. (2018) Unpacking the push-pull system: assessing the contribution of companion crops along a gradient of landscape complexity. *Agriculture, Ecosystems & Environment* 268, 115–123. DOI: 10.1016/j.agee.2018.09.012

Kenis, M. (2023) Prospects for classical biological control of *Spodoptera frugiperda* (Lepidoptera: Noctuidae) in invaded areas using parasitoids from the Americas. *Journal of Economic Entomology* 116, 331–341. DOI: 10.1093/jee/toad029

Kenis, M., Du Plessis, H., Van den Berg, J., Ba, M.N., Goergen, G. *et al.* (2019) *Telenomus remus*, a candidate parasitoid for the biological control of *Spodoptera frugiperda* in Africa, is already present on the continent. *Insects* 10, 92. DOI: 10.3390/insects10040092

Kotey, D.A., Assefa, Y., Obi, A. and Van den Berg, J. (2016) Disseminating genetically modified (GM) maize technology to smallholder farmers in the Eastern Cape province of South Africa: extension personnel's awareness of stewardship requirements and dissemination practices. *South African Journal of Agricultural Extension* 44, 59–74.

Kotey, D.A., Assefa, Y. and Van den Berg, J. (2021) Comparative incidence of maize stem borers on GM and non-GM maize under two smallholder cropping systems in the Eastern Cape province of South Africa. *Asian Journal of Agriculture and Biology* 2021, 202010550. DOI: 10.35495/ajab.2020.10.550

Krüger, W., Van den Berg, J. and Van Hamburg, H. (2008) The relative abundance of maize stem borers and their parasitoids at the Tshiombo irrigation scheme in Venda, South Africa. *South African Journal of Plant and Soil* 25, 144–151. DOI: 10.1080/02571862.2008.10639910

Kruger, M., Van Rensburg, J.B.J. and Van den Berg, J. (2012) Transgenic Bt maize: farmers' perceptions, refuge compliance and reports of stem borer resistance in South Africa. *Journal of Applied Entomology* 136, 38–50. DOI: 10.1111/j.1439-0418.2011.01616.x

Kumela, T., Simiyu, J., Sisay, B., Likhayo, P., Mendesil, E. *et al.* (2019) Farmers' knowledge, perceptions, and management practices of the new invasive pest, fall armyworm (*Spodoptera frugiperda*) in Ethiopia and Kenya. *International Journal of Pest Management* 65, 1–9. DOI: 10.1080/09670874.2017.1423129

Kuyah, S., Sileshi, G.W., Nkurunziza, L., Chirinda, N., Ndayisaba, P.C. *et al.* (2021) Innovative agronomic practices for sustainable intensification in sub-Saharan Africa. A review. *Agronomy for Sustainable Development* 41, 16. DOI: 10.1007/s13593-021-00673-4

Lacambra, C., Molloy, D., Lacambra, J., Leroux, I., Klossner, L. *et al.* (2020) *Factsheet – Resilience Solutions for the Maize Sector in South Africa.* Washington, DC: Inter-American Development Bank. Available at: https://publications.iadb.org/en/factsheet-resilience-solutions-for-the-maize-sector-in-south-africa (accessed April 20, 2024).

Langyintuo, A.S., Mwangi, W., Diallo, A.O., MacRobert, J., Dixon, J. *et al.* (2010) Challenges of the maize seed industry in eastern and southern Africa: a compelling case for private–public intervention to promote growth. *Food Policy* 35, 323–331. DOI: 10.1016/j.foodpol.2010.01.005

Lay, J., Nolte, K. and Sipangule, K. (2018) *Large-scale Farms and Smallholders: Evidence from Zambia.* Working Paper 310. Hamburg, Germany: German Institute of Global and Area Studies (GIGA). Available at: www.jstor.org/stable/resrep16500 (accessed January 1, 2024).

Maass, B.L., Midega, C.A.O., Mutimura, M., Rahetlah, V.B., Salgado, P. *et al.* (2015) Homecoming of *Brachiaria*: improved hybrids prove useful for African animal agriculture. *East African Agricultural and Forestry Journal* 81, 71–78. DOI: 10.1080/00128325.2015.1041263

Makale, F., Mugambi, I., Kansiime, M.K., Yuka, I., Abang, M. *et al.* (2021) Fall armyworm in Botswana: impacts, farmer management practices and implications for sustainable pest management. *Pest Management Science* 78, 1060–1070. DOI: 10.1002/ps.6717

Maphumulo, N., du Plessis, H. and Van den Berg, J. (2023) The effect of wood ash and soil applications on the behaviour and survival of *Spodoptera frugiperda* (Lepidoptera: Noctuidae) larvae on maize. *Insects* 14, 813. DOI: 10.3390/insects14100813

Matova, P.M., Kamutando, C.N., Magorokosho, C., Kutywayo, D., Gutsa, F. *et al.* (2020) Fall-armyworm invasion, control practices and resistance breeding in Sub-Saharan Africa. *Crop Science* 60, 2951–2970. DOI: 10.1002/csc2.20317

McGrath, D., Huesing, J.E., Beiriger, R., Nuessly, G., Tepa-Yotto, T.G. *et al.* (2018) Monitoring, surveillance, and scouting for fall armyworm. In: Prasanna, B.M., Huesing, J.E., Eddy, R. and Peschke, V.M. (eds) *Fall Armyworm in Africa: A Guide for Integrated Pest Management*, 1st edn. Mexico, CDMX: CIMMYT, pp. 11–28. Available at: http://hdl.handle.net/10883/19204 (accessed January 1, 2024).

Mendesil, E., Tefera, T., Blanco, C.A., Paula-Moraes, S.V., Huang, F. *et al.* (2023) The invasive fall armyworm, *Spodoptera frugiperda*, in Africa and Asia: responding to the food security challenge, with priorities for integrated pest management research. *Journal of Plant Diseases and Protection* 130, 1175–1206. DOI: 10.1007/s41348-023-00777-x

Midega, C.A.O., Khan, Z.R., Van den Berg, J. and Ogol, C.K.P.O. (2005) Habitat management and its impact on maize stemborer colonization and crop damage levels in Kenya and South Africa. *African Entomology* 13, 333–340.

Midega, C.A.O., Khan, Z.R., Van den Berg, J., Ogol, C.K.P.O., Pickett, J.A. *et al.* (2006) Maize stemborer predator activity under "push–pull" system and Bt-maize: a potential component in managing Bt resistance. *International Journal of Pest Management* 52, 1–10. DOI: 10.1080/09670870600558650

Midega, C.A.O., Pittchar, J.O., Pickett, J.A., Hailu, G.W. and Khan, Z.R. (2018) A climate-adapted push-pull system effectively controls fall armyworm, *Spodoptera frugiperda* (J E Smith), in maize in East Africa. *Crop Protection* 105, 10–15. DOI: 10.1016/j.cropro.2017.11.003

Midingoyi, S.G., Affognon, H.D., Macharia, I., Ong'amo, G., Abonyo, E. *et al.* (2016) Assessing the long-term welfare effects of the biological control of cereal stemborer pests in East and Southern Africa: evidence from Kenya, Mozambique and Zambia. *Agriculture, Ecosystems & Environment* 230, 10–23. DOI: 10.1016/j.agee.2016.05.026

Moolman, H.J., Van den Berg, J., Conlong, D., Cugala, D., Siebert, S.J. *et al.* (2014) Species diversity and distribution of lepidopteran stem borers in South Africa and Mozambique. *Journal of Applied Entomology* 138, 152–166. DOI: 10.1111/jen.12085

Mukarumbwa, P. and Taruvinga, A. (2023) Landrace and GM maize cultivars' selection choices among rural farming households in the Eastern Cape Province, South Africa. *GM Crops and Food* 14, 1–15. DOI: 10.1080/21645698.2023.2215146

Mwalusepo, S., Massawe, E.S., Johansson, T., Abdel-Rahman, E., Gathara, M. *et al.* (2018) Modelling the distributions of maize stem borers at local scale in East African mountain gradients using climatic and edaphic variables. *African Entomology* 26, 458–470. DOI: 10.4001/003.026.0458

Nagoshi, R.N. (2019) Evidence that a major subpopulation of fall armyworm found in the Western Hemisphere is rare or absent in Africa, which may limit the range of crops at risk of infestation. *PLoS ONE* 14, e0208966. DOI: 10.1371/journal.pone.0208966

Nagoshi, R.N., Goergen, G., Koffi, D., Agboka, K., Adjevi, A.K.M. *et al.* (2022) Genetic studies of fall armyworm indicate a new introduction into Africa and identify limits to its migratory behavior. *Scientific Reports* 12, 1941. DOI: 10.1038/s41598-022-05781-z

Ndolo, D., Njuguna, E., Adetunji, C.O., Harbor, C., Rowe, A. *et al.* (2019) Research and development of biopesticides: challenges and prospects. *Outlooks on Pest Management* 30, 267–276. DOI: 10.1564/v30_dec_08

Niassy, S., Agbodzavu, M.K., Kimathi, E., Mutune, B., Abdel-Rahman, E.F.M. *et al.* (2021) Bioecology of fall armyworm *Spodoptera frugiperda* (JE Smith), its management and potential patterns of seasonal spread in Africa. *PLoS ONE* 16, e0249042. DOI: 10.1371/journal.pone.0249042

Niassy, S., Agbodzavu, M.K., Mudereri, B.T., Kamalongo, D., Ligowe, I. *et al.* (2022) Performance of push–pull technology in low-fertility soils under conventional and conservation agriculture farming systems in Malawi. *Sustainability* 14, 2162. DOI: 10.3390/su14042162

Njuguna, E., Nethononda, P., Maredia, K., Mbabazi, R., Kachapulula, P. *et al.* (2021) Experiences and perspectives on *Spodoptera frugiperda* (Lepidoptera: Noctuidae) management in Sub-Saharan Africa. *Journal of Integrated Pest Management* 12, 7. DOI: 10.1093/jipm/pmab002

Ntiri, E.S., Calatayud, P.A., Van den Berg, J., Schulthess, F. and Le Ru, B.P. (2016) Influence of temperature on intra- and interspecific resource utilization within a community of lepidopteran maize stemborers. *PLoS ONE* 11, e0148735. DOI: 10.1371/journal.pone.0148735

Ntiri, E.S., Calatayud, P.-A., Musyoka, B., Van den Berg, J. and Le Ru, B.P. (2019) Spatio-temporal interactions between maize lepidopteran stemborer communities and possible implications from the recent invasion of *Spodoptera frugiperda* (Lepidoptera: Noctuidae) in sub-Saharan Africa. *Environmental Entomology* 48, 573–582. DOI: 10.1093/ee/nvz024

Nyamutukwa, S., Mvumi, B.M. and Chinwada, P. (2022) Sustainable management of fall armyworm, *Spodoptera frugiperda* (JE Smith): challenges and proposed solutions from an African perspective. *International Journal of Pest Management* 2022, 2027549. DOI: 10.1080/09670874.2022.2027549

Oben, E.O., Ntonifor, N.N., Kekeunou, S. and Abbeytakor, M.N. (2015) Farmers knowledge and perception on maize stem borers and their indigenous control methods in south western region of Cameroon. *Journal of Ethnobiology and Ethnomedicine* 11, 77. DOI: 10.1186/s13002-015-0061-z

Ong'amo, G.O., Le Ru, B., Calatayud, P.-A., Ogol, C.K.P.O. and Silvain, J.-F. (2013) Composition of stem borer communities in selected vegetation mosaics in Kenya. *Arthropod Plant Interactions* 7, 267–275. DOI: 10.1007/s11829-012-9241-0

Osae, M.Y., Frimpong, J.O., Sintim, J.O., Offei, B.K., Marri, D. *et al.* (2022) Evaluation of different rates of Ampligo insecticide against fall armyworm (*Spodoptera frugiperda* (JE Smith); Lepidoptera: Noctuidae) in the coastal savannah agroecological zone of Ghana. *Advances in Agriculture* 22, 5059865. DOI: 10.1155/2022/5059865

Overton, K., Maino, J.L., Day, R., Umina, P.A., Bett, B. *et al.* (2021) Global crop impacts, yield losses and action thresholds for fall armyworm (*Spodoptera frugiperda*): a review. *Crop Protection* 145, 105641. DOI: 10.1016/j.cropro.2021.105641

Prasanna, B.M., Bruce, A., Beyene, Y., Makumbi, D., Gowda, M. *et al.* (2022) Host plant resistance for fall armyworm management in maize: relevance, status and prospects in Africa and Asia. *Theoretical and Applied Genetics* 135, 3897–3916. DOI: 10.1007/s00122-022-04073-4

Pretty, J., Toulmin, C. and Williams, S. (2011) Sustainable intensification in African agriculture. *International Journal of Agricultural Sustainability* 9, 5–24. DOI: 10.3763/ijas.2010.0583

Rapsomanikis, G. (2015) *The Economic Lives of Smallholder Farmers: An Analysis based on Household Data from Nine Countries*. Rome, Italy: Food and Agriculture Organization of the United Nations. Available at: https://www.fao.org/3/i5251e/i5251e.pdf (accessed January 1, 2024).

Ratnadass, A., Avelino, J., Fernandes, P., Letourmy, P., Babin, R. *et al.* (2020) Synergies and tradeoffs in natural regulation of crop pests and diseases under plant species diversification. *Crop Protection* 146, 105658. DOI: 10.1016/j.cropro.2021.105658

Rwomushana, I., Bateman, M., Beale, T., Beseh, P., Cameron, K. *et al.* (2018) *Fall Armyworm: Impacts and Implications for Africa*. Wallingford, UK: CAB International. Available at: https://www.invasive-species.org/wp-content/uploads/sites/2/2019/02/FAW-Evidence-Note-October-2018.pdf (accessed 24 January 2021).

Sanou, M.E.R.E., Compaoré, I. and Sanon, A. (2023) Emergency response to the *Spodoptera frugiperda* invasion in Africa: what do maize producers in Burkina Faso think and do? *African Journal of Agricultural Research* 19, 101–112. DOI: 10.5897/AJAR2022.16219

Santpoort, R. (2020) The drivers of maize area expansion in Sub-Saharan Africa. How policies to boost maize production overlook the interests of smallholder farmers. *Land* 9, 68. DOI: 10.3390/land9030068

Sebastian, K. (2009) *Agro-ecological Zones of Africa (datasets)*. Washington, DC: International Food Policy Research Institute. Available at: http://hdl.handle.net/1902.1/22616 (accessed January 1, 2024).

Senay, S.D., Pardey, P.G., Chai, Y., Doughty, L. and Day, R. (2022) Fall armyworm from a maize multi-peril pest risk perspective. *Frontiers in Insect Science* 2, 971396. DOI: 10.3389/finsc.2022.971396

Shew, A.M., Tack, J.B., Nalley, L.L., Chaminuka, P. and Maali, S. (2021) Yield gains larger in GM maize for human consumption than livestock feed in South Africa. *Nature Food* 2, 104–109. DOI: 10.1038/s43016-021-00231-x

Sikora, R., Van den Berg, J. and Oercke, E. (2019) The big giveaway – farmers and biological constraints. In: Sikora, R.A., Terry, E., Vlek, P. and Chitja, J. (eds) *Transforming Agriculture in Southern Africa – Constraints, Technologies, Policies and Processes*. London, UK: Rutledge. Available at: https://www.taylorfrancis.com/chapters/oa-edit/10.4324/9780429401701-8/big-giveaway-richard-sikora-johnnie-van-den-berg-erich-christian-oerke (accessed April 20, 2024).

Silva, J.V., Baudron, F., Ngoma, H., Nyagumbo, I., Simutowe, E. *et al.* (2023) Narrowing maize yield gaps across smallholder farming systems in Zambia: what interventions, where, and for whom? *Agronomy for Sustainable Development* 43, 26. DOI: 10.1007/s13593-023-00872-1

Sisay, B., Simiyu, J., Malusi, P., Likhayo, P., Mendesil, E. *et al.* (2018) First report of the fall armyworm, *Spodoptera frugiperda* (Lepidoptera: Noctuidae), natural enemies from Africa. *Journal of Applied Entomology* 142, 800–804. DOI: 10.1111/jen.12534

Sisay, B., Simiyu, J., Mendesil, E., Likhayo, P., Ayalew, G. *et al.* (2019) Fall armyworm, *Spodoptera frugiperda* infestations in East Africa: assessment of damage and parasitism. *Insects* 10, 195. DOI: 10.3390/insects10070195

Sokame, B.M., Subramanian, S., Kilalo, D.C., Juma, G. and Calatayud, P.A. (2020) Larval dispersal of the invasive fall armyworm, *Spodoptera frugiperda*, the exotic stemborer *Chilo partellus*, and indigenous maize stemborers in Africa. *Entomologia Experimentalis et Applicata* 168, 322–331. DOI: 10.1111/eea.12899

Statista (2023) *Production of Maize in South Africa from 2000 to 2022*. New York: Statista. Available at: www.statista.com/statistics/1134833/production-of-maize-in-south-africa/ (accessed July 25, 2023).

Strydom, E., Erasmus, A., Du Plessis, H. and Van den Berg, J. (2019) Resistance status of *Busseola fusca* (Lepidoptera: Noctuidae) populations to single-and stacked-gene Bt maize in South Africa. *Journal of Economic Entomology* 112, 305–315. DOI: 10.1093/jee/toy306

Suguiyama, L., Haggblade, S., Huesing, J.E., Eddy, R., Whiten, S.R. *et al.* (2020) *Pesticide Emergency Use Authorization: An Underutilized Tool for Controlling Invasive Pests in Africa*, Policy Research Brief 124. East Lansing, MI: Michigan State University. Available at: https://www.canr.msu.edu/fsp/publications/policy-research-briefs/PB_124.pdf (accessed January 1, 2024).

Tabashnik, B.E., Van Rensburg, J.B.J. and Carrière, Y. (2009) Field-evolved insect resistance to Bt crops: definition, theory, and data. *Journal of Economic Entomology* 102, 2011–2025. DOI: 10.1603/029.102.0601

Tabashnik, B.E., Fabrick, J.A. and Carrière, Y. (2023) Global patterns of insect resistance to transgenic Bt crops: the first 25 years. *Journal of Economic Entomology* 116, 297–309. DOI: 10.1093/jee/toac183

Tadele, Z. (2017) Raising crop productivity in Africa through intensification. *Agronomy* 7, 22. DOI: 10.3390/agronomy7010022

Tambo, J.A., Kansiime, M.K., Mugambi, I., Rwomushana, I., Kenis, M. *et al.* (2020) Understanding smallholders' responses to fall armyworm (*Spodoptera frugiperda*) invasion: evidence from five African countries. *Science of the Total Environment* 740, 140015. DOI: 10.1016/j.scitotenv.2020.140015

Tambo, J.A., Mugambi, I., Onyango, D.O., Uzayisenga, B. and Romney, D. (2023) Using mass media campaigns to change pesticide use behaviour among smallholder farmers in East Africa. *Journal of Rural Studies* 99, 79–91. DOI: 10.1016/j.jrurstud.2023.03.001

Tamò, M., Glitho, I., Tepa-Yotto, G. and Muniappan, R. (2022) How does IPM 3.0 look like (and why do we need it in Africa)? *Current Opinion in Insect Science* 53, 100961. DOI: 10.1016/j.cois.2022.100961

Ten Berge, H.F., Hijbeek, R., van Loon, M.P., Rurinda, J., Tesfaye, K. *et al.* (2019) Maize crop nutrient input requirements for food security in sub-Saharan Africa. *Global Food Security* 23, 9–21. DOI: 10.1016/j.gfs.2019.02.001

Tepa-Yotto, G.T., Chinwada, P., Rwomushana, I., Goergen, G. and Subramanian, S. (2022) Integrated management of *Spodoptera frugiperda* 6 years post detection in Africa: a review. *Current Opinion in Insect Science* 52, 100928. DOI: 10.1016/j.cois.2022.100928

Thierfelder, C., Baudron, F., Setimela, P., Nyagumbo, I., Mupangwa, W. *et al.* (2018a) Complementary practices supporting conservation agriculture in southern Africa. A review. *Agronomy for Sustainable Development* 38, 16. DOI: 10.1007/s13593-018-0492-8

Thierfelder, C., Niassy, S., Midega, C., Sevgan, S., van der Berg, J. *et al.* (2018b) Low-cost agronomic practices and landscape management approaches to control FAW. In: Prasanna, B.M., Huesing, J.E., Eddy, R. and Peschke, V.M. (eds) *Fall Armyworm in Africa: A Guide for Integrated Pest Management*, 1st edn. Mexico, CDMX: CIMMYT, pp. 89–96. http://hdl.handle.net/10883/19204 (accessed January 1, 2024).

Tscharntke, T., Klein, A.M., Kruess, A., Steffan-Dewenter, I. and Thies, C. (2005) Landscape perspectives on agricultural intensification and biodiversity – ecosystem service management. *Ecology Letters* 8, 857–874. DOI: 10.1111/j.1461-0248.2005.00782.x

Uzayisenga, B., Bizimana, J.P., Dusengemungu, L., Karangwa, P. and Rukundo, P. (2020) Farmers' perceptions and preferences on pesticide use in the management of fall armyworm in Rwanda. In: Niassy, S., Ekesi, S., Migiro, L. and Oteino, W. (eds) *Sustainable Management of Invasive Pests in Africa*, Vol. 14. Cham, Switzerland: Springer, pp. 159–168. DOI: 10.1007/978-3-030-41083-4_13

Van den Berg, J. (2013) Socio-economic factors affecting adoption of improved agricultural practices by small scale farmers in South Africa. *African Journal of Agricultural Research* 8, 4490–4500. DOI: 10.5897/AJAR12.1025

Van den Berg, J. and du Plessis, H. (2022) Chemical control and insecticide resistance in *Spodoptera frugiperda* (Lepidoptera: Noctuidae). *Journal of Economic Entomology* 115, 1761–1771. DOI: 10.1093/jee/toac108

Van den Berg, J. and Nur, A.F. (1998) Chemical control. In: Polaszek, A. (ed.) *Cereal Stem Borers in Africa: Economic Importance, Taxonomy, Natural Enemies and Control*. Wallingford, UK: International Institute of Entomology/CABI Publications, pp. 319–332.

Van den Berg, J. and Van Hamburg, H. (2015) Trap cropping with Napier grass, *Pennisetum purpureum* (Schumach), decreases damage by maize stem borers. *International Journal of Pest Management* 61, 73–79. DOI: 10.1080/09670874.2014.999733

Van den Berg, J., Nur, A.F. and Polaszek, A. (1998) Cultural control. In: Polaszek, A. (ed.) *Cereal Stem Borers in Africa: Economic Importance, Taxonomy, Natural Enemies and Control*. Wallingford, UK: International Institute of Entomology/CABI Publications, pp. 333–347.

Van den Berg, J., Rebe, M., De Bruyn, J. and Van Hamburg, H. (2001) Developing habitat management systems for gramineous stemborers in South Africa. *International Journal of Tropical Pest Management* 21, 381–388. DOI: 10.1017/S1742758400008493

Van den Berg, J., Erasmus, A. and Van Rooyen, M. (2014) Maize. In: Prinsloo, G.L. and Uys, V.M. (eds) *Insects of Cultivated Plants and Natural Pastures in Southern Africa*. Pretoria, South Africa: Entomological Society of Southern Africa, pp. 88–118.

Van den Berg, J., Britz, C. and Du Plessis, H. (2021) Maize yield response to chemical control of *Spodoptera frugiperda* at different plant growth stages in South Africa. *Agriculture* 11, 826. DOI: 10.3390/agriculture11090826

Van den Berg, J., Greyvenstein, B. and du Plessis, H. (2022) Insect resistance management facing African smallholder farmers under climate change. *Current Opinion in Insect Science* 50, 100894. DOI: 10.1016/j.cois.2022.100894

Van Ittersum, M.K., Van Bussel, L.G., Wolf, J., Grassini, P., Van Wart, J. *et al.* (2016) Can sub-Saharan Africa feed itself? *Proceedings of the National Academy of Sciences* 113, 14964–14969. DOI: 10.1073/pnas.1610359113

Van Wyk, A., Van den Berg, J. and Van Hamburg, H. (2008) Diversity and comparative phenology of Lepidoptera on Bt and non-Bt maize in South Africa. *International Journal of Pest Management* 54, 77–87. DOI: 10.1080/09670870701523074

Visser, A. and Van den Berg, J. (2020) Bigger, faster, stronger: implications of inter-species interactions for IRM of lepidopteran pests of Bt maize in Africa. *Journal of Integrated Pest Management* 11, 16. DOI: 10.1093/jipm/pmaa014

Westbrook, J.K., Nagoshi, R.N., Meagher, R.L., Fleischer, S.J. and Jairam, S. (2016) Modeling seasonal migration of fall armyworm moths. *International Journal of Biometeorology* 60, 255–267. DOI: 10.1007/s00484-015-1022-x

Westbrook, J., Fleischer, S., Jairam, S., Meagher, R. and Nagoshi, R. (2019) Multigenerational migration of a pest insect. *Ecosphere* 10, e02919. DOI: 10.1002/ecs2.2919

Yigezu, G. and Wakgari, M. (2020) Local and indigenous knowledge of farmers management practice against fall armyworm (*Spodoptera frugiperda*) (JE Smith) (Lepidoptera: Noctuidae): a review. *Journal of Entomology and Zoology Studies* 8, 765–770.

5 Aphid Biological Control by Parasitoids in Winter Wheat and Canola in Heterogeneous Agricultural Landscapes

Norman C. Elliott,[1,*] Kristopher Giles,[2] Kristen Baum,[3] Michael J. Brewer,[4] Sarah Elzay,[5] and Tom Royer[2]

[1]US Department of Agriculture, Agricultural Research Service, Peanut and Small Grains Research Unit, Stillwater, OK, USA; [2]Department of Entomology and Plant Pathology, Oklahoma State University, Stillwater, OK, USA; [3]Department of Integrative Biology, Oklahoma State University, Stillwater, OK, USA; [4]Texas A&M AgriLife Research, Corpus Christi, TX, USA; [5]Department of Biological Sciences, Fort Hays State University, Hays, KS, USA

Abstract

In the United States (US) Great Plains winter wheat is the dominant crop grown in terms of total hectarage with winter canola grown in rotation with winter wheat. In Oklahoma there has been substantial research on effects of landscape structure on aphid parasitism in winter wheat and canola. Three hymenopteran parasitoid species, *Diaeretiella rapae*, *Aphelinus nigritus*, and *Lysiphlebus testaceipes*, parasitize aphids in wheat and canola. *Diaeretiella rapae* is the dominant parasitoid of the aphids *Brevicoryne brassicae*, *Myzus persicae*, and *Lipaphis pseudobrassicae* in winter canola. *Lysiphlebus testaceipes* is the dominant parasitoid of the aphids *Schizaphis graminum* and *Rhopalosiphum padi* in winter wheat. Aphid parasitism in winter wheat and canola is dependent on landscape structure and landcover by semi-natural lands (e.g. wetlands) and cultivated lands (e.g. summer crops). The effects of landcover types on aphid parasitism are parasitoid species specific. For example, increasing wetland landcover was associated with increasing aphid parasitism by *D. rapae* in canola but had no effect on aphid parasitism by *L. testaceipes* in wheat. Among the landscape metrics investigated only patch density was a predictor of aphid parasitism in both wheat and canola. The landcover of summer crops was a predictor of aphid parasitism by *L. testaceipes*, suggesting bridging of parasitoids from summer crops to winter wheat in late summer through early autumn. Greater parasitism of *R. padi* by *L. testaceipes* and *A. nigritus* in winter wheat where wheat was double-cropped with sorghum, compared to fields in continuous wheat production, highlights the importance of summer crops for bridging parasitoids to winter wheat. Overall, landscape management can improve aphid biological control in winter wheat and canola in Oklahoma and likely other areas of the Great Plains, with the proviso that there is no single recipe for landscape structure to enhance biological control by parasitoids.

Introduction

The North American Great Plains is approximately bounded by the Rio Grande river to the south, the confluence of the Mackenzie River and the Arctic Ocean to the north, the interior lowland and the Canadian Shield to the east, and the Rocky Mountains to the west. All or part

*Corresponding author: norman.elliott@usda.gov

 Arthropod Management and Landscape Considerations in Large-Scale Agroecosystems (eds M.J. Brewer and G.L. Hein)
DOI: 10.1079/9781800622777.0005

of ten US states (Montana, Wyoming, North Dakota, South Dakota, Nebraska, Kansas, Colorado, Oklahoma, Texas, and New Mexico) are included in the Great Plains, which constitutes about a third of the continental US land mass (Dietz and Robinson, 2023). Winter wheat, *Triticum aestivum* (L.) (Poales: Poaceae), is a major cash crop in the Great Plains, where approximately 8 million ha were planted annually from 2018 to 2020 in the Southern and Central Great Plains states of Colorado, Kansas, Nebraska, New Mexico, Oklahoma, and Texas (USDA-NASS, 2021).

Winter wheat is particularly important as a rainfed crop in the Southern and Central Great Plains where low and variable precipitation levels predominate and make production of rainfed crops with greater moisture requirements problematic (Peterson *et al.*, 1996). Winter wheat is typically planted in late summer through early autumn, grows from autumn through spring and is harvested in early summer. In the semi-arid Western Great Plains, winter wheat has historically been grown using a wheat-fallow rotation with 1 year of wheat followed by a year of fallow and then returning to wheat the next year. In eastern higher rainfall areas (generally east of the 100th meridian), winter wheat has been grown in year after year continuous production systems. Wheat only production in this area provides adequate grain yields with limited risk of crop failure. However, negative consequences of wheat only production are poor water- and land-use efficiency, high soil erosion, and high costs for controlling insect pests and weeds, especially winter annual grasses (Peterson *et al.*, 1996; Bushong *et al.*, 2012). Changing climate is also an increasing risk to wheat-only production because of changes in the mean and variability of precipitation and temperature, potentially creating a less predictable environment for wheat production and more common episodes of drought (Hatfield and Dold, 2018).

Winter canola, *Brassica napus* (L.) (Brassicales: Brassicaceae), is grown in the Southern and Central Great Plains and typically is planted in October, grows from autumn through spring, and is harvested in early summer. Over the past two decades winter canola has become an alternative winter crop for use in rotation with winter wheat. There often are increased profits associated with a wheat–canola rotation compared to continuous wheat production (DeVuyst *et al.*, 2009) and less expensive and more effective methods for weed control (Bushong *et al.*, 2011, 2012). All other factors being equal, reducing yield variability over time and making pest control more effective and less expensive are outcomes of wheat–canola crop production systems that lead to greater profitability (Bushong *et al.*, 2011, 2012).

This chapter focuses on winter wheat and winter canola grown in northcentral Oklahoma, where the majority of these crops are grown in the state. Aphid pests and their parasitoids are introduced. We then explore research on aphid biological control by parasitoids in winter wheat and canola from the perspective of agroecosystem-focused integrated pest management (IPM) (see Chapter 2, this volume). We discuss parasitoid ecology and biological control in Oklahoma winter wheat and canola and assess the prospects for affecting and predicting agronomically meaningful biological control of aphids by parasitoids in winter wheat and canola fields through knowledge of landscape structure.

Aphid Pests of Winter Crops in the Great Plains

The most common and economically important aphid species in Oklahoma winter wheat are the greenbug, *Schizaphis graminum* (Rondani) (Hemiptera: Aphididae), and bird cherry-oat aphid, *Rhopalosiphum padi* (L.) (Hemiptera: Aphididae) (Royer *et al.*, 2020). The English grain aphid, *Sitobion avenae* (F.) (Hemiptera: Aphididae), is also present (French *et al.*, 2001) but rarely causes economic injury to wheat (Royer *et al.*, 2020). In the semi-arid Western Great Plains, the Russian wheat aphid, *Diuraphis noxia* (Mordvilko) (Hemiptera: Aphididae), occurs (Michels *et al.*, 1994; Brewer *et al.*, 2005) and can be a significant pest of wheat (Chander *et al.*, 2006).

In canola the cabbage aphid, *Brevicoryne brassicae* (L.), green peach aphid, *Myzus persicae* (Sulzer), and turnip aphid, *Lipaphis pseudobrassicae* (Davis) (Hemiptera: Aphididae) all infest the crop in the Great Plains (Elliott *et al.*, 2014) and can be serious pests (Royer and Giles, 2017).

Grasslands (pasture and rangeland) are the most abundant non-cultivated landcover in the Southern Great Plains and often form the matrix within which agricultural fields are embedded. Agriculture is dominated by winter crops, primarily winter wheat with usually smaller areas of winter canola, and summer crops such as corn, *Zea mays* (L.) (Poales: Poaceae), cotton, *Gossypium hirsutum* (L.) (Malvales: Malvaceae), soybean, *Glycine max* (L.) Merr. (Fabales: Fabaceae), and sorghum, *Sorghum bicolor* (L.) Moench (Poales: Poaceae). Grasslands range from semi-natural lands having no history of cultivation and high plant species diversity to lands dominated by a single, sometimes non-native, grass species. Riparian areas, woodlands, wetlands, and other semi-natural lands are also present (Elliott *et al.*, 2018). Aphid pests of wheat and canola also occur on non-cultivated hosts. For example, *D. noxia* occurs on several widely distributed cool season grasses (Armstrong *et al.*, 1991; Pucherelli *et al.*, 2012). *Schizaphis graminum* also occurs on non-cultivated host grasses (Stoner and Kieckhefer, 1979; Wiktelius, 1987; Barro and Wallwork, 1992; Moore and Borer, 2012), as do *R. padi* and *S. avenae* (Stoner and Kieckhefer, 1979; Anstead *et al.*, 2003). Aphid species that infest canola occur on wild mustard species that exist in farm fields, roadsides, pastures, and semi-natural lands (Tamaki, 1975; Horn, 1981; Norris and Kogan, 2000). The predominance of crops and grasslands shift along the large latitudinal gradient of the Great Plains. The aphid species follow the local cropping mix. While this chapter's focus is on the mix of crops and aphids in northcentral Oklahoma, the concepts apply more broadly to the expansive cereal grain production system of the Great Plains (Fig. 5.1).

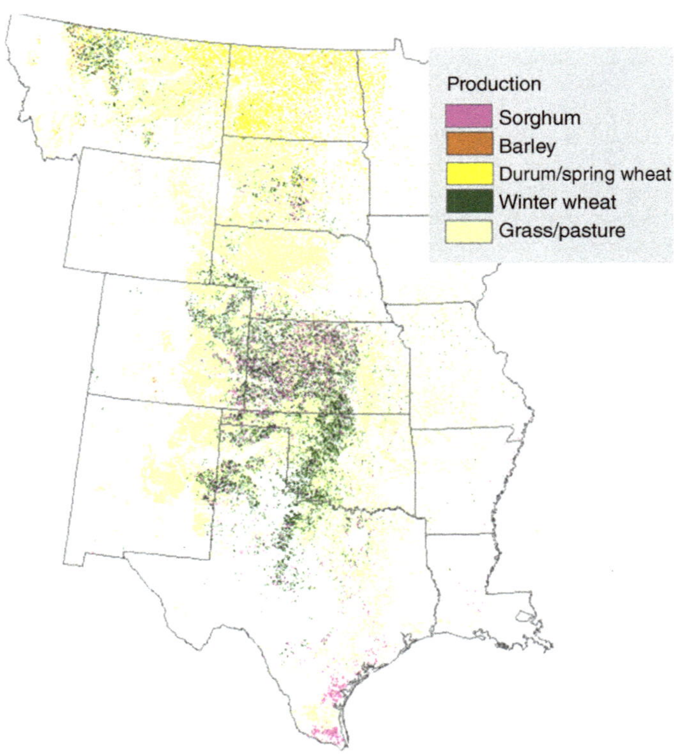

Fig. 5.1. A classified image of cropland of the North American Great Plains. Small grain production is widespread. This image was filtered to show five primary agricultural land uses and grasslands/pasture (see legend) as derived from CropScape (USDA-NASS, 2023, public domain resource).

Parasitoids of Aphids in Oklahoma Winter Wheat and Canola

A significant body of research has been completed in Oklahoma on aphid parasitism by hymenopteran parasitoids in winter crops and the role of extra-field habitats on aphid parasitism in these crops. While parasitoids can be instrumental in maintaining aphid populations in winter wheat below economically damaging densities in Oklahoma (Giles *et al.*, 2003) and elsewhere (Plećaš *et al.*, 2014), species diversity is low with a single species usually dominating the parasitoid species complex in winter wheat fields (Kring and Gilstrap, 1983; French *et al.*, 2001; Giles *et al.*, 2003; Elliott *et al.*, 2019). As is the case for winter wheat, the aphid parasitoid species complex in canola consists of only a few species with a single species consistently dominating the species complex (Elliott *et al.*, 2014, 2021, 2023).

Lysiphlebus testaceipes (Cresson) (Hymenoptera: Braconidae) is the dominant aphid parasitoid in winter wheat fields in Oklahoma and elsewhere on the Southern and Central Great Plains (Kring and Gilstrap, 1983; French *et al.*, 2001; Giles *et al.*, 2003; Colares *et al.*, 2015; Elliott *et al.*, 2018, 2021). Other aphid parasitoids are present and typically at low abundance, and several predators typically inhabit winter wheat fields (Rice and Wilde, 1991; Elliott *et al.*, 2021). During five wheat-growing seasons in Oklahoma, *L. testaceipes* and *Aphelinus nigritus* (Howard) (Hymenoptera: Aphelinidae) were the most consistently present and abundant parasitoids in wheat fields with *L. testaceipes* dominant (Elliott *et al.*, 2021).

Until recently *A. nigritus* was uncommon in wheat and sorghum and had not been observed in canola (Langston, 1970; Kring and Gilstrap, 1983; Gilstrap *et al.*, 1984; Rice and Wilde, 1991; Elliott *et al.*, 2014). However, with invasion of sorghum aphid, *Melanaphis sorghi* (Theobald) (Hemiptera: Aphididae), into the Southern and Central Great Plains in 2013 (Bowling *et al.*, 2016), abundance of *A. nigritus* on sorghum has increased (Elliott *et al.*, 2019; Maxson *et al.*, 2019; Giles *et al.*, 2021). This appears to have resulted in spillover of the species into winter wheat (Giles *et al.*, 2021; see Chapter 6, this volume).

Several predators occur in wheat in the Southern and Central Great Plains including species of Coccinellidae, Chrysopidae, Hemerobiidae, Nabidae, and Syrphidae, but they are less consistently present and often less abundant than parasitoids throughout the Great Plains (Elliott *et al.*, 2021; Brewer *et al.*, 2022; Faris *et al.*, 2022; see Chapter 8, this volume).

French *et al.* (2001) observed that the parasitoid *Diaeretiella rapae* (McIntosh) (Hymenoptera: Braconidae) occurred in canola in Oklahoma and was the dominant parasitoid of aphids in the crop. They reared *L. testaceipes* only infrequently from *L. pseudobrassicae* mummies from canola, but it was never reared from *M. persicae* mummies. Boyd and Lentz (1994) in a field study in western Tennessee (~500 km east of study areas in Oklahoma) found that *D. rapae* was the only aphid parasitoid of aphids in canola. *Diaeretiella rapae* and *A. nigritus* were the only aphid parasitoids found in canola fields in central Oklahoma (Elliott *et al.*, 2014, 2023). *Diaeretiella rapae* has a broad host and habitat range (Mackauer and Stary, 1967; Pike *et al.*, 1997; Stary, 2006). However, recent research has indicated that *D. rapae* may be more accurately described as being comprised of multiple biotypes each with a limited host range (Vaughn and Antolin, 1998; Antolin *et al.*, 2006; Navasse *et al.*, 2018; Derocles *et al.*, 2020). If this is true for *D. rapae* parasitizing aphids in canola in Oklahoma, the effect of alternate habitats on parasitism of aphids in canola might be limited. This could be exemplified in semi-natural habitats where Navasse *et al.* (2018) observed lack of host switching by *D. rapae* from aphids on the non-cultivated plant lambsquarter, *Chenopodium album* (L.) (Caryophyllale: Amaranthaceae), to aphids on cultivated Cruciferae. There are no cruciferous crops other than canola commonly grown in Oklahoma, but there are numerous species of wild crucifers (Folley, 2011), some of which are common in agroecosystems, e.g. wild mustard, *Brassica kaber* (D.C.) (Brassicales: Brassicaceae). Wild mustards grow in a variety of habitats and soil types in Oklahoma and elsewhere in the Great Plains states and are present in cultivated fields, grasslands, riparian areas, wetlands, roadsides, and disturbed sites (Folley, 2011). Some wild Cruciferae species may be higher quality hosts than canola for aphids such as *M. persicae* (Le Guigo *et al.*, 2012); therefore, presumably higher quality resources for *D. rapae*.

Sentinel plants and aphid biological control by parasitoids

Figure 5.2 illustrates some of the materials used to measure parasitism levels in studies of aphid parasitism in winter wheat and canola fields. Sentinel plants have been used to measure parasitism of aphids, for which the number of parasitized hosts on sentinel plants is assumed to be a function of the density of parasitoids searching for hosts in the field (e.g. Brewer *et al.*, 2008; Elliott *et al.*, 2018, 2023). The assumption was verified in winter wheat fields in Oklahoma by positive correlations between estimates of parasitism obtained from parasitoid accessible cages to those from sentinel plants in the same wheat field (Elliott *et al.*, 2024). The available evidence indicates that sentinel plants are effective for measuring the magnitude of aphid parasitism for comparison among fields and for comparing the parasitism at different times within a field.

Empirical studies in wheat fields have demonstrated that a threshold of roughly 20% parasitism prior to cereal aphids reaching economic injury levels will result in suppression of aphid populations by parasitoids and curtail economic losses (Giles *et al.*, 2003; Plećaš *et al.*, 2014). Elliott *et al.* (2024) in a 3-year study of winter wheat fields in northcentral Oklahoma observed that apparent parasitism (i.e. the percentage of aphids parasitized based on the ratio of mummified aphids to aphids plus mummies) in parasitoid accessible cages during spring in wheat fields differed among fields and was sometimes below and sometimes above the 20% threshold (Giles *et al.*, 2003; Plećaš *et al.*, 2014). During autumn the percentage parasitism was on average much lower than 20% (Elliott *et al.*, 2024). However, apparent parasitism may underestimate actual percentage parasitism because parasitized but still living aphids would not be accounted for. We are not aware of a known threshold for parasitism in canola at which parasitoids maintain aphid infestations below economically damaging densities. However, it is safe to assume that the number of parasitoids on sentinel canola plants is related to percentage parasitism of aphids.

Landscape Structure and Aphid Parasitoids in Winter Wheat and Canola

Influence of semi-natural lands

The spatial and temporal composition and configuration of landscape elements that contribute to agroecosystem diversity generally influence populations of natural enemies and the biological control services they provide by determining the availability of resources such as hosts and food for adults, and protection from biotic and abiotic threats (e.g. Chaplin-Kramer and Kremen, 2012; Kremen and Miles, 2012; Martin *et al.*, 2015; Redlich *et al.*, 2018; Haan *et al.*, 2020). In some agricultural landscapes, landcover of semi-natural lands is an important determinant of the level of biological control provided by insect parasitoids (e.g. Chaplin-Kramer and Kremen, 2012; Bosem Baillod *et al.*, 2017). Research in the US and elsewhere has focused on the role of extra-field habitats in the landscape matrix on parasitism of aphids in agricultural fields (e.g. Thies *et al.*, 2005; Roschewitz *et al.*, 2005; Vollhardt *et al.*, 2008; Jonsson *et al.*, 2012; Elliott *et al.*, 2018). Generally, these studies demonstrate that the landscape matrix within which an agricultural field is embedded affects parasitism of aphids in that field and that these effects can vary over time and space.

Specific to aphids in cruciferous crops, landcover of semi-natural lands has been investigated in several locations across the world with variable results. In Germany, Scheiner and Martin (2020) did not detect an effect of landcover of non-cultivated and semi-natural land on aphid parasitism by *D. rapae* in cabbage. However, in the Australian state of Victoria, Ward *et al.* (2021) found that *D. rapae* was present at low density in edge habitat surrounding canola fields from which they colonized canola. While in New South Wales in Australia *D. rapae* frequently moved from native vegetation into canola fields (Macfadyen and Muller, 2013). In Oklahoma, Elliott *et al.* (2023) found that the landcover of wetlands in the landscape nearby canola fields increased aphid parasitism by *D. rapae*.

The loss and fragmentation of natural lands is extensive in Oklahoma and is most extreme

Fig. 5.2. Materials used to obtain aphid parasitism data in winter wheat and canola fields. (A) Diagram of an emergence funnel of the type used to collect parasitoids from sentinel plants. (B) Adult parasitoids from sentinel plants in emergence funnel collection vials. (C) A sentinel wheat plant in a winter wheat field. (D) Natural enemy exclusion and parasitoid accessible cages stationed in a winter wheat field. (Artwork in the public domain (A) and photographs are the author's own: N.C. Elliott.)

where local geography, topography, and soils make widespread crop production possible. Figure 5.3 illustrates this phenomenon in northcentral Oklahoma. In Oklahoma, Elliott *et al.* (2018, 2024) demonstrated that both the composition and configuration of landscape elements surrounding wheat fields influenced parasitism of cereal aphids in wheat by *L. testaceipes*. In particular, landcover of summer crops and semi-natural lands in fine-grained landscapes were associated with greater parasitism of cereal aphids by *L. testaceipes* in winter wheat fields.

The method used by Elliott *et al.* (2018, 2023, 2024) for extracting landscape data surrounding study fields of wheat and canola is illustrated in Fig. 5.4. In northcentral Oklahoma,

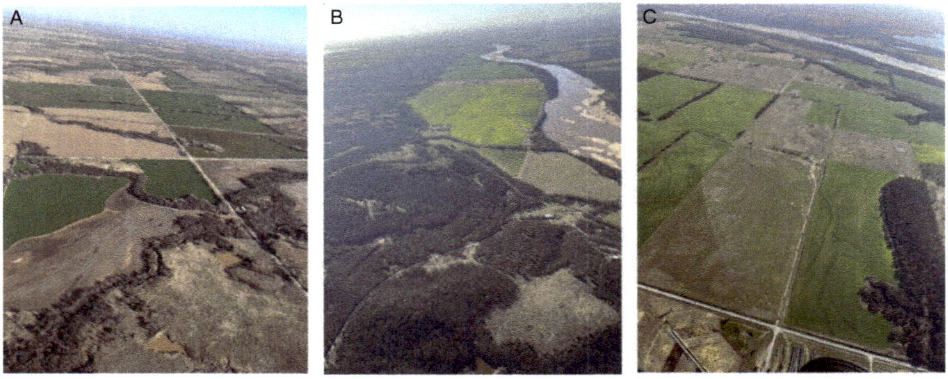

Fig. 5.3. Aerial images acquired on January 11, 2024, of the landscape 10–20 km east of the city of Perry in northcentral Oklahoma. The images illustrate a range of landcovers, from (A) land dedicated primarily to a mix of winter crops and fallow fields destined to be planted to summer crops, (B) crops inserted into primarily semi-natural land, and (C) land dedicated primarily to winter crops. (Photo credit: N.C. Elliott.)

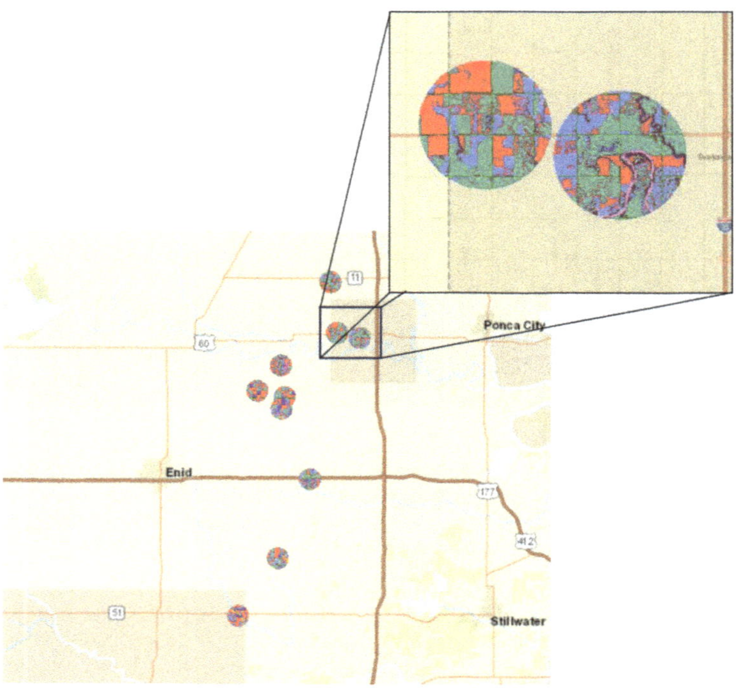

Fig. 5.4. An illustration of the cookie cutter method for extracting landscape data from circular buffers of a defined radius surrounding fields from CropScape digital maps (a background map was substituted for the CropScape base data layer to provide geographical context) (USDA-NASS, 2023, public domain resource).

particular attributes of landscapes within circular buffers of 1.5–2.5 km radius surrounding canola fields were important for determining aphid parasitism rates by *D. rapae* in canola (Elliott *et al.*, 2023). In our analysis of landscape data, wetlands were important semi-natural habitats associated with increased aphid parasitism in canola by *D. rapae*, but other semi-natural habitats (e.g. grasslands and woodlands) did not affect aphid parasitism levels in canola. Even though wetlands made up a small fraction of the total area in the circular buffers surrounding canola fields, increasing wetland landcover was associated with increasing parasitism of aphids by *D. rapae* in canola fields (Elliott *et al.*, 2023). We speculate that wetlands and their adjacent low-lying areas (which may not always be categorized as wetlands in US Department of Agriculture National Agricultural Statistics Service landcover maps) maintain green plants during the typically hot and dry summers in Oklahoma including wild mustard species, on which aphids parasitized by *D. rapae* persist. Thus, wetlands may act as source habitat for *D. rapae* from which it disperses to nearby habitats, including canola fields.

Influence of crop diversity

The importance of crop diversity to aphid biological control in crops has been demonstrated in European and midwestern US grain-producing landscapes where multiple crops can be grown contemporaneously (e.g. Gardiner *et al.*, 2009; Bosem Baillod *et al.*, 2017; Redlich *et al.*, 2018). Delayed colonization by natural enemies relative to pests in annual crops limits potential for successful biological control (Wiedenmann and Smith, 1997; Wissinger, 1997), and more specifically for biological control of aphids by *D. rapae* in crucifers (Chua, 1977; Neuville *et al.*, 2016). The role of crop diversity for increasing biological control in a particular crop is at least partially due to parasitoids colonizing and increasing in population density in fields of other crops grown in the agricultural landscape from which they disperse to the crop of interest.

In northcentral Oklahoma crop diversity occurs mainly through planting summer crops, which have limited overlap in their growing seasons with winter wheat and canola. Winter canola typically is planted in October through early November in Oklahoma (Royer and Giles, 2017) after most summer crops have matured. Therefore, there is limited potential for parasitoids to disperse directly from summer crops to canola. In fact, crop diversity was not correlated to aphid parasitism by *D. rapae* in winter canola fields (Elliott *et al.*, 2023). An increase in parasitoid recruitment in winter canola fields as the result of parasitoids dispersing from summer crops would require dispersal of adult female parasitoids to non-crop habitats where they establish and maintain populations that then bridge to winter canola. As mentioned earlier, with the exception of the percentage area of wetlands in the landscape, there was no evidence for a relationship of aphid parasitism by *D. rapae* in canola to abundance of non-cultivated landcover types in the matrix surrounding canola fields (Elliott *et al.*, 2023). Furthermore, there was no correlation between the percentage area of summer crops to the percentage area of wetlands in the landscape (Elliott *et al.*, 2023), suggesting that the positive correlation of wetlands to aphid parasitism in canola was unrelated to the presence of summer crops. Thus, there was no evidence for bridging of parasitoids from summer crops to non-cultivated vegetation and then to canola.

Unlike the situation in winter canola, crop diversity is positively correlated to aphid parasitism by *L. testaceipes* in winter wheat fields (Elliott *et al.*, 2018, 2024). About 50% of the wheat planted in Oklahoma is dual-purpose wheat planted for cattle grazing and grain production and is planted about 1 month earlier (late August through late September) than wheat used only for grain production (Edwards *et al.*, 2011). Due to an earlier planting date, the growing season of dual-purpose wheat often does overlap with the growing season of summer crops. There typically are aphid hosts for *L. testaceipes* and *A. nigritus* in both wheat and in some summer crops. For example, the aphids *S. graminum* and *R. padi* infest winter wheat and host *L. testaceipes* and *A. nigritus* (Gilstrap *et al.*, 1984; French *et al.*, 2001; Giles *et al.*, 2021). The aphids *S. graminum* and *M. sorghi* infest sorghum. *Schizaphis graminum* hosts both *L. testaceipes* and *A. nigritus* (Kring and Gilstrap, 1983; Rice and Wilde, 1991), while *M. sorghi* hosts *A. nigritus* (Maxson *et al*., 2019; Brewer *et al.*, 2022; Faris *et al.*, 2022).

Factors that contribute to early colonization of winter wheat fields by aphid parasitoids have been shown to increase their effectiveness for aphid biological control (Powell *et al.*, 1986; Vorley and Wratten, 1987; Elliott *et al.*, 2018, 2024). Colonization of Oklahoma winter wheat by aphids usually occurs soon after wheat plants emerge from the soil, and colonization by parasitoids obviously occurs after aphid colonization (Giles *et al.*, 2003; Elliott *et al.*, 2018, 2019, 2021). Both aphids and *L. testaceipes* in wheat fields continue to reproduce over winter although mortality can exceed recruitment (Giles *et al.*, 2003; Jones *et al.*, 2007; Elliott *et al.*, 2021). Because aphids and parasitoids survive and continue to reproduce in wheat fields over winter, further immigration from other habitats in spring is not necessarily required for parasitoids to effect biological control of aphids. Thus, factors that enhance earlier and greater colonization by parasitoids during autumn have potential to improve aphid biological control in winter wheat. High landcover of summer crops increases the magnitude of colonization by *L. testaceipes* and *A. nigritus* during autumn in Oklahoma winter wheat (Elliott *et al.*, 2018, 2024) and probably also leads to earlier colonization as well. As discussed previously, adult females of *D. rapae*, *L. testaceipes*, and *A. nigritus* colonize wheat fields from grassy habitats soon after aphids colonize the fields (Elliott *et al.*, 2019), and parasitoids from summer crops increase immigration levels in autumn (Elliott *et al.*, 2018, 2024).

Wheat–sorghum double-cropping and aphid biological control

As discussed earlier in this chapter and elsewhere (Powell *et al.*, 1986; Men *et al.*, 2004; Mensah and Sequeira, 2004; Tillman and Cottrell, 2012; Lopes *et al.*, 2016), potential exists for enhancing biological pest control in winter wheat by creating spatiotemporal continuity of habitats for arthropod natural enemies using alternate crops grown in winter wheat-based agroecosystems. One objective of double-cropping, and intercropping in general, is to provide alternate habitat for natural enemies to reproduce and then disperse into wheat to enhance biological control of wheat pests. Most wheat intercropping studies to date (including double-cropping) have been conducted in relatively small field plots where spatial scale is much finer than that typical of agricultural production (Lopes *et al.*, 2016). Detecting effects of adjacent habitats on recruitment of natural enemies and biological control in small experimental plantings could be masked by high rates of dispersal of natural enemies among plots. For example, adult Coccinellidae are highly mobile and exhibit short residence times in habitats (Frazer and Gilbert, 1976; Phoofolo *et al.*, 2010). Dispersal by natural enemies among habitats is an essential attribute for enhancing biological control of many insect pests including aphids. Therefore, the scale of an intercropping system must be compatible with dispersal capability and behavioral attributes of natural enemies for increased natural enemy recruitment in crops in the system to occur. This is true for northcentral Oklahoma and much of the Great Plains where agricultural fields are typically large, sometimes as large as 260 ha.

In a 2-year study in northcentral Oklahoma in autumn of 2020 and 2021, wheat was double-cropped into recently harvested production grain sorghum fields (N.C. Elliott, unpublished data). Three pairs (2020) and four pairs (2021) of wheat fields, one double-cropped with sorghum and one in continuous wheat production, were studied. To our knowledge, there was no sorghum growing within 3 km of any of the continuous wheat fields. Pairs of fields studied were located 10 km or more from other pairs, and the continuous wheat field in a pair was located 4 km or more from the sorghum–wheat double-cropped field. Parasitism was measured by placing sentinel wheat plants infested with *R. padi* in each field for 3 days and recording the number of parasitoids reared from aphids on each sentinel plant.

Aphid parasitism was greater in sorghum–wheat double-cropped fields than in continuous wheat fields. Parasitism by *L. testaceipes* was greater in sorghum–wheat double-cropped fields in both years, whereas parasitism by *A. nigritus* was greater in double-cropped fields in 2020. It is not clear whether the increased parasitism in wheat resulted from parasitoids existing on aphids on green vegetation remaining in the sorghum field after harvest, such as remnant and regenerated sorghum tillers, or from parasitoids immigrating into the fields from adjacent habitats such as Johnsongrass, *Sorghum halepense* (Poales: Poaceae), which harbored increased numbers of parasitoids that had immigrated from the adjacent sorghum field at an earlier date. Johnsongrass stands are ubiquitous in Oklahoma and commonly border agricultural fields. Johnsongrass hosts several aphid pest species of wheat, such as *M. sorghi* and *S. graminum*, both hosts of parasitoids that also parasitize aphids in wheat. The study indicated that planting wheat into the harvested sorghum resulted in greater temporal continuity of parasitoid populations to relay parasitoids from sorghum to winter wheat (Table 5.1). The main take-away is that there is potential to improve aphid biological control in winter wheat fields in the large-scale crop production systems typical of Oklahoma by appropriate crop rotation.

Figure 5.5 illustrates results of an analysis of parasitism data from winter wheat and canola in northcentral Oklahoma (Elliott *et al.*, 2023, 2024). Constrained redundancy analysis (constrained RDA) of response variables (number of parasitoids per sentinel plant) using the same predictors (percentage woodlands, percentage wetlands, percentage summer crops, Shannon's patch diversity index (SHDI), and patch density) and co-variables (season and planting year) was accomplished. The constrained RDAs for canola and wheat were significant (Fig. 5.5). The landscape variables associated with parasitism level on sentinel plants were mostly different for canola and wheat. Patch density (patch density was highly correlated with percentage wetlands which was discussed earlier) had a strong positive association with parasitism by *D. rapae* in canola and a weak positive association with parasitism by *L. testaceipes* in wheat. In contrast, SHDI had a weak positive association with parasitism by *D. rapae* in canola and a strong positive association to parasitism by *L. testaceipes* in wheat. There were no other landscape variables with similar associations to aphid parasitism in both crops. The analyses indicate that there is limited potential for a single prescription for managing landscape composition and configuration that will simultaneously improve aphid biological control by parasitoids in both canola and wheat. However, focusing management to improve biological control in a single crop, such as winter wheat, might achieve success with no apparent decrease and perhaps a slight increase in aphid biological control by parasitoids in winter canola.

Summary and Conclusions

We discussed the potential for agroecosystem-focused aphid pest management incorporating biological control by parasitoids as pest management system components. The main conclusion of this chapter is that potential exists for managing agricultural landscapes to improve aphid biological control in winter wheat and canola grown in northcentral Oklahoma, with the proviso that there is no single recipe for landscape composition and configuration to enhance biological control by parasitoids in both crops. While much of the information reviewed in this chapter comes from studies conducted in northcentral Oklahoma, there is reasonable expectation that similar relationships of landscape structure to aphid parasitism in winter wheat and canola will exist in environmentally similar areas of the Great Plains.

Aphid parasitism in both winter wheat and canola are dependent on landscape structure with both agricultural lands (such as the area of summer crops) and non-agricultural lands (such as the area of wetlands) being predictors

Table 5.1. Parasitism in wheat fields double-cropped with sorghum versus fields in continuous wheat production. (Content is the author's own: N.C. Elliott.)

Species/cropping system	2020	2021
Lysiphlebus testaceipes		
Sorghum–wheat	64.3 ± 14.6*	28.7 ± 13.1*
Continuous wheat	29.6 ± 7.8	2.83 ± 1.1
Aphelinus nigritus		
Sorghum–wheat	5.00 ± 1.61*	2.26 ± 1.21
Continuous wheat	0.13 ± 0.09	1.39 ± 0.61

Parasitism was measured on 12 sentinel plants stationed in each study field. Means were calculated for three study fields of each type (wheat with summer fallow and double-cropped wheat following sorghum) in 2020 and four study fields of each type in 2021. *, significantly greater mean parasitism rate for a particular crop pair (for each parasitoid and year).

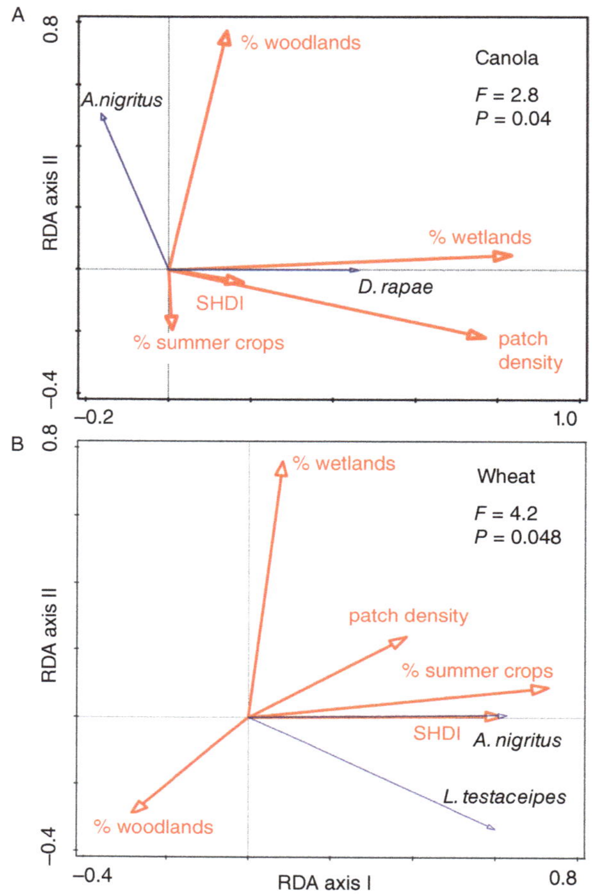

Fig. 5.5. Redundancy analysis (RDA) of aphid parasitism in (A) winter canola and (B) winter wheat on predictors measuring attributes of landscape composition and configuration using Shannon's patch diversity index (SHDI) ($P < 0.05$). (Content is the author's own: N.C. Elliott.)

of aphid parasitism in winter crops. Effects on parasitism related to the composition (the amount of various landcover types) and configuration (one or more features of the size, distribution, and shape of patches of landcover types) of the landscape matrix were specific to the parasitoid species and crop. Among the landscape metrics investigated only patch density, which measures the average area of patches of all landcover types in a landscape, was a predictor of parasitism level in both wheat and canola. Even for patch density the strength of relationships to parasitism was greater for *D. rapae* in canola than for *L. testaceipes* in wheat.

A final point is that studies conducted at coarse spatial and temporal scales, like most of those described in this chapter, serve the purpose of identifying attributes of landscapes that contribute to particular ecological phenomena of interest. To progress to a state where research has its greatest value to agroecosystem-focused IPM, additional research should focus on identifying the ecological processes occurring in key landscape elements, such as wetlands and summer crops, that lead to the patterns observed in the broad-scale studies. For example, wetlands warrant detailed study of the plant, aphid, and *D. rapae* associations that lead to greater *D. rapae* populations of canola fields. Aphid parasitoids respond to the spatial and temporal availability of resources within and among landscape elements. For example, it is the temporal pattern of host availability and perhaps of other resources in wetlands that determine their positive effect on aphid parasitism by *D. rapae* in canola. Results will help guide management of agricultural landscapes to improve resilience to aphid pests of winter crops.

References

Anstead, J.A., Burd, J.D. and Shufran, K.A. (2003) Over-summering and biotypic diversity of *Schizaphis graminum* (Homoptera: Aphididae) populations on noncultivated grass hosts. *Environmental Entomology* 32, 662–667. DOI: 10.1603/0046-225X-32.3.662

Antolin, M.F., Bjorksten, T.A. and Vaughn, T.T. (2006) Host-related fitness trade-offs in a presumed generalist parasitoid, *Diaeretiella rapae* (Hymenoptera: Aphidiidae). *Ecological Entomology* 31, 242–254. DOI: 10.1111/j.1365-2311.2006.00769.x

Armstrong, J.S., Porter, M.R. and Peairs, F.B. (1991) Alternate hosts of the Russian wheat aphid (Homoptera: Aphididae) in northeastern Colorado. *Journal of Economic Entomology* 86, 1691–1694. DOI: 10.1093/jee/84.6.1691

Barro, P.J.D. and Wallwork, H. (1992) The role of annual grasses in the phenology of *Rhopalosiphum padi* in the low rainfall belt of South Australia. *Annals of Applied Biology* 121, 455–467. DOI: 10.1111/j.1744-7348.1992.tb03456.x

Bosem Baillod, A., Tscharntke, T., Clough, Y. and Batáry. P. (2017) Landscape-scale interactions of spatial and temporal cropland heterogeneity drive biological control of cereal aphids. *Journal of Applied Ecology* 54, 1804–1813. DOI: 10.1111/1365-2664.12910

Bowling, R.D., Brewer, M.J., Kerns, D.L., Gordy, J., Seiter, N. *et al.* (2016) Sugarcane aphid (Hemiptera: Aphididae): a new pest on sorghum in North America. *Journal of Integrated Pest Management* 7, 1–13. DOI: 10.1093/jipm/pmw011

Boyd, M.L. and Lentz, G.L. (1994) Seasonal incidence of aphids and the aphid parasitoid *Diaeretiella rapae* (M'Intosh) (Hymenoptera: Aphidiidae) on rapeseed in Tennessee. *Environmental Entomology* 23, 349–353. DOI: 10.1093/ee/23.2.349

Brewer, M.J., Noma, T. and Elliott, N.C. (2005) Hymenopteran parasitoids and dipteran predators of the invasive aphid *Diuraphis noxia* after enemy introductions: temporal variation and implication for future aphid invasions. *Biological Control* 33, 315–323. DOI: 10.1016/j.biocontrol.2005.03.013

Brewer, M.J., Noma, T., Elliott, N.C., Kravchenko, A.N. and Hild, A.L. (2008) A landscape view of cereal aphid parasitoid dynamics reveals sensitivity to farm- and region-scale vegetation structure. *European Journal of Entomology* 105, 503–511. DOI: 10.14411/eje.2008.066

Brewer, M.J., Elliott, N.C., Esquivel, I.L., Jacobson, A.L., Faris, A.M. *et al.* (2022) Natural enemies, mediated by landscape and weather conditions, shape response of the sorghum agroecosystem of North America to the invasive aphid *Melanaphis sorghi*. *Frontiers in Insect Science* 2, 830997. DOI: 10.3389/finsc.2022.830997

Bushong, J., Peeper, T., Boyles, M. and Stone, A. (2011) Italian ryegrass (*Lolium perenne*), feral cereal rye (*Secale cereale*), and volunteer wheat (*Triticum aestivum*) control in winter canola. *Weed Technology* 25, 344–349. DOI: 10.1614/WT-D-10-00078.1

Bushong, J.A., Griffith, A.P., Peeper, T.F. and Epplin, F.M. (2012) Continuous winter wheat versus a winter canola–winter wheat rotation. *Agronomy Journal* 104, 324–330. DOI: 10.2134/agronj2011.0244

Chander, S., Ahuja, L.R., Peairs, F.B., Aggarwal, P.K. and Kalra, N. (2006) Modeling the effect of Russian wheat aphid, *Diuraphis noxia* (Mordvilko) and weeds in winter wheat as guide to management. *Agricultural Systems* 88, 494–513. DOI: 10.1016/j.agsy.2005.07.009

Chaplin-Kramer, R. and Kremen, C. (2012) Pest control experiments show benefits of complexity at landscape and local scales. *Ecological Applications* 22, 1936–1948. DOI: 10.1890/11-1844.1

Chua, T.H. (1977) Population studies of *Brevicoryne brassicae* (L.), its parasites and hyperparasites in England. *Population Ecology* 19, 125–139. DOI: 10.1007/BF02510944

Colares, F., Michaud, J.P., Bain, C.L. and Torres, J.B. (2015) Indigenous aphid predators show high levels of preadaptation to a novel prey, *Melanaphis sacchari* (Hemiptera: Aphididae). *Journal of Economic Entomology* 108, 2546–2555. DOI: 10.1093/jee/tov235

Derocles, S.A., Navasse, Y., Buchard, C., Plantegenest, M. and Le Ralec, A. (2020) "Generalist" aphid parasitoids behave as specialists at the agroecosystem scale. *Insects* 11, 6. DOI: 10.3390/insects11010006

DeVuyst, E.A., Epplin, F., Peeper, T.F. and Boyles, M.C. (2009) *Oklahoma Canola Systems vs. Continuous Wheat Budget Comparison*. AGEC-257. Stillwater, OK: Oklahoma State University Extension.

Dietz, J.L. and Robinson, E.B. (2023) Great Plains. *Encyclopedia Britannica*. Available at: https://www.britannica.com/place/Great-Plains (accessed January 1, 2024).

Edwards, J.T., Carver, B.F., Horn, G.W. and Payton, M.E. (2011) Impact of dual-purpose management on wheat grain yield. *Crop Science* 51, 2181–2185. DOI: 10.2135/cropsci2011.01.0043

Elliott, N.C., Backoulou, G.F., Giles, K.L. and Royer, T.A. (2014) Parasitism of aphids in canola fields in central Oklahoma. *Journal of Agricultural and Urban Entomology* 30, 59–64. DOI: 10.3954/1523-5475-30.0.59

Elliott, N.C., Brewer, M.J. and Giles, K.L (2018) Landscape context affects aphid parasitism by *Lysiphlebus testaceipes* (Hymenoptera: Aphidiinae) in wheat fields. *Environmental Entomology* 47, 803–811. DOI: 10.1093/ee/nvy035

Elliott, N.C., Giles, K.L., Brewer, M.J. and Jessie, C.N. (2019) Early season parasitism of cereal aphids in wheat fields and field borders. *Southwestern Entomologist* 44, 11–19. DOI: 10.3958/059.044.0102

Elliott, N.C., Giles, K.L., Baum, K.A. and Elzay, S.D. (2021) Quantitative assessment of aphid parasitoids and predators in central Oklahoma wheat fields during five growing seasons. *Southwestern Entomologist* 46, 833–842. DOI: 10.3958/059.046.0404

Elliott, N.C., Giles, K.L., Baum, K.A. and Elzay, S.D. and Backoulou, G.F. (2023) Role of parasitoids and landscape structure in aphid population dynamics in winter canola. *Biological Control* 186, 105330. DOI: 10.1016/j.biocontrol.2023.105330

Elliott, N.C., Giles, K.L., Baum, K.A. and Elzay, S.D. and Backoulou, G.F. (2024) Aphid parasitoids and aphid suppression in Oklahoma winter wheat fields are dependent on landscape structure. *Journal of Economic Entomology* 117, toae073. DOI: 10.1093/jee/toae073

Faris, A.M., Elliott, N.C. and Brewer, M.J. (2022) Suppression of the sugarcane aphid, *Melanaphis sacchari* Zehntner (Hemiptera: Aphididae), by resident natural enemies on susceptible and resistant sorghum hybrids. *Environmental Entomology* 51, 332–339. DOI: 10.1093/ee/nvab147

Folley, P. (2011) *The Guide to Oklahoma Wildflowers*. Iowa City, IA: University of Iowa Press.

Frazer, B.D. and Gilbert, N. (1976) Coccinellids and aphids: a quantitative study of the impact of adult ladybirds (Coleoptera: Coccinellidae) preying on field populations of pea aphids (Homoptera: Aphididae). *Journal of the Entomological Society of British Columbia* 73, 33–56.

French, B.W., Elliott, N.C., Kindler, S.D. and Arnold, D.C. (2001) Seasonal occurrence of aphids and natural enemies in wheat and associated crops. *Southwestern Entomologist* 26, 49–62.

Gardiner, M.M., Landis, D.A., Gratton, C., DiFonzo, C.D., O'Neal, M. *et al.* (2009) Landscape diversity enhances biological control of an introduced crop pest in the north-central USA. *Ecological Applications* 19, 143–154. DOI: 10.1111/j.1472-4642.2009.00563.x

Giles, K.L., Jones, D.B., Royer, T.A., Elliott, N.C. and Kindler, S.D. (2003) Development of a sampling plan in winter wheat that estimates cereal aphid parasitism levels and predicts population suppression. *Journal of Economic Entomology* 96, 975–982. DOI: 10.1093/jee/96.3.975

Giles, K.L., Elliott, N.C., Butler, H.E. and Baum, K.A. (2021) Increase in importance of *Aphelinus nigritus* (Howard) on winter crops in Oklahoma coincides with invasion of sugarcane aphid on sorghum in Oklahoma. *Southwestern Entomologist* 46, 59–68. DOI: 10.3958/059.046.0105

Gilstrap, F.E., Kring, T.J. and Brooks, G.W. (1984) Parasitism of aphids (Homoptera: Aphididae) associated with Texas sorghum. *Environmental Entomology* 13, 1613–1617. DOI: 10.1093/ee/13.6.1613

Haan, N.L., Zhang, Y. and Landis, D.A. (2020) Predicting landscape configuration effects on agricultural pest suppression. *Trends in Ecology & Evolution* 35, 175–186. DOI: 10.1016/j.tree.2019.10.003

Hatfield, J.L. and Dold, C. (2018) Agroclimatology and wheat production: coping with climate change. *Frontiers in Plant Science* 9, 224. DOI: 10.3389/fpls.2018.00224

Horn, D.J. (1981) Effect of weedy backgrounds on colonization of collards by green peach aphid, *Myzus persicae*, and its major predators. *Environmental Entomology* 10, 285–289. DOI: 10.1093/ee/10.3.285

Jones, D.B., Giles, K.L., Elliott, N.C. and Payton, M.E. (2007) Parasitism of greenbug, *Schizaphis graminum,* by the parasitoid *Lysiphlebus testaceipes,* at winter temperatures. *Environmental Entomology* 36, 1–8. DOI: 10.1603/0046-225X(2007)36[1:POGSGB]2.0.CO;2

Jonsson, M., Buckley, H.L., Case, B.S., Wratten, S.D., Hale, R.J. *et al.* (2012) Agricultural intensification drives landscape-context effects on host–parasitoid interactions in agroecosystems. *Journal of Applied Ecology* 49, 706–714. DOI: 10.1111/j.1365-2664.2012.02130.x

Kremen, C. and Miles, A. (2012) Ecosystem services in biologically diversified versus conventional farming systems: benefits, externalities, and trade-offs. *Ecology and Society* 17, 40. DOI: 10.5751/ES-05035-170440

Kring, T.J. and Gilstrap, F.E. (1983) Within-field distribution of greenbug (Homoptera: Aphididae) and its parasitoids in Texas winter wheat. *Journal of Economic Entomology* 76, 57–62. DOI: 10.1093/jee/76.1.57

Langston, D.T. (1970) Laboratory studies on the aphid parasite, *Aphelinus nigritus* Howard (Hymenoptera: Aphelinidae). MSc dissertation. Stillwater, OK: Oklahoma State University.

Le Guigo, P., Maingeneau, A. and Le Corff, J. (2012) Performance of an aphid *Myzus persicae* and its parasitoid *Diaeretiella rapae* on wild and cultivated Brassicaceae. *Journal of Plant Interactions* 7, 326–332. DOI: 10.1080/17429145.2011.628417

Lopes, T., Hatt, S., Xu, Q., Chen, J., Liu, Y. *et al.* (2016) Wheat (*Triticum aestivum* L.)-based intercropping systems for biological pest control – a review. *Pest Management Science* 72, 2193–2202. DOI: 10.1002/ps.4332

Macfadyen, S. and Muller, W. (2013) Edges in agricultural landscapes: species interactions and movement of natural enemies. *PLoS ONE* 8, e59659. DOI: 10.1371/journal.pone.0059659

Mackauer, M. and Stary, P. (1967) *World Aphidiidae (Hym. Ichneumonoidea).* Paris, France: Le Frangois.

Martin, E.A., Reineking, B., Seo, B. and Steffan-Dewenter, I. (2015) Pest control of aphids depends on landscape complexity and natural enemy interactions. *PeerJ* 3, e1095. DOI: 10.7717/peerj.1095

Maxson, E.L., Brewer, M.J., Rooney, W.L. and Woolley, J.B. (2019) Species composition and abundance of the natural enemies of sugarcane aphid, *Melanaphis sacchari* (Zehnter) (Hemiptera: Aphididae), on sorghum in Texas. *Proceedings of the Entomological Society of Washington* 121, 657–680. DOI: 10.4289/0013-8797.121.4.657

Men, X., Ge, F., Yardim, E. and Parajulee, M. (2004) Evaluation of winter wheat as a potential relay crop for enhancing biological control of cotton aphids. *BioControl* 49, 701–714. DOI: 10.1007/s10526-004-5278-z

Mensah, R.K. and Sequeira, R.V. (2004) Habitat manipulation for insect pest management in cotton cropping systems. In: Gurr, G.M., Wratten, S.D. and Altieri, M.A. (eds) *Ecological Engineering for Pest Management: Advances in Habitat Manipulation for Arthropods.* Wallingford, UK: CAB International, pp. 187–198.

Michels, G.J., Whitaker-Deerberg, R.L., Gilstrap, F.E. and McKinnon, L.K. (1994) *Releases of Exotic Parasitoids and Predators of the Russian Wheat Aphid, Diuraphis noxia (Kurdjumov), in the Texas Panhandle, 1987–1993.* Bulletin 1720. College Station, TX: Texas Agricultural Experiment Station.

Moore, S.M. and Borer, E.T. (2012) The influence of host diversity and composition on epidemiological patterns at multiple spatial scales. *Ecology* 93, 1095–1105. DOI: 10.1890/11-0086.1

Navasse, Y., Derocles, S.A.P., Plantegenest, M. and Le Ralec, A. (2018) Ecological specialization in *Diaeretiella rapae* (Hymenoptera: Braconidae: Aphidiinae) on aphid species from wild and cultivated plants. *Bulletin of Entomological Research* 108, 175–184. DOI: 10.1017/S0007485317000657

Neuville, S., Le Ralec, A., Outreman, Y. and Jaloux, B. (2016) The delay in arrival of the parasitoid *Diaeretiella rapae* influences the efficiency of cabbage aphid biological control. *BioControl* 61, 115–126. DOI: 10.1007/s10526-015-9702-3

Norris, R.F. and Kogan, M. (2000) Interactions between weeds, arthropod pests, and their natural enemies in managed ecosystems. *Weed Science* 48, 94–158. DOI: 10.1614/0043-1745(2000)048[0094:IBWAPA]2.0.CO;2

Peterson, G.A., Schlegel, A.J., Tanaka, D.L. and Jones, O.R. (1996) Precipitation use efficiency as affected by cropping and tillage systems. *Journal of Production Agriculture* 9, 180–186. DOI: 10.2134/jpa1996.0180

Phoofolo, M.W., Giles, K.L. and Elliott, N.C. (2010) Effects of relay-intercropping sorghum with winter wheat, alfalfa, and cotton on lady beetle (Coleoptera: Coccinellidae) abundance and species composition. *Environmental Entomology* 39, 763–774. DOI: 10.1603/EN09122

Pike, K.S., Starý, P., Miller, T., Allison, D., Boydston, L. *et al.* (1997) Small-grain aphid parasitoids (Hymenoptera: Aphelinidae and Aphidiidae) of Washington: distribution, relative abundance, seasonal occurrence, and key to known North American species. *Environmental Entomology* 26, 1299–1311. DOI: 10.1093/ee/26.6.1299

Plećaš, M., Gagić, V., Janković, M., Petrović-Obradović, O., Kavallieratos, N.G. *et al.* (2014) Landscape composition and configuration influence cereal aphid–parasitoid–hyperparasitoid interactions and biological control differentially across years. *Agriculture, Ecosystems & Environment* 183, 1–10. DOI: 10.1016/j.agee.2013.10.016

Powell, W., Dean, G.J. and Wilding, N. (1986) The influence of weeds on aphid-specific natural enemies in winter wheat. *Crop Protection* 5, 182–189. DOI: 10.1016/0261–2194(86)90100–6

Pucherelli, S.F., Peairs, F.B., Merrill, S.C. and Randolph, T.L. (2012) Russian wheat aphid (Hemiptera: Aphididae) reproduction and development on five noncultivated grass hosts. *Arthropod-Plant Interactions* 6, 67–73. DOI: 10.1007/s11829-011-9152-5

Redlich, S., Martin, E.A. and Steffan-Dewenter, I. (2018) Landscape-level crop diversity benefits biological pest control. *Journal of Applied Ecology* 55, 2419–2428. DOI: 10.1111/1365-2664.13126

Rice, M.E. and Wilde, G.E. (1991) Aphid predators associated with conventional- and conservation-tillage winter wheat. *Journal of the Kansas Entomological Society* 64, 245–250.

Roschewitz, I., Hücker, M., Tscharntke, T. and Thies, C. (2005) The influence of landscape context and farming practices on parasitism of cereal aphids. *Agriculture, Ecosystems & Environment* 108, 218–227. DOI: 10.1016/j.agee.2005.02.005

Royer, T.A. and Giles, K.L. (2017) The OKANOLA project: challenges in managing insect pests of canola in the southern plains. In: Reddy, G.V. (ed.) *Integrated Management of Insect Pests on Canola and Other Brassica Oilseed Crops*. Wallingford, UK: CAB International, pp. 147–156.

Royer, T.A., Giles, K., Elliott, N.C. and Zarrabi, A.A. (2020) *Small Grain Aphids in Oklahoma and their Management*. EPP-7099. Stillwater, OK: Oklahoma State University Extension.

Scheiner, C. and Martin, E.A. (2020) Spatiotemporal changes in landscape crop composition differently affect density and seasonal variability of pests, parasitoids and biological pest control in cabbage. *Agriculture, Ecosystems & Environment* 301, 107051. DOI: 10.1016/j.agee.2020.107051

Stary, P. (2006) *Aphid Parasitoids of the Czech Republic*. Praha, Czech Republic: Akademie VedCeské Republiky.

Stoner, W.N. and Kieckhefer, R.W. (1979) Survival and reproduction of four cereal aphids on certain range grasses. *Environmental Entomology* 8, 694–695. DOI: 10.1093/ee/8.4.694

Tamaki, G. (1975) Weeds in orchards as important alternate sources of green peach aphids in late spring. *Environmental Entomology* 4, 958–960. DOI: 10.1093/ee/4.6.958

Thies, C., Roschewitz, I. and Tscharntke, T. (2005) The landscape context of cereal aphid–parasitoid interactions. *Proceedings of the Royal Society B: Biological Science* 272, 203–210. DOI: 10.1098/rspb.2004.2902

Tillman, P.G. and Cottrell, T.E. (2012) Incorporating a sorghum habitat for enhancing lady beetles (Coleoptera: Coccinellidae) in cotton. *Psyche: A Journal of Entomology* 2012, 1–6. DOI: 10.1155/2012/150418

USDA-NASS (2021) *Agricultural Statistics 2021*. US Department of Agriculture, National Agricultural Statistics Service. Washington, DC: US Government Printing Office. Available at: https://downloads.usda.library.cornell.edu/usda-esmis/files/j3860694x/gq67m183k/np194m208/2021_Agricultural_Statistics_Book.pdf (accessed May 10, 2024).

USDA-NASS (2023) *CropScape – Cropland Data Layer*. Washington, DC: US Department of Agriculture, National Agricultural Statistics Service. Available at: https://nassgeodata.gmu.edu/CropScape (accessed July 20, 2023).

Vaughn, T.T. and Antolin, M.F. (1998) Population genetics of an opportunistic parasitoid in an agricultural landscape. *Heredity* 80, 152–162. DOI: 10.1046/j.1365-2540.1998.00248.x

Vollhardt, I.M., Tscharntke, T., Wäckers, F.L., Bianchi, F.J. and Thies, C. (2008) Diversity of cereal aphid parasitoids in simple and complex landscapes. *Agriculture, Ecosystems & Environment* 126, 289–292. DOI: 10.1016/j.agee.2008.01.024

Vorley, V.T. and Wratten, S.D. (1987) Migration of parasitoids (Hymenoptera: Braconidae) of cereal aphids (Hemiptera: Aphididae) between grassland, early-sown cereals and late-sown cereals in southern England. *Bulletin of Entomological Research* 77, 555–568. DOI: 10.1017/S0007485300012050

Ward, S.E., Umina, P.A., Macfadyen, S. and Hoffmann, A.A. (2021) Hymenopteran parasitoids of aphid pests within Australian grain production landscapes. *Insects* 12, 44. DOI: 10.3390/insects12010044

Wiedenmann, R.N. and Smith, J.W., Jr (1997) Attributes of natural enemies in ephemeral crop habitats. *Biological Control* 10, 16–22. DOI: 10.1006/bcon.1997.0544

Wiktelius, S. (1987) The role of grasslands in the yearly life-cycle of *Rhopalosiphum padi* (Homoptera: Aphididae) in Sweden. *Annals of Applied Biology* 110, 9–15. DOI: 10.1111/j.1744-7348.1987.tb03227.x

Wissinger, S.A. (1997) Cyclic colonization in predictably ephemeral habitats: a template for biological control in annual crop systems. *Biological Control* 10, 4–15. DOI: 10.1006/bcon.1997.0543

6 Ecology of Aphid Parasitoids in Winter Wheat Habitats of the Southern Plains: How Latitude and Crop Diversity Influence Pest Management

Kristopher Giles,[1,*] Norman C. Elliott,[2] Tom Royer,[1] Haley Butler,[1] and Nina Rudin[1]

[1]*Department of Entomology and Plant Pathology, Oklahoma State University, Stillwater, Oklahoma, USA;* [2]*US Department of Agriculture, Agricultural Research Service, Peanut and Other Field Crops Research Unit, Stillwater, OK, USA*

Abstract

Cropping systems of the United States Southern Plains have been designed over time to optimize productivity during climatic conditions that include relatively mild winters and hot dry summers. Mild winters allow producers to grow winter wheat as the primary crop, but these winters can also allow for season-long infestations of insect pests, primarily aphids. A long-standing and continuing body of research conducted in this region has focused on management of aphids in wheat, with particular focus on ecological interactions between aphids and their parasitoids. Findings from these studies revealed that (i) aphid parasitoids are susceptible to the freezing temperatures common at higher latitudes in the region, (ii) winter survival and foraging activity drive local aphid–parasitoid dynamics, and (iii) greater levels of agricultural landscape diversity in time and space, conserve parasitoids, and increase aphid regulation in winter crops.

Introduction

The United States (US) Great Plains has been a focal region for the study of cereal aphids and biological control for decades and much of the research on these topics has occurred within the vast expanses of winter wheat, *Triticum aestivum* (L.), landscapes. In particular, researchers have made great strides in describing in detail the ecology of aphid parasitoids and their role in pest suppression in the Southern Great Plains (Giles *et al.*, 2008; Royer *et al.*, 2015; Brewer *et al.*, 2019; Elliott *et al.*, 2023). In this chapter we define winter wheat cropping systems of the Southern Great Plains, aphid and parasitoid ecology within these cropping systems, and how research findings have allowed for development of more holistic management programs for aphids.

Winter wheat is one of the most important US crops generating up to US$10 billion in revenue annually, and the US Great Plains is the largest wheat-producing region (Vitale *et al.*, 2020; USDA-NASS, 2022). Winter wheat varieties are adapted to resist damage from freezing temperatures

*Corresponding author: kris.giles@okstate.edu

 Arthropod Management and Landscape Considerations in Large-Scale Agroecosystems (eds M.J. Brewer and G.L. Hein)
DOI: 10.1079/9781800622777.0006

via increases in cellular sucrose concentrations and dormancy mechanisms until warmer temperatures in spring promote tillering and jointing through grain fill (Wise *et al.*, 2011; Edwards *et al.*, 2015). A unique set of climatic conditions in the Southern Great Plains, from central Texas into the panhandle region of Texas and Oklahoma, and through all of Oklahoma into Kansas, allow winter wheat to be grown as forage for cattle, for dual purpose (cattle forage and grain production), or for grain production only (Marburger *et al.*, 2018; Panyi *et al.*, 2022; USDA-NASS, 2022). In this region approximately 6.5 million ha of winter wheat are planted annually from September through November in conventional or no-till systems. When grain is produced, harvest occurs from May through June. The success of this crop in this region is directly attributable to: (i) advanced wheat breeding programs in each state that develop adapted cultivars, (ii) research-based site-specific agronomic recommendations and enterprise budget models, and (iii) well-established pest management programs that allow producers to significantly reduce inputs while managing pests (Royer *et al.*, 2015, 2020; Vitale *et al.*, 2020; Royer and Giles, 2022a).

Production Systems: Winter Canola and Summer Sorghum

Although a large percentage of wheat in this region is planted as continuous monoculture, rotations with winter and summer crops do occur in time and space (Hawkins *et al.*, 2017; Lofton *et al.*, 2017). Winter canola, *Brassica napus* (L.), is the primary annual winter rotational crop for the region, whereas sorghum, *Sorghum bicolor* (L.), is one of the major summer rotational crops (USDA-NASS, 2022). These two crops, along with winter wheat, are the focus of this chapter primarily because they provide vital habitats and aphid host species that are utilized in time and space by aphid parasitoids important for pest suppression (Elliott *et al.*, 2014, 2018, 2023; Elkins *et al.*, 2022).

Winter canola is grown almost exclusively in Oklahoma and southern Kansas, and represents a group of rapeseed varieties selectively bred to survive prolonged low temperatures and limited moisture via vernalization in compact rosettes. Most cultivars planted are glyphosate resistant to facilitate weed control and planted from late September into October (Boyles *et al.*, 2012; Lofton *et al.*, 2017; USDA-NASS, 2022). Following winter survival, rosettes begin to bolt and produce elongate racemes of clustered flowers followed by seedpod formation. Harvest occurs from May through June like that for wheat (Lofton *et al.*, 2017). In the Southern Great Plains, winter-adapted canola is now grown in rotation with winter wheat every 1–3 years to diversify continuous wheat systems and increase on-farm net profits (DeVuyst *et al.*, 2009; Franke *et al.*, 2009; Lofton *et al.*, 2017; Elliott *et al.*, 2023). Canola acreage has increased in this region from 0 ha in 2003 to over 137,000 ha in 2014, but since it has declined to a current 7300 ha (USDA-NASS, 2022). This decline was due primarily to inconsistent production issues, reduced demand, and the lack of local oilseed processing facilities (Arnason, 2020). Currently, aphids remain the major pest problem. Aphid management in winter canola is limited to preventative insecticidal seed treatments and curative foliar insecticide applications which are used to prevent aphids from reaching economically damaging levels (Franke *et al.*, 2009; Royer and Giles, 2017, 2022b).

Sorghum is a primary summer crop in the Southern Great Plains, and approximately 80% of sorghum in the US is planted annually in Texas, Oklahoma, and Kansas (Hawkins *et al.*, 2017; Brewer *et al.*, 2019; USDA-NASS, 2022). Sorghum can be planted in the spring after the last freeze in fallow ground as a stand-alone summer crop. However, it is also commonly grown in rotation with wheat; therefore, it is planted in the early summer following wheat harvest (Hawkins *et al.*, 2017). Sorghum is a drought-tolerant, resource-conserving crop, traits that make it an essential summer grain crop in the hot and dry Southern Great Plains. Indeed, sorghum is one of the few crops that reliably produces acceptable yields despite harsh environmental conditions within the region (Elliott *et al.*, 2021; Elkins *et al.*, 2022).

Winter Wheat: Aphid Pests and Crop Damage

Stakeholders involved in winter wheat production in the Southern Great Plains have developed an extensive list of research and extension priorities related to management of insect pests, and aphids have been and remain a major concern (Keenan *et al.*, 2007; Giles and Walker, 2009; Royer *et al.*, 2015; Elliott *et al.*, 2018). Because winter wheat is one of the few sources of green vegetation throughout winter months in this region (also winter canola), it is frequently utilized by arthropods as a food source and shelter. Primary pests of winter wheat include aphids (Hemiptera: Aphididae), cutworms and armyworms (Lepidoptera: Noctuidae), Hessian fly (*Mayetiola destructor* [Say] [Diptera: Cecidomyiidae]), wire worms (Coleoptera: Elateridae), grasshoppers (Orthoptera), and mites (Acari) (Keenan *et al.*, 2007; Giles *et al.*, 2008, 2021; Chen *et al.*, 2009; Royer *et al.*, 2015; Schmid *et al.*, 2018; Royer and Giles, 2022a).

The most frequent and damaging are the aphids *Schizaphis graminum* (Rondani) (greenbug) and *Rhopalosiphum padi* (L.) (bird-cherry oat aphid (BCOA)). Severe local outbreaks are sporadic and occur approximately every 10 years (Royer *et al.*, 2020). These aphids are more likely to reach pest status when wheat is grown as expansive monocultures or when insecticides disrupt natural enemy activity (Brewer and Elliott, 2004; Giles *et al.*, 2008). Aphids are phloem feeders and divert plant nutrients for their growth and reproduction, but they can also act as vectors of plant viruses (see Chapter 7, this volume), and their honeydew can facilitate growth of sooty molds that impede photosynthesis (Royer *et al.*, 2015).

Greenbugs are unique in how they injure plants. They produce enzymes that break down cell walls and chloroplasts, causing chlorotic lesions. Continued feeding can cause leaf and root death, yield loss, and plant death (Kindler *et al.*, 2002; Royer *et al.*, 2015). In the Southern Great Plains, greenbugs are monoecious (living on the same summer and winter host plant), anholocyclic (asexual reproduction), and overwinter as nymphs or adults (Mittler and Gorder, 1991). Although not fully supported by movement studies, greenbugs are thought to survive in southern areas of the Great Plains and move northward each winter wheat growing season during fall and spring (Royer *et al.*, 2015). Indeed, Jones *et al.* (2008) documented that greenbugs will die at approximately −26°C; therefore, winter survival would not be possible in geographical areas when and where these temperatures occur. These temperatures occur during winter in the Texas Panhandle and north of Oklahoma (Giles *et al.*, 2003; Mesonet, 2023).

Economic damage caused by greenbugs in the Great Plains has been estimated to exceed US$100 million annually (Burton *et al.*, 1985; Webster, 1995; Kindler *et al.*, 2002; Giles *et al.*, 2008; Royer *et al.*, 2011). Both Burton *et al.* (1985) and Kindler *et al.* (2002) documented losses in small-plot experimental trials in Oklahoma. Kindler *et al.* (2002) documented between 0.51 and 1.17 kg/ha loss in yield per aphid-day (aphid feeding over time) depending on precipitation levels. The relationship between losses from small-plot experiments and whole-field infestations is often difficult to establish; however, these findings are currently utilized for calculation of economic thresholds for all susceptible wheat varieties (Kindler *et al.*, 2003; Royer *et al.*, 2015; Royer and Giles, 2022a).

BCOA is very common in the region, but outbreaks are rare though it is thought to be consistently more damaging to winter wheat than greenbugs (Ismail *et al.*, 2003; Hunger *et al.*, 2012; Riley-Collett, 2017). In the Southern Great Plains, BCOA is thought to be monoecious and anholocyclic, overwintering as nymphs or adults like greenbugs (Riley-Collett, 2017). Injury in this region is difficult to observe, as tissue remains green; however, the transmission of barley yellow dwarf virus is the primary concern with BCOA populations in winter wheat. This disease can cause significant damage to young plants in the fall (Hunger *et al.*, 2012; Riley-Collett, 2017; see Chapter 7, this volume).

Several studies have demonstrated the negative relationship between BCOA infestations and wheat productivity. Increasing BCOA populations inhibit forage growth, reduce grain yields, and ultimately net returns in other regions of the US (Riedell *et al.*, 1999; Ismail *et al.*, 2003). Small-plot data from Oklahoma demonstrated that grain yields were significantly influenced by BCOA infestation level over time (aphid-days) (Fig. 6.1) and suggested that moderate-heavy infestations would result in economic losses similar to greenbug losses (Kindler *et al.*, 2002, 2003; K. Giles, unpublished data).

Aphid management in wheat in the Southern Great Plains has received considerable attention (Giles *et al.*, 2000a, b, 2003, 2021; Kindler *et al.*, 2002; Royer *et al.*, 2002a, b, 2005, 2015).

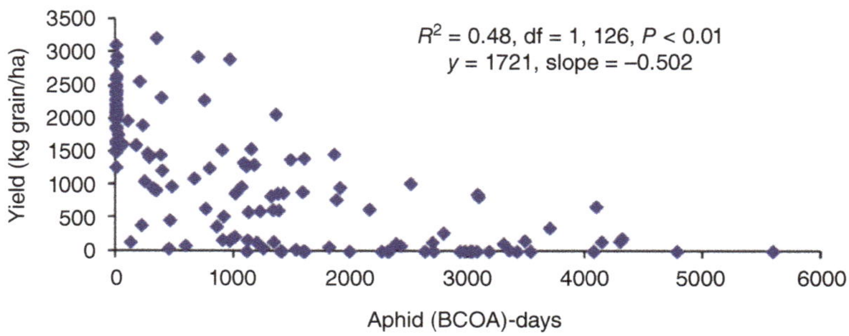

Fig. 6.1. Yield versus *Rhopalosiphum padi* (L.) (bird-cherry oat aphid (BCOA)) aphid-days for small wheat plots indicating losses similar to *Schizaphis graminum* (Rondani) (greenbug) under typical growing conditions (cultivar "Custer"). (Content is the author's own: K. Giles.)

Much of this information is incorporated into web-based or application-based decision support systems for managing aphids in winter wheat (Royer *et al.*, 2015; Giles *et al.*, 2017; Boswell-Gore, 2022). Studies on natural enemy ecology and incorporating their impact on aphids have been critical to the development of aphid management programs (Giles *et al.*, 2017).

Natural Enemies — Aphid Parasitoids in Wheat

A diversity of natural enemies is found feeding on or attacking aphids in wheat in the Southern Great Plains (Elliott *et al.*, 2021). The focus here is on aphid parasitoids that have regularly been demonstrated to maintain populations below economic injury levels (Jones, 2001; Giles *et al.*, 2003, 2017, 2021; Elliott *et al.*, 2018). Three parasitoids, *Lysiphlebus testaceipes* (Cr.), *Aphelinus nigritus* (Howard), and *Diaeretiella rapae* (McIntosh), are regularly found attacking greenbugs and *R. padi* during the fall and spring months.

Lysiphlebus testaceipes (Hymenoptera: Braconidae) is a solitary aphidiine endoparasitoid with a wide host range, and its biology and ecology are well described (Kring and Gilstrap, 1983; Royer *et al.*, 2001; Jones *et al.*, 2003, 2007, 2008; Hughes *et al.*, 2010, 2011; Jessie, 2017; Elliott *et al.*, 2018). Parasitized aphids become swollen tan-colored "mummies" once the parasitoid larva begins pupation; the developed adult parasitoid subsequently cuts a circular hole in the mummy and emerges. This wasp is the region's dominant parasitoid in winter wheat (Elliott *et al.*, 2018; Giles *et al.*, 2021), representing over 90% of parasitized aphids. The frequency of its mummies have been incorporated into aphid management guides (Giles *et al.*, 2017). The importance of *L. testaceipes* in winter wheat has been attributed to its rapid dispersal abilities among fields, its functional and numerical responses, and also its cold weather foraging activity and tolerance (Jones, 2005; Jones *et al.*, 2008; Elliott *et al.*, 2018).

The importance of *L. testaceipes* in the regulation of cereal aphids has been highlighted by several studies examining parasitism rates and ovipositional activities. In the Southern Great Plains, *L. testaceipes* can respond to low or moderate populations of cereal aphids throughout the winter wheat growing season (Jones, 2001, 2005; Giles *et al.*, 2003; Jones *et al.*, 2008). In field cage studies, *L. testaceipes* was able to suppress greenbug populations in winter wheat when initial adult female parasitoid-to-aphid ratios were less than 1 (parasitoid):50 (aphid) (Jones, 2001). Interestingly, *L. testaceipes* adults can remain active and attack aphids throughout winter months and survive temperatures as low as –8°C for 12 h, which is lower than the developmental threshold of their cereal aphid hosts (Royer *et al.*, 2001; Jones, 2005). This interesting cold weather dynamic is also observed in Europe (Hughes *et al.*, 2011). Population regulation is shifted toward the functional response by providing reliable biological control because parasitism rates increase over cold winter months when aphid populations are very slowly increasing (Giles *et al.*, 2003, Jones, 2005; Jones *et al.*, 2007).

Aphelinus nigritus (Hymenoptera: Aphelinidae) is a solitary endoparasitoid with a wide host range and is regularly found on sorghum during the summer (Gilstrap *et al.*, 1984; Maxson *et al.*, 2019; Elliott *et al.*, 2021) and at low levels on winter wheat during the fall and spring (Kring and Gilstrap, 1983; Elliott *et al.*, 2019; Giles *et al.*, 2021). This species has long been observed to cause significant mortality of cereal aphids throughout the Southern Great Plains (Gilstrap *et al.*, 1984; Maxson *et al.*, 2019; Elliott *et al.*, 2021; Brewer *et al.*, 2022; Faris *et al.*, 2022). *Aphelinus nigritus* mummies are identified by their black shiny color and elongate slender body form. The last larval instar will pupate, and the adult will emerge after cutting a circular hole in the aphid exuviae. Adult females cause mortality by both host feeding and parasitism (Maxson *et al.*, 2019).

Little is known about the life history traits of *A. nigritus*, including its functional and numerical responses. Studies on their thermal requirements for development and seasonal dynamics, including overwintering strategies, are ongoing and described below. Interestingly, in the Central Great Plains (Kansas, eastern Colorado, and southern Nebraska), *A. nigritus* is rarely found actively parasitizing aphids in winter wheat until late spring (Rice and Wilde, 1988; Giles *et al.*, 2008; Elliott *et al.*, 2019, 2021; Rudin, 2023), suggesting the possibility that this species overwinters in a dormant state, with delayed emergence, versus migrating north each spring. We hypothesize that *Aphelinus nigritus*, similar to other Aphelinidae (Hopper *et al.*, 2018), is less mobile than *L. testaceipes* (Elliott *et al.*, 2018, 2019, 2023) and must utilize dormancy strategies to survive extreme winter temperatures (Rudin, 2023). Further studies on how *A. nigritus* moves among habitats in space and time will help to describe population dynamics at increasing spatial scales including the large expanses of the Great Plains.

Diaeretiella rapae (Hymenoptera: Braconidae) is found worldwide and can regulate aphid populations in several cruciferous crops (Brassicaceae) (Elliott *et al.*, 2023). This species produces a mummy similar in appearance to *L. testaceipes*, depending upon host species, but adults are easily distinguished by wing venation. This species is also known to attack aphids in cereal crops, but levels are often extremely low or absent altogether (Elliott *et al.*, 2014, 2021; Ferguson *et al.*, 2018; Giles *et al.*, 2021). When given the choice between winter canola and winter wheat, *D. rapae* exhibits a strong preference for winter canola (Ferguson *et al.*, 2018). Prior to statewide winter canola, *Brassica napus* (L.), production in Oklahoma, *D. rapae* was only occasionally found parasitizing cereal aphids in sorghum and winter wheat, primarily BCOA and *Rhopalosiphum maidis* (Fitch) (corn leaf aphid) (Gilstrap *et al.*, 1984; Giles *et al.*, 2003, 2021; Elliott *et al.*, 2023).

Aphid–parasitoid dynamics in winter wheat

Several experimental and quantitative field studies conducted in Oklahoma from 1997 to 2002 revealed that *L. testaceipes* and *A. nigritus* effectively prevent aphid populations from reaching economic injury levels in winter wheat. In recent field studies, *L. testaceipes* and to a lesser extent *A. nigritus* were found in most Oklahoma winter wheat fields providing aphid suppression below economic injury levels (Giles *et al.*, 2021; Elliott *et al.*, 2021). The key to this impact appears to be early colonization in fall and as previously described continued foraging activity by *L. testaceipes* during relatively mild winters (Giles *et al.*, 2003; Jones, 2005; Jones *et al.*, 2007, 2008; Hughes *et al.*, 2010; Elliott *et al.*, 2018, 2019). Numbers of *D. rapae* can be found in wheat fields with proximity to canola fields where preferred aphid hosts are abundant, but this species is thought to provide little to no impact on aphid populations in wheat (Elliott *et al.*, 2014, 2023; Giles *et al.*, 2021).

Parasitoid–aphid dynamics vary dramatically throughout the Great Plains. Work conducted from Texas to Nebraska from 2002 to 2006 indicated that parasitoids provided little to no impact on aphid populations north of the Texas Panhandle and Oklahoma (Giles *et al.*, 2008). Indeed, aphid parasitoid numbers were very low in Nebraska and Colorado (Table 6.1), and they arrived in the spring well after common insect predators which have been shown to maintain aphids below economic injury levels in

Table 6.1. Average mummies per wheat tiller ± SE from 2002 to 2006 in the Central (Nebraska and Colorado) and Southern Great Plains (Texas Panhandle and Oklahoma) wheat fields. (Content is the author's own: K. Giles.)

	Nebraska and Colorado	Texas Panhandle	Oklahoma
Mummies ± SE	0.0008 ± 0.0003	0.0155 ± 0.0075	0.0218 ± 0.057
Fields sampled	130	213	147

SE, standard error

winter wheat in Kansas (Rice and Wilde, 1988; Giles *et al.*, 2008). This data can be misleading with such low overall ratios of parasitoid mummies per tiller (Table 6.1), but these samples from a large number of fields include 4 years of very low densities of aphids and parasitoids on a large number of fields which is common in the region (Giles *et al.*, 2003). Indeed, even very low ratios of parasitoids to aphids in Oklahoma result in population regulation below economic thresholds (Jones, 2001; Giles *et al.*, 2003).

Latitudinal differences in natural enemies' assemblages are important, particularly when predicting local temporal dynamics during winter wheat growing seasons. How competition among parasitoids that attack cereal aphids in the Southern Great Plains influences aphid regulation has been little studied. Jessie (2017) revealed that *L. testaceipes* is a superior competitor to *D. rapae* and is not impacted by this competition. Interestingly, recent experimental work indicates that when together in time and space, *A. nigritus* appears to regularly outcompete *L. testaceipes* for host resources (Butler *et al.*, 2022). However, these species are rarely found together on the same host patch in cereal fields and appear to avoid each other to optimize coexistence (Giles *et al.*, 2022).

Winter survival and diapause

A set of laboratory studies conducted by Jones (2005) revealed important information about survival of *L. testaceipes* at winter temperatures. This information is of particular importance throughout the latitudes of the Great Plains and where average and extreme temperatures dictate parasitoid survival. Supercooling points, the lowest temperature for survival, for young *L. testaceipes* adults and those in a late mummy stage were similar to those for greenbugs at about −26°C (Jones *et al.*, 2008). However, adults more than 12 h old had significantly higher supercooling points (−20°C to −22°C) and were thus more susceptible to dying at these temperatures. These extreme winter temperatures occur north of Oklahoma on a regular basis (Mesonet, 2023) and likely cause widespread mortality of regional aphid parasitoid populations, thus limiting survival and aphid parasitism in Kansas and more northern states (Giles *et al.*, 2008).

Just as important to survival of parasitoids are long periods of average low winter temperatures among the latitudes of the Great Plains. Jones (2005) demonstrated a lethal temperature over a 12-h period for *L. testaceipes* adults of −8°C. Interestingly, this average low regularly occurs during January just north of the Oklahoma border. Altogether, both extreme lows and average lows in winter wheat fields north of Oklahoma are significant barriers to *L. testaceipes* winter survival, thus impacting typical winter dynamics and aphid suppression in wheat. Interestingly, average winter temperatures appeared to have increased over the past 50 years at most latitudes of the Southern Great Plains (Mesonet, 2023) (Fig. 6.2) and perhaps this change will allow for greater survival and impact of *L. testaceipes* in wheat fields north of Oklahoma.

Understanding how seasonality affects biological functions is critical for describing local and regional population dynamics of important aphid parasitoids in wheat, and it can also enhance our understanding of dormancy strategies utilized by insects to survive adverse seasonal conditions (Rudin, 2023). Insect dormancy falls into one of three categories: diapause, oligopause, or quiescence. These categories represent a sequence of evolutionary adaptations for development commonly dependent upon cyclic abiotic conditions in the organism's environment and are characterized by varying degrees of arrested growth and development. The category a species falls into depends upon the nature and extent of deviation of the environmental factors

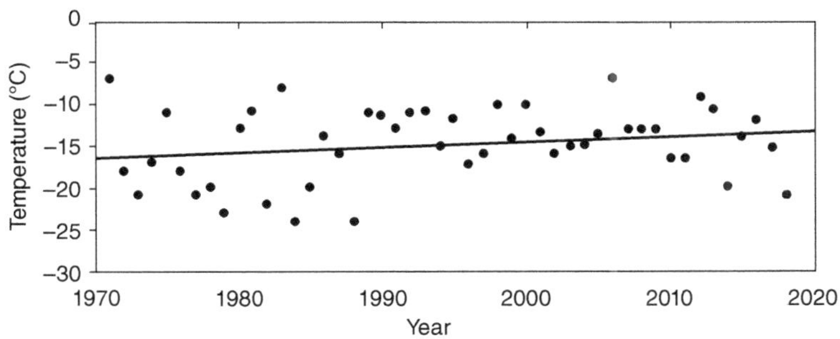

Fig. 6.2. 50-year minimum low January temperatures for Blackwell in the far north of Oklahoma (Mesonet, 2023). (Public domain resource.)

from optimal conditions due to variations in weather (Mansingh, 1971; Tatsumi and Takada, 2005, 2006).

Diapause is the most extreme category of dormancy and is characterized by an extended period of cessation of feeding, complex biochemical changes, arrested growth, and is under complete endocrine control. It is often present in populations exposed to extreme adversity, such as long, cold winters or hot, dry summers (also called aestivation). Quiescence is the least extreme category of dormancy, and is often characterized as sudden, unanticipated, and non-cyclic. It is usually short-term and may occur at any stage of the life cycle. There are only slight setbacks in growth, feeding is essential, and there are often very simple biochemical adjustments. Between these two extremes is oligopause, which can be characterized by some growth arrestment, slight and fast biochemical adjustments, and periodic dependence on food. It is generally seen in populations in areas of mild winters or summers (Mansingh, 1971; Tatsumi and Takada, 2006).

Most cases of dormancy are induced by seasonal changes in temperature and photoperiod (Mansingh, 1971), but dormancy also can be induced by other factors (Mansingh, 1971; Brodeur and McNeil, 1989). For example, Brodeur and McNeil (1989) demonstrated that diapause induction in some aphid parasitoids can be influenced by complex interactions among parasitoids, aphids, and environmental factors over multiple generations. Diapause in aphid parasitoids is induced primarily by environmental changes experienced by the developing larva (Trimble *et al.*, 1990; Bernal and González, 1996). Tatsumi and Takada (2005, 2006) reported evidence of both oligopause and diapause induction when Aphelinid larvae were exposed to lower temperatures and reduced day lengths. Interestingly, results from field studies of *A. nigritus* have suggested that winter conditions in central and northern parts of the Great Plains may be inducing delayed emergence (Giles *et al.*, 2008, 2021; Elliott *et al.*, 2019). Studies on *L. testaceipes* seasonal biology, however, have revealed no indication of diapause for this species (Hughes *et al.*, 2011).

A recent set of studies described (i) the relationship between temperature and development for *A. nigritus* and (ii) how temperature and photoperiod changes influence development and diapause induction (Rudin, 2023). The total temperature development requirement of *A. nigritus* was estimated at 216 degree-days with an average lower developmental threshold of 9.1°C. Follow-up diapause studies assessed the effect of exposure to environmental change on immatures. Results indicated that exposure to lower fluctuating temperatures and shorter days cause delays in development greater than those expected (216 degree-days) for *A. nigritus*. These temperature changes are typical of fall conditions north of Oklahoma, and delayed development associated with the onset of average winter conditions indicates that *A. nigritus* can enter oligopause. This finding is critical and may reveal why this parasitoid is rarely found or found at very low levels during the early spring; well after the accumulation of 216 degree-days required for *A. nigritus* development (Giles *et al.*, 2021). Perhaps this is a delayed emergence strategy utilized by *A. nigritus* to survive extreme winter temperatures further north in the Great Plains (Rudin, 2023).

Spatial and temporal dynamics

Giles *et al.* (2003) and Elliott *et al.* (2018) demonstrated highly variable aphid infestations among wheat fields in the Southern Great Plains, but parasitism by *L. testaceipes* was remarkably independent of host density. It was suggested that this parasitoid is conserved in agricultural landscapes, rapidly colonizes new wheat habitats, and could be described as an *r*-selected natural enemy (Elliott *et al.*, 2018). Follow-up field studies revealed that both *L. testaceipes* and *A. nigritus* rapidly colonize newly planted wheat fields and surrounding vegetation, and they are both present when initial aphid populations are at low levels (Elliott *et al.*, 2019). These parasitoid–aphid dynamics during fall appear critical for winter-long aphid suppression, particularly during mild winters (Giles *et al.*, 2003, 2017, 2021; Jones *et al.*, 2007, 2008).

It has been hypothesized that the local overlap between winter wheat, winter canola, and sorghum in space and time creates a "green bridge" for aphid parasitoids to survive year-round (Giles *et al.*, 2021; Elliott *et al.*, 2023; see also Chapter 7, this volume). Indeed, wheat and canola occur nearby in time and space, and recent studies of insect activity in winter canola revealed the most common natural enemies attacking aphids are *D. rapae* and *A. nigritus* (Jessie, 2017; Elliott *et al.*, 2023). Aphid populations act as resources in winter canola and likely attract parasitoids, and this crop may subsequently serve as a parasitoid source crop in rotational landscapes. In particular, *D. rapae* and *A. nigritus* are expected to respond positively to increasing proportions of winter canola in the landscape in Oklahoma, as these species utilize brassica aphids and disperse shorter distances compared with *L. testaceipes* (Elliott *et al.*, 2023). However, the currently limited acreage of winter canola (USDA-NASS, 2022) and its multi-year rotation with winter wheat result in highly fragmented habitats for both *D. rapae* and *A. nigritus* (Elliott *et al.*, 2023).

Winter wheat and sorghum occasionally overlap in time and space before wheat harvest in early summer but more frequently in early fall before sorghum harvest. They share many of the same aphid pests, plus a few additional crop-specific species utilized by common parasitoids (Elliott *et al.*, 2014, 2021; Maxson *et al.*, 2019; Brewer *et al.*, 2022). The most common parasitoids in sorghum are *L. testaceipes* and *A. nigritus* (Elliott *et al.*, 2021; Brewer *et al.*, 2022; Faris *et al.*, 2022). Historically, *L. testaceipes* has been the most abundant parasitoid in these systems (Kring and Gilstrap, 1983; Gilstrap *et al.*, 1984; Elliott *et al.*, 2019). However, following the invasion of the sorghum aphid, *A. nigritus* has increased in abundance in sorghum (Faris *et al.*, 2022) and wheat (Giles *et al.*, 2021).

A few studies in the region have provided important insights into how aphid parasitism in wheat and canola is influenced by landscape complexity. Elliott *et al.* (2018; N.C. Elliott unpublished data) conducted separate 3-year studies describing how local landscapes with summer crops (primarily sorghum) affect aphid parasitism in wheat during fall and spring. A consistent finding was that the presence of sorghum in landscapes, at spatial scales up to a 3 km radius, was significantly correlated with higher levels of aphid parasitism by *L. testaceipes* in wheat fields. The overlap of vegetation from late planted sorghum and winter wheat in the fall appears to provide a local green bridge for these parasitoids allowing for regulation of aphids below economic injury levels (see Chapter 5, this volume) (Fig. 6.3).

A recent 3-year study by Elliott *et al.* (2023) investigated how landscape structure affects aphid parasitism in winter canola and this study further described the biology of *D. rapae* and *A. nigritus*. Interestingly, the presence of wetlands, and perhaps aphids on wild brassicas associated with these wetlands, were found to be positively correlated with higher parasitism by *D. rapae* and aphid population suppression in canola. Local landscapes with summer crops (sorghum), however, were associated with higher parasitism and aphid population suppression by *A. nigritus*, a finding consistent with wheat studies. These studies increase our understanding of aphid parasitism in the Southern Great Plains on winter crops, and they provide the foundation for incorporation of landscape complexity and natural enemy biology into aphid management decisions (Giles *et al.*, 2017; Keasar *et al.*, 2023; see Chapter 5, this volume) (Fig. 6.3).

Aphid Parasitism and Wheat Integrated Pest Management Programs: Development and Future Goals

Wheat integrated pest management (IPM) programs in the Southern Great Plains are comprehensive and in constant change addressing multiple weed,

Fig. 6.3. Aerial view of an example agricultural landscape of winter crops (green cropland vegetation), summer crops (cropland parcels in fallow at the time of the photo), and associated wetlands. (Photo credit: N.C. Elliott.)

pathogen, and insect pests (Giles and Walker, 2009; Bynum *et al.*, 2018; Royer and Giles, 2022a; Zukoff *et al.*, 2023). Relative to insect pests, resistant varieties have successfully targeted Hessian outbreaks (Schmid *et al.*, 2018), but much of the progress has focused on aphid pests (Giles *et al.*, 2017). Indeed, collaborations within the Southern Great Plains have resulted in advancements for aphid management in wheat including development of economic thresholds (Kindler *et al.*, 2002, 2003), efficient aphid sampling plans (Giles *et al.*, 2000a, b), documenting the suppressive effects of parasitoids (Giles *et al.*, 2003), and development of smartphone applications (e.g. "Glance-N-Go") (Boswell-Gore, 2022) that are dynamic and integrate parasitism into management recommendations (Royer *et al.*, 2015; Giles *et al.*, 2017).

Use of dynamic economic thresholds are rare because they must be adaptable to stakeholder needs and achieve an acceptable level of reliability (error rates) which is based on long-term knowledge of pest–natural enemy dynamics in changing agricultural systems (Naranjo *et al.*, 2015; Giles *et al.*, 2017; Keasar *et al.*, 2023). Measured adoption rates and impacts are also difficult to access, but several US Department of Agriculture National Agricultural Statistics Service surveys (2009, 2012, 2015, 2017, 2019) (USDA-NASS, 2022) have included a question for winter wheat growers in Oklahoma on whether natural enemies were considered prior to making an insecticide application. The only available farmer resource for this decision is the "Glance-N-Go" aphid + parasitism program, an aphid sampling and threshold program that incorporates parasitoid presence and predicted impact (Boswell-Gore, 2022). Based on survey results, farmers managing 46% of wheat in Oklahoma likely utilize the dynamic thresholds from the "Glance-N-Go" program. Looking at 1995–1998, 2002 and 2004 survey data prior to delivery of "Glance-N-Go" in the Southern Great Plains (USDA-NASS, 2022), and incorporating reductions in insecticide use through 2022, a conservative cumulative estimate of US$150 million in savings since 2004 was calculated based on use of the dynamic aphid + parasitism thresholds in the Southern Great Plains.

These savings are calculated from wheat farms in Oklahoma and are not representative of other regions of the Southern Great Plains where "Glance-N-Go" is not as commonly utilized. Based on our discussion previously concerning latitude and aphid–parasitoid dynamics, we expect usefulness of dynamic aphid + parasitism thresholds in central and northern Texas wheat fields, but less use with increased latitude into Kansas where aphid parasitoids provide little aphid suppression (Giles *et al.*, 2008). Because of substantial annual climatic differences among latitudes and challenges in region-wide coordination of these programs, adaptation and adoption of dynamic thresholds, represented by the limited use of "Glance-N-Go" beyond Oklahoma remains a barrier in the Southern Great Plains (Royer *et al.*, 2015; Giles *et al.*, 2017).

Currently, because of typical climatic conditions, aphid parasitoids are unlikely to maintain consistently high populations in the northern portions of the Southern Great Plains. Therefore, wheat producers in these areas should expect little to no aphid suppression and would be advised to utilize sampling and management plans for aphids alone. However, we believe there is opportunity to improve reliability and utilization of dynamic aphid + parasitism thresholds within local winter wheat landscapes of Oklahoma and Texas. Systems such as "Glance-N-Go" (Boswell-Gore, 2022) allow for integration of insect count data, parasitism levels, and management recommendations from several wheat fields within an agricultural landscape (farm level or larger) that can be summarized over time (Giles *et al.*, 2017). This summarized data combined with results from previous landscape studies in the region (Elliott *et al.*, 2018, 2021, 2023) allow for predictions of aphid suppression within any georeferenced wheat field. Future hypothesis testing studies need to incorporate annual landscape context data into predictions about aphid parasitism and risks of economic aphid infestations in individual wheat fields. Findings from these studies would allow for production planning and estimates of pesticide costs, but also provide an opportunity to alter farm landscape design to conserve aphid parasitoids if these strategies fit within producer farm goals.

Summary and Conclusions

The wide range of climatic variability within the Southern Great Plains creates challenges for wheat producers who must manage insect pests to grow this crop of low value per unit area, even though its cumulative value across the Great Plains is substantial (Giles *et al.*, 2008). Aphid pests regularly infest wheat, and producers have utilized findings from regional research on aphid–parasitoid interactions and predictions on aphid control to optimize their local pest management strategies (Keenan *et al.*, 2007). Indeed, advanced web and application-based sampling and decision tools are available and used for management decisions in individual winter wheat fields (Giles *et al.*, 2017). However, research has revealed that aphid–parasitoid dynamics are often dictated by temperature extremes among the latitudes of the Southern Great Plains. Furthermore, recent research has also described how agricultural landscape diversity in time and space at increasing spatial scales influences local aphid–parasitoid dynamics and pest suppression. Clearly, more advanced aphid management research and implementation programs are needed that incorporate short and long-term temperature variation effects on aphid–parasitoid dynamics as well as predictions of pest suppression based on landscape diversity. These advancements will likely require innovative data management platforms that integrate weather data and cropping system diversity at increasing spatial scales allowing for dynamic pest management recommendations (see Chapter 13, this volume). As with any system, changing stakeholder goals and adoption will dictate platform use (Giles *et al.*, 2017).

References

Arnason, R. (2020) *Bloom Goes Bust on Oklahoma Canola Prospects*. The Western Producer. Available at: https://www.producer.com/crops/bloom-goes-bust-on-oklahoma-canola-prospects/ (accessed December 1, 2023).

Bernal, J.S. and González, D. (1996) Thermal requirements of *Aphelinus albipodus* (Hayat and Fatima) (Hym., Aphelinidae) on *Diuraphis noxia* (Mordwilko) (Hom., Aphididae) hosts. *Journal of Applied Entomology* 120, 631–637. DOI: 10.1111/j.1439-0418.1996.tb01663.x

Boswell-Gore, A. (2022) *Glance-N-Go App the Best Tool for Aphid Infestation*. Stillwater, OK: Oklahoma State University Extension. Available at: https://extension.okstate.edu/articles/2022/app-the-best-tool-for-aphid-infestation.html (accessed December 1, 2023)

Boyles, M., Bushong, J., Sanders, H. and Stamm, M. (2012) *Great Plains Canola Production Handbook*. MF-2734. Manhattan, KS: Agricultural Experiment Station and Cooperative Extension Service.

Brewer, M.J. and Elliott, N.C. (2004) Biological control of cereal aphids in North America and mediating effects of host plant and habitat manipulations. *Annual Review of Entomology* 49, 219–242. DOI: 10.1146/annurev.ento.49.061802.123149

Brewer, M.J., Peairs, F.B. and Elliott, N.C. (2019) Invasive cereal aphids of North America: ecology and pest management. *Annual Review of Entomology* 64, 73–93. DOI: 10.1146/annurev-ento-011118-111838

Brewer, M.J., Elliott, N.C. Esquivel, I.L. Jacobson, A.L., Faris, A.M. *et al.* (2022) Natural enemies, mediated by landscape and weather conditions, shape response of the sorghum agroecosystem of North America to the invasive aphid *Melanaphis sorghi*. *Frontiers in Insect Science* 2, 830997. DOI: 10.3389/finsc.2022.830997

Brodeur, J. and McNeil, J.N. (1989) Biotic and abiotic factors involved in diapause induction of the parasitoid, *Aphidius nigripes* (Hymenoptera: Aphidiidae). *Journal of Insect Physiology* 35, 969–974. DOI: 10.1016/0022-1910(89)90020-6

Burton, R.L., Simon, D.D., Starks, K.J. and Morrison, R.D. (1985) Seasonal damage by greenbugs (Homoptera: Aphididae) to a resistant and a susceptible variety of wheat. *Journal of Economic Entomology* 78, 395–401. DOI: 10.1093/jee/78.2.395

Butler, H.E., Giles, K. and Elliott, N. (2022) Inter- and intraspecific competition between solitary endoparasitoids drives host consumption and community structure. Presented at: *Ecology of Aphidophaga 15, Lleida, Spain.*

Bynum, E., Knutson, A., Reed, B. and Gonzales, J.D. (2018) *Managing Insect and Mite Pests of Texas Small Grains*. ENTO-084. College Station, TX: Texas A&M AgriLife Extension.

Chen, M.S., Echegaray, E., Whitworth, R.J., Wang, H., Sloderbeck, P.E. *et al.* (2009) Virulence analysis of Hessian Fly (*Mayetiola destructor*) populations from Texas, Oklahoma, and Kansas. *Journal of Economic Entomology* 102, 774–780. DOI: 10.1603/029.102.0239

DeVuyst, E.A., Epplin, F., Peeper, T. and Boyles, M.C. (2009) *Oklahoma Canola Systems vs. Continuous Wheat Budget Comparison*. AGEC-257. Stillwater, OK: Oklahoma State University Extension.

Edwards, J., Calhoun, R., Lollato, R., Cruppe, G., Carver, B. *et al.* (2015) *2015 Small Grains Variety Performance Tests*. CR-2143. Stillwater, OK: Oklahoma State University Extension.

Elkins, B.H., Eubanks, M.D., Faris, A.M., Waing, H.H. and Brewer, M.J. (2022) Landscape complexity has mixed effects on an invasive aphid and its natural enemies in sorghum agroecosystems. *Environmental Entomology* 51, 660–669. DOI: 10.1093/ee/nvac036

Elliott, N.C., Backoulou, G.F., Giles, K.L. and Royer, T.A. (2014) Aphids and parasitoids in wheat and nearby canola fields in central Oklahoma. *Southwestern Entomologist* 39, 23–28. DOI: 10.3958/059.039.0103

Elliott, N.C., Brewer, M.J. and Giles, K.L. (2018) Landscape context affects aphid parasitism by *Lysiphlebus testaceipes* in wheat fields. *Environmental Entomology* 47, 803–811. DOI: 10.1093/ee/nvy035

Elliott, N.C., Giles, K.L., Brewer, M.J. and Jessie, C.N. (2019) Early season parasitism of cereal aphids in wheat fields and field borders. *Southwestern Entomologist* 44, 11–19. DOI: 10.3958/059.044.0102

Elliott, N.C., Giles, K.L., Baum, K.A. and Elzay, S.D. (2021) Quantitative assessment of aphid parasitoids and predators in central Oklahoma wheat fields during five growing seasons. *Southwestern Entomologist* 46, 833–842. DOI: 10.3958/059.046.0404

Elliott, N.C., Giles, K.L., Baum, K.A., Elzay, S.D. and Backoulou, G.F. (2023) Role of parasitoids and landscape structure in aphid population dynamics in winter canola. *Biological Control* 186, 105330. DOI: 10.1016/j.biocontrol.2023.105330

Faris, A.M., Elliott, N.C. and Brewer, M.J. (2022) Suppression of the sugarcane aphid, *Melanaphis sacchari* (Hemiptera: Aphididae), by resident natural enemies on susceptible and resistant sorghum hybrids. *Environmental Entomology* 51, 332–339. DOI: 10.1093/ee/nvab147

Ferguson, M.E., Giles, K.L., Payton, M. and Royer, T.A. (2018) Behavioral and ovipositional response of *Diaeretiella rapae* (Hymenoptera: Braconidae) to *Rhopalosiphum padi* and *Brevicoryne brassicae* in winter wheat and winter canola. *Environmental Entomology* 47, 1517–1524. DOI: 10.1093/ee/nvy151

Franke, T.C., Kelsey, K.D. and Royer, T.A. (2009) *Pest Management Needs Assessment for Oklahoma Canola Producers*. EPP-7085. Stillwater, OK: Oklahoma State University Extension. Available at: http://pods.dasnr.okstate.edu/docushare/dsweb/Get/Document-6235/EPP-7085web.pdf (accessed December 1, 2023).

Giles, K.L. and Walker, N.W. (2009) Dissemination and impact of IPM programs in US agriculture. In: Peshin, R. and Dhawan, A.K (eds) *Integrated Pest Management: Innovation-Development Process*, Vol. 2. Dordrecht, Netherlands: Springer, pp. 481–506.

Giles, K.L., Royer, T.A., Elliott, N.C. and Kindler, S.D. (2000a) Development and validation of a binomial sequential sampling plan for the Greenbug (Homoptera: Aphididae) infesting winter wheat in the Southern Plains. *Journal of Economic Entomology* 93, 1522–1530. DOI: 10.1603/0022-0493-93.5.1522

Giles, K.L., Royer, T.A., Elliott, N.C. and Kindler, S.D. (2000b) Binomial sequential sampling of *Rhopalosiphum padi* in winter wheat in the Southern Plains. *Southwestern Entomologist* 25, 191–199.

Giles, K.L., Jones, D.B., Royer, T.A., Elliott, N.C. and Kindler, S.D. (2003) Development of a sampling plan in winter wheat that estimates cereal aphid parasitism levels and predicts population suppression. *Journal of Economic Entomology* 96, 975–982. DOI: 10.1093/jee/96.3.975

Giles, K., Hein, G. and Peairs, F. (2008) Areawide pest management of cereal aphids in dryland wheat systems of the Great Plains, USA. In: Koul, O., Cuperus, G.W. and Elliott, N.C. (eds) *Area-wide Pest Management: Theory to Implementation*. Wallingford, UK: CAB International, pp. 441–466.

Giles, K.L., McCornack, B.P., Royer, T.A. and Elliott, N.C. (2017) Incorporating biological control into IPM decision making. *Current Opinion in Insect Science* 20, 84–89. DOI: 10.1016/j.cois.2017.03.009

Giles, K.L., Elliott, N.C., Butler, H.E. and Baum, K.A. (2021) Increase in importance of *Aphelinus nigritus* (Howard) on winter crops in Oklahoma coincides with invasion of sugarcane aphid on sorghum in Oklahoma. *Southwestern Entomologist* 46, 59–68. DOI: 10.3958/059.046.0105

Giles, K.L., Elliott, N.C., Brewer, M.J., Bliss, K., Butler, H.E. *et al.* (2022) The spatial scale of co-occurrence of aphid parasitoids in sorghum: summary of multi-year population data in the US Southern Plains. Presented at: *Ecology of Aphidophaga 15, Lleida, Spain*.

Gilstrap, F.E., Kring, T.J. and Brooks, G.W. (1984) Parasitism of aphids (Homoptera: Aphididae) associated with Texas sorghum. *Environmental Entomology* 13, 1643–1617. DOI: 10.1093/ee/13.6.1613

Hawkins, S., Massey, B., Williams, E. and Hodges, M. (2017) *Grain Sorghum Production Calendar*. PSS-2113. Stillwater, OK: Oklahoma State University Extension.

Hopper, K.R., Oppenheim, S.J., Kuhn, K.L., Lanier, K., Hoelmer, K.A. *et al.* (2018) Counties not countries: variation in host specificity among populations of an aphid parasitoid. *Evolutionary Applications* 12, 815–829. DOI: 10.1111/eva.12759

Hughes, G.E., Owen, E., Sterk, G. and Bale, J.S. (2010) Thermal activity thresholds of the parasitic wasp *Lysiphlebus testaceipes* and its aphid prey: implications for the efficacy of biological control. *Physiological Entomology* 35, 373–378. DOI: 10.1111/j.1365-3032.2010.00754.x

Hughes, G.E., Sterk, G. and Bale, J.S. (2011) Thermal biology and establishment potential in temperate climates of the aphid parasitoid, *Lysiphlebus testaceipes*. *BioControl* 56, 19–33. DOI: 10.1007/s10526-010-9314-x

Hunger, B., Edwards, J., Royer, T. and Giles, K. (2012) *Effect of Planting Date and Seed Treatment on Diseases and Insect Pests of Wheat*. CR-7088. Stillwater, OK: Oklahoma State University Extension.

Ismail, E.A., Giles, K.L., Coburn, L., Royer, T.A., Hunger, R.M. *et al.* (2003) Effects of aphids, barley yellow dwarf, and grassy weeds in grazed winter wheat. *Southwestern Entomologist* 28, 121–130.

Jessie, W.P. (2017) Effects of interspecific competition and narrow-spectrum insecticides on the survival and development of *Diaeretiella rapae* in Oklahoma winter canola. PhD dissertation. Stillwater, OK: Oklahoma State University.

Jones, D.B. (2001) Natural enemy thresholds for greenbug, *Schizaphis graminum* Rondani, on winter wheat. MS thesis. Stillwater, OK: Oklahoma State University.

Jones, D.B. (2005) Winter ecology of the parasitoid *Lysiphlebus*. PhD dissertation. Stillwater, OK: Oklahoma State University.

Jones, D.B., Giles, K.L., Berberet, R.C., Royer, T.A., Elliott, N.C. *et al.* (2003) Functional responses of an introduced parasitoid and an indigenous parasitoid on greenbug at four temperatures. *Environmental Entomology* 32, 425–432. DOI: 10.1603/0046-225X-32.3.425

Jones, D.B., Giles, K.L., Elliott, N.C. and Payton, M.E. (2007) Parasitism of greenbug, *Schizaphis graminum*, by the parasitoid *Lysiphlebus testaceipes*, at winter temperatures. *Environmental Entomology* 36, 1–8. DOI: 10.1603/0046-225X(2007)36[1:POGSGB]2.0.CO;2

Jones, D.B., Giles, K.L. and Elliott, N.C. (2008) Supercooling points of *Lysiphlebus testaceipes* and its host *Schizaphis graminum*. *Environmental Entomology* 37, 1063–1068. DOI: 10.1093/ee/37.5.1063

Keasar, T., Wajnberg, E., Heimpel, G., Hardy, I.C.W., Harpaz, L.S. *et al.* (2023) Dynamic economic thresholds for insecticide applications against agricultural pests: importance of pest and natural enemy migration. *Journal of Economic Entomology* 116, 321–330. DOI: 10.1093/jee/toad019

Keenan, S.P., Giles, K.L., Burgener, P.A. and Elliott, N. (2007) Collaborating with wheat growers in demonstrating areawide integrated pest management. *Journal of Extension* 45(1), 1FEA7.

Kindler, S.D., Elliott, N.C., Giles, K.L., Royer, T.A., Fuentes-Granados, R. *et al.* (2002) Effect of greenbugs on winter wheat yield. *Journal of Economic Entomology* 95, 89–95. DOI: 10.1603/0022-0493-95.1.89

Kindler, S.D., Elliott, N.C., Giles, K.L. and Royer, T.A. (2003) Economic injury levels for the greenbug, *Schizaphis graminum*, in winter wheat. *Southwestern Entomology* 28, 163–166.

Kring, T.J. and Gilstrap, F.E. (1983) Within-field distribution of greenbug (Homoptera: Aphididae) and its parasitoids in Texas winter wheat. *Journal of Economic Entomology* 76, 57–62. DOI: 10.1093/jee/76.1.57

Lofton, J., Arnall, B. and Bushong, J. (2017) *Managing Winter Canola in Oklahoma*. PSS-2130. Stillwater, OK: Oklahoma State University Extension.

Mansingh, A. (1971) Physiological classification of dormancies in insects. *The Canadian Entomologist* 103, 983–1009. DOI: 10.4039/Ent103983-7

Marburger, D., Hunger, B., Carver, B. and Royer, T. (2018) *Wheat Variety Comparison*. PSS-2142. Stillwater, OK: Oklahoma State University Extension.

Maxson, E.L., Brewer, M.J., Rooney, W.L. and Woolley, J.B. (2019) Species composition and abundance of the natural enemies of sugarcane aphid, *Melanaphis sacchari* (Zehnter) (Hemiptera: Aphididae), on sorghum in Texas. *Proceedings of the Entomological Society of Washington* 121, 657–680. DOI: 10.4289/0013-8797.121.4.657

Mittler, T.E. and Gorder, N.K.N. (1991) Variation between clones of *Schizaphis graminum* (Homoptera: Aphididae) in the photoperiodic induction of sexual morphs. *Environmental Entomology* 20, 433–440. DOI: 10.1093/ee/20.2.433

Naranjo, S.E., Ellsworth, P.C. and Frisvold, G.B. (2015) Economic value of biological control in integrated pest management of managed plant systems. *Annual Review of Entomology* 60, 621–645. DOI: 10.1146/annurev-ento-010814-021005

Mesonet (2023) *Mesonet: Oklahoma's Weather Network*. Available at: https://www.mesonet.org/ (accessed December 1, 2023).

Panyi, A.F., Biermacher, J.T., Wade Brorsen, B., Reuter, R. and Rogers, J.K. (2022) Economics of no-tilling winter cereal pasture for growing beef cattle in Oklahoma. *Agronomy Journal* 115, 833–843. DOI: 10.1002/agj2.21281

Rice, M.E. and Wilde, G.E. (1988) Experimental evaluation of predators and parasitoids in suppressing greenbugs (Homoptera: Aphididae) in sorghum and wheat. *Environmental Entomology* 17, 836–841. DOI: 10.1093/ee/17.5.836

Riedell, W.E., Kieckhefer, R.W., Haley, S.D., Langham, M.A. and Evenson, P.D. (1999) Winter wheat responses to bird cherry-oat aphids and barley yellow dwarf virus infection. *Crop Science* 39, 158–163. DOI: 10.2135/cropsci1999.0011183X003900010025x

Riley-Collett, D.P. (2017) Factors influencing bird cherry-oat aphid (*Rhopalosiphum padi* L.) feeding behavior and fitness. PhD dissertation. Stillwater, OK: Oklahoma State University.

Royer, T.A. and Giles, K.L. (2017) The OKANOLA project: challenges in managing insect pests of canola in the southern plains. In: Reddy, G.V.P. (ed.) *Integrated Management of Insect Pests on Canola and Other Brassica Oilseed Crops*. Boston, MA: CAB International, pp. 147–156.

Royer, T.A. and Giles, K.L. (2022a) *Management of Insect and Mite Pests in Small Grains*. CR-7194. Stillwater, OK: Oklahoma State University Extension.

Royer, T.A. and Giles, K.L. (2022b) *Management of Insect and Mite Pests in Canola*. CR-7667. Stillwater, OK: Oklahoma State University Extension.

Royer, T.A., Giles, K.L., Kindler, S.D. and Elliott, N.C. (2001) Developmental response of three geographic isolates of *Lysiphlebus testaceipes* (Hymenoptera: Aphididae) to temperature. *Environmental Entomology* 30, 637–641. DOI: 10.1603/0046-225X-30.4.637

Royer, T.A., Giles, K.L. and Elliott, N.C. (2002a) *Glance 'n Go Sampling for Greenbugs in Winter Wheat: Spring Edition*. L-306. Stillwater, OK: Oklahoma State University Extension.

Royer, T.A., Giles, K L. and Elliott, N.C. (2002b) *Glance 'n Go Sampling for Greenbugs in Winter Wheat: Fall Edition*. L-307. Stillwater, OK: Oklahoma State University Extension.
Royer, T.A., Giles, K.L., Nyamanzi, T., Hunger, R.M., Krenzer, E.G. *et al.* (2005) Economic evaluation of the effects of planting date and application rate of imidacloprid for management of cereal aphids and barley yellow dwarf in winter wheat. *Journal of Economic Entomology* 98, 95–102. DOI: 10.1093/jee/98.1.95
Royer, T.A., Elliott, N.C. and Giles, K.L. (2011) Field efficacy of wintertime insecticide applications against greenbugs, *Schizaphis graminum* (Rondani) (Hemiptera: Aphididae), on winter wheat (*Triticum aestivum* L.). *Crop Protection* 30, 826–832. DOI: 10.1016/j.cropro.2011.03.002
Royer, T.A., Pendleton, B.B., Elliott, N.C. and Giles, K.L. (2015) Greenbug (Hemiptera: Aphididae) biology, ecology and management in wheat and sorghum. *Journal of Integrated Pest Management* 6, 19. DOI: 10.1093/jipm/pmv018
Royer, T.A., Giles, K.L. and Elliott, N.C. (2020) *Small Grains Aphids in Oklahoma and their Management*. EPP-7099. Stillwater, OK: Oklahoma State University Extension.
Rudin, N. (2023) Effects of temperature and light cycle on the development of *Aphelinus nigritus* (Howard) in the Southern Great Plains. MS thesis. Stillwater, OK: Oklahoma State University.
Schmid, R.B., Knutson, A., Giles, K.L. and McCornack, B.P. (2018) Hessian fly (Diptera: Cecidomyiidae) biology and management in wheat. *Journal of Integrated Pest Management* 9, 14. DOI: 10.1093/jipm/pmy008
Tatsumi, E. and Takada, H. (2005) Effects of photoperiod and temperature on adult oligopause of *Aphelinus asychis* and larval diapause of *A. albipodus* (Hymenoptera: Aphelinidae). *Applied Entomology and Zoology* 40, 447–456. DOI: 10.1303/aez.2005.447
Tatsumi, E. and Takada, H. (2006) Overwintering of the aphid parasitoids *Aphelinus asychis* and *A. albipodus* (Hymenoptera: Aphelinidae) under natural conditions in Kyoto, Japan. *Applied Entomology and Zoology* 41, 139–144. DOI: 10.1303/aez.2006.139
Trimble, R.M., Blommers, L.H.M. and Helsen, H.H.M. (1990) Diapause termination and thermal requirements for postdiapause development in *Aphelinus mali* at constant and fluctuating temperatures. *Entomologia Experimentalis et Applicata* 56, 61–69. DOI: 10.1111/j.1570-7458.1990.tb01381.x
USDA-NASS (2022) *Home Page*. Washington, DC: US Department of Agriculture, National Agricultural Statistics. Available at: https://www.nass.usda.gov/ (accessed December 1, 2023).
Vitale, J.D., Vitale, P.P., Epplin, F., Giles, K.L., Elliott, N.C. *et al.* (2020) Farm management practices used by wheat producers in the western Great Plains: estimating their productivity and profitability. *Journal of Applied Farm Economics* 3, 4. DOI: 10.7771/2331-9151.1027
Webster, J.A. (1995) *Economic Impact of the Greenbug in the Western United States: 1992–1993*. Publication 155. Stillwater, OK: Great Plains Agricultural Council.
Wise, K., Johnson, B., Mansfield, C. and Krupke, C. (2011) *Managing Wheat by Growth Stage*. ID-422. West Lafayette, IN: Purdue Extension.
Zukoff, A., McCornack, B.P. and Whitworth, R.J. (2023) *Wheat Insect Pest Management*. MF745. Manhattan, KS: KSU Cooperative Extension.

7 Cereal Aphids and Yellow Dwarf Viruses Among Wheat and Other Large-scale Agroecosystems of Central North America

Louis S. Hesler* and Shannon L. Osborne
US Department of Agriculture, Agricultural Research Service, Integrated Cropping Systems Unit, Brookings, SD, USA

Abstract

Yellow dwarf disease (YDD) of wheat and other small grains significantly reduces grain yield in all production regions, including the Great Plains of central North America. Multiple cereal aphid species (Hemiptera: Aphididae) transmit the viruses responsible for YDD, but four species—*Rhopalosiphum padi* (L.), *R. maidis* (Fitch), *Schizaphis graminum* (Rondani), and *Sitobion avenae* (F.)—are principal vectors in central North America. These aphids may be present year-round in the Southern Great Plains and annually extend their range northward in late spring by riding favorable wind currents. Model development for large-scale prediction of the severity of aphid infestation and virus transmission in wheat is ongoing but currently limited by the complexity of wheat–vector–virus dynamics. Lacking accurate predictive tools, wheat producers may scout fields to determine the need for foliar insecticide applications to limit cereal aphids and virus transmission in individual fields. Alternatively, wheat producers may proactively utilize a range of preventative measures such as insecticidal seed treatments, optimization of planting dates to limit cereal aphid infestation and virus transmission, and elimination of grassy weeds and volunteer grain plants that serve as "green bridge" hosts for cereal aphids and yellow dwarf viruses (YDVs). Producers may also use wheat cultivars, to the extent they are available, that are resistant to cereal aphids and those that are resistant or tolerant to YDD. Many of these practices are best applied regionally to achieve disease reduction in a growing region. The wheat agroecosystem hosts a rich set of aphid natural enemies that includes parasitoids, entomopathogens, and predators such as lady beetles, ground beetles, and damsel bugs. Additional research is needed to develop predictive models of aphid dispersal in relation to YDD and to enhance practices, applied within individual fields and regionally, that fortify the natural enemy complex for timely exploitation of biocontrol services against nascent cereal aphid infestations of wheat fields and thereby curtail YDV transmission.

Introduction

Many arthropod pests of agriculture are notorious for dispersing among different crops within a large geographic region (Irwin *et al.*, 2007; Pimentel, 2007). Some also spread plant pathogens during their movement among crops (Reynolds *et al.*, 2006). Aphids (Hemiptera: Aphididae) represent one group of insects that are well known for their ability to transmit a wide variety of plant-pathogenic viruses to crops (Fingu-Mabola and Francis, 2021).

Yellow dwarf disease (YDD) of wheat and other small-grain crops is a major, worldwide

*Corresponding author: louis.hesler@usda.gov

 Arthropod Management and Landscape Considerations in Large-Scale Agroecosystems (eds M.J. Brewer and G.L. Hein)
DOI: 10.1079/9781800622777.0007

plant disease syndrome (Walls *et al.*, 2019; Aradottir and Crespo-Herrera, 2021). It is ecologically and epidemiologically complex because it involves a series of viruses, various aphid vectors, and a wide range of both crop and non-crop hosts (Irwin and Thresh, 1990; Walls *et al.*, 2019). Consequently, YDD management is also complex (Walls *et al.*, 2019; Van den Eynde *et al.*, 2020).

YDD is one of the most significant diseases of wheat and other cereal crops grown in central North America, composed primarily of the expansive Great Plains prairie which has been modified for animal grazing and crop production (Pike, 1990; Hewings and Eastman, 1995; Rotenberg *et al.*, 2016; Walls *et al.*, 2019). This chapter will focus on the ecology, impact, and management of YDD in that region. In doing so, it will outline the complexities of this plant disease syndrome by discussing the multiplicity of yellow dwarf viruses (YDVs), their host plants and aphid vectors, and cover various approaches for managing the disease, from protection of individual fields to disruption of the virus threat regionally.

The Crop–Virus–Vector Pathosystem

Crops

This overview will focus principally on wheat and other small-grain crops. Wheat is the dominant small-grain crop in central North America, with a production value worth several billion US dollars each year (USDA-NASS, 2021; USDA-FAS, 2023a, b). Its production extends from Mexico and throughout the Great Plains of North America, which run approximately from the Texas, USA–Mexico border in the south, through the central US states, and northward into eastern Montana and southcentral Canada.

The classes of wheat vary by region (USDA-ERS, 2023) (Fig. 7.1). The Southern and Central Great Plains grow a preponderance of hard red and soft red winter wheat classes that are planted in the fall and harvested in late spring. The wheat belt of the Northern Great Plains is predominated by durum and hard red spring wheat classes, which are planted in late spring and harvested in late summer. In the transition between these regions, both winter and spring types of hard red winter wheat are common. In Mexico, durum and soft white wheat are planted in late fall and harvested in spring (Mulvaney *et al.*, 2014; USDA-FAS, 2023b).

Other small-grain crops in central North America include barley, oats, and triticale (USDA-NASS, 2021). These are often grown in the same areas where wheat is cultivated. Though they can predominate locally, overall acreages of oats, barley, and triticale in central North America are relatively small compared to wheat (USDA-NASS, 2021).

Corn is also a dominant crop in the Great Plains. In many areas of the Great Plains, wheat and corn may be co-dominant. In addition, sorghum can account for substantial acreage in the Southern and Central Great Plains. For this chapter, wheat is the focal crop of economic interest in terms of YDVs and cereal aphids, and corn and sorghum will be regarded as alternative hosts for the viruses and their vectors. The large-scale range and extent of production of wheat and the alternate crop and non-crop hosts and potential virus reservoirs in the landscape will be considered in the section "Management of cereal aphids and yellow dwarf viruses" after consideration of the biology and ecological interactions of the YDVs and their cereal aphid vectors.

Yellow dwarf viruses

YDVs of grasses are one of the most widespread plant-pathogenic viruses globally (Rybicki, 2015; Ali *et al.*, 2018). YDVs are single-stranded RNA viruses and are phloem limited (Irwin and Thresh, 1990; Miller and Lozier, 2022). The taxonomic classification of YDVs is dynamic (Miller and Lozier, 2022; Wei *et al.*, 2023). Currently, two genera comprise 11 known species of YDVs that infect grasses (Sõmera *et al.*, 2021; Miller and Lozier, 2022; Erickson *et al.*, 2023), and three other species have not been assigned to a genus (Walker *et al.*, 2022) (Fig. 7.2). The genus *Luteovirus* (Family *Tombusviridae*) includes seven species that are collectively known under the name barley YDV (BYDV). The genus *Polerovirus* (Family *Poleroviridae*) includes four species that are classified under the name cereal YDV (CYDV), with

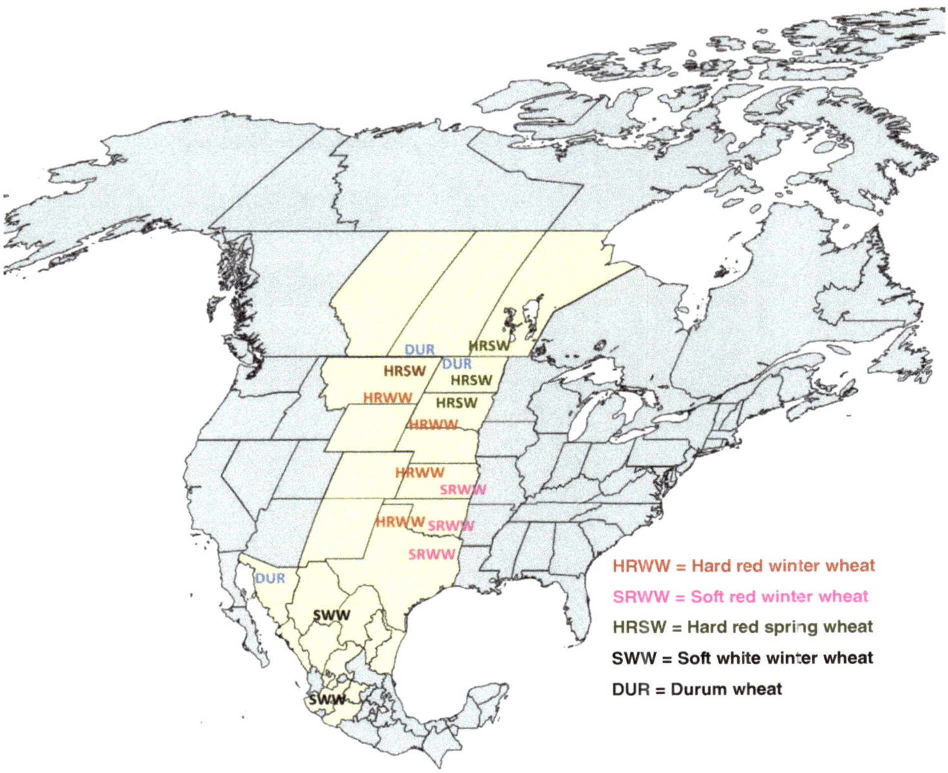

Fig. 7.1. Main wheat-producing states and provinces of central North America and the classes of wheat grown. (Content is the author's own: L.S. Hesler.)

maize YDV (MYDV)-RMV) retaining the acronym that was established before its classification as a *Polerovirus* (Miller and Lozier, 2022).

YDVs are widespread geographically and have a very broad host range. Collectively, YDVs infect small grains, corn, sorghum, many grassy weeds, and over 150 species of non-crop grasses (D'Arcy, 1995). Other viruses with "yellow dwarf" in their name (e.g. potato yellow dwarf, onion yellow dwarf) do not impact small-grain crops (Miller and Lozier, 2022).

YDVs can produce various symptoms (Ali *et al.*, 2018; Walls *et al.*, 2019). As the name implies, they typically cause a yellowing of leaf tissue (Fig. 7.3) and other discoloring and streaking, although leaf reddening often occurs in oats and in some wheat varieties. These viruses also cause dwarfing or stunting of plants. Smaller plants typically produce fewer tillers. In turn, these tillers may also have less seed and those seeds may have reduced seed mass. Symptoms of YDD generally intensify with cooler weather (Ali *et al.*, 2018; Walls *et al.*, 2019).

Currently, there is no curative or antiviral treatment for YDV infections in plants. Once infected, a plant may suffer grain yield loss and serve as a source of inoculum for aphid vectors (Hewings and Eastman, 1995; Ali *et al.*, 2018). Yield losses in wheat have been recorded up to 80% and are greatest when plants are infected as seedlings, but generally yield loss is much less (Pike, 1990). Obtaining accurate data on yield loss from YDVs is challenging because of simultaneous infection with other disease organisms (Rotenberg *et al.*, 2016). Yield loss due to YDVs in the Great Plains is estimated to peak at about 15–22% but is typically 5% or less in any given year (Pike, 1990; Rotenberg *et al.*, 2016; Walls

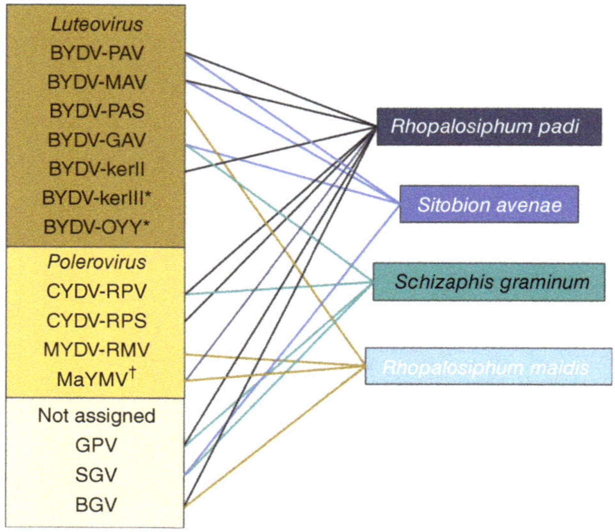

Fig. 7.2. Schematic depicting the currently known associations among yellow dwarf viruses (YDVs; see main text) (left) and the four main aphid vectors (right). *, The vector of *Luteovirus* BYDV-kerIII is unknown (Aradottir and Crespo-Herrera, 2021), and the vector of *Luteovirus* BYDV-OYY is unknown (Sõmera *et al.*, 2021). †, maize yellow mosaic virus (MaYMV) (also known as MYDV-RMV-II). (Content is the author's own: L.S. Hesler.)

Fig. 7.3. Wheat plants showing yellow dwarf disease symptoms. (A) Early infection symptoms showing leaf yellowing progressing from leaf tip downward. (B) Cluster of tillers showing leaf tip yellowing in canopy. (C) Chlorotic wheat leaves of later infection contrast sharply with the deep green of normal, healthy wheat leaves. (Photo credits: (A, B) Stephen Wegulo, University of Nebraska; (C) Keith Weller, USDA, Agricultural Research Service. Photographs are in public domain.)

et al., 2019). Occasionally, plants positive for YDV infection do not show yield loss compared with non-infected plants (Rotenberg *et al.*, 2016).

Vectors of yellow dwarf viruses

Aphids (Hemiptera: Aphididae) are the only known vectors of YDVs, although the vector of BYDV-kerIII is unknown (Aradottir and Crespo-Herrera, 2021). Many of the YDV vectors are known generally as "cereal aphids" because they use cereal crops and other grasses as hosts for most (holocyclic species) or perhaps all of their life cycle (anholocyclic species). Many species of cereal aphids cause economic damage in cereals through their feeding, irrespective of YDV (Hesler *et al.*, 2018; Aradottir and Crespo-Herrera, 2021).

More than 20 species of cereal aphids infest small-grain plants and other cereals in central North America (Halbert and Voegtlin, 1995). Four of these are the main vectors of YDVs in central North America (Hewings and Eastman, 1995; Wegulo and Hein, 2013; Hesler *et al.*, 2018): *Rhopalosiphum padi* (L.), the bird cherry-oat aphid (Fig. 7.4); *Sitobion avenae* (F.), the English grain aphid; *Schizaphis graminum* (Rondani), the greenbug; and *Rhopalosiphum maidis* (Fitch), the corn leaf aphid. Figure 7.2 depicts the complex relationships among these four aphids and YDV species. In particular, notice the higher degree of virus associations with *R. padi*, suggesting its centrality in YDV epidemiology (Walls *et al.*, 2019; Aradottir and Crespo-Herrera, 2021; Jones *et al.*, 2021).

Diuraphis noxia (Kurdjumov), the Russian wheat aphid, is a major pest of wheat in the West-central Great Plains (Porter *et al.*, 2009). However, it is not an efficient YDV vector (Damsteegt *et al.*, 1992; Halbert *et al.*, 1992) and thus not considered important in the epidemiology of YDV in North America (Halbert *et al.*, 1992). Two other cereal aphids are newly invasive to central North America. *Melanaphis sorghi* (Theobald), the sorghum aphid, is a major pest of grain, forage, and sweet sorghum. It also infests several weedy grass species (Brewer *et al.*, 2019; see Chapter 8, this volume). However, it is not known as a pest of wheat in North America (Harris-Shultz *et al.*, 2019; Armstrong *et al.*, 2021). *Sipha maydis* (Passerini), the hedgehog grain aphid, infests wheat and other small-grain plants (Puterka *et al.*, 2019), but its role as a YDV vector is unclear.

Aphids feed from the phloem with piercing-sucking mouthparts, setting them up as ideal vectors for YDVs. YDVs are termed "persistent," meaning that they are retained indefinitely within the body of the aphids and transmitted via the saliva as the aphid feeds on a grassy plant (Aradottir and Crespo-Herrera, 2021). Aphids may acquire YDVs during their various developmental stages

Fig. 7.4. *Rhopalosiphum padi*, the bird cherry-oat aphid: (A) late-stage nymph with wing buds, (B) winged viviparous female, and (C) wingless viviparous female. Images of aphids are at different scales. (Photo credit: Eric Beckendorf, USDA, Agricultural Research Service. Photographs are in the public domain.)

(Burgess *et al.*, 1999) and require a minimum acquisition access period for YDV that ranges between 15 min and 3 h (Gray *et al.*, 1991). YDV transmission may be quite rapid. Jiménez *et al.* (2020) discovered that BYDV can be inoculated during brief intracellular punctures in phloem cells of barley by *R. padi* before the aphid's mouthparts reach the sieve element during the continuous salivation phase. Once infective, a cereal aphid remains infective, even across molts, for the rest of its life (Colvin *et al.*, 2006; Ali *et al.*, 2018).

Aphid dispersal and virus spread

The spread of YDV depends on cereal aphid dispersal. However, the proportion of viruliferous individuals within cereal aphid populations in assorted regions varies greatly, but it is typically <30% and often <10% (Halbert *et al.*, 1992; Fabre *et al.*, 2005). Thus, the degree of YDV spread may differ considerably from cereal aphid dispersal. The scale of aphid dispersal may also vary greatly, from aphids simply dispersing to a neighboring plant to long-range dispersal that traverses a continent (Irwin *et al.*, 2007).

Cereal aphids and YDD are perennial in wheat and other small-grain agroecosystems of central North America (Michaud, 2010; Hesler *et al.*, 2018). Cereal aphid viviparae persist throughout the year in areas without prolonged freezing winter temperatures (i.e. Mexico through Texas and some years as far north as Nebraska), but normally they cannot survive harsh winter conditions typical of latitudes north of Kansas (Kieckhefer *et al.*, 1974; Royer *et al.*, 2005; Wegulo and Hein, 2013; Crespo-Herrera *et al.*, 2015). However, climatic warming is expected to increase overwintering success and increase virus transmission in northern regions (Michaud, 2010). In southern regions, cereal aphids typically infest wheat in southcentral North America from early autumn through mid-spring and may continue to reproduce and transmit YDV throughout the winter (Royer *et al.*, 2005). In contrast, in northern areas cereal aphid viviparae infest wheat and other small grains from late spring through summer and may go on to infest newly planted winter wheat through mid-autumn (Kieckhefer and Lytle, 1976; Hesler *et al.*, 2018).

Two of the major YDV vectors are known to overwinter (holocyclically as eggs) in northern areas of central North America: *S. graminum* on various grasses and *R. padi* on *Prunus* spp. (Webster and Phillips, 1912; Wadley, 1931; Kieckhefer *et al.*, 1974). However, these overwintering populations are considered to play a minor role in epidemiology of YDD in the region (Gill, 1970; Kieckhefer *et al.*, 1974).

Thus, each spring, re-establishment of cereal aphids in the Northern Great Plains appears largely dependent upon an influx wind-aided long-distance movement of winged aphids originating from maturing cereal fields in the Southern Great Plains (Kieckhefer *et al.*, 1974; Michaud, 2010) (Fig. 7.5). It is plausible that cereal aphids may also be replenished from wheat and other small grains even farther south in Mexico. The likelihood of this movement theory is supported by the movement of *M. sorghi*. Seasonal wind patterns and planting of spring

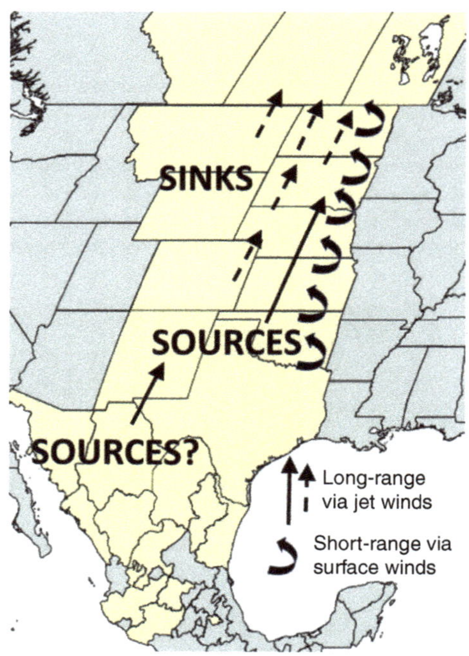

Fig. 7.5. Generalized depiction of northward long-range and short-range movement of cereal aphids from source fields in southcentral US into Mexico to sink fields in central North America. (Content is the author's own: L.S. Hesler.)

sorghum facilitate long-distance wind-aided dispersion of *M. sorghi* from far southern Texas (that is also representative of the northeastern Mexico cereal-growing season) based on aeroecological modeling populated with weather, *M. sorghi*, and sorghum data (Wang *et al.*, 2020). In years with the suitable biotic and abiotic conditions, the influx of cereal aphids may be quite dramatic, resulting in extremely high incidence of virus-infected plants (Irwin *et al.*, 1988). The seasonal appearance of cereal aphids in wheat and other small grains at northern latitudes has been documented for over 100 years (Webster and Phillips, 1912; Wadley, 1931). However, it was not until the 1960s that the appearance of cereal aphids and YDD in northern areas was linked to (i) their prevalence in southern wheat fields and (ii) southerly jet winds associated with synoptic weather systems (Hodson and Cook, 1960; Wallin *et al.*, 1967; Wallin and Loonan, 1971).

A low-level jet wind is defined as a stream of fast-moving air in the layer from ground level to about 1000 m above (Irwin *et al.*, 1988; USDA-NWS, 2023). Low-level jet winds often originate on cloudless nights that follow bright sunny afternoons when high pressure lies over the eastern or southeastern states and low pressure dominates the southern or central Rocky Mountains (Irwin *et al.*, 1988). Jet winds can be substantial, sometimes blowing up to 80 km/h for 1–3 days at a time.

The buildup of cereal aphid populations in wheat fields in southcentral North America is also required for significant transport of aphids via jet winds. Populations of cereal aphids consist of winged and wingless individuals. The proportion of winged individuals (alate viviparae) in a population increases as aphids become crowded on leaves and stems and as virus-infected plants weaken (see Colvin *et al.*, 2006, and references therein). When these winged aphids engage in active flight on warm afternoons and evenings with low-level jet winds, they may be transported several hundred kilometers and fall out in cereal crop fields in more northerly states (Wallin *et al.*, 1967; Wallin and Loonan, 1971; Irwin *et al.*, 1988; Irwin and Thresh, 1990). The pattern of fallout is considered random and may be localized to a few counties or extensive enough to include portions of several states (Irwin *et al.*, 1988). In areas of fallout, the incidence of aphid-infested plants can rise dramatically, sometimes followed by a high incidence of YDD. Heavy aphid infestations and high YDD incidence in cereal crop fields in the Northern Great Plains occur relatively infrequently, perhaps only once in 10 years or so on average (Hewings and Eastman, 1995). Particularly large peaks in aphids and viral diseases in cereal crops of the Northern Great Plains include outbreaks in Minnesota and several other states in 1959 and 1977, Iowa in 1975, Illinois and other upper Midwest states in the US in the early 1980s (Irwin *et al.*, 1988), and portions of Minnesota, North Dakota and South Dakota in 1999 (L.S. Hesler and S.L. Osborne, personal observations). It is unknown whether there is a reverse movement of cereal aphids from northern regions, where grains are maturing in later summer to early autumn, into some regions of more subtropical climates of Mexico where grains are in cultivation in the winter.

In some years the appearance of cereal aphids may be sudden and striking, but in most years, aphid population buildup does not coincide with transport via low-level jet winds (Kieckhefer *et al.*, 1974). Cereal aphids may still be carried northward each spring by lighter, shorter duration southerly winds. In these cases, aphid transport would appear less dramatic, as aphid populations move among crop fields within the landscape and eventually from southern states to northern ones (Kieckhefer *et al.*, 1974; Irwin *et al.*, 2007). Notably, cereal aphids were a nearly constant part of the airborne fauna sampled in suction traps from June to mid-November in eastern South Dakota (Kieckhefer and Lytle, 1976). It is unclear whether stimuli that cause aphid flight associated with long-range, regional dispersal by low-level jet winds differ from those that result in shorter-range dispersal within the landscape (Irwin *et al.*, 2007). Research is needed to better understand the conditions underlying short and intermediate cereal aphid flights and dispersal.

If the aphids moving into wheat fields in the spring are carrying YDVs, fields will eventually show characteristic roughly circular chlorotic patches (Wegulo and Hein, 2013). These patches arise following initial colonization by a winged viruliferous cereal aphid on a particular plant. The offspring of these winged aphids may then acquire YDVs, become infective, and move radially to infect neighboring plants, ultimately giving rise to the characteristic circular yellow patches within a cereal field (Ali *et al.*, 2018; Walls *et al.*, 2019).

While the scale of recolonization of northern wheat-growing regions by cereal aphids is dramatic, evidence shows that cereal aphids recolonize newly planted winter wheat fields at a local scale in autumn following a break in the annual wheat-cropping cycle (Kieckhefer and Lytle, 1976). That is, the wheat crops eventually mature and dry down for harvest, with the exact timing varying with the type of wheat (winter or spring) and the latitude at which wheat is grown. However, at this point in the season, wheat in a particular region is no longer available or suitable as a host for cereal aphids. Cereal aphids then must find alternative hosts that enable them to persist through the break in the growing cycle until winter wheat is planted and emerges in late summer or early autumn. This is known as the "green bridge period."

The means by which cereal aphids persist during this green bridge period vary with species and region. For instance, in southcentral North America (e.g. Oklahoma), *S. graminum* oversummers largely on various non-crop grasses typically found outside of agroecosystems and undergoes sexual reproduction (Anstead *et al.*, 2003; Royer *et al.*, 2015). In contrast, in the northern wheat regions, parthenogenic forms of cereal aphids, including *S. graminum*, are most frequently found on corn, sorghum, and sudangrass, followed next by their frequency on volunteer small grains, and infrequently found on weeds such as rough barnyard grass, and various species of foxtails (Hesler and Dagel, 2010). However, contrary to southern regions, cereal aphids were not found on non-cultivated grasses in various settings, such as in and around cropland, managed grasslands, prairies, and other natural areas (Hesler and Dagel, 2010). These findings showed that cultivated crops serve as a green bridge in mid to late summer for cereal aphids in northern regions, and the contribution of crops such as corn outweighed that of volunteer grains and weeds.

Management of Cereal Aphids and Yellow Dwarf Viruses

Predictive monitoring tools and management decision support systems (DSS) appear promising for large-scale agroecosystems in central North America. An outstanding example of predictive decision support is the soybean Pest Information Platform for Extension and Education (soybean PIPE) system for predicting soybean rust in the US Midwest and other regions of North America (Hershman *et al.*, 2011; Walls *et al.*, 2019). Soybean PIPE has been credited with nearly US$300 million in annual fungicide savings (Hershman *et al.*, 2011).

Predictive models as part of DSS may be especially useful with the cereal aphid–YDD pathosystem because there is no curative or antiviral treatment for YDV infections in wheat plants—once infected, a plant will likely go on to suffer yield loss and serve as a source of inoculum for aphid vectors (Hewings and Eastman, 1995; Walls *et al.*, 2019). Advances have been made in developing predictive models of cereal aphid infestation and the severity of YDV infection in wheat fields. At least 17 DSS have been developed that predict one or more of the following: cereal aphid activity, YDV incidence or risk, efficacy of cereal aphid management options, and economic damage and yield loss (Van den Eynde *et al.*, 2020, cf. Table S4 and references therein; Leybourne *et al.*, 2023, cf. Table 1 and references therein; Van den Eynde, 2023). Individual models were developed for use in the United Kingdom (UK) and Ireland, Belgium (Flanders), US, Australia, and New Zealand. However, a DSS used in the UK and Ireland is the only one that is currently active. This particular DSS (AHDB, 2024) has combined the models of Kendall *et al.* (1992) and Morgan (2000) to predict YDD severity in the UK and Ireland based on knowledge about cereal aphid flight activity in relation to temperature.

In North America, two DSS have been developed for cereal aphids (Elliott *et al.*, 2004) or the cereal aphid–YDV pathosystem (Walls *et al.*, 2016). The DSS of Elliott *et al.* (2004) was developed for Oklahoma, factoring in abiotic, biotic, and agronomic conditions under which winter wheat is grown to predict the population growth rates of *S. graminum* and calculation of treatment thresholds, although YDV was not considered in their system. This DDS was intended for management of individual fields. Walls *et al.* (2016) developed a DSS that modeled the site-specific decision process for cereal aphid and YDV management and validated recommendations produced by the model for delayed planting of winter wheat and the use of insecticide-treated seed using historical data from central North America. However, these two DSS are no longer publicly available, as well as several of the other DSS listed by Leybourne *et al.* (2023).

Accordingly, several researchers have noted the general lack of sustained programs in forecasting aphid displacements within large-scale agricultural production systems such as wheat agroecosystems (e.g. Irwin *et al.*, 2007; Wang *et al.*, 2020; Leybourne *et al.*, 2023), notwithstanding the DSS for the management of cereal aphids and YDD in the UK and Ireland (Leybourne *et al.*, 2023). Leybourne *et al.* (2023) have argued that barriers to DSS utilization include (i) lack of engagement with end-users in system development, (ii) lack of economic and environmental quantification, (iii) disregard of consumer benefits in using DDS, and (iv) limited end-user trust in DSS recommendations. Arguably, these factors are interrelated, and increased end-user participation may be gained by engagement of end-users early in the development of a DSS, with attention especially to economic benefits of using a DSS system (Elliott *et al.*, 2004). In addition to these social and economic factors, DSS validity and reliability may depend on the ability to model abiotic and biotic factors (Irwin *et al.*, 2007; Walls *et al.*, 2016). For central North America, these may include factors such as the magnitude and timing of large air masses that move cereal aphids from source to sink; the timing, magnitude, and location of source aphids; and their viruliferousness (Irwin *et al.*, 2007; Walls *et al.*, 2016).

The development of DSS for aphids in other large-scale agroecosystems may be applicable for developing areawide DSS for the cereal aphid–YDV–wheat pathosystem in central North America. One such system involves recent efforts to effectively model long-range dispersal of *M. sorghi* for its areawide pest management in sorghum agroecosystems of the Southern to Central Great Plains of North America (= *M. sacchari* of Wang *et al.*, 2021). Through a series of simulations to forecast infestations of *M. sorghi* in sorghum, Wang *et al.* (2020, 2021) determined that modeling priorities should emphasize elements that accurately represent the dispersion and deposition of immigrating aphids and deemphasize the timing and magnitude of aphid emigrations. In a related set of model analyses, Koralewski *et al.* (2021) identified dispersal duration, take-off height, and maximum dispersal height as three critical parameters that control the long-distance, wind-borne dispersal of *M. sorghi*. Although the *M. sorghi* DSS is in research mode, the results in modeling areawide dispersal of *M. sorghi* may suggest useful approaches for modeling cereal aphid infestation and movement among wheat agroecosystems of central North America, with the caveat that critical information about types and levels of particular YDVs may need to be incorporated.

Given the lack of predictive ability and curative treatments at present and potential for YDD-induced yield loss, it is imperative to proactively manage wheat crops to minimize aphid infestations, YDV infection, and yield loss (Hesler *et al.*, 2018; Walls *et al.*, 2019). Management options against cereal aphids and yellow dwarf in wheat may be broadly classified as preventative and remedial measures (Glen, 2000; Ali *et al.*, 2018; Hesler *et al.*, 2018; Walls *et al.*, 2019). Preventative measures include cultural controls, plant resistance and tolerance, biological control, and insecticide seed dressings, whereas remedial measures are limited to foliar insecticides (Glen, 2000; Hesler *et al.*, 2018).

Cultural controls

An effective cultural control tactic is to exploit windows of phenological asynchrony between wheat and cereal aphids. This may be achieved by planting wheat such that younger, more susceptible stages are present during a relatively pest-free period (Glen, 2000). This entails a delay in planting of winter wheat (Hesler *et al.*, 2005; Royer *et al.*, 2005; Bockus *et al.*, 2016). Delayed planting of winter wheat also avoids problems with *Mayetiola destructor* (Say) (Diptera: Cecidomyiidae), the Hessian fly, and *Aceria tosichella* Keifer (Prostigmata: Eriophyidae), the wheat curl mite) (Hesler *et al.*, 2005; Wegulo and Hein, 2013). However, adoption of the practice of delayed planting of winter wheat is not widespread in the Great Plains for various reasons, such as balancing farm workload in the fall, the need to establish wheat early for livestock grazing, and the threat of winterkill if wheat is established too late (Royer *et al.*, 2005).

Analogously, early planting of spring wheat would be predicted to protect young wheat plants from cereal aphids and YDV transmission. As cereal aphids may re-appear each spring by late April in northern states such as South Dakota (Kieckhefer *et al.*, 1974) and early May in Manitoba (Gill, 1970), planting wheat in late March to early April may enable plants to reach developmental stages that would be relatively tolerant of aphid infestation and YDV infection in late April to early May. However, this practice has not been adequately tested in spring wheat regions (Hesler *et al.*, 2010). In much of central North America, spring is typically the time of year with the highest number of rainy days, and cool, wet weather may thwart timely planting of spring wheat.

The control of grassy weeds and volunteer grains in late summer may eliminate alternative hosts of cereal aphids and YDVs (Wegulo and Hein, 2013; Hesler *et al.*, 2018). Their control is achieved through tillage and application of herbicides. Elimination of these alternative hosts will also help to control *M. destructor* and *A. tosichella*, with its associated wheat viruses (Wegulo and Hein, 2013).

Cultural control practices aimed at curbing cereal aphid infestation and YDD are generally expected to be applicable to individual fields or small farmscapes. However, widespread adoption of a particular cultural control tactic across a region would be expected to produce areawide benefits in terms of aphid control and YDD management.

Insecticides

Wheat producers may scout fields to determine the need for foliar insecticide applications to limit cereal aphids and virus transmission. Thresholds for managing cereal aphids to prevent their direct damage vary depending on state or region, aphid species, and crop development stage, and typically range from 5 to 30 aphids per plant (Walls *et al.*, 2016). Sampling systems have been developed in Oklahoma that account for seasonal differences in aphid distribution, price of wheat, controls costs, and the proportion of parasitized (i.e. mummified) aphids (Giles *et al.*, 2003; Royer *et al.*, 2004; see Chapter 6, this volume).

Cereal aphid thresholds are considerably lower when the objective is limiting YDV infection, arguably around one aphid per plant, but in practice, the mere presence of aphids within wheat fields may trigger spraying. In central North America, foliar sprays have been effective in limiting cereal aphid populations within individual fields (Royer *et al.*, 2015) but inadequately tested in terms of preventing infections and yield loss from YDVs (Hesler *et al.*, 2018).

Insecticide-treated seed can effectively control aphids and minimize risk of YDV infection (Gourmet *et al.*, 1996; Gray *et al.*, 1996; Royer *et al.*, 2005; Bockus *et al.*, 2016). In some cases, systemic insecticides are more effective than contact sprays in preventing aphid feeding damage (Royer *et al.*, 2005). A neonicotinoid insecticide is typically used to treat seed before planting (Royer *et al.*, 2005). Such treatments are intended to protect wheat during early growth stages when aphids move into spring wheat fields in the spring (Gourmet *et al.*, 1996) and in the fall for winter wheat (Gray *et al.*, 1996; Royer *et al.*, 2005; Bockus *et al.*, 2016).

Given various competing interests at the farmscape level, farmers often decide not to delay planting and thus sow winter wheat during a susceptible period. In these situations, use of insecticide-treated seeds can be a preventative option to effectively manage aphids and avoid YDV infection of winter wheat (Royer *et al.*, 2005; Bockus *et al.*, 2016). However, the use of neonicotinoid-treated seeds remains controversial in terms of its relation to integrated pest management (IPM) (Douglas and Tooker, 2015) and off-target effects (McNamara *et al.*, 2020; Holland *et al.*, 2021). Moreover, the development of resistance to neonicotinoids is a concern, given their relatively widespread use in wheat, the short generation time of cereal aphids, and the development of resistance in large-scale wheat agroecosystems outside of North America (Wang *et al.*, 2018; Xu *et al.*, 2022). Ultimately, wheat growers must weigh this prophylactic approach to wheat management given the infrequency of direct economic damage from cereal aphids, their typically low proportion of viruliferousness, and the sporadic nature of YDD outbreaks (Hesler *et al.*, 2005, 2018; Royer *et al.*, 2005). The use of foliar insecticides tends to be applied remedially as needed in individual fields, and thus the benefit is largely confined to those fields (Royer *et al.*, 2004). Insecticide seed treatments are largely prophylactic in nature and reduce initial infestations of cereal aphids and YDV infection of wheat plants. Thus, their use is expected to have a greater impact on YDD epidemiology on a scale beyond the individual field (Gray *et al.*, 1996).

Plant resistance and tolerance

Both cereal aphids and YDVs may be managed through the general approach of host-plant resistance (Aradottir and Crespo-Herrera, 2021). Host-plant resistance may be categorized as resistance or tolerance to both aphids (Stout, 2013; Mitchell *et al.*, 2016) and YDVs (Pagán and García-Arenal, 2018; Walls *et al.*, 2019). Resistance limits pest reproduction and decreases the ability of pests to reach damaging levels on or in the wheat plant, whereas tolerance enables a wheat plant to yield well despite supporting a relatively high aphid population or titer of YDV in infected aphids (Stout, 2013; Mitchell *et al.*, 2016; Walls *et al.*, 2019).

Resistance and tolerance to a wide range of cereal aphid species have been identified in wheat and relatives of wheat. Several of these aphid-resistant lines have been used to breed for aphid-resistant wheat lines through conventional and modern genetic approaches (Aradottir and Crespo-Herrera, 2021, and references therein; Simon *et al.*, 2021). Resistance to cereal aphids varies in the degree to which it reduces aphid population growth, but resistant lines typically enable at least small populations to persist on individual plants. This degree of resistance may be valuable in decreasing the number of vectors and thereby curtailing spread of YDV regionally, particularly since cereal aphids move from southern to northern wheat production areas of central North America (Hesler *et al.*, 2010). On the other hand, tolerance to aphids is probably not practicable, because while it would limit yield loss from the aphids, it could have negative implications for the epidemiology of YDV. Namely, tolerance of wheat cultivars to cereal aphids would allow proliferation of vector populations that could intensify the YDV spread within individual fields and arguably to other fields in the same area and moreover from southern to northern wheat-growing regions (Hesler *et al.*, 2010).

Resistance to YDV may be important to wheat plants within individual fields, but it also has wider implications on the regional epidemiology of YDD (Colvin *et al.*, 2006; Bosque-Pérez and Eigenbrode, 2011; Ingwell *et al.*, 2012). This is due largely to the fact that YDV-infected plants are preferred, higher quality hosts for cereal aphids (*R. padi*) over non-infected plants, thus leading to greater aphid populations with a high proportion of viruliferous individuals that may then spread within fields, local areas, and regions (Jiménez-Martínez and Bosque-Pérez, 2004; Ingwell *et al.*, 2012). In contrast, YDV-resistant wheat plants have the potential to suppress YDD within fields and regionally, as YDV-resistant wheat plants harbor a lower virus titer, reduce virus transmission efficiency by *R. padi*, and negatively impact the demographics of *R. padi* (Jiménez-Martínez and Bosque-Pérez, 2004; Jiménez-Martínez *et al.*, 2004).

However, resistance to YDVs does not naturally occur in wheat, but resistance and tolerance genes have been introgressed from wild grasses into various wheat cultivars (Jarošová *et al.*, 2016; Choudhury *et al.*, 2017). As such, strong potential exists for further development of YDV-resistant wheat adapted to the Great Plains and potentially other regions. However, performance of these wheat cultivars has varied, and resistance and tolerance are not equally conferred against different strains or species of YDVs (Bockus *et al.*, 2016). Consequently, wheat-breeding programs vary in their prioritization of selecting and advancing wheat lines with YDV-resistant or tolerant cultivars and in commitment to producing lines resistant to the various YDV vectors.

Biological control

The wheat agroecosystem hosts a rich set of natural enemies that includes parasitoids, entomopathogens, and predators such as lady beetles, ground beetles, and damsel bugs (Hesler *et al.*, 2018; Brewer *et al.*, 2019). Natural enemies may suppress cereal aphid populations and thereby limit YDV infections overall, particularly in young wheat plants, which are most vulnerable to the effects of YDV infection (Ali *et al.*, 2018). Thus, conservation biological control measures that promote an early-season reservoir of natural enemies in and around wheat fields are particularly useful for cereal aphid and YDD management. Within wheat agroecosystems, conservation biological control tactics that facilitate the early presence of natural enemies include reduced tillage and no-till approaches, cover cropping, and the deployment of banker plants (i.e. alternative reservoir plants) within or adjacent to wheat fields (Sacco-Martret de Préville *et al.*, 2022). Conservation biological control that affects cereal aphids may be extended to the landscape level (Brewer *et al.*, 2019). For instance, greater abundances of aphid parasitoids and predators such as lady beetles, ground beetles, damsel bugs, and lacewings correlate positively with higher habitat diversity adjacent to cereal production fields (Elliott *et al.*, 1999, 2018; Brewer *et al.*, 2008; Sacco-Martret de Préville *et al.*, 2022; see Chapters 2, 5, and 8, this volume).

However, for many kinds of aphid natural enemies, recruitment to wheat typically occurs once cereal aphid populations have established, and therefore their maximum efficacy may not correspond temporally to effectively limit the spread of YDV infections within that field or in the local landscape (Sacco-Martret de Préville *et al.*, 2022; Roudine *et al.*, 2023). Similarly for entomopathogens, levels of natural infection may take a relatively long period to build up to adequately suppress and, thus, effectively limit cereal aphid populations and YDV transmission to prevent economic loss (Butt *et al.*, 2001). Nonetheless, natural enemies remain important in reducing within-field populations of cereal aphids and thereby minimizing direct damage by the aphids in individual fields and limiting the vector capacity of aphids in a larger landscape or region (Brewer *et al.*, 2019). Measures, such as the use of "soft" insecticides that are differentially more toxic to aphids than biocontrol agents, may also play a role in conserving aphid natural enemies (Torres and Bueno, 2018) and limiting YDD in individual wheat agroecosystems. Other conservation biological control measures, such as banker plants and managing vegetation adjacent to the crop are also relevant for aphid control at various stages of wheat development (see Chapter 5, this volume). Means to increase adoption of farming practices that promote conservation biological control, such as monetary incentives or field demonstration plots, may be needed (Brewer and Goodell, 2012). Currently in central North America, however, the functional contribution of incentivized programs that are intended to promote conservation biological control activity in wheat agroecosystems is poorly understood in terms of the suppression of cereal aphids and YDD in individual fields or in a regional context. Research is needed to increase understanding about the impact of particular farm policies on biocontrol services within the larger wheat agroecosystem.

While natural enemies may have an overall positive effect on cereal aphid suppression, they also have non-consumptive effects on cereal aphids (Power, 1991). These non-consumptive effects include altered behavior, such as increased movement of aphids and thus the potential for increased secondary transmission of YDV to other plants within a field (Power, 1991; Fingu-Mabola and Francis, 2021; Roudine *et al.*, 2023). However, the relatively lengthy transmission and acquisition times for vectors of persistently transmitted viruses such as YDVs diminish the impact of increased aphid movement due to natural enemies (Roudine *et al.*, 2023). Nonetheless, the non-consumptive effects of natural enemies on cereal aphids in relation to YDD epidemiology is under-studied and requires further research to understand the subtle effects of aphid–natural enemy interactions (Roudine *et al.*, 2023).

Summary and Conclusions

YDD remains one of the most important and challenging diseases that limits wheat production. Challenges in managing this disease in wheat agroecosystems arise from the complexity of the plant–virus–vector pathosystem. Foliar insecticides and insecticide-treated seed are effective in controlling this disease, but alternative tactics should be optimized for a more environmentally friendly approach to YDD management. Additional efforts are needed to develop predictive models of cereal aphid infestations and YDD severity, with emphases on engagement of end-users early in the development of a

DSS, attention to its economic benefits, and understanding the aeroecology involved the dispersal and deposition of immigrating aphids. In addition, further research is needed to determine practices that enhance the effectiveness of the natural enemy complex against cereal aphid vectors of YDV, including the functional contribution of incentivized programs that are intended to promote natural enemy conservation. In addition, wheat-breeding programs should prioritize and advance sources useful in developing modern YDV-resistant wheat cultivars.

Acknowledgments

Thanks very much to Mike Brewer, Gary Hein and Ashleigh Faris for inviting us to write this chapter. L.S.H., S.L.O., Marie Langham, Walter Riedell, Kurt Dagel, Cecil Tharp, Eric Beckendorf, David Schneider, Mike Volek, Jessica Smith, Malissa Mayer, Kendra Kallemeyn, and Toby Bryant contributed to some aspects of research summarized in this chapter. Figures 7.1 and 7.5 were created using MapChart. L.S.H. wrote the chapter text, and L.S.H. and S.L.O. participated in revising and editing the manuscript and approved its final version. Sharon Schneider and Lauren Hesler reviewed an earlier draft of this chapter. Funding to USDA-ARS CRIS Project Number 3080-21220-008-000-D provided support for writing this chapter. Mention of trade names or commercial products in this article is solely for the purpose of providing specific information and does not imply recommendation or endorsement by the USDA. The USDA is an equal opportunity provider and employer.

References

AHDB (2024) *Barley Yellow Dwarf Virus (BYDV) Management Tool for Cereals*. Coventry, UK: Agriculture and Horticulture Development Board. Available at: https://ahdb.org.uk/bydv (accessed January 1, 2024).

Ali, M., Anwar, S., Shuja, M.N., Tripathi, R.K. and Singh, J. (2018) The genus *Luteovirus* from infection to disease. *European Journal of Plant Pathology* 151, 841–860. DOI: 10.1007/s10658-018-1425-8

Anstead, J.A., Shufran, K.A. and Burd, J.D. (2003) Over-summering and biotypic diversity of *Schizaphis graminum* (Homoptera: Aphididae) populations on noncultivated grass hosts. *Environmental Entomology* 32, 662–667. DOI: 10.1603/0046-225X-32.3.662

Aradottir, G.I. and Crespo-Herrera, L. (2021) Host plant resistance in wheat to barley yellow dwarf viruses and their aphid vectors: a review. *Current Opinion in Insect Science* 45, 59–68. DOI: 10.1016/j.cois.2021.01.002

Armstrong, J.S., Hoback, W.W. and Springer, T.L. (2021) Host suitability for the sugarcane aphid *Melanaphis sacchari* (Hemiptera: Aphididae) on sorghum, sorghum-sudangrass, millets and other forage grass species. *Trends in Entomology* 17, 1–11.

Bosque-Pérez, N.A. and Eigenbrode, S.D. (2011) The influence of virus-induced changes in plants on aphid vectors: insights from luteovirus pathosystems. *Virus Research* 159, 201–205. DOI: 10.1016/j.virusres.2011.04.020

Bockus, W.W., De Wolf, E.D. and Todd, T.C. (2016) Management strategies for barley yellow dwarf on winter wheat in Kansas. *Plant Health Progress* 17, 122–127. DOI: 10.1094/PHP-RS-15-0050

Brewer, M.J. and Goodell, P.G. (2012) Approaches and incentives to implement integrated pest management that addresses regional and environmental issues. *Annual Review of Entomology* 57, 41–59. DOI: 10.1146/annurev-ento-120709-144748

Brewer, M.J., Noma, T. and Elliott, N.C. (2008) A landscape perspective in managing vegetation for beneficial plant–pest–natural enemy interactions: a foundation for areawide pest management. In: Koul, O., Cuperus, G.W. and Elliott, N.C. (eds) *Areawide Pest Management: Theory and Implementation*. Wallingford, UK: CAB International, pp. 81–96.

Brewer, M.J., Peairs, F.B. and Elliott, N.C. (2019) Invasive cereal aphids of North America: ecology and pest management. *Annual Review of Entomology* 64, 73–93. DOI: 10.1146/annurev-ento-011118-111838

Burgess, A.J., Harrington, R. and Plumb, R.T. (1999) Barley and cereal yellow dwarf virus epidemiology and control strategies. In: Smith, H.G. and Barker, H. (eds) *The Luteoviridae*. Wallingford, UK: CAB International, pp. 211–279.

Butt, T.M., Jackson, C. and Magan, N. (2001) *Fungi as Biocontrol Agents: Progress, Problems and Potential*. Wallingford, UK: CAB International.

Choudhury, S., Hu, H., Meinke, H., Shabala, S., Westmore, G., Larkin, P. and Zhou, M. (2017) Barley yellow dwarf viruses: infection mechanisms and breeding strategies. *Euphytica* 213, 1–22. DOI: 10.1007/s10681-017-1955-8

Colvin, J., Omongo, C.A., Govindappa, M.R., Stevenson, P.C., Maruthi, M.N. *et al.* (2006) Host-plant viral infection effects on arthropod-vector population growth, development and behaviour: management and epidemiological implications. *Advances in Virus Research* 67, 419–452. DOI: 10.1016/S0065-3527(06)67011-5

Crespo-Herrera, L.A., Singh, R.P. and Åhman, I. (2015) Field population development of bird cherry-oat aphid and greenbug (Hemiptera: Aphididae) on wheat-alien substitution and translocation lines. *Euphytica* 203, 249–260. DOI: 10.1007/s10681-014-1244-8

Damsteegt, V.D., Gildow, F.E., Hewings, A.D. and Carroll, T.W. (1992) A clone of the Russian wheat aphid (*Diuraphis noxia*) as a vector of the barley yellow dwarf, barley stripe mosaic and brome mosaic viruses. *Plant Disease* 76, 1155–1160. DOI: 10.1094/PD-76-1155

D'Arcy, C.J. (1995) Symptomology and host range of barley yellow dwarf. In: D'Arcy, C.J. and Burnett, P.A. (eds) *Barley Yellow Dwarf: 40 Years of Progress*. St Paul, MN: APS Press, pp. 9–28.

Douglas, M.R. and Tooker, J.F. (2015) Large-scale deployment of seed treatments has driven rapid increase in use of neonicotinoid insecticides and preemptive pest management in U.S. field crops. *Environmental Science and Technology* 49, 5088–5097. DOI: 10.1021/es506141g

Elliott, N.C., Kieckhefer, R.W., Lee, J.H. and French, B.W. (1999) Influence of within-field and landscape factors on aphid predator populations in wheat. *Landscape Ecology* 14, 239–252. DOI: 10.1023/A:1008002528345

Elliott, N.C., Royer, T.A., Giles, K.L., Kindler, S.D., Porter, D.R. *et al.* (2004) A web-based decision support system for managing greenbugs in wheat. *Crop Management* 3, 1–6. DOI: 10.1094/CM-2004-1006-01-MG

Elliott, N.C., Brewer, M.J. and Giles, K.L. (2018) Landscape context affects aphid parasitism by *Lysiphlebus testaceipes* (Hymenoptera: Aphidiinae) in wheat fields. *Environmental Entomology* 47, 803–811. DOI: 10.1093/ee/nvy035

Erickson, A., Jiang, J., Kuo, Y.W. and Falk, B.W. (2023) Construction and use of an infectious cDNA clone to identify aphid vectors and susceptible monocot hosts of the polerovirus barley virus G. *Virology* 579, 178–185. DOI: 10.1016/j.virol.2023.01.011

Fabre, F., Plantegenest, M., Mieuzet, L., Dedryver, C.A., Leterrier, J.-L. *et al.* (2005) Effects of climate and land use on the occurrence of viruliferous aphids and the epidemiology of barley yellow dwarf disease. *Agriculture, Ecosystems & Environment* 106, 49–55. DOI: 10.1016/j.agee.2004.07.004

Fingu-Mabola, J.C. and Francis, F. (2021) Aphid–plant–phytovirus pathosystems: influencing factors from vector behaviour to virus spread. *Agriculture* 11, 502. DOI: 10.3390/agriculture11060502

Giles, K.L., Jones, D.B., Royer, T.A., Elliott, N.C. and Kindler, S.D. (2003) Development of a sampling plan in winter wheat that estimates cereal aphid parasitism levels and predicts population suppression. *Journal of Economic Entomology* 96, 975–982. DOI: 10.1093/jee/96.3.975

Gill, C.C. (1970) Epidemiology of barley yellow dwarf in Manitoba and effect of the virus on yield of cereals. *Phytopathology* 60, 1826–1830. DOI: 10.1094/Phyto-60-1826

Glen, D.M. (2000) The effects of cultural measures on cereal pests and their role in integrated pest management. *Integrated Pest Management Reviews* 5, 25–40. DOI: 10.1023/A:1009609504464

Gourmet, C., Kolb, F.L., Smyth, C.A. and Pedersen, W.L. (1996) Use of imidacloprid as a seed-treatment insecticide to control barley yellow dwarf virus (BYDV) in oat and wheat. *Plant Disease* 80, 136–141. DOI: 10.1094/PD-80-0136

Gray, S.M., Power, A.G., Smith, D.M., Seaman, A.J. and Altman, N.S. (1991) Aphid transmission of barley yellow dwarf virus: acquisition access periods and virus concentration requirements. *Phytopathology* 81, 539–545. DOI: 10.1094/Phyto-81-539

Gray, S.M., Bergstrom, G.C., Vaughan, R., Smith, D.M. and Kalb, D.W. (1996) Insecticidal control of cereal aphids and its impact on the epidemiology of the barley yellow dwarf luteoviruses. *Crop Protection* 15, 687–697. DOI: 10.1016/S0261-2194(96)00040-3

Halbert, S. and Voegtlin, D. (1995) Biology and taxonomy of vectors of barley yellow dwarf viruses. In: D'Arcy, C.J. and Burnett, P.A. (eds) *Barley Yellow Dwarf: 40 Years of Progress*. St Paul, MN: APS Press, pp. 217–258.

Halbert, S.E., Connelly, B.J., Bishop, G.W. and Blackmer, J.L. (1992) Transmission of barley yellow dwarf virus by field collected aphids (Homoptera, Aphididae) and their relative importance in barley yellow dwarf epidemiology in southwestern Idaho. *Annals of Applied Biology* 121, 105–121. DOI: 10.1111/j.1744-7348.1992.tb03991.x

Harris-Shultz, K.R., Armstrong, J.S. and Jacobson, A. (2019) Invasive cereal aphids of North America: biotypes, genetic variation, management, and lessons learned. *Trends in Entomology* 15, 99–122.

Hershman, D.E., Sikora, E.J. and Giesler, L.J. (2011) Soybean rust PIPE: past, present, and future. *Journal of Integrated Pest Management* 2, D1–D7. DOI: 10.1603/IPM11001

Hesler, L.S. and Dagel, K.J. (2010) Grass hosts of cereal aphids (Hemiptera: Aphididae) between wheat-cropping cycles in South Dakota. *Great Lakes Entomologist* 43, 1–10.

Hesler, L.S., Riedell, W.E., Langham, M.A.C. and Osborne, S.L. (2005) Insect infestations, incidence of viral plant diseases, and yield of winter wheat in relation to planting date in the northern Great Plains. *Journal of Economic Entomology* 98, 2020–2027. DOI: 10.1093/jee/98.6.2020

Hesler, L.S., Hadi, B.A.R. and Tharp, C.I. (2010) Bird cherry-oat aphid (Hemiptera: Sternorrhyncha, Aphidinae): biology, pest status, and management in wheat. *Trends in Entomology* 6, 53–70.

Hesler, L.S., Sappington, T.W., Luttrell, R.G., Allen, K.C. and Papiernik, S.K. (2018) Selected early-season pests of wheat in the United States and factors affecting their risks of infestation. *Journal of Integrated Pest Management* 9, 17. DOI: 10.1093/jipm/pmx023

Hewings, A.D. and Eastman, C.E. (1995) Epidemiology of barley yellow dwarf in North America. In: D'Arcy, C.J. and Burnett, P.A. (eds) *Barley Yellow Dwarf: 40 Years of Progress*. St Paul, MN: APS Press, pp. 75–106.

Hodson, A.C. and Cook, E.F. (1960) Long-range aerial transport of the harlequin bug and the greenbug into Minnesota. *Journal of Economic Entomology* 53, 604–608. DOI: 10.1093/jee/53.4.604

Holland, J.M., McHugh, N.M. and Salinari, F. (2021) Field specific monitoring of cereal yellow dwarf virus aphid vectors and factors influencing their immigration within fields. *Pest Management Science* 77, 4100–4108. DOI: 10.1002/ps.6435

Ingwell, L.L., Eigenbrode, S.D. and Bosque-Pérez, N.A. (2012) Plant viruses alter insect behavior to enhance their spread. *Scientific Reports* 2, 578. DOI: 10.1038/srep00578

Irwin, M.E. and Thresh, J.M. (1990) Epidemiology of barley yellow dwarf: a study in ecological complexity. *Annual Review of Phytopathology* 28, 393–424. DOI: 10.1146/annurev.py.28.090190.002141

Irwin, M E., Thresh, J.M. and Harrison, B.D. (1988) Long-range aerial dispersal of cereal aphids as virus vectors in North America. *Philosophical Transactions of the Royal Society of London B* 321, 421–446. DOI: 10.1098/rstb.1988.0101

Irwin, M.E., Kampmeier, G.E. and Weisser, W.W. (2007) Aphid movement: process and consequences. In: Van Emden, H.F. and Harrington, R. (eds) *Aphids as Crop Pests*. Wallingford, UK: CAB International, pp. 153–186.

Jarošová, J., Beoni, E. and Kundu, J.K. (2016) Barley yellow dwarf virus resistance in cereals: approaches, strategies and prospects. *Field Crops Research* 198, 200–214. DOI: 10.1016/j.fcr.2016.08.030

Jiménez, J., Arias-Martín, M., Moreno, A., Garzo, M. and Fereres, A. (2020) Barley yellow dwarf virus can be inoculated during brief intracellular punctures in phloem cells before the sieve element continuous salivation phase. *Phytopathology* 110, 85–93. DOI: 10.1094/PHYTO-07-19-0260-FI

Jiménez-Martínez, E.S. and Bosque-Pérez, N.A. (2004) Variation in barley yellow dwarf virus transmission efficiency by *Rhopalosiphum padi* (Homoptera: Aphididae) after acquisition from transgenic and non-transformed wheat genotypes. *Journal of Economic Entomology* 97, 1790–1796. DOI: 10.1093/jee/97.6.1790

Jiménez-Martínez, E.S., Bosque-Pérez, N.A., Berger, P.H. and Zemetra, R.S. (2004) Life history of the bird cherry-oat aphid, *Rhopalosiphum padi* (Homoptera: Aphididae), on transgenic and untransformed wheat challenged with barley yellow dwarf virus. *Journal of Economic Entomology* 97, 203–212. DOI: 10.1093/jee/97.2.203

Jones, R.A.C., Sharman, M., Trębicki, P., Maina, S. and Congdon, B.S. (2021) Virus diseases of cereal and oilseed crops in Australia: current position and future challenges. *Viruses* 13, 2051. DOI: 10.3390/v13102051

Kendall, D.A., Brain, P. and Chinn, N.E. (1992) A simulation model of the epidemiology of barley yellow dwarf virus in winter sown cereals and its application to forecasting. *Journal of Applied Ecology* 29, 414–426. DOI: 10.2307/2404510

Kieckhefer, R.W. and Lytle, W.F. (1976) Summer and autumn flight of cereal aphids in South Dakota. *Journal of Economic Entomology* 69, 421–422. DOI: 10.1093/jee/69.3.421

Kieckhefer, R.W., Lytle, W.F. and Spuhler, W. (1974) Spring movement of cereal aphids into South Dakota. *Environmental Entomology* 3, 347–350. DOI: 10.1093/ee/3.2.347

Koralewski, T.E., Wang, H.H., Grant, W.E., Brewer, M.J., Elliott, N.C. *et al.* (2021) Modeling the dispersal of wind-borne pests: sensitivity of infestation forecasts to uncertainty in parameterization of long-distance airborne dispersal. *Agricultural and Forest Meteorology* 301, 108357. DOI: 10.1016/j.agrformet.2021.108357

Leybourne, D.J., Ramsden, M., White, S., Wang, R., Huang, H. *et al.* (2023) Online decision support systems, remote sensing and artificial intelligence applications for wheat pest management. In: Eigenbrode, S. and Rashed, A. (eds) *Advances in Understanding Insect Pests Affecting Wheat and Other Cereals*. Philadelphia, PA: Burleigh Dodds, pp. 411–444.

McNamara, L., Gauthier, K., Walsh, L., Thébaud, G., Gaffney, M. *et al.* (2020) Management of yellow dwarf disease in Europe in a post-neonicotinoid agriculture. *Pest Management Science* 76, 2276–2285. DOI: 10.1002/ps.5835

Michaud, J.P. (2010) Implications of climate change for cereal aphids on the Great Plains of North America. In: Kindlmann, P., Dixon, A.F.G. and Michaud, J.P. (eds) *Aphid Biodiversity Under Environmental Change: Patterns and Processes*. Dordrecht, Netherlands: Springer, pp. 69–89. DOI: 10.1007/978-90-481-8601-3_5

Miller, W.A. and Lozier, Z. (2022) Yellow dwarf viruses of cereals: taxonomy and molecular mechanisms. *Annual Review of Phytopathology* 60, 121–141. DOI: 10.1146/annurev-phyto-121421-125135

Mitchell, C., Brennan, R.M., Graham, J. and Karley, A.J. (2016) Plant defense against herbivorous pests: exploiting resistance and tolerance traits for sustainable crop protection. *Frontiers in Plant Science* 7, 1132. DOI: 10.3389/fpls.2016.01132

Morgan, D. (2000) Population dynamics of the bird-cherry oat aphid, *Rhopalosiphum padi* (L.), during the fall and winter: a modelling approach. *Agricultural and Forest Entomology* 2, 297–304. DOI: 10.1046/j.1461-9563.2000.00079.x

Mulvaney, M.J., Verhulst, N., Herrera, J.M., Mezzalama, M. and Govaerts, B. (2014) Improved wheat performance with seed treatments under dry sowing on permanent raised beds. *Field Crops Research* 164, 189–198. DOI: 10.1016/j.fcr.2014.04.017

Pagán, I. and García-Arenal, F. (2018) Tolerance to plant pathogens: theory and experimental evidence. *International Journal of Molecular Sciences* 19, 810. DOI: 10.3390/ijms19030810

Pike, K.S. (1990) A review of barley yellow dwarf virus grain yield losses. In: Burnett, P.A. (ed.) *World Perspectives on Barley Yellow Dwarf*. Texcoco, Mexico: CIMMYT, pp. 356–361.

Pimentel, D. (2007) Area-wide pest management: environmental, economic and food issues. In: Vreysen, M.J.B., Robinson, A.S. and Hendrichs, J. (eds) *Area-wide Control of Insect Pests, From Research to Field Implementation*. Dordrecht, Netherlands: Springer, pp. 35–47. DOI: 10.1007/978-1-4020-6059-5_2

Porter, D.H., Harris, M.O., Hesler, L.S. and Puterka, G.J. (2009) Insects which challenge global wheat production. In: Carver, B.F. (ed.) *Wheat Science and Trade*. Ames, IA: Wiley-Blackwell, pp. 189–201.

Power, A.G. (1991) Virus spread and vector dynamics in genetically diverse plant populations. *Ecology* 72, 232–241. DOI: 10.2307/1938917

Puterka, G.J., Hammon, R.W., Mornhinweg, D.W., Springer, T.L., Armstrong, J.S. *et al.* (2019) Distribution of a new invasive species, *Sipha maydis* Passerini (Heteroptera: Aphididae), on cereals and wild grasses in the Great Plains and Rocky Mountain states. *Journal of Economic Entomology* 112, 1713–1721. DOI: 10.1093/jee/toz068

Reynolds, D.R., Chapman, J.W. and Harrington, R. (2006) The migration of insect vectors of plant and animal viruses. *Advances in Virus Research* 67, 453–517. DOI: 10.1016/S0065-3527(06)67012-7

Rotenberg, D., Bockus, W.W., Whitfield, A.E., Hervey, K., Baker, K.D. *et al.* (2016) Occurrence of viruses and associated grain yields of paired symptomatic and nonsymptomatic tillers in Kansas winter wheat fields. *Phytopathology* 106, 202–210. DOI: 10.1094/PHYTO-04-15-0089-R

Roudine, S., Le Lann, C., Bouvaine, S., Le Ralec, A. and van Baaren, J. (2023) Can biological control be a strategy to control vector-borne plant viruses? *Journal of Pest Science* 96, 451–470. DOI: 10.1007/s10340-022-01587-0

Royer, T.A., Giles, K.L., Kindler, S.D. and Elliott, N.C. (2004) *Cereal Aphid Expert System and Glance 'n Go Sampling for Greenbugs: Questions and Answers*. CR-7191. Stillwater, OK: Oklahoma Cooperative Extension Service. Available at: https://hdl.handle.net/11244/333359 (accessed January 1, 2024).

Royer, T.A., Giles, K.L., Nyamanzi, T., Hunger, R.M., Krenzer, E.G. *et al.* (2005) Economic evaluation of the effects of planting date and application rate of imidacloprid for management of cereal aphids and barely yellow dwarf in winter wheat. *Journal of Economic Entomology* 98, 95–102. DOI: 10.1093/jee/98.1.95

Royer, T.A., Pendleton, B.B., Elliott, N.C. and Giles, K.L. (2015) Greenbug (Hemiptera: Aphididae) biology, ecology, and management in wheat and sorghum. *Journal of Intergrated Pest Management* 6, 19. DOI: 10.1093/jipm/pmv018

Rybicki, E.P. (2015) A Top Ten list for economically important plant viruses. *Archives of Virology* 160, 17–20. DOI: 10.1007/s00705-014-2295-9

Sacco-Martret de Préville, A., Ortiz-Martinez, S., Plantegenest, M. and Canard, E. (2022) Effect of conservation agriculture on aphid biocontrol by generalist (carabid beetle) and specialist (parasitoids wasp) natural enemy communities in winter wheat. *Frontiers in Ecology and Evolution* 10, 893787. DOI: 10.3389/fevo.2022.893787

Simon, A.L., Caulfield, J.C., Hammond-Kosack, K.E., Field, L.M. and Aradottir, G.I. (2021) Identifying aphid resistance in the ancestral wheat *Triticum monococcum* under field conditions. *Scientific Reports* 11, 13495. DOI: 10.1038/s41598-021-92883-9

Sõmera, M., Massart, S., Tamisier, L., Sooväli, P., Sathees, K. *et al.* (2021) A survey using high-throughput sequencing suggests that the diversity of cereal and barley yellow dwarf viruses is underestimated. *Frontiers in Microbiology* 12, 673218. DOI: 10.3389/fmicb.2021.673218

Stout, M.J. (2013) Reevaluating the conceptual framework for applied research on host-plant resistance. *Insect Science* 20, 263–272. DOI: 10.1111/1744-7917.12011

Torres, J.B. and Bueno, A.D.F. (2018) Conservation biological control using selective insecticides – a valuable tool for IPM. *Biological Control* 126, 53–64. DOI: 10.1016/j.biocontrol.2018.07.012

USDA-ERS (2023) *Wheat Data*. Washington, DC: US Department of Agriculture, Economic Research Service. Available at: https://www.ers.usda.gov/data-products/wheat-data/ (accessed January 1, 2024).

USDA-FAS (2023a) *Canada Wheat Area, Yield and Production*. Washington, DC: US Department of Agriculture, Foreign Agricultural Service, Foreign Agricultural Service. Available at: https://ipad.fas.usda.gov/countrysummary/?id=CA (accessed January 1, 2024).

USDA-FAS (2023b) *Mexico Wheat Area, Yield and Production*. Washington, DC: US Department of Agriculture, Foreign Agricultural Service. Available at: https://ipad.fas.usda.gov/countrysummary/?id=MX (accessed January 1, 2024).

USDA-NASS (2021) *Crop Production*. Washington, DC: US Department of Agriculture, National Agricultural Statistics Service. Available at: https://www.nass.usda.gov/Publications/Todays_Reports/reports/crop0822.pdf (accessed January 1, 2024).

USDA-NWS (2023) *Low-level Jet Winds*. Washington, DC: US Department of Commerce, National Weather Service. Available at: https://forecast.weather.gov/glossary (accessed January 1, 2024).

Van den Eynde, R. (2023) A model-based integrated approach to support the control of barley yellow dwarf virus in Flanders. PhD thesis. Ghent, Belgium: Ghent University. http://hdl.handle.net/1854/LU-01GV43FE4W601QEWB08P2WGZHJ (accessed January 1, 2024).

Van den Eynde, R., Van Leeuwen, T. and Haesaert, G. (2020) Identifying drivers of spatio-temporal dynamics in barley yellow dwarf virus epidemiology as a critical factor in disease control. *Pest Management Science* 76, 2548–2556. DOI: 10.1002/ps.5851

Wadley, F.M. (1931) Ecology of *Toxoptera graminum*, especially as to factors affecting importance in the northern United States. *Annals of the Entomological Society of America* 24, 325–395. DOI: 10.1093/aesa/24.2.325

Walker, P.J., Siddell, S.G., Lefkowitz, E.J., Mushegian, A.R., Adriaenssens, E.M. *et al* (2022) Recent changes to virus taxonomy ratified by the International Committee on Taxonomy of Viruses (2022). *Archives of Virology* 167, 2429–2440. DOI: 10.1007/s00705-022-05516-5

Wallin, J.R. and Loonan, D.V. (1971) Low-level jet winds, aphid vectors, local weather, and barley yellow dwarf virus outbreaks. *Phytopathology* 61, 1068–1070. DOI: 10.1094/Phyto-61-1068

Wallin, J.R., Peters, D. and Johnson, L.C. (1967) Low-level jet winds, early cereal aphid and barley yellow dwarf detection in Iowa. *Plant Disease Reporter* 51, 528–530.

Walls, J.T., III, Caciagli, P., Tooker, J.F., Russo, J.M., Rajotte, E.G. *et al.* (2016) Modeling the decision process for barley yellow dwarf management. *Computers and Electronics in Agriculture* 127, 775–786. DOI: 10.1016/j.compag.2016.08.005

Walls, J., III, Rajotte, E. and Rosa, C. (2019) The past, present, and future of barley yellow dwarf management. *Agriculture* 9, 23. DOI: 10.3390/agriculture9010023

Wang, H.H., Grant, W.E., Koralewski, T.E., Elliott, N.C., Brewer, M.J. *et al.* (2020) Where do all the aphids go? A series of thought experiments within the context of area-wide pest management. *Agricultural Systems* 185, 102957. DOI: 10.1016/j.agsy.2020.102957

Wang, H.H., Grant, W.E., Koralewski, T.E., Brewer, M.J. and Elliott, N.C. (2021) Simulating migration of wind-borne pests: "deconstructing" representation of the emigration process. *Ecological Modelling* 460, 109742. DOI: 10.1016/j.ecolmodel.2021.109742

Wang, K., Zhang, M., Huang, Y., Yang, Z., Su, S. *et al.* (2018) Characterisation of imidacloprid resistance in the bird cherry-oat aphid, *Rhopalosiphum padi*, a serious pest on wheat crops. *Pest Management Science* 74, 1457–1465. DOI: 10.1002/ps.4834

Webster, F.M. and Phillips, W.J. (1912) *The Spring Grain-Aphis or 'Greenbug'*. Bureau of Entomology Bulletin 110. Washington, DC: US Department of Agriculture.

Wegulo, S.W. and Hein, G.L. (2013) *Yellow Dwarf of Wheat, Barley, and Oats*. NebGuide 1823. Lincoln, NE: University of Nebraska–Extension. Available at: http://extensionpublications.unl.edu/assets/pdf/g1823.pdf (accessed January 1, 2024).

Wei, S., Chen, G., Yang, H., Huang, L., Gong, G. *et al.* (2023) Global molecular evolution and phylogeographic analysis of barley yellow dwarf virus based on the cp and mp genes. *Virology Journal* 20, 130. DOI: 10.1186/s12985-023-02084-1

Xu, T., Zhang, S., Liu, Y., Ma, L., Li, X. *et al.* (2022) Slow resistance evolution to neonicotinoids in field populations of wheat aphids revealed by insecticide resistance monitoring in China. *Pest Management Science* 78, 1428–1437. DOI: 10.1002/ps.6760

8 Natural Enemy Suppression Supplemented by Regional Pest Management for the Invasive *Melanaphis sorghi*, Sorghum Aphid, on Sorghum

Ashleigh M. Faris,[1,*] Michael J. Brewer,[2] and Norman C. Elliott[3]
[1]Department of Entomology and Plant Pathology, Oklahoma State University, Stillwater, OK, USA; [2]Texas A&M AgriLife Research, Corpus Christi, TX, USA; [3]US Department of Agriculture, Agricultural Research Service, Peanut and Other Field Crops Research Unit, Stillwater, OK, USA

Abstract
Sorghum in North America became at risk to *Melanaphis sorghi* (Theobald) (Hemiptera: Aphididae), sorghum aphid, after the aphid's invasion in 2013 and near-continental-scale expansion within a few years. Mixed indications of sorghum aphid outbreak severity and suppression by natural enemies were reported across the grain sorghum production system spanning the United States (US), Mexico, and Caribbean islands. This production system provided an opportunity to evaluate agroecosystem response to an invading pest species by natural enemies, how the response varied spatially across regions, and how host plant resistance may complement natural enemy suppression. The natural enemy complex was similar across four regions from the US Central Great Plains, south along the US and Mexico Gulf Coast, and to southeastern US states. Two primary parasitoids, six coccinellid species, and several syrphid fly and lacewing species were common. In some but not all regions, natural enemies were associated with reduced aphid abundance. Landscape influences on natural enemies varied by region, from negligible (southern Texas) to modest (in the more temperate conditions of the Central Great Plains) association with natural enemy abundance. Validation and estimation of aphid suppression were pursued using cage exclosures in field settings. Aphid suppression commonly exceeded 90% on aphid-susceptible and -resistant sorghums in these widely separated regions. There was a pattern of more suppression by parasitoids in the south Texas location and a mix of predator and parasitoids contributing to suppression in the Oklahoma location. Based on these results, suppression by resident parasitoids and predators is widespread, but aphid risk persists in some regions. Partially aphid-resistant sorghum is expected to reduce aphid risk directly and not interfere with aphid suppression by resident natural enemies.

Introduction

The monotypic nature of large-scale field crops promotes a homogenous response to plant stress; however, these crops experience spatially variable pest pressures that provide challenges as well as opportunities for management. Classic integrated pest management (IPM) decision making would seem appropriate to such a cropping system by emphasizing the use of remedial pest management

*Corresponding author: ashleigh.faris@okstate.edu

 Arthropod Management and Landscape Considerations in Large-Scale Agroecosystems (eds M.J. Brewer and G.L. Hein)
DOI: 10.1079/9781800622777.0008

tactics applied at the individual field level once economic thresholds are exceeded. However, the modest crop value of most field crops and regional variability of pest severity and crop sensitivity hamper IPM applied field by field. Under the current crop and pest consulting business model, IPM decision making applied field by field in large field crop systems appears cost and resource constrained in many circumstances (Brewer and Goodell, 2012). Here, we focus on the premise that pests may be suppressed well by natural enemies and the crop may be more tolerant of pest presence in some regions, while other regions may be prone to pest outbreaks and crop stress due to pest injury. A meta-analysis of over 50 studies indicated that pests and natural enemies are variable across a variety of agroecosystems and have inconsistent responses to a landscape composition metric (i.e. proportion of landcover with crops or a focal crop damaged by a pest species) (Karp *et al.*, 2018). This interpretation is consistent with the proposition that pest risk and innate pest suppression by resident natural enemies are spatially variable and conditional on agricultural and environmental conditions (i.e. landscape considerations; see Chapter 2, this volume).

Favorable pest management and crop health outcomes become more likely when innate pest suppression is coupled with preventative pest management tactics that are strategically applied to regions at high pest risk. This regionally oriented IPM approach may be more effective, provide more pest management and environmental benefits, and foster agroecosystem resilience to pests compared with IPM decision making based on pest assessment done field by field across the growing season (Brewer and Elliott, 2023). Here we provide the research basis and framework for applying this concept to management of *Melanaphis sorghi* (Theobald) (Hemiptera: Aphididae), sorghum aphid, that invaded sorghum, *Sorghum bicolor* (L.) Moench, in North America in the 2010s. Regarding works cited, many previous publications of *M. sorghi* on sorghum in North America have used the name *Melanaphis sacchari* (Zehntner). This designation was before molecular and ecological evidence that indicated *M. sorghi* recently invaded and was predominantly found on sorghum in the United States (US). It is distinctive from the morphologically similar *M. sacchari* that has been a long-established resident of sugarcane in the US (Harris-Shultz *et al.*, 2017; Nibouche *et al.*, 2021).

Invasion of the Sorghum Agroecosystem of North America

The invasion of sorghum by *M. sorghi* in North America provided an opportunity to evaluate the agroecosystem response of a widely planted grain crop that experiences a range of landscape and weather conditions. From 2012 to 2019, between 1.9 and 3.2 million ha of grain sorghum were harvested annually in the US, and production in Mexico was also extensive (Bowling *et al.*, 2016; USDA-NASS, 2020). *Melanaphis sorghi* principally damages sorghum (Blackman and Eastop, 1984). It has long been associated with sorghum in Africa and Asia, but it is suppressed by natural enemies in subtropical and tropical regions in Africa (Singh *et al.*, 2004). In 2013, outbreaks were detected on sorghum along the Texas Gulf coast, southern Oklahoma, and the Mississippi Delta area of Louisiana and Mississippi. From 2013 to 2015, *M. sorghi* spread rapidly. The near-continental scale of the invasion and subsequent rapid spread extended south into Mexico and the Caribbean islands. Northward, its range extended to the central US (southern Illinois and Kansas), and along the eastern coast from Florida to North Carolina. It was detected to lesser extents in Arizona and California (Bowling *et al.*, 2016; Harris-Shultz *et al.*, 2017). The extensive spread reached across a substantial gradient of temperature, rainfall and humidity, and geography. Four regions considered in depth in this chapter generally followed a transition from subtropical to temperate climates. The most southern region (labelled South) was the Texas coastal areas that extended from Rio Grande Valley with a subtropical climate to more temperate areas near Houston. More temperate areas included central Texas to the lower Texas Panhandle and across central Oklahoma (labelled South Great Plains), and southern and central Alabama (labelled South East (Eastern US)). The most northern temperate region included the Texas and Oklahoma Panhandles through northern Oklahoma and adjacent southern Kansas (labelled North Great Plains).

Sorghum damage results from a combination of loss of plant nutrients, increased plant water stress, induced leaf chlorosis, and reduced photosynthesis because of the buildup of sooty mold growing on aphid honeydew. Initially, sorghum foliage remains green despite the presence of numerous aphids but becomes chlorotic after several weeks of uncontrolled *M. sorghi* increase (Fig. 8.1). An uncontrolled infestation can delay plant development and maturity, reduce the number of heads and seed weight, increase head sterility, and lead to plant death (Fig. 8.1). Grain quality and harvest efficiency can be affected by heads tainted with sticky honeydew. Notably, sorghum disease vectored by *M. sorghi* has not been observed in south Texas during the current invasion (M.J. Brewer, personal observation) nor documented to any substantial degree worldwide (Singh *et al.*, 2004).

Sorghum hybrids in cultivation in North America were largely lacking in bottom-up constraints to *M. sorghi* as most were susceptible to *M. sorghi* (Bowling *et al.*, 2016). Outbreaks were widespread, with commercial hybrids often observed supporting dense *M. sorghi* populations and suffering economic loss. Gordy *et al.* (2019) estimated an economic threshold of about 40 aphids per leaf on a range of commercial sorghum hybrids planted in several US states. High reproduction in southern areas created source populations for annual spread regionally by wind-aided flights of alates (Wang *et al.*, 2019; Koralewski *et al.*, 2020). But exceptions occurred, with low *M. sorghi* populations occurring sporadically in time and space in the South region (southern Texas) (M.J. Brewer, personal observation).

Potential bottom-up aphid regulation was identified within a few years of the initial outbreaks in the form of partially aphid-resistant sorghum hybrids. These hybrids were quickly characterized as partly tolerant and antibiotic in greenhouse screening and full season field testing (Armstrong *et al.*, 2015; Gordy *et al.*, 2021) (Fig. 8.1). Notably, a few hybrids were already available in the market and grown in some areas based on other desirable traits such as shortened days to maturity (Gordy *et al.*, 2021). Other aphid-resistant sorghum hybrids progressed quickly through breeding programs as resistant germplasm was readily available (Peterson *et al.*, 2018).

Another bottom-up consideration was that *M. sorghi* occurred and thrived on sorghum regrowth in harvested grain fields, forage sorghum fields, and the alternative host *Sorghum halepense* (L.) Pers. (Johnsongrass) (Esquivel *et al.*, 2021).

Fig. 8.1. *Melanaphis sorghi* on sorghum. (A) Initial colonization and population growth near an economic threshold (ET) of 40 aphids per leaf with little visual leaf chlorosis; inset shows *M. sorghi* apterae and aphids mummified (shiny black) by the parasitoid *Aphelinus nigritus*. (B) Largely uncontrolled population growth with *M. sorghi* densities averaging 500 aphids per leaf (few natural enemies detected) developed over 3–4 weeks with signs of plant decline (leaf chlorosis, overall plant necrosis, and grain head decline); inset shows over 500 apterae on a leaf. (C) Partially resistant sorghum hybrid infested with *M. sorghi* at levels exceeding the traditionally accepted ET of 40 aphids per leaf; inset shows *M. sorghi* apterae and one alate on a healthy green leaf (Gordy *et al.*, 2019, 2021). (Photo credits: J. Gordy and M.J. Brewer, in public domain through Texas A&M AgriLife Research.)

Johnsongrass persists through winter in the South and South Great Plains regions, providing harborage to *M. sorghi* (Esquivel *et al.*, 2021) and natural enemies (Faris *et al.*, 2022a). These characteristics suggested a high risk of *M. sorghi* outbreaks in subtropical growing areas, particularly south Texas, Mexico, and the Caribbean, consistent with invasion ecology and invasive species range expansion (Jeschke, 2014; Barbet-Massin *et al.*, 2018). During the first few years of the invasion, this pattern was borne out in field surveys (Brewer *et al.*, 2017; Rodríguez-del-Bosque *et al.*, 2018).

Potential top-down forces on suppressing *M. sorghi* populations were a point of discussion during the expansion phase of the invasion. The periodic outbreaks of *M. sorghi* on sorghum overlapped in geographic range with a suite of natural enemies suppressing other aphids occurring in cereal grains grown in the North American Great Plains (Colares *et al.*, 2015; Brewer *et al.*, 2019). From a landscape perspective and broadly relevant across systems, spatial arrangement of different field crops and semi-natural habitat may increase pest suppression by natural enemies. The consistency of this impact is dependent on insect mobility across crops and semi-natural habitats that differ in resource availability for insects and plant sensitivity to insects (Bianchi and van der Werf, 2004; Karp *et al.*, 2018). But estimating potential benefits of pest suppression relative to crop pest risk requires system-specific study (Brewer and Elliott, 2023). For example, parasitism of *Diuraphis noxia* (Mordvilko) (Hemiptera: Aphididae) by *Aphelinus albipodus* (Hayat & Fatima) (Hymenoptera: Aphelinidae) was associated with declines in aphid abundance. This was documented where wheat production was nested in a relatively diverse cropping system (Brewer *et al.*, 2008) and where winter wheat was interplanted with sunflower in a strip crop system (Ahern and Brewer, 2002).

Early field observations of *M. sorghi* and its natural enemies, the availability of partially aphid-resistant sorghum hybrids, and the history of past cereal aphid invasions in the North American Great Plains stimulated a two-part *M. sorghi* research effort. This effort considered the: (i) spatially variable aphid and natural enemy abundance across regions and (ii) extent of aphid suppression by aphid-susceptible and partially aphid-resistant sorghum hybrids. The aim was to consider aphid suppression by natural enemies as an innate agroecosystem response to the *M. sorghi* invasion and whether a preventative pest management strategy, partial sorghum resistance to *M. sorghi*, would complement natural suppression when used regionally where aphid risk was persistent.

Regional Associations with Natural Enemies

In a comparison of *M. sorghi* and its natural enemies (Brewer *et al.*, 2022), sorghum fields were randomly selected for sampling from a list of probable *M. sorghi*-infested fields provided by local collaborators in 40 US counties from 2016 to 2019. These counties fell in the four regions that followed a latitudinal transition from subtropical, to temperate climates described above and in Brewer *et al.* (2022). Plant growth stage ranged from mid vegetative growth through early grain fill stages of reproductive growth. Each year, 2–12 fields were sampled in each county, and each field was sampled around 8 times by inspecting a minimum of 160 leaves per sampling event. Supplementing this coordinated effort, records on *M. sorghi* and natural enemies were available using the similar techniques before and after the organized survey in these South regions, as well as from Mexico.

The 4-year survey was fully reported in Brewer *et al.* (2022) and summarized here. Average *M. sorghi* densities typically fell within the range of 20–70 *M. sorghi* per leaf, which bridged across subeconomic to economic densities (Gordy *et al.*, 2019). In this survey and reported by others, natural enemies preying on *M. sorghi* consisted of about 19 species including solitary endoparasitoids (Hymenoptera: Aphelinidae, Braconidae), a hyperparasitoid (Hymenoptera: Encyrtidae), lady beetles (Coleoptera: Coccinellidae), hoverflies (Diptera: Syrphidae), lacewings (Neuroptera: Chrysopidae and Hemerobiidae), and pirate bugs (Hemiptera: Anthocoridae) (e.g. Colares *et al.*, 2015; Brewer *et al.*, 2017; Maxson *et al.*, 2019; Lahiri *et al.*, 2020). Photographs of adults, selected larvae, and mummified *M. sorghi* can be found in Maxson *et al.* (2019) and outreach publications (e.g. Bowling *et al.*, 2016; Faris and Brewer, 2022).

The South region experienced low *M. sorghi* densities (around 23 *M. sorghi* per leaf) with some populations exceeding the economic threshold of 40 aphids per leaf (Figs 8.2 and 8.3). Abundances of both parasitoids and lady beetle larvae were relatively high (Fig. 8.2) compared with other regions of the survey (Brewer *et al.*, 2022). A temporal shift in the natural enemy complex was seen, with *Aphelinus nigritus* (Hymenoptera: Aphelinidae), becoming numerically dominant as the years progressed (Fig. 8.2). These aphid and natural enemy population trends occurred across a range of agricultural landscapes along the Texas Gulf coast, with limited indication that abundance was tied to landscape conditions (Brewer *et al.*, 2022; Elkins *et al.*, 2022). Elsewhere, *Lysiphlebus testaceipes* (Cresson) (Hymenoptera: Braconidae) parasitized *M. sorghi* in northeastern Mexico (Rodríguez-del-Bosque *et al.*, 2018) and periodically parasitized *M. sorghi* in the South and South Great Plains regions (e.g. south and central Texas as reported in Maxson *et al.*, 2019). Coccinellids and selected other predator taxa were commonly reported from the South and South Great Plains regions (Maxson *et al.*, 2019) (Fig. 8.2) as well as further south in Mexico (Rodriguez-Velez *et al.*, 2021) and the North Great Plains region of the 2016–2019 survey (Brewer *et al.*, 2022).

In contrast, *M. sorghi* abundance commonly exceeded the economic threshold of 40 *M. sorghi* per leaf in the South East region (average of 130 *M. sorghi* per leaf) (Brewer *et al.*, 2022). In this region, natural enemy abundance was not associated with *M. sorghi* abundance and landscape conditions, but it was associated with weather, particularly rainfall. In the South Great Plains region, a shift to more common occurrence of predators was seen. Coccinellid abundance appeared associated with landscape composition, particularly the extent of sorghum production and shrubland (Brewer *et al.*, 2022).

Prior knowledge of natural enemies preying on cereal aphids and their association with landscape, agricultural, and weather conditions (Brewer *et al.*, 2019) also supported the idea that natural enemies were a top-down force in *M. sorghi* suppression and contributed to shaping the sorghum agroecosystem response to the *M. sorghi* invasion. On wheat, parasitism by *L. testaceipes* was sufficient to prevent populations of *Schizaphis graminum* (Rondani) (Hemiptera: Aphididae) from reaching damaging population levels (Giles *et al.*, 2003), and parasitism was positively correlated with plant diversity (cropland and grassland/pasture) (Elliott *et al.*, 2018; see Chapters 5 and 6, this volume). Parasitism of the aphid *D. noxia* by the parasitoid *A. albipodus* was associated with declines in *D. noxia* abundance, especially in areas where wheat production was nested in grasslands and other crops (Brewer *et al.*, 2008), including where wheat and sunflower were grown in a strip cropping system (Ahern and Brewer, 2002). Outside the North American Great Plains, Thies *et al.* (2005) studied parasitoids of cereal aphids affected by landscape diversity in Europe. They concluded that increasing landscape diversity was associated with both increasing parasitoid abundance but also increasing cereal aphid abundance.

Field suppression of *M. sorghi* was indicated by repeated measurements of *M. sorghi* and natural enemies in the South region across the growing season, revealing examples of classic predator–prey population oscillations (Fig. 8.3). *Aphelinus nigritus* was assumed to be the primary suppression agent due to its dominance in the surveys. Notably, some of the first outbreaks of the invasion occurred in this region (Bowling *et al.*, 2016). Of the regions sampled, the South region was the closest climate match to subtropical and tropical areas where *M. sorghi* has long infested sorghum in Africa and natural enemies commonly regulate *M. sorghi* (Singh *et al.*, 2004).

Overall, *M. sorghi* populations were spatially variable and declining through the four years of the survey (Fig. 8.2). But outbreaks still occurred periodically. Also, consumption rates for predators, rates of parasitism, and different generation times of predators and parasitoids does limit comparison of field aphid mortality across taxa using the survey data. For these reasons, controlled experiments guided partly by the survey results were carried out to validate and estimate aphid suppression on sorghum. An additional aim was to consider prospects for regional use of partially resistant sorghum to complement suppression by natural enemies where aphid risk persisted.

Melanaphis sorghi Suppression on Resistant Sorghum

Observations of *M. sorghi* and their natural enemies on both aphid-susceptible and -resistant hybrids

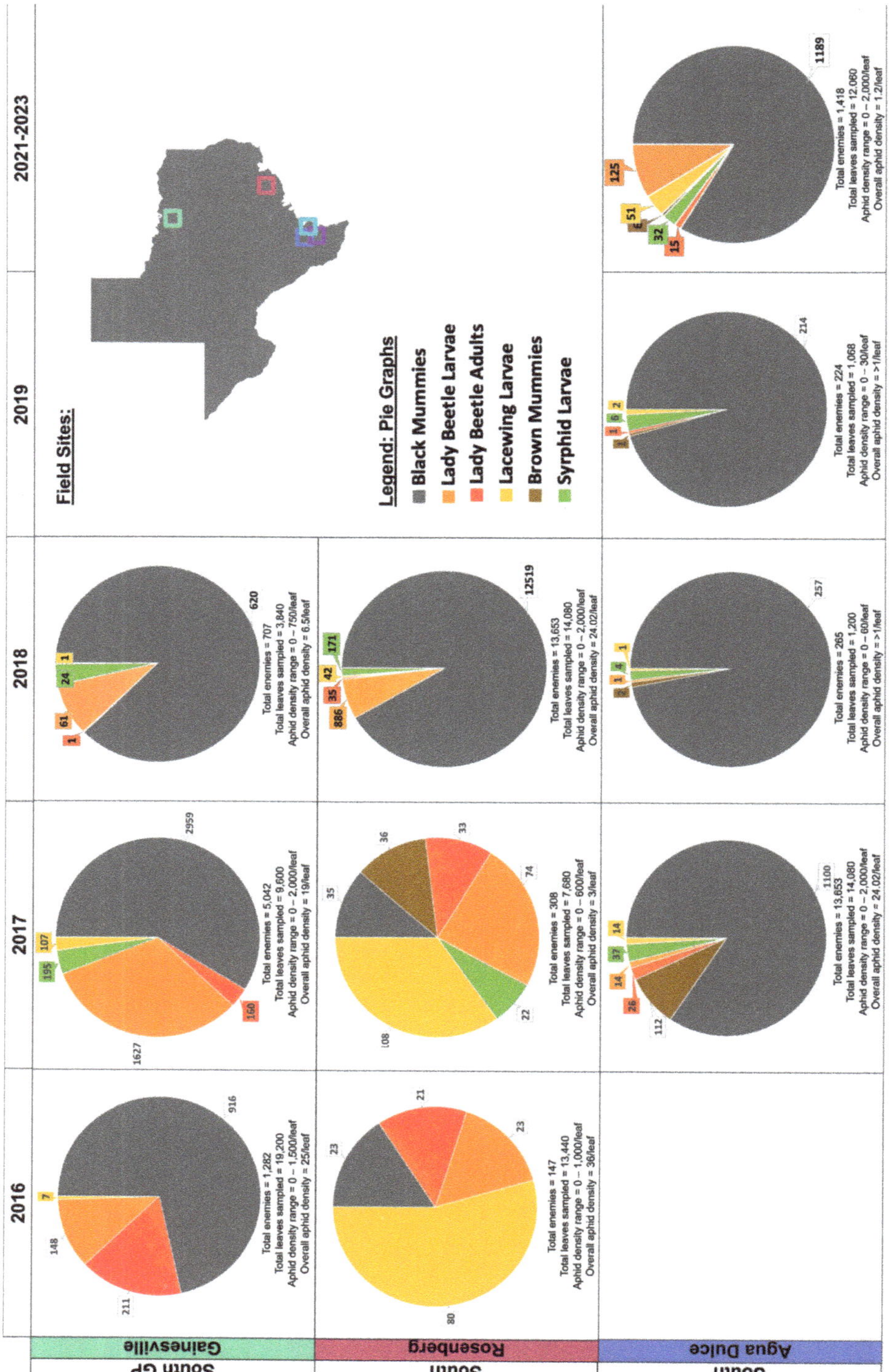
2016
2017
2018
2019
2021-2023
Field Sites:
Legend: Pie Graphs
Black Mummies
Lady Beetle Larvae
Lady Beetle Adults
Lacewing Larvae
Brown Mummies
Syrphid Larvae
South GP
Gainesville
South
Rosenberg
South
Agua Dulce

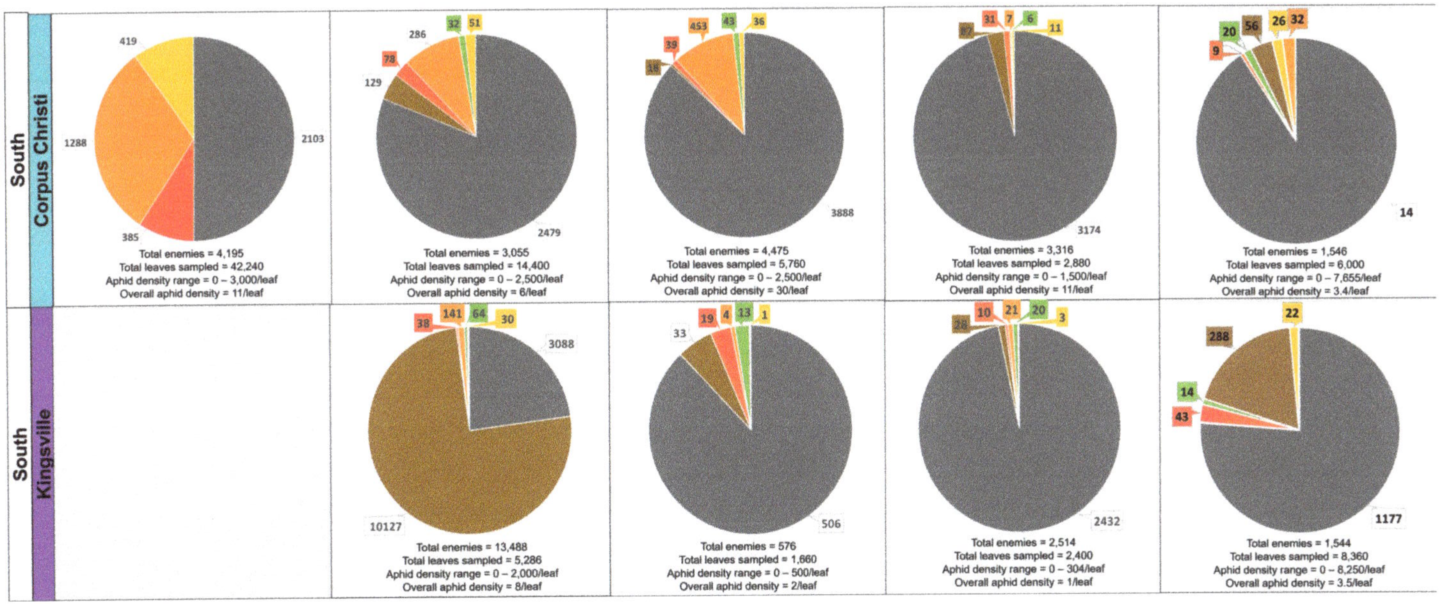

Fig. 8.2. Field survey of natural enemies of *Melanaphis sorghi* across localities in Texas during the mid 2010s through the early 2020s. These data represent multiple field visits in the specified region. The number of sampling visits range from 2 to 20 conducted in 1 to 12 fields at the various locations. These surveys overlapped with surveys conducted in the South region (Agua Dulce, Corpus Christi, Kingsville, and Rosenberg areas of south Texas) and South Great Plains GP region (Gainesville area of central Texas) of the coordinated 2016–2019 survey reported in Brewer *et al.* (2022) and including selective 2021 and 2023 updates. Pie charts display an abundance of natural enemy taxa; text below pie charts provide total enemies, total leaves sampled, *M. sorghi* density range, and overall *M. sorghi* density. (Content is the author's own: A.M. Faris.)

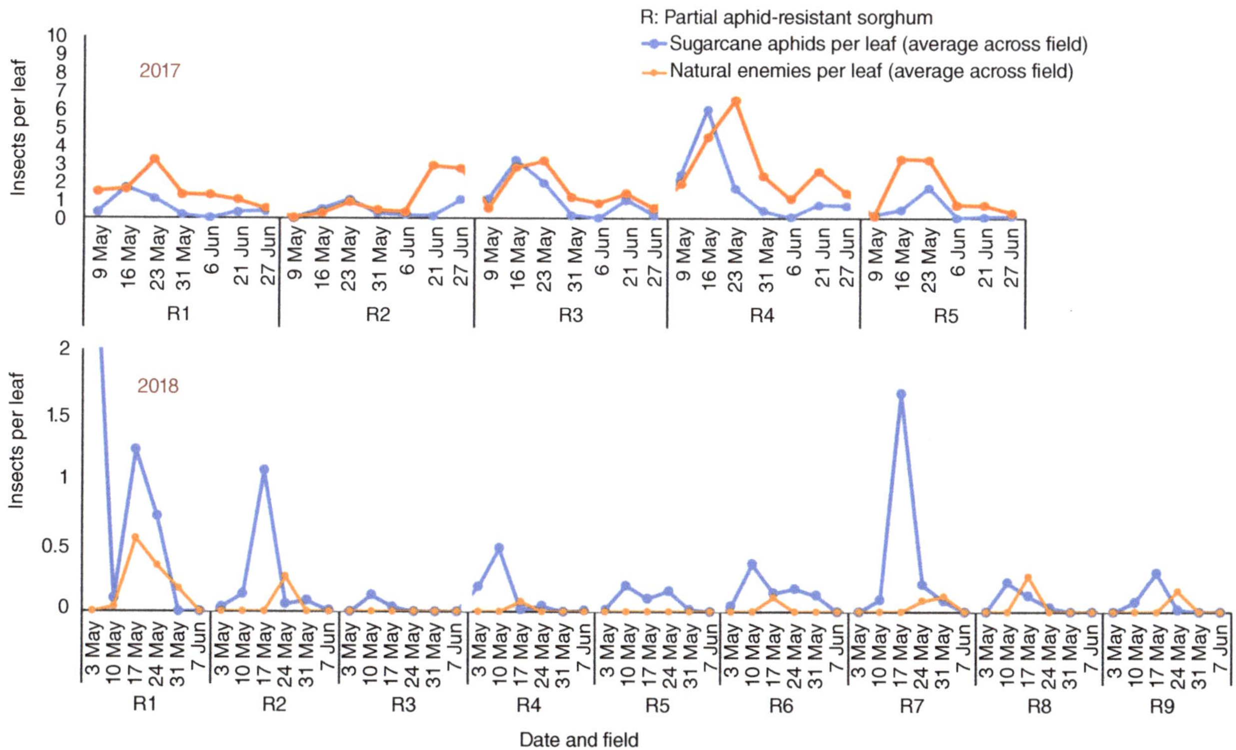

Fig. 8.3. Repeated field measurements of *Melanaphis sorghi* and natural enemy abundance across the growing season revealing examples of classic predator–prey populations oscillations on partially aphid-resistant sorghum. The dominant natural enemy species was the parasitoid *Aphelinus nigritus* (see Fig. 8.2 for relative abundance of the natural enemy complex, note change in scale of charts). The location and data came from the Agua Dulce area of south Texas as represented in aggregate in Fig. 8.2. (Content is the author's own: A.M. Faris.)

led to a controlled cage exclosure experiment to evaluate the combined effects of top-down (natural enemies) and bottom-up (aphid-resistant hybrids) control across the South and South Great Plains regions (Faris *et al.*, 2022b). Similar mechanical cage exclosure studies have been used with other cereal and soybean aphids to evaluate the impact of natural enemies on these pests on susceptible host plants (Kring *et al.*, 1985; Michels *et al.*, 2001; Lee *et al.*, 2005; Costamagna *et al.*, 2008). While some growers across the South and South Great Plains regions were electing to plant commercially available, aphid semi-resistant sorghum hybrids, other growers continued to plant aphid-susceptible sorghum. Because aphid population growth was higher in susceptible sorghum hybrids than in resistant hybrids, aphid suppression by natural enemies was not directly comparable across these hybrids (Gordy *et al.*, 2021).

Using a natural enemy exclusion cage technique, natural enemy suppression of *M. sorghi* was evaluated on partially aphid-resistant and -susceptible hybrids by comparing aphid population levels of each hybrid protected and exposed to natural enemies under similar agronomic field conditions in Stillwater, OK (representing the South Great Plains region of the 4-year survey) and Corpus Christi, TX (representing the South region).

The hybrids were similar in maturity and were planted in a small plot with eight replications at both locations. The experiment was carried out during one growing season at the South Great Plains location and two growing seasons at the South region. At the 4–5 leaf stage, plants were artificially infested with 50 colony-reared *M. sorghi* and then caged with the appropriate cage treatment that was modified from Lee *et al.* (2005). There were three natural enemy treatments: (i) an open treatment where predators and parasitoids had access to the infested aphids, (ii) a partial exclosure treatment where parasitoids had aphid access and predators did not (coarse mesh fabric), and (iii) a full exclosure treatment where predators and parasitoids could not access the aphids (fine mesh fabric). Sorghum plants were destructively sampled at the 6–7 leaf growth stage. Sampling consisted of removing whole plants from the cage, bagging the plant, and transferring to the laboratory for whole plant counts of *M. sorghi*, predators, and mummies. Predators were identified to species using Maxson *et al.* (2019) photographs. Mummies were clipped from leaves and placed in gel capsules to rear parasitoids to adults for species identification using Maxson *et al.* (2019). Data from plants sampled 3 weeks after aphid infestation are reviewed here. The complete data set and analyses can be found in Faris *et al.* (2022b).

The most abundant natural enemy across the two locations and years was the parasitoid *A. nigritus*, as indicated by black mummies (Table 8.1). The second most abundant natural enemy was coccinellids, although predators were rarely detected in 2018 in the South region (Table 8.1). The adult coccinellids were of similar species across the cage study regions and years. *Scymnus* spp. (Coleoptera: Coccinellidae) were the most abundant adult lady beetle species in Corpus Christi, TX, in 2019, and *Hippodamia convergens* (Guérin-Méneville) (Coleoptera: Coccinellidae) was the most abundant adult lady beetle species in Stillwater, OK, in 2019 (Table 8.1). There were minimal observations of chrysopids, hemerobiids, and syrphids (family identification only) at both Corpus Christi, TX and Stillwater, OK sites. The natural enemy composition of this cage study was similar to the natural enemies observed in the field during the 2017 and 2018 monitoring activities, where the parasitoid *A. nigritus* was relatively high in abundance followed by low predator numbers of coccinellids, syrphids, and chrysopids/hemerobiids (Elliott *et al.*, 2021) (Fig. 8.2). There were also similarities to the southeastern US *M. sorghi* natural enemy complex, with slight differences in lady beetle adult species (Uyi *et al.*, 2022).

No mummies were observed in the complete exclusion treatment, verifying the cages' ability to exclude natural enemies. When natural enemy abundance data were parsed out for the two cage treatments that allowed natural enemies access to the artificially infested *M. sorghi*, mummies and predators were recovered from the no exclusion and partial exclusion treatments on both the aphid-susceptible and -resistant hybrids (Fig. 8.4). As seen in Fig. 8.4, predators in low numbers were noted in the partial exclusion cages, verifying largely successful partial exclusion of predators. There were noticeably more predators relative to the number of parasitoids in the no exclusion treatment at the Stillwater, OK site than at the Corpus Christi, TX site (Fig. 8.4). The low predator numbers at the Corpus Christi,

Table 8.1. Relative abundance (n) of natural enemies collected from susceptible and resistant sorghum at the South region, Corpus Christi, TX (2018 and 2019), and the South Great Plains region, Stillwater, OK (2019). (Content is the authors' own: A.M. Faris, M.J. Brewer, and N.C. Elliott.)

	Corpus Christi, TX		Stillwater, OK
Natural enemy	2018	2019	2019
Parasitoids			
Aphelinus nigritus	85	1251	75
Adults			
Cycloneda sanguinea	0	17	0
Scymnus spp.	0	72	0
Coccinella septempunctata	1	2	1
Hippodamia convergens	0	5	5
Coleomegilla maculata	0	1	4
Harmonia axyridis	0	5	1
Juveniles			
Coccinellidae juveniles	0	58	32
Syrphidae larvae	0	30	1
Chrysopidae/Hemerobiidae larvae	0	8	9

Data shown here are from the no exclusion (open field) treatment. Juvenile coccinellids, syrphids, and chrysopids/hemerobiids were identified to taxonomic family level.

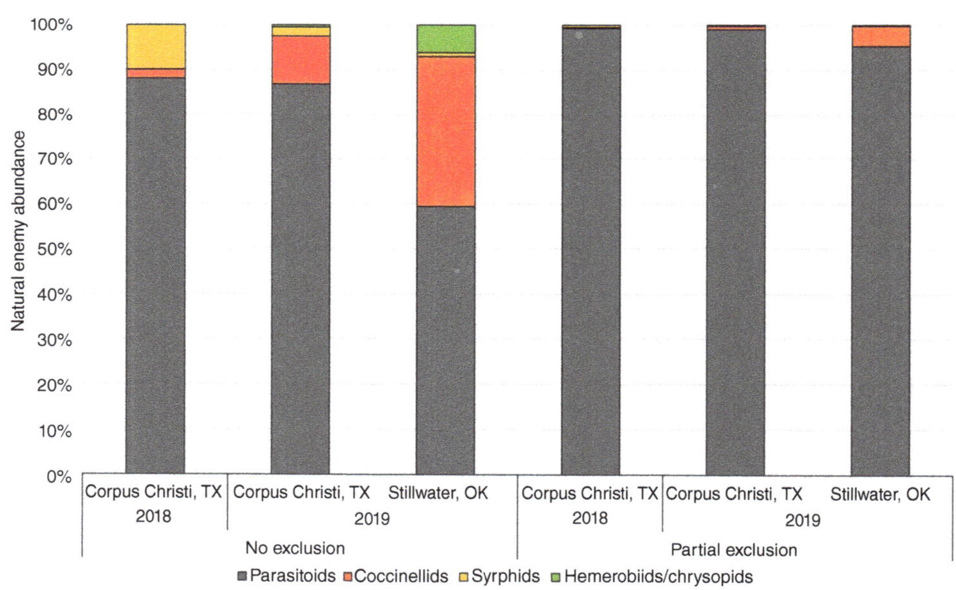

Fig. 8.4. Relative abundance of natural enemies collected from susceptible and resistant sorghum at the South region, Corpus Christi, TX (2018 and 2019), and the South Great Plains region, Stillwater, OK (2019), by location, year and cage treatment. Data shown here are from the cages that permitted natural enemy access to the artificially infested aphids: (i) the no exclusion (open field) cages which did not exclude enemies and (ii) the partial exclusion (coarse mesh) cages which excluded predators but allowed parasitoids. (Content is the authors' own: A.M. Faris, M.J. Brewer, and N.C. Elliott.)

TX site was representative of what was observed during field monitoring of *M. sorghi* and natural enemies (Fig. 8.2).

Previous studies have shown that bottom-up forces such as the planting of aphid-resistant sorghum varieties can serve as a baseline for protecting against *M. sorghi* by suppressing population growth (Sharma *et al.*, 2013; Armstrong *et al.*, 2015; Haar *et al.*, 2019; Paudyal *et al.*, 2019; Pekarcik and Jacobson, 2021). Of these studies, one was conducted in the South region where significant suppression (50–80% reduction) of *M. sorghi* was observed on an aphid-resistant sorghum variety (Armstrong *et al.*, 2015). When all natural enemies were excluded using the full exclusion cage, the *M. sorghi* populations were only limited by the aphid-resistant variety, and the number of *M. sorghi* on aphid-resistant hybrids was not significantly different from susceptible hybrids (Fig. 8.5). Taking into consideration that the economic threshold for *M. sorghi* in this region is 40 aphids per leaf and the sampled plants had seven leaves (40 aphids per leaf × 7 leaves = 280 aphids), the number of *M. sorghi* per leaf were more than double the economic threshold regardless of the hybrid type, location, and year in the full exclusion cages. This highlights the important role that natural enemies can play in the suppression of *M. sorghi* on aphid-susceptible and -resistant hybrids.

Since *A. nigritus* and coccinellids as a group were the most abundant natural enemies, they were the focus of further analyses. Estimates of percent aphid suppression were calculated when two conditions were met. First, significantly fewer *M. sorghi* were found in the no or partial exclusion cages than in the complete exclusion cages for one or both sorghum hybrid types, and one of the natural enemy measurements was significantly greater in the no or partial exclusion cages than in the complete exclusion cages for one or both sorghum hybrid types. In these cases, percent aphid suppression associated with predator and parasitoid activity = 100 × (1 − number of aphids in the no exclusion cage/number of aphids in the complete exclusion cage), and percent aphid suppression associated with parasitoid activity only = 100 × (1 − number of aphids in the partial exclusion cage/number of aphids in the complete exclusion cage). For parasitoids, apparent per cent field parasitism was also calculated for reference to past studies reporting this measurement, i.e. 100 × number of mummies/(number of mummies + aphids) (Fig. 8.6).

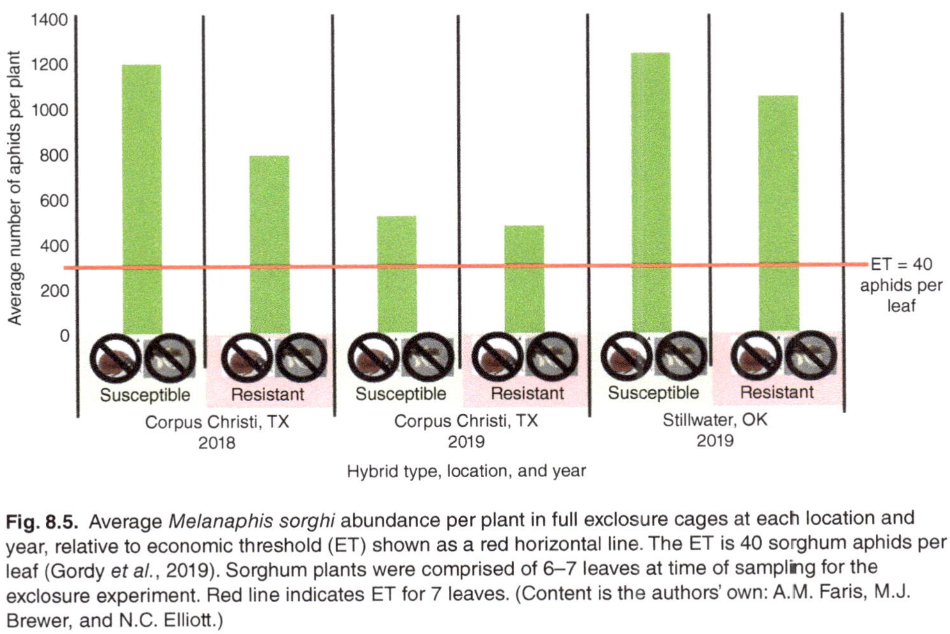

Fig. 8.5. Average *Melanaphis sorghi* abundance per plant in full exclosure cages at each location and year, relative to economic threshold (ET) shown as a red horizontal line. The ET is 40 sorghum aphids per leaf (Gordy *et al.*, 2019). Sorghum plants were comprised of 6–7 leaves at time of sampling for the exclosure experiment. Red line indicates ET for 7 leaves. (Content is the authors' own: A.M. Faris, M.J. Brewer, and N.C. Elliott.)

In the Corpus Christi, TX 2018 full exclusion cages, where *M. sorghi* had no exposure to parasitoids or predators, there were over 1600 *M. sorghi* per plant in both aphid-resistant and -susceptible hybrids (Fig. 8.6A). When *M. sorghi* were exposed to parasitoids in the partial exclosure cages, there were approximately 25 *M. sorghi* per plant. Using the above formulae, *M. sorghi* suppression by *A. nigritus*, was estimated at about 95% on the resistant hybrid and 80% on the susceptible hybrid. The ratio of black mummies to *M. sorghi* indicated that for every two *M. sorghi* observed, there was one black mummy recovered in the partial exclosure and open cages (Fig. 8.6C). The 2019 results at this location were similar.

The 2019 Stillwater, OK experiment had a contrasting story. At this location there were over 1800 *M. sorghi* in the full cage treatments in both aphid-susceptible and -resistant hybrids (Fig. 8.6B). In the partial exclosure cages, approximately 1500 *M. sorghi* were observed, indicating that some suppression was occurring on both hybrids. *Melanaphis sorghi* suppression due to *A. nigritus* was estimated to be approximately 11% on the resistant hybrid and approximately 33% on the susceptible hybrid when comparing *M. sorghi* detected in the open and partial exclusion treatments (Fig. 8.6D). Black mummies were present in the partial cages, but the parasitism by *A. nigritus* did not explain all the *M. sorghi* suppression at this location. There was a significant difference between aphid abundance in the partial and open cages (Fig. 8.6B). Suppression of *M. sorghi* in the open treatment was much higher (>90%) compared to aphids in the full exclusion treatment, supporting the conclusion of additional suppression by predators (Fig. 8.6D). This is also supported by the observation of more predators relative to the number of parasitoids at the Stillwater, OK site (Fig. 8.4).

Prior to this study, suppression of *M. sorghi* by natural enemies was assumed due to the observation within the sorghum growing season that as the number of *M. sorghi* increased, the natural enemy abundance followed a similar

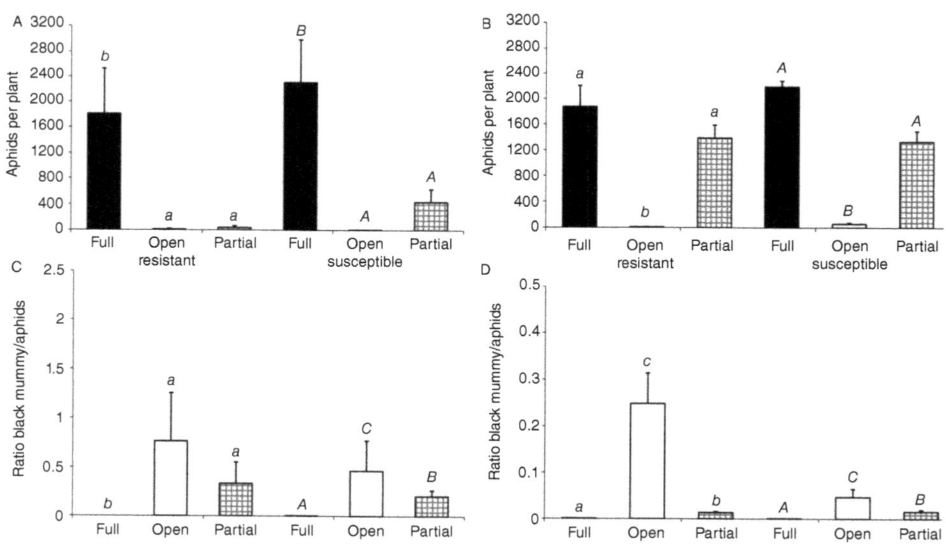

Fig. 8.6. Mean abundance of *Melanaphis sorghi* and the ratio of black mummies to *M. sorghi* recorded at the second timepoint collection in the South region, Corpus Christi, TX 2018 (A and C, respectively) and the South Great Plains region, Stillwater, OK 2019 (B and D, respectively). Tukey's means separation test was used for the comparisons across natural enemy treatments for each resistant (italic lower case letters) and susceptible (italic upper case letters) hybrid for each year. Means sharing a common italic uppercase letter did not significantly differ using the Tukey's means separation test. (Content is the authors' own: A.M. Faris, M.J. Brewer, and N.C. Elliott.)

trend, displaying classic predator–prey dynamics (Fig. 8.3), with less frequent outbreaks of *M. sorghi* on sorghum being observed (Fig. 8.2). The exclosure study validated suppression of *M. sorghi* by natural enemies. Suppression was largely due to the parasitoid, *A. nigritus*, in the 2 years at the Corpus Christi, TX location. At the Stillwater, OK location, there was a more complex picture, where suppression of *M. sorghi* was due to parasitism by *A. nigritus* and predation by coccinellids. Aphid semi-resistant hybrids have proven to be effective in suppressing *M. sorghi* population growth in the South region (Armstrong *et al.*, 2015), but not in the study presented. However, it was unknown how these hybrids may impact suppression by natural enemies. Fortunately, the aphid-resistant hybrids do not have a negative impact on natural enemy suppression of *M. sorghi*, as suppression by natural enemies was observed on both hybrid types at both locations and years. Aphid suppression appears to be due to a combination of parasitoids and predators, and semi-resistant hybrids are compatible with these resident natural enemies.

A study conducted by Uyi *et al.* (2022) evaluated the impact of sorghum resistance to *M. sorghi*, planting date, and foliar insecticide use on *M. sorghi* suppression and natural enemies in the southeastern US. This study highlighted the use of host plant resistance in combination with using insecticides with low toxicity to natural enemies. *Aphelinus nigritus* was observed parasitizing *M. sorghi* on aphid-susceptible and -resistant hybrids, with or without the use of foliar insecticides (Uyi *et al.*, 2022). In the South region (Corpus Christi, TX), *A. nigritus* was observed in the field early in the sorghum growing season where it parasitized *Rhopalosiphum maidis* (Fitch) (Hemiptera: Aphididae), corn leaf aphid, found in the whorl of the sorghum plant. Elliott *et al.* (2021) noted that the subtropical climate for the Texas Coastal Plains (South region) may allow for year-round activity of aphids and their natural enemies.

The habitat surrounding the sorghum field sites may also play a role in natural enemy colonization and recruitment to *M. sorghi* both in and out of the sorghum growing season (Elliott *et al.*, 1999, 2018; Bianchi *et al.*, 2006; Chaplin-Kramer *et al.*, 2011; Faris *et al.*, 2022a). A sentinel pot study was conducted in the South region (Corpus Christi, TX) at the same time as the exclosure cage study reviewed here (Faris *et al.*, 2022b). Before, during, and after the sorghum growing season, sentinel pots of aphid-infested susceptible sorghum were placed in Johnsongrass, riparian areas (agricultural ditches that can contain Johnsongrass), sorghum (during the growing season), and remnant sorghum (outside of the growing season) (Faris *et al.*, 2022a). Parasitoids were the most abundant natural enemy during the sorghum growing season and lady beetle adults were the most abundant natural enemy off-season when sorghum was not in cultivation. *Aphelinus nigritus* was most abundant in the riparian habitat and was present in and out of the sorghum growing season in all three habitats. This supported the cage study results indicating this parasitoid was a key player in *M. sorghi* suppression within the South region (Faris *et al.*, 2022a, b). Overall, natural enemies were found in all habitats but were most abundant in riparian areas and in-season sorghum and least abundant in remnant sorghum (Faris *et al.*, 2022a). In terms of IPM practices, destroying sorghum regrowth from fields following harvest may substantially reduce aphid population growth and movement of aphids to overwintering sites in Johnsongrass. Fortunately, natural enemies are also present in Johnsongrass outside of the cultivated sorghum season. Johnsongrass may be serving as a green bridge for *M. sorghi* and its natural enemies ahead of the next growing season, potentially preventing outbreaks in sorghum. Additional research may help illuminate whether crop and non-crop vegetation at a local scale can support natural enemies when sorghum is not present.

While *A. nigritus* is abundant and has become more widespread, the killing power and generation turnover differences between parasitoids and predators should be considered to characterize suppression by natural control. Studies have evaluated predator–prey models consisting of lady beetle larvae and adults and lacewing larvae to characterize their suppression of *M. sorghi* at different pest densities. It was found that all life stages of *Coccinella septempunctata*, *Harmonia axyridis* (Pallas) (Coleoptera: Coccinellidae), and *Chrysoperla rufilabris* (Burmeister) (Neuroptera: Chrysopidae) significantly reduced aphid population growth (Hewlett *et al.*, 2019). In the South Great Plains and South regions, where the exclosure cage study was conducted, Elliott *et al.* (2021) elucidated geographic patterns of *M. sorghi* natural enemy occurrence, abundance, and predatory potential during the sorghum growing season. They quantified the importance of natural enemy taxa using three variables: cumulative degree-days on the date of first occurrence, taxon mean density, and the relative predation rate (number of aphids killed per day) of taxon from published literature (Elliott *et al.*, 2021). Larval coccinellids had over a twofold larger taxonomic importance index ($I = 1.47$) than *A. nigritus* ($I = 0.61$), but Elliott *et al.* (2021) also found that the South region, where *A. nigritus* was the predominant species, had larger importance indices (two locations sampled, $I = 1.27$) than any location in the South Great Plains region further north in Oklahoma (three locations sampled, $I = 1.02$). The South region seems particularly well suited for *A. nigritus* to suppress *M. sorghi*, based on its high abundance (Fig. 8.4), direct measures of *M. sorghi* suppression (Fig. 8.6), and occurrence in cultivated sorghum and surrounding habitat. The landscape of the South Great Plains region north of this location (north Texas and Oklahoma) may be more suited to coccinellids. Quantifying suppression potential of natural enemy taxa in different habitats and on aphid-susceptible and -resistant sorghum should be combined with geographic variation of natural enemy abundance in regions where *M. sorghi* is an important pest. This information can be used to determine where conservation biological control efforts and planting aphid-resistant sorghum hybrids may be most effective and needed.

Summary and Conclusions

From a regional or agroecosystem-focused IPM perspective, this review of several studies adds to the evidence that parasitism and predation by resident natural enemies are well suited to aphid management in the low-input large-scale cereal agroecosystem of the North American Great Plains (Giles *et al.*, 2008; Brewer *et al.*, 2008; Bowling *et al.*, 2016). The regional-scale risk assessment as a product of combining aphid and natural enemy surveys can help guide where to plant aphid-resistant sorghum. For *M. sorghi*, use of partially aphid-resistant sorghum does not deter parasitism and predation (Brewer *et al.*, 2017; Maxson *et al.*, 2019). Planting aphid-resistant sorghum is particularly important where risk is more persistent, variable, or prone to variable natural enemy activity across landscape and weather conditions. In such cases, variable top-down forces provided by natural enemies benefit from the addition of bottom-up forces of sorghum resistant to *M. sorghi*, as well as strategic use of insecticides with low toxicity to natural enemies (Uyi *et al.*, 2022). Future studies that evaluate the impact of landscape and environmental variations on pests and pest suppression will advance the utility of regional pest management.

Acknowledgments

We thank the United Sorghum Checkoff Program for financial support. The USDA Agricultural Research Service also provided funding through the Areawide Pest Management Program, project 58-3072-6-013. We thank Ivin Lam for assistance with the figures. This research was conducted partly under the umbrella of USDA National Institute of Food and Agriculture, USA Hatch project TEX0-2-9394 assigned to M.J.B. at Texas A&M AgriLife Research. Mention of trade names or commercial products in this publication is solely for the purpose of providing specific information and does not imply recommendation or endorsement by the USDA. The USDA is an equal opportunity provider and employer.

References

Ahern, R.G. and Brewer, M.J. (2002) Effect of different wheat production systems on the presence of two parasitoids (Hymenoptera: Aphelinidae; Braconidae) of the Russian wheat aphid in the North American Great Plains. *Agriculture, Ecosystems & Environment* 92, 201–210. DOI: 10.1016/S0167-8809(01)00296-1

Armstrong, J.S., Rooney, W.L., Peterson, G.C., Villanueva, R.T., Brewer, M.J. *et al.* (2015) Sugarcane aphid (Hemiptera: Aphididae): host range and sorghum resistance including cross-resistance from greenbug sources. *Journal of Economic Entomology* 108, 576–582. DOI: 10.1093/jee/tou065

Barbet-Massin, M., Rome, Q., Villemant, C. and Courchamp, R. (2018) Can species distribution models really predict the expansion of invasive species? *PLoS ONE* 13, e0193085. DOI: 10.1371/journal.pone.0193085

Bianchi, F.J.J.A. and van der Werf, W. (2004) Model evaluation of the function of prey in non-crop habitats for biological control by ladybeetles in agricultural landscapes. *Ecological Modelling* 171, 177–193. DOI: 10.1016/j.ecolmodel.2003.08.003

Bianchi, F.J.J.A., Booij, C.J.H. and Tscharntke, T. (2006) Sustainable pest regulation in agricultural landscapes: a review on landscape composition, biodiversity and natural pest control. *Proceedings of the Royal Society B: Biological Sciences* 273, 1715–1727. DOI: 10.1098/rspb.2006.3530

Blackman, R.L. and Eastop, V.F. (1984) *Aphids on the World's Crops: An Identification and Information Guide*. London, UK: Wiley.

Bowling, R.D., Brewer, M.J., Kerns, D.L., Gordy, J., Seiter, N. *et al.* (2016) Sugarcane aphid (Hemiptera: Aphididae): a new pest on sorghum in North America. *Journal of Integrated Pest Management* 7, 12. DOI: 10.1093/jipm/pmw011

Brewer, M.J. and Elliott, N.C. (2023) Recent advances in agroecological research for increasing scope of areawide pest management of arthropods in cropping systems. *Current Opinion in Insect Science* 56, 101019. DOI: 10.1016/j.cois.2023.101019

Brewer, M.J. and Goodell, P.B. (2012) Approaches and incentives to implement integrated pest management that addresses regional and environmental issues. *Annual Review of Entomology* 57, 41–59. DOI: 10.1146/annurev-ento-120709-144748

Brewer, M.J., Noma, T., Elliott, N.C., Kravchenko, A.N. and Hild, A.L. (2008) A landscape view of cereal aphid parasitoid dynamics reveals sensitivity to farm- and region-scale vegetation structure. *European Journal of Entomology* 105, 503–511. DOI: 10.14411/eje.2008.066

Brewer, M.J., Gordy, J.W., Kerns, D.L., Woolley, J.B., Rooney, W.L. *et al.* (2017) Sugarcane aphid population growth, plant injury, and natural enemies on selected grain sorghum hybrids in Texas and Louisiana. *Journal of Economic Entomology* 110, 2109–2118. DOI: 10.1093/jee/tox204

Brewer, M.J., Peairs, F.B. and Elliott, N.C. (2019) Invasive cereal aphids of North America: ecology and pest management. *Annual Review of Entomology* 64, 73–93. DOI: 10.1146/annurev-ento-011118-111838

Brewer, M.J., Elliott, N.C., Esquivel, I.L., Jacobson, A.L., Faris, A.M. *et al.* (2022) Natural enemies, mediated by landscape and weather conditions, shape response of the sorghum agroecosystem of North America to the invasive aphid *Melanaphis sorghi*. *Frontiers in Insect Science* 2, 830997. DOI: 10.3389/finsc.2022.830997

Chaplin-Kramer, R., O'Rourke, M.E., Blitzer, E.J. and Kremen, C. (2011) A meta-analysis of crop pest and natural enemy response to landscape complexity. *Ecology Letters* 14, 922–932. DOI: 10.1111/j.1461-0248.2011.01642.x

Colares, F., Michaud, J.P., Bain, C.L. and Torres, J.B. (2015) Recruitment of aphidophagous arthropods to sorghum plants infested with *Melanaphis sacchari* and *Schizaphis graminum* (Hemiptera: Aphididae). *Biological Control* 90, 6–24. DOI: 10.1016/j.biocontrol.2015.05.009

Costamagna, A.C., Landis, D.A. and Brewer, M.J. (2008) The role of natural enemy guilds in *Aphis glycines* suppression. *Biological Control* 45, 368–379. DOI: 10.1016/j.biocontrol.2008.01.018

Elkins, B.H., Eubanks, M.D., Faris, A.M., Wang, H.-H. and Brewer, M.J. (2022) Landscape complexity has mixed effects on an invasive aphid and its natural enemies in sorghum agroecosystems. *Environmental Entomology* 51, 660–669. DOI: 10.1093/ee/nvac036

Elliott, N.C., Kieckhefer, R.W. and Lee, J.H. (1999) Influence of within-field and landscape factors on aphid predator populations in wheat. *Landscape Ecology* 14, 239–252. DOI: 10.1023/A:1008002528345

Elliott, N.C., Brewer, M.J. and Giles, K.L. (2018) Landscape context affects aphid parasitism by *Lysiphlebus testaceipes* (Hymenoptera: Aphidiinae) in wheat fields. *Environmental Entomology* 47, 803–811. DOI: 10.1093/ee/nvy035

Elliott, N., Giles, K., Brewer, M., Szczepaniec, A., Knutson, A. *et al.* (2021) Recruitment of natural enemies of the invasive sugarcane aphid vary spatially and temporally in sorghum fields in the Southern Great Plains of the USA. *Southwestern Entomologist* 46, 357–372. DOI: 10.3958/059.046.0207

Esquivel, I.L., Faris, A.M. and Brewer, M.J. (2021) Sugarcane aphid, *Melanaphis sacchari* (Hemiptera: Aphididae), abundance on sorghum and johnsongrass in a laboratory and field setting. *Crop Protection* 148, 105715. DOI: 10.1016/j.cropro.2021.105715

Faris, A.M. and Brewer, M.J. (2022) *Natural Enemies of the Sugarcane Aphid on Sorghum in South Texas.* ENTO-PU-175. College Station, TX: Texas A&M Extension Service.

Faris, A.M., Brewer, M.J. and Elliott, N.C. (2022a) Parasitoids and predators of the invasive aphid *Melanaphis sorghi* found in sorghum and non-crop vegetation of the sorghum agroecosystem. *Insects* 13, 606. DOI: 10.3390/insects13070606

Faris, A.M., Elliott, N.C. and Brewer, M.J. (2022b) Suppression of the sugarcane aphid, *Melanaphis sacchari* Zehntner (Hemiptera: Aphididae), by resident natural enemies on susceptible and resistant sorghum hybrids. *Environmental Entomology* 51, 332–339. DOI: 10.1093/ee/nvab147

Giles, K.L., Jones, D.B., Royer, T.A., Elliott, N.C. and Kindler, S.D. (2003) Development of a sampling plan in winter wheat that estimates cereal aphid parasitism levels and predicts population suppression. *Journal of Economic Entomology* 96, 975–982. DOI: 10.1603/0022-0493-96.3.975

Giles, K., Hein, G. and Peairs, F. (2008) Areawide pest management of cereal aphids in dryland wheat systems of the Great Plains, USA. In: Koul, O., Cuperus, G.W. and Elliott., N.C. (eds) *Areawide Pest Management: Theory and Implementation*. Wallingford, UK: CAB International, pp. 441–466.

Gordy, J.W., Brewer, M.J., Bowling, R.D., Buntin, G.D., Seiter, N.J. *et al.* (2019) Development of economic thresholds for sugarcane aphid (Hemiptera: Aphididae) in susceptible grain sorghum hybrids. *Journal of Economic Entomology* 112, 1251–1259. DOI: 10.1093/jee/toz028

Gordy, J.W., Seiter, N.J., Kerns, D.L., Reay-Jones, F.P.F., Bowling, R.D. *et al.* (2021) Field assessment of aphid doubling time and yield of sorghum susceptible and partially resistant to sugarcane aphid (Hemiptera: Aphididae). *Journal of Economic Entomology* 11, 2076–2087. DOI: 10.1093/jee/toab135

Haar, P.J., Buntin, G.D., Jacobson, A., Pekarcik, A., Way, M.O. and Zarrabi, A. (2019) Evaluation of tactics for management of sugarcane aphid (Hemiptera: Aphididae) in grain sorghum. *Journal of Economic Entomology* 112, 2719–2730. DOI: 10.1093/jee/toz215

Harris-Shultz, K., Ni, X., Wadl, P.A., Wang, X., Huang, F. *et al.* (2017) Microsatellite markers reveal a predominant sugarcane aphid (Homoptera: Aphididae) clone is found on sorghum in seven states and one territory of the USA. *Crop Science* 57, 2064–2072. DOI: 10.2135/cropsci2016.12.1010

Hewlett, J.A., Szczepaniec, A. and Eubanks, M.D. (2019) The effects of sugarcane aphid density in sorghum on predation by lady beetles and lacewings. *Biological Control* 129, 171–177. DOI: 10.1016/j.biocontrol.2018.10.015

Jeschke, J.M. (2014) General hypotheses in invasion ecology. *Diversity and Distributions* 20, 1229–1234. DOI: 10.1111/ddi.12258

Karp, D.S., Chaplin-Kramer, R., Meehan, T.D., Martin, E.A., DeClerck, F. *et al.* (2018) Crop pests and predators exhibit inconsistent responses to surrounding landscape composition. *Proceedings of the National Academy of Sciences* 115, E7863–E7870. DOI: 10.1073/pnas.1800042115

Koralewski, T.E., Wang, H.-H., Grant, W.E., Brewer, M.J., Elliott, N.C. *et al.* (2020) Integrating models of atmospheric dispersion and crop-pest dynamics: linking detection of local aphid infestations to forecasts of region-wide invasion of cereal crops. *Annals of the Entomological Society of America* 113, 79–87. DOI: 10.1093/aesa/saz047

Kring, T.J., Gilstrap, F.E. and Michels, G.J., Jr (1985) Role of indigenous coccinellids in regulating greenbugs (Homoptera: Aphididae) on Texas grain sorghum. *Journal of Economic Entomology* 78, 269–273. DOI: 10.1093/jee/78.1.269

Lahiri, S., Ni, X., Buntin, G.D. and Toews, M.D. (2020) Parasitism of *Melanaphis sacchari* (Hemiptera: Aphididae) by *Lysiphlebus testaceipes* (Hymenoptera: Braconidae) in the greenhouse and field. *Journal of Entomological Science* 55, 14–24. DOI: 10.18474/0749-8004-55.1.14

Lee, J.H., Elliott, N.C., Kindler, S.D., French, B.W., Walker, C.B. *et al.* (2005) Natural enemy impact of the Russian wheat aphid in southeastern Colorado. *Environmental Entomology* 34, 115–123. DOI: 10.1603/0046-225X-34.1.115

Maxson, E.L., Brewer, M.J., Rooney, W.L. and Woolley, J.B. (2019) Species composition and abundance of the natural enemies of sugarcane aphid, *Melanaphis sacchari* (Zehntner) (Hemiptera: Aphididae), on sorghum in Texas. *Proceedings of the Entomological Society of Washington* 121, 657–680. DOI: 10.4289/0013-8797.121.4.657

Michels, G.J., Elliott, N.C., Romero, R.A., Owings, D.A. and Bible, J.B. (2001) Impact of indigenous coccinellids on Russian wheat aphids and greenbugs (Homoptera: Aphididae) infesting winter wheat in the Texas Panhandle. *Southwestern Entomologist* 26, 97–114.

Nibouche, S., Costet, L., Medina, R.F., Holt, J.R., Sadeyen, J. *et al.* (2021) Morphometric and molecular discrimination of the sugarcane aphid, *Melanaphis sacchari*, (Zehntner, 1897) and the sorghum aphid *Melanaphis sorghi* (Theobald, 1904). *PLoS ONE* 16, e0241881. DOI: 10.1371/journal.pone.0241881

Paudyal, S., Armstrong, J.S., Harris-Shultz, K.R., Wang, H., Giles, K.L. *et al.* (2019) Evidence of host plant specialization among the U.S. sugarcane aphid (Hemiptera: Aphididae) genotypes. *Trends in Entomology* 15, 47–58.

Pekarcik, A.J. and Jacobson, A.L. (2021) Evaluating sugarcane aphid, *Melanaphis sacchari* (Hemiptera: Aphididae), population dynamics, feeding injury, and grain yield among commercial sorghum varieties in Alabama. *Journal of Economic Entomology* 114, 757–768. DOI: 10.1093/jee/toab013

Peterson, G.C., Armstrong, J.S., Pendleton, B.B., Stelter, M. and Brewer, M.J. (2018) Registration of RTx3410 through RTx3428 sorghum germplasm resistant to sugarcane aphid [*Melanaphis sacchari* (Zehntner)]. *Journal of Plant Registrations* 12, 391–398. DOI: 10.3198/jpr2018.02.0007crg

Rodríguez-del-Bosque, L.A., Rodríguez-Vélez, B., Sarmiento-Cordero, M.A. and Arredondo-Bernal, H.C. (2018) Natural enemies of Melanaphis sacchari on grain sorghum in Northeastern Mexico. *Southwestern Entomologist* 43, 277–279. DOI: 10.3958/059.043.0103

Rodriguez-Velez, J.M., Uribe-Mu, C.A., Sarmiento-Cordero, M.A., Rodriguez-Velez, B., Cruz-Rodriguez, J.A. *et al.* (2021) Relationships among *Melanaphis sacchari*, its predator insect communities, and weather conditions in sorghum crops at Colima, Mexico. *Southwestern Entomologist* 46, 47–58. DOI: 10.3958/059.046.0104

Sharma, H.C., Sharma, S.P. and Munghate, R.S. (2013) Phenotyping for resistance to the sugarcane aphid *Melanaphis sacchari* (Hemiptera: Aphididae) in *Sorghum bicolor* (Poaceae). *International Journal of Tropical Insect Science* 33, 227–238. DOI: 10.1017/S1742758413000271

Singh, B.U., Padmaja, P.G. and Seetharama, N. (2004) Biology and management of the sugarcane aphid, *Melanaphis sacchari* (Zehntner) (Homoptera: Aphididae), in sorghum: a review. *Crop Protection* 23, 739–755. DOI: 10.1016/j.cropro.2004.01.004

Thies, C., Roschewitz, I. and Tscharntke, T. (2005) The landscape context of cereal aphid-parasitoid interactions. *Proceedings of the Royal Society B* 272, 203–210. DOI: 10.1098/rspb.2004.2902

USDA-NASS (2020) *Statistics by Subject*. Washington, DC: US Department of Agriculture, National Agricultural Statistics. Available at: https://www.nass.usda.gov/Statistics_by_Subject (accessed January 23, 2020).

Uyi, O., Lahiri, S., Ni, X., Buntin, D., Jacobson, A. *et al.* (2022) Host plant resistance, foliar insecticide application and natural enemies play a role in the management of *Melanaphis sorghi* (Hemiptera: Aphididae) in grain sorghum. *Frontiers in Plant Science* 13, 1006225. DOI: 10.3389/fpls.2022.1006225

Wang, H.-H., Grant, W.E., Elliott, N.C., Brewer, M.J., Koralewski, T.E. *et al.* (2019) Integrated modelling of the life cycle and aeroecology of wind-borne pests in temporally variable spatially heterogeneous environment. *Ecological Modelling* 399, 23–38. DOI: 10.1016/j.ecolmodel.2019.02.014

9 Western Corn Rootworm Resistance to Bt Maize within the Midwestern Agricultural Landscape

Aaron J. Gassmann[1,*] and Lance J. Meinke[2]
[1]*Department of Plant Pathology, Entomology and Microbiology, Iowa State University, Ames, IA, USA;* [2]*Department of Entomology, University of Nebraska – Lincoln, Lincoln, NE, USA*

Abstract

Western corn rootworm, *Diabrotica virgifera virgifera* (LeConte) (Coleoptera: Chrysomelidae), is among the most serious pests of maize in the United States (US) Corn Belt, with larval feeding on maize roots causing most of the yield loss from this pest. Genetically engineered maize that produces insecticidal toxins from *Bacillus thuringiensis* Berliner (Bt) has been used to manage rootworm larvae for more than two decades. However, field-evolved resistance to Bt maize by western corn rootworm is now prevalent in the western Corn Belt. Several factors likely contributed to field-evolved resistance, including limited adult dispersal and the propensity of adults to mate within natal fields. These factors facilitate the development of high pest abundance in continuous maize fields and the evolution of resistance when the same management tactic is used repeatedly. The high level of mortality caused by Bt maize imposes strong selection for resistance, a pattern also observed for some broadcast-applied conventional insecticides, with several parallels between field-evolved resistance to Bt maize and conventional insecticides. Patterns of spatial heterogeneity in Bt resistance highlight the importance of local management as a key driver of resistance. In general, fields in continuous maize cultivation with continuous use of the same Bt traits are at the greatest risk of resistance while those fields that use a greater diversity of management approaches, in particular crop rotation, are at lower risk. This difference in risk of resistance is reflected in fine-scale patterns of spatial heterogeneity in resistance, and in larger, regional difference in resistance. As future transgenic traits are developed and applied for management of western corn rootworm, it will be important to apply insights gained from these cases of resistance to Bt maize. In particular, management efforts should focus on reducing pest abundance and delaying resistance evolution through the application of integrated pest management and planting of non-Bt refuges.

Introduction

The western corn rootworm, *Diabrotica virgifera virgifera* (LeConte) (Coleoptera: Chrysomelidae), is one of the most significant insect pests of maize in the United States (US) (Gray *et al.*, 2009). Annually, this species is responsible for US$2 billion in yield losses and associated management costs (Sappington *et al.*, 2006; Wechsler and Smith, 2018). Western corn rootworm is univoltine, developing through three larval instars while maize is actively growing during the

*Corresponding author: aaronjg@iastate.edu

 Arthropod Management and Landscape Considerations in Large-Scale Agroecosystems (eds M.J. Brewer and G.L. Hein)
DOI: 10.1079/9781800622777.0009

spring and early summer (Chiang, 1973). Larvae only survive on roots of a small number of grasses while adults feed on above-ground plant tissues and pollen (Fig. 9.1) (Branson and Ortman, 1967; Branson and Ortman, 1970; Clark and Hibbard, 2004; Oyediran *et al.*, 2004; Moeser and Hibbard, 2005; Campbell and Meinke, 2006). Maize is the primary host in modern agro-ecosystems and is strongly attractive to adults for feeding and oviposition (Branson and Krysan, 1981; Moeser and Hibbard, 2005). Because of this, most adults oviposit in the soil around maize plants. Eggs enter an obligatory diapause and overwinter (Meinke *et al.*, 2009). Significant larval feeding injury can reduce growth of maize plants by interfering with water and nutrient uptake, in addition to decreasing plant stability and reducing grain yield (Spike and Tollefson, 1991; Godfrey *et al.*, 1993; Hou *et al.*, 1997; Gray and Steffey, 1998; Urías-López *et al.*, 2000; Urías-López and Meinke, 2001; Tinsley *et al.*, 2013). The western corn rootworm is a highly invasive pest, with its geographic range expanding across the US Corn Belt from southwestern Nebraska in the 1920s to New Jersey along the Atlantic coast by the 1980s (Gray *et al.*, 2009; Meinke *et al.*, 2009). This species also has been inadvertently introduced into Europe on a number of occasions, which has led to establishment and geographic spread through a number of countries over time (Bažok, 2001; Miller *et al.*, 2005; Bažok *et al.*, 2021).

Several cropping sequences are commonly practiced by farmers in the US Corn Belt, which impact western corn rootworm population dynamics. Crop rotation from maize to a non-host

Fig. 9.1. (A) Lateral view of an adult western corn rootworm. (B) First, second, and third instar of western corn rootworm larvae. (C) Dorsal view of an adult western corn rootworm. (D) Maize roots without feeding injury from larval corn rootworm. (E) Maize roots that have suffered heavy feeding injury from larval corn rootworm. (Photo credits: (A) Lance J. Meinke, University of Nebraska; (B, C) Jim Kalisch, University of Nebraska; (D, E) Aaron J. Gassmann, Iowa State University.)

crop (e.g. soybean) is a recommended management tactic because larvae that eclose from eggs laid the previous season will starve in the absence of maize root tissue, essentially resetting the population density in a field to zero (Bare, 1930; Cullen *et al.*, 2013). Continuous maize (maize planted consecutively for ≥2 years) is a profitable agronomic practice that supports the annual demand for maize as feed for the large confined-livestock industry and ethanol production. This agronomic system facilitates the buildup of large western corn rootworm populations over time and increased larval injury, making management an annual challenge (Meinke *et al.*, 2009; Souza *et al.*, 2019). Because maize is the optimal host for the western corn rootworm, the proportion of the agricultural landscape planted to continuous maize will be one of the key determinants of overall pest density at the field and farm levels (Meinke *et al.*, 2009). The use of continuous maize versus crop rotation varies within and among states, but in general there is more annual maize/soybean rotation in the eastern Corn Belt and more continuous maize in the western Corn Belt.

In this chapter, we review the evolution of resistance by western corn rootworm to maize that is genetically engineered to produce insecticidal toxins from the bacterium *Bacillus thuringiensis* Berliner (Bt). We address how the biology of western corn rootworm and interactions of western corn rootworm with cropping practices and the agricultural landscape influenced the evolution of resistance to Bt maize. We also discuss parallels between resistance to Bt maize and conventional insecticides and insights that can be gained by comparing the response of western corn rootworm to these two management approaches. The chapter concludes with a consideration of how our current knowledge may be applied to improve future management of this serious agricultural pest.

Adult Dispersal Behavior

Dispersal and migratory behaviors of insects can influence both population dynamics and the geographic spread of a species (Johnson, 1969; Dingle, 1972). For western corn rootworm, like many insect species, the adult stage is the primary stage that engages in movement across the landscape, with larvae typically moving less than 1 m (Hibbard *et al.*, 2003; Rudeen and Gassmann, 2013). Adult rootworm display a bimodal distribution for adult dispersal (Coats *et al.*, 1986; Naranjo, 1990; Yu *et al.*, 2019). Western corn rootworm tethered flight studies have revealed that the distribution of flight duration is heavily skewed toward short duration or trivial flights (many <1 min), but some individuals engage in longer duration or sustained flights (>20–30 min) (Coats *et al.*, 1986; Naranjo, 1990; Stebbing *et al.*, 2005; Yu *et al.*, 2019). In addition, mean tethered flight time for females has been reported to be longer than for males, and while both males and females undertake sustained flight, this is more commonly exhibited by females (Coats *et al.*, 1986; Naranjo, 1990; Stebbing *et al.*, 2005). Sustained flight is often exhibited by mated females during the pre-ovipositional period (3–10 days old), although Naranjo (1990) reported females as old as 30 days exhibiting sustained flight. Studies of in-field movement suggest that western corn rootworm adults not engaged in long-distance dispersal move less than 40 m/day (Spencer *et al.*, 2003; Hughson and Spencer, 2015). Western corn rootworm tethered flight activity increases during early-mid morning and evening, which is consistent with the diurnal periodicity of activity documented in the field (Witkowski *et al.*, 1975; Coats *et al.*, 1986; Grant and Seevers, 1990; Naranjo, 1990; Isard *et al.*, 2000; Stebbing *et al.*, 2005).

Many factors can interact with innate western corn rootworm flight characteristics to determine spatial and temporal patterns of movement (Grant and Seevers, 1989; Beckler *et al.*, 2004; Toepfer *et al.*, 2007; Meinke *et al.*, 2009). Adult emigration and immigration rates can significantly affect the population density in a maize field during specific periods (Darnell *et al.*, 2000; Meinke *et al.*, 2009). Because adult western corn rootworms are strongly attracted to maize pollen and reproductive plant tissues as food sources, differences in maize phenology within and among fields of maize can have a substantial influence on local spatial distributions (Hill and Mayo, 1974; Prystupa *et al.*, 1988; Darnell *et al.*, 2000; Moeser and Hibbard, 2005; Campbell and Meinke, 2006). The ability of western corn rootworm to utilize ear tissues of maize as a food source from pollination through the early reproductive stages facilitates retention of adults within fields. Late season movement of western corn rootworm adults to alternate pollen sources (e.g. blooming weeds) also is observed commonly when maize becomes a suboptimal host during later maize reproductive growth stages (Campbell and Meinke, 2006). However, despite these alternative sources of food, western corn rootworm maintains a strong affinity for maize as the primary oviposition site regardless of adult feeding site (Siegfried and Mullin, 1990; Levine and Oloumi-Sadeghi, 1991). The ability of adult western corn rootworm females to recolonize first-year maize fields at the local level and to colonize larger geographic areas over time is well documented (Godfrey and Turpin, 1983; Gray *et al.*, 2009; Meinke *et al.*, 2009).

Despite the mobility of adult western corn rootworms, the number of adults emerging within a large continuous maize field is a primary contributor to the total population density present within that field (Meinke *et al.*, 2009). Females mate in the natal field upon emergence (Hill, 1975; Spencer *et al.*, 2009), and the highest density of adults within a field is typically associated with the peak adult emergence period (Darnell *et al.*, 1999; Bažok, 2001; Meinke *et al.*, 2009). Collectively, these biological characteristics suggest that, even though local trivial movement can be very high and some individuals exhibit sustained flight and long-distance movement, many individuals stay in or around the natal field, ovipositing locally and exhibiting relatively limited net dispersal from the point of origin. The limited net dispersal displayed by many adults has important implications for population dynamics, resulting in high pest densities for fields that are in continuous maize production for multiple, consecutive growing seasons. Furthermore, as discussed later in the chapter, limited adult dispersal and mating within natal fields facilitate field-evolved resistance to Bt maize and conventional insecticides when these practices are used repeatedly in the same field.

Bt Traits Targeting Rootworm and Approaches for Resistance Management

Since 2003, transgenic Bt maize has played a central role in management of western corn rootworm larvae (Gassmann and Reisig, 2023). The first Bt trait targeting western corn rootworm produced Bt toxin Cry3Bb1, which was followed by the US Environmental Protection Agency (EPA) registration and subsequent commercialization of Bt traits producing mCry3A, Gpp34/Tpp35Ab1 (formerly classified as Cry34/35Ab1) (Crickmore *et al.*, 2021), and eCry3.1Ab between 2005 and 2012 (Gassmann, 2021). With the exception of eCry3.1Ab, these Bt traits were released singly and later combined in Bt pyramids, and these Bt pyramids included Cry3Bb1 with Gpp34/Tpp35Ab1, mCry3A with Gpp34/Tpp35Ab1, and mCry3A with eCry3.1Ab (Gassmann, 2021). More recently transgenic maize that produces double-stranded RNA (dsRNA) has been combined with Bt traits for management of rootworm larvae. This trait (DvSnf7) was released as a pyramid with either Cry3Bb1 alone or Cry3Bb1 and Gpp34/Tpp35Ab1 (EPA, 2017; Head *et al.*, 2017).

In all cases, the US EPA registration of a Bt trait requires an accompanying resistance management strategy (Gould, 1998; EPA, 2022). Resistance management for Bt traits targeting western corn rootworm has been based on the refuge strategy and the pyramid/refuge strategy (Gould, 1998; Roush, 1998). The refuge strategy delays resistance by reducing the frequency of homozygous resistant individuals within a population (Carrière *et al.*, 2010; Tabashnik and Carrière, 2017). Under the refuge strategy, the cultivation of a Bt crop is accompanied by the presence of non-Bt host plants, which serve as a source of Bt-susceptible individuals. Mating between Bt-susceptible individuals from refuge plants with Bt-resistant individuals that survive on a Bt crop results in the production of individuals that are heterozygous for resistance (Gould, 1998; Carrière *et al.*, 2010; Tabashnik and Carrière, 2017). To the extent that these heterozygous individuals have lower fitness on a Bt crop than homozygous resistant individuals, resistance will be delayed compared to a scenario in which refuges are absent (Tabashnik *et al.*, 2008). Importantly, the inheritance of resistance plays a central role in determining the effectiveness of refuges in delaying resistance. The greatest delays in resistance will be achieved when resistance is recessive, but resistance will evolve most rapidly when resistance is inherited as a dominant trait (Tabashnik and Carrière, 2017).

In the case of the pyramid/refuge strategy, resistance is delayed by redundant killing (Roush, 1998). Bt pyramids contain two Bt traits that target the same pest, and delays in resistance are achieved when individuals that are resistant to one trait in a pyramid are killed by the second trait (Carrière *et al.*, 2015). The presence of non-Bt refuges is central to the success of Bt pyramids in delaying resistance because mating of Bt-susceptible individuals from refuges with individuals surviving on a Bt pyramid acts to reduce the frequency of individuals that harbor alleles for resistance to both Bt traits (Roush, 1998). Additionally, the delays in resistance achieved by the pyramid/refuge strategy are sensitive to the presence of cross-resistance between Bt toxins present in a pyramid, and to the frequency of alleles for resistance to either Bt trait in a pyramid (Gould, 1998). Resistance is expected to evolve more rapidly if there is cross-resistance between Bt traits or if there is an elevated frequency of alleles for resistance to either Bt trait, compared to a scenario when these factors are absent (Gould, 1998; Carrière *et al.*, 2015).

For both a refuge strategy and the pyramid/refuge strategy, the risk of resistance is inversely related to the proportion of the landscape planted to non-Bt refuges, with rate of resistance evolution increasing as the proportion of the landscape planted to refuges decreases (Gould, 1998; Roush, 1998; Tabashnik *et al.*, 2008). When the first Bt traits were released for management of rootworm, Bt maize contained one Bt trait targeting rootworm and was accompanied by a 20% non-Bt refuge (Gassmann *et al.*, 2011; Andow *et al.*, 2016). These refuges were spatially structured and consisted of blocks of non-Bt maize adjacent to Bt maize (Gassmann *et al.*, 2011; Andow *et al.*, 2016). This refuge structure followed requirements for spatially structured refuges that were already in place for European corn borer, *Ostrinia nubilalis* (Lepidoptera: Crambidae), and other lepidopteran pests (Andow *et al.*, 2016).

Later, single-traited Bt maize was made available with a blended refuge, in which bags of maize seed contained a mixture of Bt and non-Bt seeds (Deitloff *et al.*, 2016). Because the closer spatial proximity of Bt and non-Bt plants was expected to facilitate mating between refuge individuals and Bt-selected individuals, the required non-Bt refuge was reduced to 10%. Once maize hybrids with a pyramid of Bt traits targeting rootworm became available, the required refuge size was further reduced to 5% (either as a blended refuge or spatially structured refuge) because the risk of resistance to a Bt pyramid was presumed to be less than the risk of resistance to a single trait (Andow *et al.*, 2016; EPA, 2022; Gassmann and Reisig, 2023).

Temporal and Geographic Patterns in the Evolution of Resistance to Bt Maize

When Bt maize was first made available to farmers, it was characterized by high levels of efficacy in killing larval rootworm and reducing the associated larval feeding injury to roots (Andow *et al.*, 2016). Consequently, it was rapidly adopted by farmers and widely used (Gassmann *et al.*, 2011; Dunbar *et al.*, 2016; St. Clair and Gassmann, 2021). This in turn resulted in strong selection for resistance. Reports by farmers and crop consultants of high levels of feeding injury to Cry3Bb1 maize by western corn rootworm provided the first indications of resistance to Bt maize (Gassmann *et al.*, 2011, 2012, 2014). For example, injury to Cry3Bb1 maize in some portions of the affected fields often exceeded two nodes of root tissue pruned by larval rootworm feeding, which equates to a reduction in yield of approximately 30% (Dun *et al.*, 2010; Tinsley *et al.*, 2013). Additionally, fields with high levels of feeding injury to Bt maize also harbored large populations of western corn rootworm (Dunbar *et al.*, 2016). These first cases of resistance to Cry3Bb1 were documented in Iowa from western corn rootworm populations that were sampled in 2009, 6 years after Cry3Bb1 maize was first grown by farmers. However, over the next few years, instances of resistance to Cry3Bb1 maize were found more broadly within the western Corn Belt, with cases of resistance also documented in Illinois, Minnesota, and Nebraska (Wangila *et al.*, 2015; Schrader *et al.*, 2016; Zukoff *et al.*, 2016). In all instances, these first cases of resistance were documented by visiting fields with high levels of injury to Bt maize, rather than sampling populations more broadly in the landscape, suggesting the presence of spatial structure in the distribution of resistance.

Concurrent with the documentation of Cry3Bb1 resistance, high levels of feeding injury by western corn rootworm to mCry3A maize also were observed. Subsequent bioassays demonstrated that this arose from cross-resistance between mCry3A and Cry3Bb1, with western corn rootworm populations demonstrating elevated survival on both Cry3Bb1 maize and mCry3A maize, even if there was no history of exposure to mCry3A (Gassmann *et al.*, 2014; Wangila *et al.*, 2015; Jakka *et al.*, 2016; Zukoff *et al.*, 2016). Subsequent studies found that cross-resistance also extended to eCry3.1Ab; however, the level of cross-resistance between either Cry3Bb1 or mCry3A and eCry3.1Ab was less pronounced than the level of cross-resistance between Cry3Bb1 and mCry3A (Jakka *et al.*, 2016). Initial studies documenting resistance by western corn rootworm to these various Cry3 toxins found that populations were susceptible to Bt maize producing Gpp34/Tpp35Ab1, pointing to an absence of cross-resistance between Cry3 traits and Gpp34/Tpp35Ab1 (Gassmann *et al.*, 2011, 2012, 2014; Wangila *et al.*, 2015; Jakka *et al.*, 2016). Beginning in 2013, high levels of feeding injury to maize producing Gpp34/Tpp35Ab1 were observed, and subsequent bioassays indicated the presence of resistance to this Bt trait in Iowa (Gassmann *et al.*, 2016). Additional studies found resistance to Gpp34/Tpp35Ab1 more broadly in the western Corn Belt, including Minnesota and Nebraska, and additional fields with resistant populations in Iowa (Ludwick *et al.*, 2017; Gassmann *et al.*, 2020; Reinders and Meinke, 2022; Reinders *et al.*, 2022).

Factors Affecting the Evolution of Resistance to Bt Maize

As described earlier, western corn rootworm adults may disperse only short distances, often remaining within their natal field, especially if that field is of larger size. Additionally, most mating occurs among adults that emerge from the same field, an effect that is caused by limited adult movement but also by mating behavior. Western corn rootworm is protandrous, with adult males emerging, on average, approximately 5 days before females (Spencer *et al.*, 2009). After emerging, males are unable to mate for several days; however, females can mate shortly after emerging as adults, and while still teneral (Spencer *et al.*, 2009).

These patterns of localized dispersal and within-field mating have important implications for pest adaptation to Bt maize (Meinke *et al.*, 2009). First, the extent to which refuges are spatially structured (i.e. physically separated from portions of the field planted to Bt maize) will reduce mating between refuge individuals and Bt-selected individuals, and act to accelerate the evolution of resistance (Spencer *et al.*, 2003; Spencer *et al.*, 2012; Hughson and Spencer, 2015). This effect may be exacerbated by delayed emergence of western corn rootworm from Bt maize compared to non-Bt maize, although this asynchrony in emergence may diminish as Bt resistance evolves within a population (Petzold-Maxwell *et al.*, 2013a, b; Shrestha and Gassmann, 2019; Reinders *et al.*, 2021). Second, management practices that are used within a field can impose localized selection pressure for resistance. More specifically, if a field is planted to the same Bt-traited maize for three or more growing seasons, the western corn rootworm population in that field has an increased potential to evolve resistance because the population will experience multiple generations of selection for resistance (Gassmann *et al.*, 2011). The importance of this dynamic is supported by empirical observations of a positive relationship between the number of years a field was planted to Bt maize and the magnitude of Bt resistance (Gassmann *et al.*, 2011; Reinders *et al.*, 2018).

While patterns of within-field movement and mating produced a scenario that favored the evolution of resistance to Bt maize by western corn rootworm, there were likely several additional factors that heightened the risk of resistance. First, estimates of initial resistance allele frequency suggest values between 0.05 and 0.20, which enables a rapid response to selection from Bt maize because resistance alleles would be prevalent in most populations (Onstad and Meinke, 2010). Second, a number of studies have found that resistance to Bt maize by western corn rootworm is inherited in a non-recessive manner, enabling individuals that are heterozygous for resistance to survive on Bt maize (Meihls *et al.*, 2008; Petzold-Maxwell *et al.*, 2012; Ingber and Gassmann, 2015; Paolino and Gassmann, 2017; Smith *et al.*, 2023). Finally, fitness costs of resistance to Bt maize by western corn rootworm (i.e. the reduction in fitness of resistant individuals versus susceptible individuals in the absence of Bt) tend to be limited (Petzold-Maxwell *et al.*, 2012; Hoffmann *et al.*, 2015; Ingber and Gassmann, 2015; Paolino and Gassmann, 2017; Smith *et al.*, 2023). While available evidence supports the hypothesis that resistant individuals suffer fitness costs of resistance, these effects can be minor. For example, although resistance to Cry3Bb1 maize declined significantly in the absence of selection, populations were still found to have significantly greater survival on Cry3Bb1 maize than a susceptible strain after six generations without selection (St. Clair *et al.*, 2020a). All these factors favor a rapid response to selection, and when coupled with a strong localized adaptation due to limited net adult dispersal, should lead to localized areas in the landscape with resistant populations.

In addition to these biological factors, the manner in which Bt technology was developed and brought to the market also likely contributed to evolution of resistance. In particular, three of the four Bt traits targeting rootworm (i.e. Cry3Bb1, Gpp34/Tpp35Ab1 and mCry3A) were first deployed singly before being released as a pyramid, with the evolution of resistance to single traits compromising the capacity of pyramids to delay resistance (Gassmann, 2021; Gassmann and Reisig, 2023). Furthermore, data on potential cross-resistance among traits were not gathered prior to commercial cultivation of Bt maize targeting rootworm. All Bt pyramids using Gpp34/Tpp35Ab1 included mCry3A or Cry3Bb1 which are cross-resistant. Earlier evaluations of cross-resistance for future transgenic traits should help to provide a more accurate assessment of the risk of resistance. Additionally, release of traits as pyramids, rather than using traits singly prior to pyramiding, should lead to a more robust approach for resistance management.

Resistance to Bt Maize Within the Agricultural Landscape

Spatial patterns in resistance range from fine-scale, field-level patterns to broader geographic patterns that span counties, states, and regions. Research on the spatial structure of resistant populations has focused primarily on Iowa and Nebraska. In general, fine-scale spatial patterns in resistance appear to be more prevalent when resistance first evolves, before pest dispersal acts to homogenize resistance within the landscape, but even several years after resistance is first detected, evidence of spatial structure is still present (Gassmann *et al.*, 2011; Reinders *et al.*, 2018; Shrestha *et al.*, 2018).

Examples of fine-scale spatial patterning include the first cases of resistance to Cry3Bb1 and Gpp34/Tpp35Ab1 in Iowa. When resistance to Cry3Bb1 was first documented in Iowa, populations that were either resistant or susceptible to Cry3Bb1 were found within the agricultural landscape, and in some cases, in the same county (Gassmann *et al.*, 2011). Similarly, with resistance to Gpp34/Tpp35Ab1, resistance was found to be a mosaic within the landscape, with both resistant and susceptible populations co-occurring within the state (Gassmann *et al.*, 2016, 2020). Research in Nebraska also found that spatial patterns in resistance were present for both Cry3Bb1 and Gpp34/Tpp35Ab1 (Reinders *et al.*, 2018, 2022). For Cry3Bb1 resistance in both states, a significant association was found between the use of Cry3Bb1 maize within a field and the level of resistance in that field, with the magnitude of resistance showing a positive association with the number of years Cry3Bb1-traited maize was grown (Gassmann *et al.*, 2011; Reinders *et al.*, 2018). These spatial patterns in resistance point to the importance of local selection (i.e. within-field management) in driving patterns of resistance evolution and provide evidence for the independent evolution of resistance among populations.

For resistance to Cry3Bb1 maize in Iowa, studies conducted 4–5 years after resistance was documented found evidence of spatial variation in resistance; however, some level of Cry3Bb1 resistance was present in all populations studied (Shrestha *et al.*, 2018). Specifically, resistance was found in fields planted continuously to maize for 7 years or more and fields that were recently rotated out of maize production (i.e. those only in maize production for 2 years), in addition to fields where high levels of injury to Cry3 (i.e. mCry3A or Cry3Bb1) maize were observed that year or in previous years (Shrestha *et al.*, 2018). However, fields where crop rotation had been used had a lower overall level of resistance than the other types of fields examined in the study. Similarly, in Nebraska, research conducted several years after Cry3Bb1 resistance was first documented found evidence of resistance in most field populations, but also that the magnitude of resistance was affected by within-field management actions, in particular the use of Cry3Bb1-traited maize and crop rotation (Wangila *et al.*, 2015; Reinders *et al.*, 2018).

The general pervasiveness of Cry3Bb1-resistant western corn rootworm in the agricultural landscape of Iowa and Nebraska was confirmed with subsequent studies that examined resistance in broader geographic regions spanning groups of neighboring fields to individual counties. One study in Iowa examined the level of Cry3Bb1 resistance in fields with a history of high levels of feeding injury to Cry3 maize, and in neighboring fields (St. Clair *et al.*, 2020b). This study found similar levels of resistance between these types of fields (St. Clair *et al.*, 2020b). Additionally, a study comparing Cry3Bb1 resistance in counties of Iowa with and without reports of high levels of feeding injury to Cry3 maize found similar levels of Cry3Bb1 resistance between these categories of counties (St. Clair *et al.*, 2020a). In Nebraska, fields sampled from several counties in the northeastern portion of the state were found to contain western corn rootworm populations with some level of Cry3Bb1 resistance in almost all cases (Reinders *et al.*, 2022).

Additional research on broader landscape-level effects on resistance comes from a study comparing geospatial data from Iowa on the planting of maize and instances where high levels of feeding injury to Cry3 maize by western corn rootworm were observed (St. Clair and Gassmann, 2021). Specifically, this research found that the occurrence of high levels of feeding injury to Cry3 maize by western corn rootworm, and the presence of Cry3Bb1 resistance, were associated with a higher prevalence of continuous maize cultivation in the surrounding landscape (St. Clair and Gassmann, 2021). Within a 3.2 km radius of fields where high levels of feeding injury to Cry3 maize were observed, there was significantly more maize planted than in randomly selected points across Iowa's agricultural landscape. It is also likely that fields planted to maize were predominantly Cry3Bb1 maize during the period of time examined in the study (Dunbar *et al.*, 2016). In this scenario, adjacent fields with similar management approaches generated a larger landscape where resistance could evolve, with western corn rootworm moving among fields that were in close spatial proximity, and substantial pest populations arising in some cases. In these instances, farmers observed high levels of feeding injury to Cry3 maize, and subsequent bioassays provided evidence of Cry3Bb1 resistance as a causal factor for this injury (Gassmann *et al.*, 2011, 2012, 2014, 2016; Jakka *et al.*, 2016; St. Clair and Gassmann, 2021). A complementary study by Carrière *et al.* (2020) examined geospatial data from Illinois, Iowa, and Minnesota and found that instances of severe feeding injury to Cry3Bb1 maize by western corn rootworm within a crop reporting district (i.e. a region spanning approximately 10–15% of a state's area) were inversely related to the prevalence of crop rotation. Additionally, this study provided data showing the high prevalence of Cry3Bb1 maize cultivation within the agricultural landscape (Carrière *et al.*, 2020). Taken together these studies illustrate the larger, landscape-level patterns in resistance and high levels of feeding injury to Bt maize by western corn rootworm that can arise from continuous maize cultivation coupled with extensive use of Bt-traited maize.

A common approach used by farmers to mitigate the presence of Cry3Bb1 resistance was to plant maize containing a pyramid of Cry3Bb1 and Gpp34/Tpp35Ab1 (Dunbar *et al.*, 2016; Carrière *et al.*, 2020). Due to the presence of Cry3Bb1 resistance, this approach increased the risk of resistance to Gpp34/Tpp35Ab1 compared to the use of a pyramid in which resistance to either Bt trait was absent (Andow *et al.*, 2016). The first cases of resistance to Gpp34/Tpp35Ab1 were in Iowa and consisted of populations sampled in 2013 from fields where high levels of feeding injury were observed to single-traited Gpp34/Tpp35Ab1 maize and maize that contained a pyramid of Cry3 and Gpp34/Tpp35Ab1 (Gassmann *et al.*, 2016). These western corn rootworm populations showed incomplete resistance to Gpp34/Tpp35Ab1, but complete resistance to Cry3Bb1 (Gassmann *et al.*, 2016). While complete resistance indicates no difference in fitness of insects on Bt versus non-Bt maize, incomplete resistance arises when Bt maize imposes some reductions in fitness (e.g. survival), but fitness of the resistant population on Bt maize is still greater than that observed for a known susceptible population (Tabashnik *et al.*, 2014).

Additionally, this study found an absence of Gpp34/Tpp35Ab1 resistance for populations sampled from fields that did not have high levels of feeding injury to maize containing Gpp34/Tpp35Ab1, indicating the presence of landscape-level heterogeneity in resistance to this Bt trait. A follow-up study of populations sampled in Iowa during 2017 found evidence of complete resistance to Gpp34/Tpp35Ab1 in a field with high levels of feeding injury to maize containing a pyramid of Gpp34/Tpp35Ab1 and Cry3Bb1, suggesting that the magnitude of resistance to Gpp34/Tpp35Ab1 in the landscape had increased (Gassmann *et al.*, 2020). In Nebraska, a study of 20 field populations sampled in 2018 and 2019 found a mosaic of resistance to maize containing a pyramid of Cry3Bb1 and Gpp34/Tpp35Ab1 (Reinders *et al.*, 2022). While all populations displayed some level of resistance to this Bt pyramid, in many cases survival was lower than on non-Bt maize (i.e. resistance was incomplete) (Reinders *et al.*, 2022). A second study conducted in the same region and during the same time found that most, but not all, populations sampled were resistant to Gpp34/Tpp35Ab1 maize, but again, resistance was incomplete (Reinders and Meinke, 2022).

In contrast to the presence of pervasive resistance to Cry3Bb1 and Gpp34/Tpp35Ab1 in regions of Iowa and Nebraska, states along the eastern edge of the Corn Belt (i.e. Indiana and Ohio) and states further east, appear to have limited issues with Bt resistance by western corn rootworm. As discussed previously, continuous maize cultivation and continuous use of the same Bt traits is an important driver of resistance evolution, and contrasting patterns of maize cultivation between these regions is likely a key factor affecting regional patterns in Bt resistance, with the lesser area planted to maize in the eastern region contributing to fewer observations of resistance evolution (USDA-NASS, 2022).

Connections Between Management With Bt Traits and Conventional Insecticides

There are interesting parallels between western corn rootworm resistance to conventional insecticides and Bt maize. In the western US Corn Belt, where continuous maize is commonly grown in large, irrigated monocultures, insecticides have been extensively used over time to manage annual densities of western corn rootworm and other secondary insect pests (Meinke *et al.*, 2021). Since the 1950s, this has contributed to the sequential occurrence of western corn rootworm resistance evolution to broadcast-applied cyclodiene soil insecticides (Ball and Weekman, 1962, 1963) and aerial-applied organophosphate (Meinke *et al.*, 1998; Zhou *et al.*, 2002), carbamate (Scharf *et al.*, 1999), and pyrethroid insecticides (Pereira *et al.*, 2015; Souza *et al.*, 2019). In each case, resistance evolution occurred when most of the western corn rootworm population was exposed to a toxin over time because there was no structured refuge (see Case Study 9.1). These patterns of insecticide resistance are in contrast to the lack of resistance evolution by western corn rootworm larvae to soil insecticides that are applied in a banded pattern, and these contrasting results have been attributed to western corn rootworm larvae feeding on maize roots growing outside of the insecticide-treated zone (i.e. a refuge). This feeding pattern appears to often result in an inconsistent effect of soil insecticides on adult emergence, with the band or in-furrow placement protecting the main root system but doing little to reduce adult rootworm abundance (Gray *et al.*, 1992; Boetel *et al.*, 2003; Petzold-Maxwell *et al.*, 2013b; Souza *et al.*, 2020).

These observations parallel patterns for field-evolved resistance by western corn rootworm to Bt maize. Prior to resistance evolution, both broadcast-applied insecticides and Bt maize caused high mortality for western corn rootworm, which imposed strong selection for resistance. For example, Bt maize can impose substantial larval mortality, reducing adult abundance by 65% to over 95% (Petzold-Maxwell *et al.*, 2013b; Head *et al.*, 2014, 2017; Keweshan *et al.*, 2015; Campbell *et al.*, 2017). When resistance of western corn rootworm to conventional insecticides and Bt maize was initially reported, a similar pattern was observed across the landscape. Case histories revealed significant variation in susceptibility among western corn rootworm populations at local and regional levels reinforcing the importance of local rootworm management as a key driver of resistance evolution (Meinke *et al.*, 1998;

Case Study 9.1: Field-evolved resistance to methyl parathion by western corn rootworm and geographic spread of resistance in Nebraska

The history of resistance to the insecticide methyl parathion in Nebraska during the 1990s provides an interesting parallel to observations of field-evolved resistance to Bt maize by western corn rootworm at local and landscape levels. Methyl parathion resistance evolved in western corn rootworm populations after intensive repeated use of adult management over multiple years. Field failures and subsequent bioassays confirmed methyl parathion resistance in two counties separated by about 150 km (Meinke *et al.*, 1998; Miller *et al.*, 2009). Management of western corn rootworm was more diversified in the landscape between the resistance focal points (i.e. use of soil insecticides, crop rotation, and less adult management) and populations in these areas were generally still susceptible to methyl parathion (Meinke *et al.*, 1997, 1998; Miller *et al.*, 2009). Between 1998 and 2021, the area of susceptibility between the two initial foci was greatly reduced, and many populations had become resistant or showed an increase in the magnitude of resistance (Miller *et al.*, 2009; Meinke *et al.*, 2021). Key factors facilitating the spread of resistance across the landscape appeared to be the semi-dominant inheritance of resistance and similar dispersal capacity between methyl parathion-resistant and -susceptible populations (i.e. few fitness costs of resistance) (Parimi *et al.*, 2003; Stebbing *et al.*, 2005). Additionally, selection pressure from adult management programs continued in a patchwork pattern across the landscape until programs were abandoned in the early 2000s (Meinke *et al.*, 2021).

Zhou *et al.*, 2002; Gassmann *et al.*, 2011, 2020; Reinders *et al.*, 2018).

Other similarities include few fitness costs of resistance and often non-recessive inheritance associated with resistance (Gassmann, 2021; Meinke *et al.*, 2021), both of which may facilitate resistance evolution and contribute to the retention of resistance for substantial periods of time in the absence of selection (Parimi *et al.*, 2006; Wang *et al.*, 2013; Wangila and Meinke, 2017; St. Clair *et al.*, 2020a; Paddock *et al.*, 2021). The parallels between resistance by western corn rootworm to conventional insecticides and Bt traits raises the hypothesis that the spread and persistence of resistance to current Bt traits may follow a pattern similar to that of resistance to methyl parathion and other conventional insecticides.

Summary and Conclusions

The first Bt trait for management of corn rootworm (Cry3Bb1) has been available for more than two decades, and all other Bt traits have been used for more than a decade (Gassmann, 2012; Tabashnik and Carrière, 2017). The long duration these traits have been used, coupled with the high risk of resistance, has resulted in Bt resistance in many areas of the western US Corn Belt. Recently, transgenic corn that kills corn rootworm using RNA interference (RNAi) was commercialized (Head *et al.*, 2017; Reinders *et al.*, 2023). This novel transgenic trait (DvSnf7) uses dsRNA encoding an essential protein in rootworm. Ingestion of dsRNA results in silencing of this gene, which in turn kills the insect (Bolognesi *et al.*, 2012; Ramaseshadri *et al.*, 2013). Additional transgenic events that use RNAi to target other rootworm genes also may be developed (USDA-APHIA, 2020). Other, future, transgenic traits are likely to use insecticidal proteins derived from *B. thuringiensis* in addition to proteins derived from other organisms (Gassmann and Reisig, 2023). For example, a rootworm-active protein toxin derived from *Pseudomonas chlororaphis* (Guignard and Sauvageau) has been shown to kill rootworm larvae when produced in transgenic maize (Schellenberger *et al.*, 2016).

As new transgenic technologies become available for management of corn rootworm, it will be important to refine approaches for managing pest resistance based on insights gained from the first generation of transgenic traits. Factors that contributed to widespread resistance by western corn rootworm to the first generation of Bt traits were likely the lack of a high dose for these Bt traits, minimal fitness costs that accompany resistance, elevated initial resistance allele frequency, high rates of mating within natal fields, and limited net dispersal by adult western corn rootworm. Because of these risk factors, fields that were in continuous maize production, with the same Bt traits used for multiple seasons, provided an environment where resistance could readily evolve.

It will also be important to identify cases where heightened risks of resistance are present for future transgenic traits. For example, the current RNAi trait targeting corn rootworm (DvSnf7) is used in a pyramid with Cry3Bb1 alone or Cry3Bb1 and Gpp34/Tpp35Ab1. Because of resistance to these Bt traits, there is an elevated risk of RNAi resistance compared to a scenario in which an RNAi trait is used in a pyramid with Bt traits for which rootworm has not evolved resistance. Furthermore, past research has shown that western corn rootworm can readily evolve resistance to RNAi, at a rate similar to resistance to Bt traits, and that this resistance tends to be broad-spectrum, extending to other RNAi traits (Meihls *et al.*, 2008; Deitloff *et al.*, 2016; Khajuria *et al.*, 2018).

Even when resistance is inherited in a non-recessive manner, refuges will act to delay resistance, albeit less effectively than when resistance is recessive (Tabashnik *et al.*, 2008). The use of larger refuges will help delay resistance to future transgenic rootworm traits (Gassmann *et al.*, 2011; Tabashnik and Gould, 2012). These refuges could include planting a larger portion of a field to maize that lacks traits targeting corn rootworm, or refuges could be temporal, with fields rotated over sequential years between rootworm-traited maize and maize that lacks rootworm traits (Gassmann and Reisig, 2023). Additionally, greater use of blended refuges, which consist of Bt and non-Bt plants interspersed in a field, could benefit efforts to manage resistance because blended refuges promote mating between Bt-selected individuals and refuge individuals.

Ideally, a management plan for western corn rootworm should focus on reducing larval injury to an acceptable level within a growing season while also slowing the rate of resistance evolution. In general, applying a greater diversity of management approaches within an integrated pest management (IPM) framework will manage root injury and reduce selection for resistance, which consequently, should reduce the rate at which resistance evolves (Andow *et al.*, 2016; EPA, 2016; Martinez and Caprio, 2016; Reinders *et al.*, 2018; Gassmann, 2021). This approach is most effective when novel transgenic technologies are introduced or when resistance to existing technologies is low because pest susceptibility can be maintained with an IPM strategy. Examples of current tactics that can be used in a complementary way with Bt traits in IPM programs include rotation from maize to a non-host crop (e.g. soybean) (Bare, 1930; Cullen *et al.*, 2013), foliar insecticides targeting adult females to reduce oviposition (Meinke, 2014; Dang *et al.*, 2023), use of non-rootworm-traited maize with soil-applied insecticide (Cullen *et al.*, 2013; Gassmann, 2021), and potentially, the use of insect-pathogenic nematodes for biological control (Shields *et al.*, 2018; Shields, 2019). Rotation of maize to a non-host crop is an effective agronomic practice in many areas of the US Corn Belt because it resets or reduces the existing western corn rootworm density to near zero. This enables a non-rootworm Bt hybrid to be planted in first-year maize and reduces selection pressure compared to continuous planting of the same Bt trait over time. The density and Bt susceptibility of western corn rootworm in areas surrounding a rotated field will determine the rate at which western corn rootworm recolonize that field and the extent to which the recolonizing population is resistant or susceptible to Bt traits, and this in turn will affect future management strategies needed in the rotated field.

Well-timed broadcast foliar insecticide applications targeting adult western corn rootworm can be used to reduce pest density and complement other tactics that may be deployed the following season in continuous maize (Meinke *et al.*, 2021; Dang *et al.*, 2023). This approach is most commonly used in the western Corn Belt because of the availability of chemigation and a well-developed aerial application industry, plus the necessary agricultural consultant base to appropriately time applications (Meinke *et al.*, 2021). Caution must be used when including adult management in an IPM program due to the heightened risk of resistance, with western corn rootworm having previously evolved resistance in adults to multiple insecticidal classes (Meinke *et al.*, 2021).

Until recently, biological control of western corn rootworm has not been available for large-scale commercial maize fields. However, the use of persistent entomopathogenic nematodes shows promise and may be an additional complementary tool for pest management in the future (Shields *et al.*, 2018; Shields, 2019).

In areas where longer-term (4–10 years or more) continuous maize is planted (e.g. in support of confined-livestock industry), an IPM approach also can be applied to manage larval injury and mitigate existing resistance. However, managing western corn rootworm is an ongoing challenge in continuous maize. Rootworm-active Bt traits provide excellent root protection when little to no Bt resistance is present, and this effect is independent of pest density (Reinders *et al.*, 2023; Meinke *et al.*, 2024). However, when moderate to high levels of resistance are present, pest density and resistance interact leading to increasing levels of feeding injury to Bt maize as pest abundance increases (Reinders *et al.*, 2023; Meinke *et al.*, 2024). In areas with high levels of resistance in the landscape, resistance cannot be remediated, but density can be managed to reduce this interaction, such that Bt maize suffers lower levels of feeding injury. Reducing the duration of continuous maize to 2–3 years, and including complementary tactics to manage pest density, can lead to acceptable levels of pest injury, even in the presence of resistance, which may help bridge to future technologies.

The history of resistance by western corn rootworm to insecticides and Bt traits clearly illustrates that resistance management needs to be part of any IPM program. This includes local and landscape considerations of the biological, behavioral, and genetic factors that affect the risk of resistance evolution (Gassmann, 2021; Meinke *et al.*, 2021). When a pest management technology is deployed as a single tactic to manage western corn rootworm, and this tactic is used repeatedly, there is a high probability of resistance evolving. More programs are needed that provide economic incentives for farmers to try or implement IPM strategies to break the long-term behavioral pattern of reliance on repeated use of single tactics to manage western corn rootworm.

References

Andow, D.A., Pueppke, S.G., Schaafsma, A.W., Gassmann, A.J., Sappington, T.W. *et al.* (2016) Early detection and mitigation of resistance to Bt maize by western corn rootworm (Coleoptera: Chrysomelidae). *Journal of Economic Entomology* 109, 1–12. DOI: 10.1093/jee/tov238

Ball, H.J. and Weekman, G.T. (1962) Insecticide resistance in the adult western corn rootworm in Nebraska. *Journal of Economic Entomology* 55, 439–441. DOI: 10.1093/jee/55.4.439

Ball, H.J. and Weekman, G.T. (1963) Differential resistance of corn rootworms to insecticides in Nebraska and adjoining states. *Journal of Economic Entomology* 56, 553–555. DOI: 10.1093/jee/56.5.553

Bare, O.S. (1930) *Corn Rootworm does Damage in Southwestern Nebraska*. Annual Report of Cooperative Extension Work in Agriculture and Home Economics, Entomology. Lincoln, NE: University of Nebraska-Lincoln.

Bažok, R. (2001) Istrazivanje biologije i ekologije *Diabrotica virgifera virgifera* LeConte – Novog clana entomofaune Hrvatske (Research on the biology and ecology of *Diabrotica virgifera virgifera* LeConte – new member of entomofauna in Croatia). PhD dissertation. Zagreb, Croatia: University of Zagreb.

Bažok, R., Lemić, D., Chiarini, F. and Furlan, L. (2021) Western corn rootworm (*Diabrotica virgifera virgifera* LeConte) in Europe: current status and sustainable pest management. *Insects* 12, 195. DOI: 10.3390/insects12030195

Beckler, A.A., French, B.W. and Chandler, L.D. (2004) Characterization of western corn rootworm (Coleoptera: Chrysomelidae) population dynamics in relation to landscape attributes. *Agricultural and Forest Entomology* 6, 129–139. DOI: 10.1111/j.1461-9563.2004.00213.x

Boetel, M.A., Fuller, B.W. and Evenson, P.D. (2003) Emergence of adult northern and western corn rootworms (Coleoptera: Chrysomelidae) following reduced soil insecticide applications. *Journal of Economic Entomology* 96, 714–729. DOI: 10.1093/jee/96.3.714

Bolognesi, R., Ramaseshadri, P., Anderson, J., Bachman, P., Clinton, W. *et al.* (2012) Characterizing the mechanism of action of double-stranded RNA activity against western corn rootworm (*Diabrotica virgifera virgifera* LeConte). *PLoS ONE* 7, e47534. DOI: 10.1371/journal.pone.0047534

Branson, T.F. and Krysan, J.L. (1981) Feeding and oviposition behavior and life cycle strategies of *Diabrotica*: an evolutionary view with implications for pest management. *Environmental Entomology* 10, 826–831. DOI: 10.1093/ee/10.6.826

Branson, T.F. and Ortman, E.E. (1967) Host range of larvae of the western corn rootworm. *Journal of Economic Entomology* 60, 201–203. DOI: 10.1093/jee/60.1.201

Branson, T.F. and Ortman, E.E. (1970) The host range of larvae of the western corn rootworm: further studies. *Journal of Economic Entomology* 63, 800–803. DOI: 10.1093/jee/63.3.800

Campbell, L.A. and Meinke, L.J. (2006) Seasonality and adult habitat use by four *Diabrotica* species at prairie–corn interfaces. *Environmental Entomology* 35, 922–936. DOI: 10.1603/0046-225X-35.4.922
Campbell, L.A., Prasifka, P.L., Storer, N.P., Rule, D.M. and Hendrix, W.H. (2017) Effects of Bt corn and egg density on western corn rootworm (Coleoptera: Chrysomelidae) adult emergence and estimation of effective Bt dose. *Journal of Economic Entomology* 110, 607–614. DOI: 10.1093/jee/tow286
Carrière, Y., Crowder, D.W. and Tabashnik, B.E. (2010) Evolutionary ecology of insect adaptation to Bt crops. *Evolutionary Applications* 3, 561–573. DOI: 10.1111/j.1752-4571.2010.00129.x
Carrière, Y., Crickmore, N. and Tabashnik, B.E. (2015) Optimizing pyramided transgenic Bt crops for sustainable pest management. *Nature Biotechnology* 33, 161–168. DOI: 10.1038/nbt.3099
Carrière, Y., Brown, Z., Aglasanb, S., Dutilleul, P., Carroll, M. *et al.* (2020) Crop rotation mitigates impacts of corn rootworm resistance to transgenic Bt corn. *Proceedings of the National Academy of Sciences* 117, 18385–18392. DOI: 10.1073/pnas.2003604117
Chiang, H.C. (1973) Bionomics of the northern and western corn rootworms. *Annual Review of Entomology* 18, 47–72. DOI: 10.1146/annurev.en.18.010173.000403
Clark, T.L. and Hibbard, B.E. (2004) Comparison of nonmaize hosts to support western corn rootworm (Coleoptera: Chrysomelidae) larval biology. *Environmental Entomology* 33, 681–689. DOI: 10.1603/0046-225X-33.3.681
Coats, S.A., Tollefson, J.J. and Mutchmor, J.A. (1986) Study of migratory flight in the western corn rootworm (Coleoptera: Chrysomelidae). *Environmental Entomology* 15, 620–625. DOI: 10.1093/ee/15.3.620
Crickmore, N., Berry, C., Panneerselvam, S., Mishra, R., Connor, T.R. *et al.* (2021) A structure-based nomenclature for *Bacillus thuringiensis* and other bacteria-derived pesticidal proteins. *Journal of Invertebrate Pathology* 186, 107438. DOI: 10.1016/j.jip.2020.107438
Cullen, E.M., Gray, M.E., Gassmann, A.J. and Hibbard, B.E. (2013) Resistance to Bt corn by western corn rootworm (Coleoptera: Chrysomelidae) in the U.S. Corn Belt. *Journal of Integrated Pest Management* 4(3), D1–D6. DOI: 10.1603/IPM13012
Dang, T.M., Velez, A.M., Valencia-Jimenez, A., Reinders, J.D., Stricklin, E.E. *et al.* (2023) Characterization of western corn rootworm (Coleoptera: Chrysomelidae) susceptibility to foliar insecticides in northeastern Nebraska. *Journal of Economic Entomology* 116, 945–955. DOI: 10.1093/jee/toad067
Darnell, S.J., Meinke, L.J., Young, L.J. and Gotway, C.A. (1999) Geostatistical investigation of the small-scale spatial variation of western corn rootworm (Coleoptera: Chrysomelidae) adults. *Environmental Entomology* 28, 266–274. DOI: 10.1093/ee/28.2.266
Darnell, S.J., Meinke, L.J. and Young, L.J. (2000) Influence of corn phenology on adult western corn rootworm (Coleoptera: Chrysomelidae) distribution. *Environmental Entomology* 29, 587–595. DOI: 10.1603/0046-225X-29.3.587
Deitloff, J., Dunbar, M.W., Ingber, D.A., Hibbard, B.E. and Gassmann, A.J. (2016) Effects of refuges on the evolution of resistance to transgenic corn by the western corn rootworm, *Diabrotica virgifera virgifera* LeConte. *Pest Management Science* 72, 190–198. DOI: 10.1002/ps.3988
Dingle, H. (1972) Migration strategies of insects. *Science* 175, 1327–1334. DOI: 10.1126/science.175.4028.1327
Dun, Z., Mitchell, P.D. and Agosti, M. (2010) Estimating *Diabrotica virgifera virgifera* damage functions with field trial data: applying an unbalanced nested error component model. *Journal of Applied Entomology* 134, 409–419. DOI: 10.1111/j.1439-0418.2009.01487.x
Dunbar, M.W., O'Neal, M.E. and Gassmann, A.J. (2016) Effects of field history on corn root injury and adult abundance of northern and western corn rootworm (Coleoptera: Chrysomelidae). *Journal of Economic Entomology* 109, 2096–2104. DOI: 10.1093/jee/tow163
EPA (2016) *Framework to Delay Corn Rootworm Resistance*. Washington, DC: US Environmental Protection Agency. Available at: https://www.epa.gov/regulation-biotechnology-under-tsca-and-fifra/framework-delay-corn-rootworm-resistance (accessed September 25, 2023).
EPA (2017) *MON 89034 × MIR162 × MON87411 Insect-protected, Herbicide-tolerant Corn.* Washington, DC: US Environmental Protection Agency. Available at: https://www3.epa.gov/pesticides/chem_search/ppls/000524-00635-20170301.pdf (accessed September 21, 2023).
EPA (2022) *Insect Resistance Management for Bt Plant-incorporated Protectants*. Washington, DC: US Environmental Protection Agency. Available at: https://www.epa.gov/regulation-biotechnology-under-tsca-and-fifra/insect-resistance-management-bt-plant-incorporated (accessed September 21, 2023).
Gassmann, A.J. (2012) Field-evolved resistance to Bt maize by western corn rootworm: predictions from the laboratory and effects in the field. *Journal of Invertebrate Pathology* 110, 287–293. DOI: 10.1016/j.jip.2012.04.006
Gassmann, A.J. (2021) Resistance to Bt maize by western corn rootworm: effects of pest biology, the pest–crop interaction and the agricultural landscape on resistance. *Insects* 12, 136. DOI: 10.3390/insects12020136

Gassmann, A.J. and Reisig, D.D. (2023) Management of insect pests with Bt crops in the United States. *Annual Review of Entomology* 68, 31–49. DOI: 10.1146/annurev-ento-120220-105502

Gassmann, A.J., Petzold-Maxwell, J.L., Keweshan, R.S. and Dunbar, M.W. (2011) Field-evolved resistance to Bt maize by western corn rootworm. *PLoS ONE* 6, e22629. DOI: 10.1371/journal.pone.0022629

Gassmann, A.J., Petzold-Maxwell, J.L., Keweshan, R.S. and Dunbar, M.W. (2012) Western corn rootworm and Bt maize: challenges of pest resistance in the field. *GM Crops and Food* 3, 235–244. DOI: 10.4161/gmcr.20744

Gassmann, A.J., Petzold-Maxwell, J.L., Clifton, E.H., Dunbar, M.W., Hoffmann, A.M. *et al.* (2014) Field-evolved resistance by western corn rootworm to multiple *Bacillus thuringiensis* toxins in transgenic maize. *Proceedings of the National Academy of Sciences* 111, 5141–5146. DOI: 10.1073/pnas.1317179111

Gassmann, A.J., Shrestha, R.B., Jakka, S.R.K., Dunbar, M.W., Clifton, E.H. *et al.* (2016) Evidence of resistance to Cry34/35Ab1 corn by western corn rootworm (Coleoptera: Chrysomelidae): root injury in the field and larval survival in plant-based bioassays. *Journal of Economic Entomology* 109, 1872–1880. DOI: 10.1093/jee/tow110

Gassmann, A.J., Shrestha, R.B., Kropf, A.L., St. Clair, C.R. and Brenizer, B.D. (2020) Field-evolved resistance by western corn rootworm to Cry34/35Ab1 and other *Bacillus thuringiensis* traits in transgenic maize. *Pest Management Science* 76, 268–276. DOI: 10.1002/ps.5510

Godfrey, L.D. and Turpin, F.T. (1983) Comparison of western corn rootworm (Coleoptera: Chrysomelidae) adult populations and economic thresholds in first-year and continuous corn fields. *Journal of Economic Entomology* 76, 1028–1032. DOI: 10.1093/jee/76.5.1028

Godfrey, L.D., Meinke, L.J. and Wright, R.J. (1993) Vegetative and reproductive biomass accumulation in field corn: response to root injury by western corn rootworm (Coleoptera: Chrysomelidae). *Journal of Economic Entomology* 86, 1557–1573. DOI: 10.1093/jee/86.5.1557

Gould, F. (1998) Sustainability of transgenic insecticidal cultivars: integrating pest genetics and ecology. *Annual Review of Entomology* 43, 701–726. DOI: 10.1146/annurev.ento.43.1.701

Grant, R.H. and Seevers, K.P. (1989) Local and long range movements of adult western corn rootworm (Coleoptera: Chrysomelidae) as evidenced by washup along southern Lake Michigan shores. *Environmental Entomology* 18, 266–272. DOI: 10.1093/ee/18.2.266

Grant, R.H. and Seevers, K.P. (1990) The vertical movement of adult western corn rootworms (*Diabrotica virgifera virgifera*) relative to the transport of momentum and heat. *Agricultural and Forest Meteorology* 49, 191–203. DOI: 10.1016/0168-1923(90)90032-2

Gray, M.E. and Steffey, K.L. (1998) Corn rootworm (Coleoptera: Chrysomelidae) larval injury and root compensation of 12 maize hybrids: an assessment of the economic injury index. *Journal of Economic Entomology* 91, 723–740. DOI: 10.1093/jee/91.3.723

Gray, M.E., Felsot, A.S., Steffey, K L. and Levine, E. (1992) Planting time application of soil insecticides and western corn rootworm (Coleoptera: Chrysomelidae) emergence: implications for long-term management programs. *Journal of Economic Entomology* 85, 544–553. DOI: 10.1093/jee/85.2.544

Gray, M.E., Sappington, T.W., Miller, N.J., Moeser, J. and Bohn, M.O. (2009) Adaptation and invasiveness of western corn rootworm: intensifying research on a worsening pest. *Annual Review of Entomology* 54, 303–321. DOI: 10.1146/annurev.ento.54.110807.090434

Head, G., Carroll, M., Clark, T., Galvan, T., Huckaba, R.M. *et al.* (2014) Efficacy of SmartStax® insect-protected corn hybrids against corn rootworm: the value of pyramiding the Cry3Bb1 and Cry34/35Ab1 proteins. *Crop Protection* 57, 38–47. DOI: 10.1016/j.cropro.2013.11.025

Head, G.P., Carroll, M.W., Evans, S.P., Rule, D.M., Willse, A.R. *et al.* (2017) Evaluation of SmartStax and SmartStax PRO maize against western corn rootworm and northern corn rootworm: efficacy and resistance management. *Pest Management Science* 73, 1883–1899. DOI: 10.1002/ps.4554

Hibbard, B.E., Duran, D.P., Ellersieck, M.R. and Ellsbury, M.M. (2003) Post establishment movement of western corn rootworm larvae (Coleoptera: Chrysomelidae) in central Missouri corn. *Journal of Economic Entomology* 96, 599–608. DOI: 10.1603/0022-0493-96.3.599

Hill, R.E. (1975) Mating, oviposition patterns, fecundity and longevity of the western corn rootworm. *Journal of Economic Entomology* 68, 311–315. DOI: 10.1093/jee/68.3.311

Hill, R.E. and Mayo, Z.B. (1974) Trap corn to control corn rootworms. *Journal of Economic Entomology* 67, 748–750. DOI: 10.1093/jee/67.6.748

Hoffmann, A.M., French, B.W., Hellmich, R.L., Lauter, N. and Gassmann, A.J. (2015) Fitness costs of resistance to Cry3Bb1 maize by western corn rootworm. *Journal of Applied Entomology* 139, 403–415. DOI: 10.1111/jen.12209

Hou, X., Meinke, L.J. and Arkebauer, T.J. (1997) Soil moisture and larval western corn rootworm injury: influence on gas exchange parameters in corn. *Agronomy Journal* 89, 709–717. DOI: 10.2134/agronj1997.00021962008900050001x

Hughson, S.A. and Spencer, J.L. (2015) Emergence and abundance of western corn rootworm (Coleoptera: Chrysomelidae) in Bt cornfields with structured and seed blend refuges. *Journal of Economic Entomology* 108, 114–125. DOI: 10.1093/jee/tou029

Ingber, D.A. and Gassmann, A.J. (2015) Inheritance and fitness costs of resistance to Cry3Bb1 corn by western corn rootworm (Coleoptera: Chrysomelidae). *Journal of Economic Entomology* 108, 2421–2432. DOI: 10.1093/jee/tov199

Isard, S.A., Spencer, J.L., Nasser, M.A. and Levine, E. (2000) Aerial movement of western corn rootworm (Coleoptera: Chrysomelidae): diel periodicity of flight activity in soybean fields. *Environmental Entomology* 29, 226–234. DOI: 10.1603/0046-225X(2000)029[0226:AMOWCR]2.0.CO;2

Jakka, S.R.K., Shrestha, R.B. and Gassmann, A.J. (2016) Broad-spectrum resistance to *Bacillus thuringiensis* toxins by western corn rootworm (*Diabrotica virgifera virgifera*). *Scientific Reports* 6, 27860. DOI: 10.1038/srep27860

Johnson, C.G. (1969) *Migration and Dispersal of Insects by Flight*. London: Methuen.

Keweshan, R.S., Head, G.P. and Gassmann, A.J. (2015) Effects of pyramided Bt corn and blended refuges on western corn rootworm and northern corn rootworm (Coleoptera: Chrysomelidae). *Journal of Economic Entomology* 108, 720–729. DOI: 10.1093/jee/tov005

Khajuria, C., Ivashuta, S., Wiggins, E., Flagel, L., Moar, W. *et al.* (2018) Development and characterization of the first dsRNA-resistant insect population from western corn rootworm, *Diabrotica virgifera virgifera LeConte*. *PLoS ONE* 13, e0197059. DOI: 10.1371/journal.pone.0197059

Levine, E. and Oloumi-Sadeghi, H. (1991) Management of diabroticite rootworms in corn. *Annual Review of Entomology* 36, 229–255. DOI: 10.1146/annurev.en.36.010191.001305

Ludwick, D.C., Meihls, L.N., Ostlie, K.R., Potter, B.D., French, L. *et al.* (2017) Minnesota field population of western corn rootworm (Coleoptera: Chrysomelidae) shows incomplete resistance to Cry34Ab1/Cry35Ab1 and Cry3Bb1. *Journal of Applied Entomology* 141, 28–40. DOI: 10.1111/jen.12377

Martinez, J.C. and Caprio, M.A. (2016) IPM use with the deployment of a non-high dose Bt pyramid and mitigation of resistance for western corn rootworm (*Diabrotica virgifera virgifera*). *Environmental Entomology* 45, 747–761. DOI: 10.1093/ee/nvw015

Meihls, L.N., Higdon, M.L., Siegfried, B.D., Miller, N.J., Sappington, T.W. *et al.* (2008) Increased survival of western corn rootworm on transgenic corn within three generations of on-plant greenhouse selection. *Proceedings of the National Academy of Sciences* 105, 19177–19182. DOI: 10.1073/pnas.0805565105

Meinke, L.J. (2014) *Adult Corn Rootworm Suppression: Corn Rootworm Management in the Transgenic Era*. Grow: Plant Health Exchange. St Paul, MN: American Phytopathological Society. Available at: http://www.plantmanagementnetwork.org/edcenter/seminars/corn/AdultRootworm/ (accessed September 26, 2023).

Meinke, L.J., Siegfried, B.D., Wright, R.J. and Chandler, L.D. (1997) Western corn rootworm resistance to insecticides: current situation in Nebraska. In: *Proceedings of the Illinois Agricultural Pesticides Conference, Champaign, IL*, pp. 88–92.

Meinke, L.J., Siegfried, B.D., Wright, R.J. and Chandler, L D. (1998) Adult susceptibility of Nebraska western corn rootworm (Coleoptera: Chrysomelidae) populations to selected insecticides. *Journal of Economic Entomology* 91, 594–600. DOI: 10.1093/jee/91.3.594

Meinke, L.J., Sappington, T.W., Onstad, D.W., Guillemaud, T., Miller, N.J. *et al.* (2009) Western corn rootworm (*Diabrotica virgifera virgifera* LeConte) population dynamics. *Agricultural and Forest Entomology* 11, 29–46. DOI: 10.1111/j.1461-9563.2008.00419.x

Meinke, L.J., Souza, D. and Siegfried, B.D. (2021) The use of insecticides to manage the western corn rootworm, *Diabrotica virgifera virgifera*, LeConte: history, field-evolved resistance, and associated mechanisms. *Insects* 12, 112. DOI: 10.3390/insects12020112

Meinke, L.J., Reinders, J.D., Dang, T.B., Krumm, J.T., Pilcher, C.D. *et al.* (2024) Resistance management and integrated pest management insights from deployment of a Cry3Bb1 + Gpp34Ab1/Tpp35Ab1 pyramid in a resistant western corn rootworm landscape. *PLoS ONE* 19, 0299483. DOI: 10.1371/journal.pone.0299483

Miller, N., Estoup, A., Toepfer, S., Bourguet, D., Lapchin, L. *et al.* (2005) Multiple transatlantic introductions of the western corn rootworm. *Science* 310(5750), 992. DOI: 10.1126/science.1115871

Miller, N.J., Guillemaud, T., Giordano, R., Siegfried, B.D., Gray, M.E. *et al.* (2009) Genes, gene flow and adaptation of *Diabrotica virgifera virgifera*. *Agricultural and Forest Entomology* 11, 47–60. DOI: 10.1111/j.1461-9563.2008.00398.x

Moeser, J. and Hibbard, B.E. (2005) A synopsis of the nutritional ecology of larvae and adults of *Diabrotica virgifera virgifera* (LeConte) in the new and old world – nouvelle cuisine for the invasive maize pest *Diabrotica virgifera virgifera* in Europe? In: Vidal, S., Kuhlmann, U. and Edwards, C.R. (eds) *Western Corn Rootworm: Ecology and Management.* Wallingford, UK: CAB International, pp. 41–65. DOI: 10.1079/9780851998176.0041

Naranjo, S.E. (1990) Comparative flight behavior of *Diabrotica virgifera virgifera* and *Diabrotica barberi* in the laboratory. *Entomologia Experimentalis et Applicata* 55, 79–90. DOI: 10.1007/BF00188230

Onstad, D.W. and Meinke, L.J. (2010) Modeling evolution of *Diabrotica virgifera virgifera* (Coleoptera: Chrysomelidae) to transgenic corn with two insecticidal traits. *Journal of Economic Entomology* 103, 849–860. DOI: 10.1603/ec09199

Oyediran, I.O., Hibbard, B.E. and Clark, T.L. (2004) Prairie grasses as hosts of the western corn rootworm (Coleoptera: Chrysomelidae). *Environmental Entomology* 33, 740–747. DOI: 10.1603/0046-225x(2008) 37[247:pgahot]2.0.co;2

Paddock, K.J., Hibbard, B.E., Barry, J., Sethi, A., Mueller, A.L. *et al.* (2021) Restoration of susceptibility following removal of selection for Cry34/35Ab1 resistance documents fitness costs in resistant population of western corn rootworm, *Diabrotica virgifera virgifera*. *Pest Management Science* 77, 2385–2394. DOI: 10.1002/ps.6266

Paolino, A.R. and Gassmann, A.J. (2017) Assessment of inheritance and fitness costs associated with field-evolved resistance to Cry3Bb1 maize by western corn rootworm. *Toxins* 9, 159. DOI: 10.3390/toxins9050159

Parimi, S., Scharf, M.E., Meinke, L.J., Chandler, L.D. and Siegfried, B.D. (2003) Inheritance of methyl-parathion resistance in Nebraska western corn rootworm populations (Coleoptera: Chrysomelidae). *Journal of Economic Entomology* 96, 131–136. DOI: 10.1603/0022-0493-96.1.131

Parimi, S., Meinke, L.J., French, W.B., Chandler, L.D. and Siegfried, B.D. (2006) Stability and persistence of aldrin and methyl-parathion resistance in western corn rootworm populations (Coleoptera: Chrysomelidae). *Crop Protection* 25, 269–274. DOI: 10.1016/j.cropro.2005.04.017

Pereira, A.E., Wang, H., Zukoff, S.N., Meinke, L.J., French, B.W. *et al.* (2015) Evidence of field-evolved resistance to bifenthrin in western corn rootworm (*Diabrotica virgifera virgifera* LeConte) populations in western Nebraska and Kansas. *PLoS ONE* 10, e0142299. DOI: 10.1371/journal.pone.0142299

Petzold-Maxwell, J.L., Cibils-Stewart, X., French, B.W. and Gassmann, A.J. (2012) Adaptation by western corn rootworm (Coleoptera: Chrysomelidae) to Bt maize: inheritance, fitness costs and feeding preference. *Journal of Economic Entomology* 105, 1407–1418. DOI: 10.1603/ec11425

Petzold-Maxwell, J.L., Alves, A.P., Estes, R.E., Gray, M.E., Meinke, L. *et al.* (2013a) Applying an integrated refuge to manage western corn rootworm (Coleoptera: Chrysomelidae): effects on survival, fitness and selection pressure. *Journal of Economic Entomology* 106, 2195–2207. DOI: 10.1603/EC13088

Petzold-Maxwell, J.L., Meinke, L.J., Gray, M.E., Estes, R.E. and Gassmann, A.J. (2013b) Effect of Bt maize and soil insecticides on yield, injury, and rootworm survival: implications for resistance management. *Journal of Economic Entomology* 106, 1941–1951. DOI: 10.1603/ec13216

Prystupa, B., Ellis, C.R. and Teal, P.E.A. (1988) Attraction of adult *Diabrotica* (Coleoptera: Chrysomelidae) to corn silks and analysis of the host-finding response. *Journal of Chemical Ecology* 14, 635–651. DOI: 10.1007/BF01013912

Ramaseshadri, P., Segers, G., Flannagan, R., Wiggins, E., Clinton, W. *et al.* (2013) Physiological and cellular responses caused by RNAi-mediated suppression of Snf7 orthologue in western corn rootworm (*Diabrotica virgifera virgifera*) larvae. *PLoS ONE* 8, e54270. DOI: 10.1371/journal.pone.0054270

Reinders, J.D. and Meinke, L.J. (2022) Reduced susceptibility of western corn rootworm (*Diabrotica virgifera virgifera* LeConte) populations to Cry34/35Ab1-expressing maize in northeast Nebraska. *Scientific Reports* 12, 19221. DOI: 10.1038/s41598-022-23755-z

Reinders, J.D., Hitt, B.D., Stroup, W.W., French, B.W. and Meinke, L.J. (2018) Spatial variation in western corn rootworm (Coleoptera: Chrysomelidae) susceptibility to Cry3 toxins in Nebraska. *PLoS ONE* 13, e0208266. DOI: 10.1371/journal.pone.0208266

Reinders, J.D., Wangila, D.S., Robinson, E.A., French, B.W. and Meinke, L.J. (2021) Characterizing the relationship between western corn rootworm (Coleoptera: Chrysomelidae) larval survival on Cry3Bb1-expressing corn and larval development metrics. *Journal of Economic Entomology* 114, 2096–2107. DOI: 10.1093/jee/toab151

Reinders, J.D., Reinders, E.E., Robinson, E.A., French, W.B. and Meinke, L.J. (2022) Evidence of western corn rootworm (*Diabrotica virgifera virgifera* LeConte) field-evolved resistance to Cry3Bb1 + Cry34/35Ab1 maize in Nebraska. *Pest Management Science* 78, 1356–1366. DOI: 10.1002/ps.6752

Reinders, J.D., Moar, W.J., Head, G.P., Hassan, S. and Meinke, L.J. (2023) Effects of SmartStax® and SmartStax® PRO maize on western corn rootworm (*Diabrotica virgifera virgifera* LeConte) larval feeding injury and adult life history parameters. *PLoS ONE* 18, e0288372. DOI: 10.1371/journal.pone.0288372

Roush, R.T. (1998) Two-toxin strategies for management of insecticidal transgenic crops: can pyramiding succeed where pesticide mixtures have not? *Philosophical Transactions of the Royal Society of London. Series B, Biological Sciences* 353, 1777–1786. DOI: 10.1098/rstb.1998.0330

Rudeen, M.L. and Gassmann, A.J. (2013) Effects of Cry34/35Ab1 corn on survival and development of western corn rootworm, *Diabrotica virgifera virgifera*. *Pest Management Science* 69, 709–716. DOI: 10.1002/ps.3425

Sappington, T.W., Siegfried, B.D. and Guillemaud, T. (2006) Coordinated Diabrotica genetics research: accelerating progress on an urgent insect pest problem. *American Entomologist* 52, 90–97. DOI: 10.1093/ae/52.2.90

Scharf, M.E., Meinke, L.J., Siegfried, B.D., Wright, R.J. and Chandler, L.D. (1999) Carbaryl susceptiblty, diagnostic concentration determination, and synergism for U.S. populations of western corn rootworm. *Journal of Economic Entomology* 92, 33–39. DOI: 10.1093/jee/92.1.33

Schellenberger, U., Oral, J., Rosen, B.A., Wei, J.-Z., Zhu, G. *et al.* (2016) A selective insecticidal protein from *Pseudomonas* for controlling corn rootworms. *Science* 354, 634–637. DOI: 10.1126/science.aaf6056

Schrader, P.M., Estes, R.E., Tinsley, N.A., Gassmann, A.J. and Gray, M.E. (2016) Evaluation of adult emergence and larval root injury for Cry3Bb1-resistant populations of the western corn rootworm. *Journal of Applied Entomology* 141, 41–52. DOI: 10.1111/jen.12329

Shields, E.J. (2019) *Invasion of the Rootworm Snatchers*. Burnsville, MN: Progressive Farmer. Available at: https://www.dtnpf.com/agriculture/web/ag/news/article/2020/02/03/nematodes-eat-insect-guts-breakfast (accessed September 25, 2023).

Shields, E.J., Testa, A.M. and O'Neil, W.J. (2018) Long-term persistence of native New York entomopathogenic nematode isolates across crop rotation. *Journal of Economic Entomology* 111, 2592–2598. DOI: 10.1093/jee/toy258

Shrestha, R.B., Dunbar, M.W., French, B.W. and Gassmann, A.J. (2018) Effects of field history on resistance to Bt maize by western corn rootworm, *Diabrotica virgifera virgifera* LeConte (Coleoptera: Chrysomelidae). *PLoS ONE* 13, e0200156. DOI: 10.1371/journal.pone.0200156

Shrestha, R.B. and Gassmann, A.J. (2019) Field and laboratory studies of resistance to Bt corn by western corn rootworm (Coleoptera: Chrysomelidae). *Journal of Economic Entomology* 112, 2324–2334. DOI: 10.1093/jee/toz151

Siegfried, B.D. and Mullin, C.A. (1990) Effects of alternative host plants on longevity, oviposition, and emergence of western and northern corn rootworms (Coleoptera: Chrysomelidae). *Environmental Entomology* 19, 474–480. DOI: 10.1093/ee/19.3.474

Smith, E.M., Shrestha, R.B. and Gassmann, A.J. (2023) Inheritance and fitness costs of laboratory-selected resistance to Gpp34/Tpp35Ab1 corn in western corn rootworm. *Journal of Economic Entomology* 116, 565–573. DOI: 10.1093/jee/toad022

Souza, D., Vieira, B.C., Fritz, B.K., Hoffmann, W.C., Peterson, J.A. *et al.* (2019) Western corn rootworm pyrethroid resistance confirmed by aerial application simulations of commercial insecticides. *Scientific Reports* 9, 6713. DOI: 10.1038/s41598-019-43202-w

Souza, D., Peterson, J.A., Wright, R.J. and Meinke, L.J. (2020) Field efficacy of soil insecticides on pyrethroid-resistant western corn rootworms (*Diabrotica virgifera virgifera* LeConte). *Pest Management Science* 76, 827–833. DOI: 10.1002/ps.5586

Spencer, J., Onstad, D., Krupke, C., Hughson, S., Pan, Z. *et al.* (2012) Isolated females and limited males: evolution of insect resistance in structured landscapes. *Entomologia Experimentalis et Applicata* 146, 38–49. DOI: 10.1111/eea.12022

Spencer, J.L., Mabry, T.R. and Vaughn, T.T. (2003) Use of transgenic plants to measure insect herbivore movement. *Journal of Economic Entomology* 96, 1738–1749. DOI: 10.1093/jee/96.6.1738

Spencer, J.L., Hibbard, B.E., Moeser, J. and Onstad, D.W. (2009) Behaviour and ecology of the western corn rootworm (*Diabrotica virgifera virgifera* LeConte). *Agricultural and Forest Entomology* 11, 9–27. DOI: 10.1111/j.1461-9563.2008.00399.x

Spike, B.P. and Tollefson, J.J. (1991) Response of western corn rootworm infested corn to nitrogen fertilization and plant density. *Crop Science* 31, 776–785. DOI: 10.2135/cropsci1991.0011183X003100030048x

St. Clair, C.R. and Gassmann, A.J. (2021) Linking land use patterns and pest outbreaks in Bt maize. *Ecological Applications* 31, e02295. DOI: 10.1002/eap.2295

St. Clair, C.R., Clifton, E.H., Dunbar, M.W., Masloski, K.E., Paolino, A.R. *et al.* (2020a) Applying a selection experiment to test for fitness costs of Bt resistance in western corn rootworm and the effect of density on fitness costs. *Journal of Economic Entomology* 113, 2473–2479. DOI: 10.1093/jee/toaa168

St. Clair, C.R., Head, G.P. and Gassmann, A.J. (2020b) Western corn rootworm abundance, injury to corn, and resistance to Cry3Bb1 in the local landscape of previous problem fields. *PLoS ONE* 15, e0237094. DOI: 10.1371/journal.pone.0237094

Stebbing, J.A., Meinke, L.J., Naranjo, S.E., Siegfried, B.D., Wright, R.J. *et al.* (2005) Flight behavior of methyl-parathion resistant and susceptible western corn rootworm (Coleoptera: Chrysomelidae) populations from Nebraska. *Journal of Economic Entomology* 98, 1294–1304. DOI: 10.1603/0022-0493-98.4.1294

Tabashnik, B.E. and Carrière, Y. (2017) Surge in insect resistance to transgenic crops and prospects for sustainability. *Nature Biotechnology* 35, 926–935. DOI: 10.1038/nbt.3974

Tabashnik, B.E. and Gould, F. (2012) Delaying corn rootworm resistance to Bt corn. *Journal of Economic Entomology* 105, 767–776. DOI: 10.1603/EC12080

Tabashnik, B.E., Gassmann, A.J., Crowder, D.W. and Carrière, Y. (2008) Insect resistance to *Bt* crops: evidence versus theory. *Nature Biotechnology* 26, 199–202. DOI: 10.1038/nbt1382

Tabashnik, B.E., Mota-Sanchez, D., Whalon, M.E., Hollingworth, R.M. and Carrière, Y. (2014) Defining terms for proactive management of resistance to Bt crops and pesticides. *Journal of Economic Entomology* 107, 496–507. DOI: 10.1603/EC13458

Tinsley, N.A., Estes, R.E. and Gray, M.E. (2013) Validation of a nested error component model to estimate damage caused by corn rootworm larvae. *Journal of Applied Entomology* 137, 161–169. DOI: 10.1111/j.1439-0418.2012.01736.x

Toepfer, S., Ellsbury, M.M., Eschen, R. and Kuhlmann, U. (2007) Spatial clustering of *Diabrotica virgifera virgifera* and *Agriotes ustulatus* in small-scale maize fields without topographic relief drift. *Entomologia Experimentalis et Applicata* 124, 61–75. DOI: 10.1111/j.1570-7458.2007.00549.x

Urías-López, M.A. and Meinke, L.J. (2001) Influence of western corn rootworm (Coleoptera Chrysomelidae) larval injury on yield of different types of maize. *Journal of Economic Entomology* 94, 106–111. DOI: 10.1603/0022-0493-94.1.106

Urías-López, M.A., Meinke, L.J., Higley, L.G. and Haile, F.J. (2000) Influence of western corn rootworm (Coleoptera: Chrysomelidae) larval injury on photosynthetic rate and vegetative growth of different types of maize. *Environmental Entomology* 29, 861–867. DOI: 10.1603/0022-0493-94.1.106

USDA-APHIA (2020) *Pioneer Hi-Bred International, Inc.: Availability of a Petition for the Determination of Nonregulated Status for Insect Resistant and Herbicide-tolerant Maize*. Washington, DC: US Department of Agriculture, Animal Plant Health Inspection Agency. Available at: https://www.federalregister.gov/documents/2020/11/03/2020-24267/pioneer-hi-bred-international-incavailability-of-a-petition-for-the-determination-of-nonregulated (accessed September 27, 2023).

USDA-NASS (2022) *Corn for all Purposes 2022 Planted Acres by County for Selected States*. Washington, DC: US Department of Agriculture, National Agricultural Statistical Service. Available at: https://www.nass.usda.gov/Charts_and_Maps/Crops_County/cr-pl.php (accessed September 21, 2023).

Wang, H., Coates, B.S., Chen, H., Sappington, T.W., Guillemaud, T. *et al.* (2013) Role of a gamma-aminobutryic acid (GABA) receptor mutation in the evolution and spread of *Diabrotica virgifera virgifera* resistance to cyclodiene insecticides. *Insect Molecular Biology* 22, 473–484. DOI: 10.1111/imb.12037

Wangila, D.S. and Meinke, L.J. (2017) Effects of adult emergence timing on susceptibility and fitness of Cry3Bb1-resistant western corn rootworms. *Journal of Applied Entomology* 141, 372–383. DOI: 10.1111/jen.12346

Wangila, D.S., Gassmann, A.J., Petzold-Maxwell, J.L., French, B.W. and Meinke, L.J. (2015) Susceptibility of Nebraska western corn rootworm populations (Coleoptera: Chrysomelidae) populations to Bt corn events. *Journal of Economic Entomology* 108, 742–751. DOI: 10.1093/jee/tou063

Wechsler, S. and Smith, D. (2018) Has resistance taken root in U.S. corn fields? Demand for insect control. *American Journal of Agricultural Economics* 100, 1136–1150. DOI: 10.1093/ajae/aay016

Witkowski, J.F., Owens, J.C. and Tollefson, J.J. (1975) Diel activity and vertical flight distribution of adult western corn rootworms in Iowa cornfields. *Journal of Economic Entomology* 68, 351–352. DOI: 10.1093/jee/68.3.351

Yu, E.Y., Gassmann, A.J. and Sappington, T.W. (2019) Effects of larval density on dispersal and fecundity of western corn rootworm *Diabrotica virgifera virgifera* LeConte (Coleoptera: Chrysomelidae). *PLoS ONE* 14, e0212696. DOI: 10.1371/journal.pone.0212696

Zhou, X., Scharf, M.E., Parimi, S., Meinke, L.J., Wright, R.J. *et al.* (2002) Diagnostic assays based on esterase-mediated resistance mechanisms in western corn rootworms (Coleoptera: Chrysomelidae). *Journal of Economic Entomology* 95, 1261–1266. DOI: 10.1603/0022-0493-95.6.1261

Zukoff, S.N., Ostlie, K.R., Potter, B., Meihls, L.N., Zukoff, A.L. *et al.* (2016) Multiple assays indicate varying levels of cross resistance in Cry3Bb1-selected field populations of the western corn rootworm to mCry3A, eCry3.1Ab, and Cry34/35Ab1. *Journal of Economic Entomology* 109, 1387–1398. DOI: 10.1093/jee/tow073

10 Linking Corn Earworm Populations and Management to Landscapes Across North America

Anders S. Huseth,[1,*] Dominic D. Reisig,[2] and William D. Hutchison[3]
[1]Department of Entomology and Plant Pathology and the North Carolina Plant Sciences Initiative, North Carolina State University, Raleigh, NC, USA; [2]Department of Entomology and Plant Pathology, North Carolina State University, Plymouth, NC, USA; [3]Department of Entomology, University of Minnesota, St Paul, MN, USA

Abstract

Accurately modeling highly mobile pests in agricultural landscapes is a considerable challenge that limits population prediction in support of management. For some specialist insect herbivores, a close association to a limited host range enables reasonably accurate prediction of infestation risk. Moreover, simplified agricultural systems have generated successful tools that improve integrated pest management (IPM) practices due to a very predictable sequence of host utilization. In contrast, accurately modeling generalist pests with broad host ranges in dynamic cropping systems requires a deep understanding of ecological interactions and dispersal patterns among host plant patches in space and time. Corn earworm, *Helicoverpa zea* (Boddie) (Lepidoptera: Noctuidae) is an example of a challenging polyphagous pest that cycles through multiple crops in agricultural systems, but also has different movement patterns from local dispersal to long-distance migration. *Helicoverpa zea* has been one of the most well studied agricultural pests in North America and, in turn, has generated many approaches to monitoring (i.e. pheromone and black light networks) and modeling. Unfortunately, these data-driven tools are vulnerable to major changes in production practices that include, but are not limited to, the adoption of genetically engineered crops expressing *Bacillus thuringiensis* toxins and an ever-shifting set of production practices. The focus of this chapter is a targeted review of fundamental *H. zea* literature that has spurred mechanistic and statistical modeling approaches to improve understanding of this pest beyond the field edge. Recent advancements in geospatial data analytics have enabled a rapid radiation in *H. zea* analytic approaches and improved risk prediction capabilities. The motivation for this chapter is to generate new interest and approaches to study this pest's ecology and distribution using contemporary methods. We also highlight key knowledge gaps that, provided with innovative spatial data, could improve inference to motivate *H. zea* management improvements in agroecosystems.

Introduction

Landscape structure and composition are two drivers of corn earworm, *Helicoverpa zea* (Boddie) (Lepidoptera: Noctuidae), populations in row crop agroecosystems. In these landscapes, the sequential availability of crop and non-crop hosts is an important factor that helps to amplify *H. zea* population size over several generations (Kennedy and Margolies, 1985). The capability

*Corresponding author: ashuseth@ncsu.edu

 Arthropod Management and Landscape Considerations in Large-Scale Agroecosystems (eds M.J. Brewer and G.L. Hein)
DOI: 10.1079/9781800622777.0010

of *H. zea* adults to disperse among hosts in the landscape is a key factor that has allowed this pest to thrive in agroecosystems (Fitt, 1989; Kennedy and Storer, 2000). Knowledge about these landscape linkages is a key aspect of management efforts that integrate spatiotemporal host utilization for risk prediction. Row crops are key components of this ecologically based management approach because they are abundant hosts in the landscape that have different periods of suitability during the growing season. This facilitates rapid population increase and straightforward predictions about where and when economic damage may occur (Rabb, 1978) (Fig. 10.1). Discovering information about this crop–pest relationship combined with a long history of areawide monitoring for *H. zea* have been critical steps toward developing diverse integrated pest management (IPM) tactics that minimize reliance on chemical interventions. Moreover, these predictable sequences of *H. zea* population movement through genetically engineered corn, *Zea mays* (L.), and cotton, *Gossypium hirsutum* (L.), that express insecticidal *Bacillus thuringiensis* (Bt) toxins is important ecological information used to implement insecticide resistance management (IRM) practices and policy. Over time, advancements in *H. zea* IPM and IRM have required the integration of multi-scale pest ecology research to improve management outcomes, while alleviating reliance on broad-spectrum insecticides. *Helicoverpa zea* research in North America represents one of the most comprehensive examples of how the increasing availability of geospatial data can advance

Fig. 10.1. Corn earworm or bollworm, *Helicoverpa zea* (Boddie): (A) adult on a sorghum leaf, (B) late instar larvae infesting sweet corn, and (C) adult in a cotton bloom. (Photo credits: A.S. Huseth.)

pest management through a union of natural history, population dynamics, landscape ecology, and spatial data sciences.

Along with boll weevil, *Anthonomus grandis grandis* (Boheman) (Coleoptera: Curculionidae), and tobacco budworm, *Chloridea virescens* (F.) (Lepidoptera; Noctuidae), *H. zea* has been a major area of focus for economic entomologists in the southern United States (US) for more than a century that, in turn, has produced new approaches to systems-based pest management in dynamic agroecosystems (Quaintance and Brues, 1905; Bradley, 1996; Kennedy and Storer, 2000). Decades of research have focused on the spatial structure of *H. zea* populations and their relationships to key host crops in local- and regional-scale landscapes, particularly the temporal availability of early-season corn followed by upland cotton and soybean, *Glycine max* (L.), later in the growing season. The configuration of host fields and proximity to semi-natural habitats can also affect *H. zea* source–sink dispersal dynamics, often leading to differential infestation and crop injury (Stinner, 1979). This research effort generated significant fundamental knowledge about *H. zea* ecology and its interaction with the surrounding agroecosystems before farmers transitioned to intensive, insecticide-based management programs (Bradley, 1996).

Elements of spatial interactions are woven throughout early *H. zea* research with descriptions of multi-scale population variation, crop infestation patterns, and resultant damage. In a formative paper more than a century ago, Quaintance and Brues (1905) reported foundational components of geospatial knowledge about *H. zea* dispersal and infestation patterns in multiple US agroecosystems. These authors described notable differences in *H. zea* infestation and crop injury in different agricultural regions, with a specific focus on the US Cotton Belt where this insect is a severe economic pest. Before the turn of the 20th century, variable crop diversity and field configurations were included as two possible explanations for differential *H. zea* pressure across the Cotton Belt (Quaintance and Brues, 1905). Farms in the mid-south to the western cotton production region (e.g. Mississippi, Louisiana, and Texas) tended to grow larger monocultures of cotton and corn. These cropping areas were reported to have more intensive *H. zea* pressure and resultant crop injury. In contrast, farmers in the eastern US planted smaller irregularly shaped fields with a larger diversity of crops, including intercropped cowpea, which was an important subsistence crop and *H. zea* host. The increased diversity of hosts and agronomic practices in the east were one explanation for less *H. zea* cotton injury when compared to western cotton production regions (Quaintance and Brues, 1905). In retrospect, these reports of *H. zea* host and landscape associations are conceptually similar to contemporary landscape research efforts that model the response of many different cotton pest species to agroecosystem composition in space and time (Fitt, 1989; Allen and Luttrell, 2009; Carrière *et al.*, 2012; Meisner *et al.*, 2017; Sivakoff *et al.*, 2013; Huseth *et al.*, 2018; Dorman *et al.*, 2020). An important recent innovation has been the aggregation of large cotton scouting data sets to uncover the generality of pest–landscape associations using observational data analyses (Karp *et al.*, 2018; Gagic *et al.*, 2021; Rosenheim *et al.*, 2022). Undoubtedly, the observations of early cotton entomologists set the stage for decades of research relating agricultural landscapes to *H. zea* and other crop pests.

The motivation for this chapter is to connect contemporary data-driven *H. zea* research to spatial and landscape concepts presented in foundational research. It is important to acknowledge the conceptual contributions of early entomologists to our current understanding of the *H. zea* movement and management within agroecosystems (e.g. Rabb, 1978; Stinner, 1979; Kennedy and Margolies, 1985; Fitt, 1989). Because the majority of early landscape ecology research focused on *H. zea* occurred in southern US row crop agroecosystems, we used this region as a case study to discuss pest ecology in complex and temporally unstable cropping systems (Kennedy and Storer, 2000). Key to advancements of *H. zea* research has been the rapid increase in spatial data products and proliferation of accessible geospatial science software. These tools have enabled testing of fundamental ecological principles governing *H. zea* in actual landscapes. Here, we describe a selection of recent advancements in the spatial ecology of *H. zea* and discuss opportunities to integrate innovative data science to solve outstanding issues with population prediction and management of this pest. Additionally, our goal is to motivate a new generation of ecologists to explore historical insect pest literature and leverage a growing number of analytical tools to refine prediction and management of mobile lepidopteran pests at multiple spatial and temporal scales.

Host Use in Agricultural Landscapes

Research in the late 1900s included detailed observations about *H. zea* ecology, but also geographic concepts related to distribution and crop damage. Specifically, Quaintance and Brues (1905) highlight elements of spatial structure when discussing host suitability, crop diversity, dispersal, and this pest's relationship to different land use beyond the field edge. Decades later, many of these relationships were explored with both mechanistic- and simulation-based *Helicoverpa* spp. models using growing computing power and accessible programming languages (e.g. FORTRAN) (Stinner *et al.*, 1974, 1975; Gutierrez *et al.*, 1975; Getz and Gutierrez, 1982; Fitt *et al.*, 1995; Feng *et al.*, 2010). The goal of these modeling exercises was to explore the underlying ecological principles driving cycles of *Helicoverpa* spp. populations in agricultural fields and landscapes, a key step toward risk prediction. Key information for these models is a fundamental component of this pest's ecology reported in the early literature. This information includes, but is not limited to, temporal host plant use, life table analyses, overwintering conditions, and weather patterns (Fig. 10.2).

In the southern US, *H. zea* pupae overwinter in soil chambers across a variety of soil types and overlying vegetation. Winter soil temperatures are lethal to *H. zea* pupae and limit northward overwintering range expansion beyond approximately 40°N latitude (Hardwick, 1965; Diffenbaugh *et al.*, 2008; Lawton *et al.*, 2022). Recent experimental cold hardiness research determined that soil temperatures less than 5°C resulted in significant *H. zea* pupal mortality (Morey *et al.*, 2012); however, confidence intervals around median lethal temperature estimates suggest more moderate winter temperatures also affects pupal survival for some individuals in a population. Although the lower limits for overwintering survival have been established, we do know that a large proportion of individuals in more favorable environments also die during overwintering. In North Carolina, an overwintering region, Caron *et al.* (1978) and Dillard *et al.* (2023a) documented that less than 10% of overwintering pupae survive.

Vegetation and vegetation residue interact with soil type that could further complicate spatial assessment of lethal soil temperatures. The effect of vegetation on mediating soil temperature through shading and the influence of roots on soil porosity has long been recognized as an important mediator of soil temperature (Phillips and Barber, 1929). However, the magnitude of this interaction and the resultant impact on soil-borne arthropods remains understudied. Recent work has also shown that soil type and moisture interact to affect pupal survival both during the growing season and over the winter (Dillard *et al.*, 2023a, b). Hence, the impact of soil conditions in mediating survival is a potentially important factor regulating local populations across highly variable soil types in the southern US overwintering zones. In the future, developing research that examines the effect of soil conditions on *H. zea* overwintering survival in space and time could be a useful contribution to our understanding of annual population dynamics at broad spatial scales.

As a multivoltine and highly polyphagous pest, host plant associations in both crop and semi-natural habitats are the most well studied and important aspects of *H. zea* ecology. Tabulations of *H. zea* host diversity highlight the broad host range, reporting that this insect can colonize an estimated 123 plant species in 29 different families (Kogan *et al.*, 1989; Hardwick, 1996; Blanco *et al.*, 2007; Cunningham and Zalucki, 2014). Several weed species found in semi-natural habitats are key early-season resources throughout the southeastern US. These include, but are not limited to, *Geranium carolinianum* (L.), *Linaria* spp., *Trifolium incarnatum* (L.), *Vica villosa* (Roth) (Graham *et al.*, 1972; Roach, 1975; Stadelbacher, 1981; Stadelbacher *et al.*, 1986). An important caveat to these host lists is that phenological or geographic mismatches may occur between the pest and host plants. We do know that *H. zea* prefers to oviposit on flowering hosts (Johnson *et al.*, 1975). Often the insect and the flowering stage of these hosts do not co-occur in space or time. Consequently, documenting the cycle of generations passing through semi-natural weeds and cultivated host plants is important to

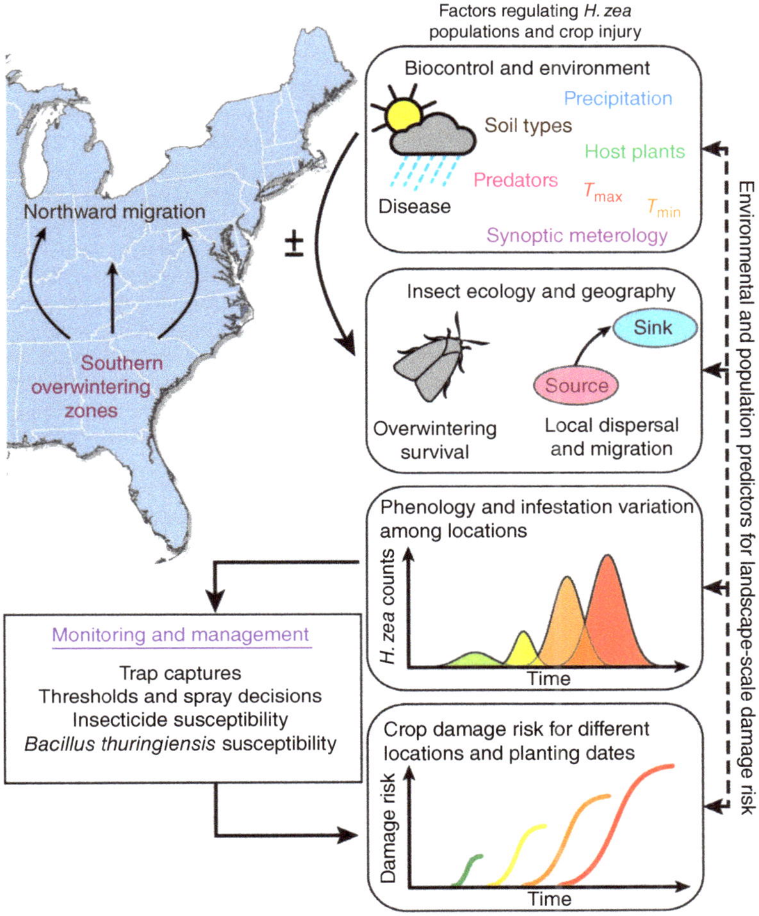

Fig. 10.2. Multiple biotic and abiotic factors drive *Helicoverpa zea* populations and infestation risk. Here, several general factors regulating population size and dispersal are presented as a conceptual framework. Management practices are also included as a process that regulates *H. zea* populations at local to regional scales. These factors have ultimate outcomes for management of crop injury, evolution of pesticide resistance, and economic damage for farmers, including northern and southern states. T_{max}, maximum temperature; T_{min}, minimum temperature. (Original artwork by A.S. Huseth.)

assess populations and management outcomes at the landscape scale (Kennedy and Margolies, 1985; Kennedy and Storer, 2000; Jackson *et al.*, 2008).

Several crop hosts play an important role in *H. zea* population dynamics. Larvae typically infest corn and then cotton or soybean later in the summer (Stinner *et al.*, 1974). This predictable sequence of seasonal host use is an important factor in annual pest management recommendations across the US Cotton Belt. It was once a widely accepted view that large numbers of larvae developed effectively to adulthood feeding on cotton. Perhaps this view was because a few larvae could negatively impact yield relative to other crops, sometimes with 100% loss (Del Pozo-Valdivia *et al.*, 2021; Reisig and Goldsworthy, 2023). However, Gould *et al.* (2002) used stable carbon isotopes to generate the first evidence suggesting that cotton is not a major developmental host of

H. zea moths. In a subsequent study, Head *et al.* (2010) further refined this relationship by testing for gossypol, a defensive compound expressed in cotton, in stable isotope studies of adult *H. zea* moths. The authors found that the majority of *H. zea* moths caught adjacent to cotton fields did not develop on cotton as larvae. These initial studies provided new knowledge about the variable contribution of common row crop hosts to *H. zea* populations. This result has been important to guide resistance management recommendations for cotton and corn expressing Bt toxins (see "Resistance management: southern row crop perspective" section below).

Host associations are key factors in recent geospatial analyses focused on the relationship between *H. zea* adult abundance to landscape composition. Early work presented hypothetical landscapes that contained different crops contained in a mosaic of non-crop habitats (Stinner, 1979). These hand-drawn maps of a typical cropping landscape found in the southeastern US, based on field observations, illustrate crop-specific variation in *H. zea* oviposition preference, larval establishment, pupal survival, and adult production (Stinner, 1979). These images communicated key life table components using a cartographic product, which provided a basis for developing knowledge about specific landscape patches that affect the life history of this pest. Since the development of this agroecosystem perspective on *H. zea* life cycles, rapid development of remotely sensed crop data has enabled researchers to replicate these plant-insect host scenarios from the field to national scale. An example of crop composition and oviposition preference was generated using contemporary remotely sensed crop data (Fig. 10.3). Here, differences in North Carolina crop field sizes are apparent, and these patch area differences are important factors to consider when estimating population dynamics in the landscape (Fig. 10.3A).

Using geospatial crop data, metrics describing landscape structure are also accessible using different spatial analysis software programs (e.g. ArcGIS, GRASS, QGIS). In our example landscape, field configuration differences are estimated using the perimeter-to-area ratio calculated with field-level attributes (Fig. 10.3B). Larger values indicate greater edge complexity, which results from more irregularly shaped crop fields and likely more non-crop habitat that harbors early-season weedy hosts for *H. zea*. Perimeter-to-area and other patch aggregation/connectivity metrics characterizing landscape structure can more effectively account for the complex structure and connectivity of patches nested within agroecosystems than the summed or proportional measures of host patch composition and abundance (Turner and Gardner 2015). Including landscape metrics that integrate spatial connectivity can be useful for pest dispersal/infestation because simple summed or proportional measures of abundance in the local landscape cannot account for structure and connectivity of patches nested in differently arranged agroecosystems. A full overview of landscape metrics is beyond the scope of this chapter; however, Turner and Gardner (2015) provide a comprehensive overview of landscape pattern measurement and include important commentary on the relationships among common metrics used in landscape analysis. An important consideration of these derived landscape metrics is that they are often highly correlated predictors (Hargis *et al.*, 1998; Peng *et al.*, 2010). Correlation between landscape predictors should be considered when testing independent variables calculated by common landscape characterization software (e.g. FRAGSTATS) (McGarigal, 1995) in regression analyses connecting *H. zea* and other organisms to agroecosystem structure.

Temporal host availability is an important concept to consider when modeling relationships between *H. zea* abundance and landscapes. In our example, host availability varies as a function of crop composition, with unequal host patch availability over time as the growing season progresses (Fig. 10.3). Increasing availability of host patches during the growing season is one contributing factor for concomitant *H. zea* population growth (Stinner *et al.*, 1982). Together, these factors contribute to an important distribution of suitable oviposition sites across the landscape in space and time. These ecological concepts are important components in any model assessment of *H. zea* activity as it relates to crop composition at multiple spatial scales.

Early field observations and mechanistic modeling described the critical relationships between crop phenology and *H. zea* oviposition (Stinner *et al.*, 1974; Johnson *et al.*, 1975; Logan *et al.*, 1979). Jackson *et al.* (2008) used intensive trapping and larval monitoring studies across five southern states to estimate spatial and

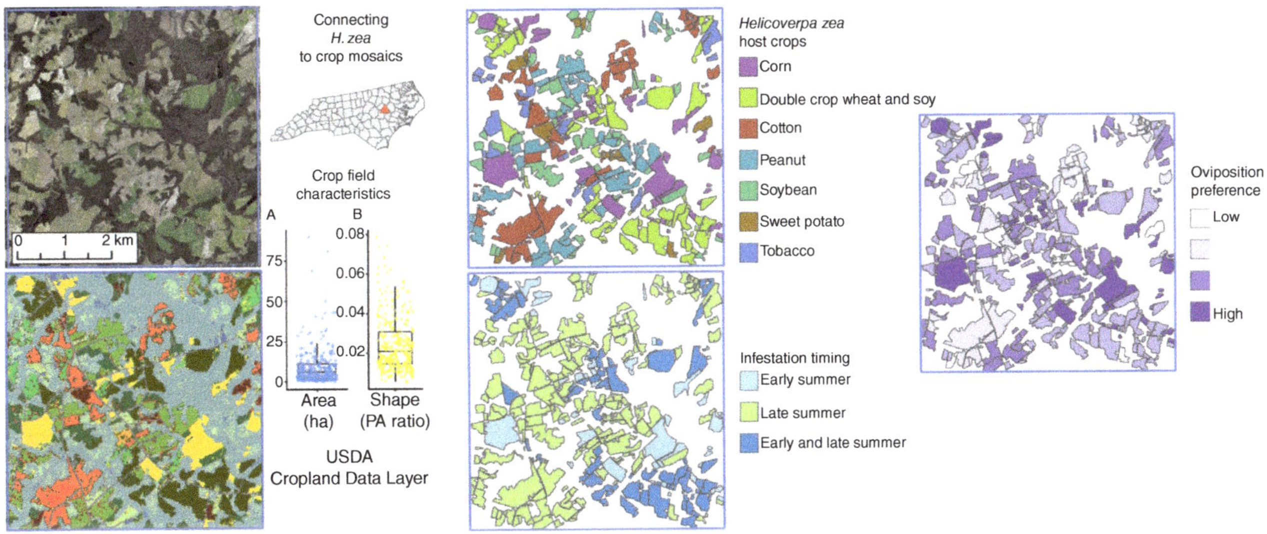

Fig. 10.3. Availability of remotely sensed crop data enables researchers to translate quantitative *Helicoverpa zea* life history studies into realistic cropscapes. Here, we present an actual crop landscape that includes field size (ha) (A) and field shape (perimeter-to-area (PA) ratio) (B) variation in areas where *H. zea* is a major economic pest. Reclassification of crop types into ecologically relevant pest-centric groups provides an informative visual assessment of the timing of *H. zea* infestation. Note, winter wheat followed by soybean in the same growing season (double crop) provides two infestation opportunities. Oviposition preference by crop was adapted from Johnson *et al.* (1975) and Stinner (1979). (Original artwork by A.S. Huseth.)

temporal variation in activity relative to local host crops in the landscape. To measure host abundance beyond the field edge, the authors used hand-digitized aerial photography to qualify relationships between *H. zea* and abundance of focal crops at 1.61 and 16.1 km radius from observation points. A similar approach was used to assess *H. zea* and *C. virescens* abundance at the interface of crops with differing reproductive suitability over time (Allen and Luttrell, 2011). The authors took an important step toward temporal assessments by accounting for phenology of the host crop, specifically integrating different soybean maturity groups with varying flowering times to quantify host synchrony with *H. zea* adult activity. In Florida, researchers related year-round *H. zea* adult counts in different cropping regions to weather and host plant factors (Santos *et al.*, 2023). This study measured adult activity at several field locations but did not directly quantify the relationship between abundance and landscape composition or configuration. Importantly, this paper illustrates that far southern latitudes could be a source of *H. zea* adults year-round. However, we do not currently know what ecological implications exist for *H. zea* that are active during winter months or how their persistent activity could affect practical aspects of management. This previously underappreciated cycle of low-level activity could become more important as climate change continues to favor earlier periods of activity and dispersal.

These temporal host concepts are important because inference can be improved when evaluating landscape-level risk for pests by considering host phenology. One important data resource for these pest–crop analyses is the National Agricultural Statistics Service Cropland Data Layer (CDL) and related National Land Cover Database (USDA-NASS, 2022). The CDL provides an annual assessment of crop and non-crop land use for the conterminous US, which is useful to estimate the abundance of different crop hosts in the environment (Fig. 10.3). However, these categorical classifications are unable to recover important nuances between temporal host availability and insect activity. The interaction between crop suitability and multivoltine life history of *H. zea* complicates development of accurate population abundance estimates because static geospatial crop data does not encompass differences in temporal suitability within the same crop classification. In practice, one of the most common spatial research approaches is to relate classified crop and non-crop categorical data derived from multi-spectral satellite imagery (i.e. CDL) to predict *H. zea* abundance. This approach may work reasonably well in temperate climates where a single *H. zea* larval generation occurs in one crop, but likely introduces unanticipated variability where the suitability of host patches varies in time.

Changing soybean agronomics in the southern US is a good example of why researchers should be cautious leveraging some remotely sensed products to predict population abundance. Indirect evidence suggests that soybean is the most important crop to serve as a reproductive host in this region for *H. zea* after they pass a generation in corn (Head *et al.*, 2010; Dorman *et al.*, 2021). Over time, soybeans have generally trended toward a broader planting window that includes many planting dates and relative maturity groups. For example, in North Carolina the recent availability of a wide range of relative maturity group soybean varieties adapted to the state has extended the period of reproductive stage soybeans (Vann *et al.*, 2021). We do know that *H. zea* prefer to oviposit on early reproductive growth stages of soybean when flowers are present (R1–R3) compared to later pod-filling stages (R4–R6) (Fig. 10.4). This short period of flowering translates into a narrow window of high reproductive preference for *H. zea* females (Johnson *et al.*, 1975; Fehr and Caviness, 1977). Studies of soybean growth can be used to illustrate the effective period of high oviposition preference. Fehr and Caviness (1977) found that an average of 22 days elapsed between R1 and R4 (range 10–37 days) which coincides with the period of highest oviposition preference (Fig. 10.4). Several variables can affect soybean flowering duration variation including, but not limited to, planting date, maturity group, soybean growth habit, and environmental conditions. The interaction between these variables can contribute to different flowering times among fields and, in turn, a wide range of highly attractive host patches for *H. zea* oviposition. As a result, field-to-field agronomic differences can lead to considerable heterogeneity in soybean flowering times, often at fine spatial scales. Because there can be substantial phenological mismatch between reproductive activity of discrete *H. zea* generations and flowering

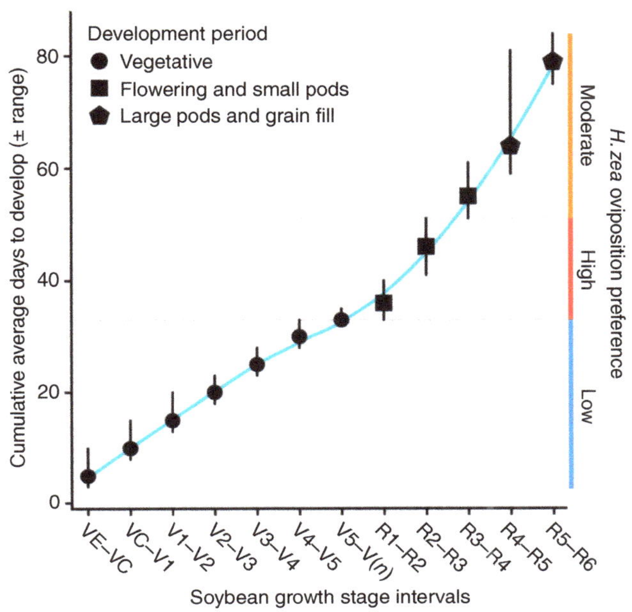

Fig. 10.4. Average cumulative development days for vegetative and reproductive stage soybeans. Range estimates illustrate the minimum and maximum days observed for each period (Fehr and Caviness, 1977). Periods of stage-specific *Helicoverpa zea* oviposition preference were reported in Johnson *et al.* (1975). (Original artwork by A.S. Huseth.)

times, best-fit models that assume soybean is a static patch may generate ecologically inaccurate risk assessments and misleading management recommendations.

To address this temporal concern, future work connecting infestation risk for multivoltine pests may need to include crop phenology into statistical models to resolve unexplained variation, particularly in complex agroecosystems where many different hosts occur or variable suitability within host type exists. Analytical improvements will likely require very large data sets that contain nuanced crop and pest management data (i.e. spatial distribution of different Bt toxin pyramids) to help validate basic drivers of *H. zea* populations in contemporary agroecosystems. These approaches have been used successfully to understand pest risk in other row crop systems (Rosenheim *et al.*, 2011, 2022), but they have not been applied widely to *H. zea* analyses at the farm or landscape scale. Next steps will require the assembly of large, spatially explicit multivariate data sets describing key features of temporally heterogeneous agroecosystems to improve predictions of *H. zea* life cycles and abundance (Stinner *et al.*, 1982; Kennedy and Storer, 2000).

Leveraging Monitoring Networks to Understand Migration

It is important to recognize that *H. zea* has two geographic regions of ecological importance in North America: the southern latitudes where the local populations can effectively overwinter and northern regions where the insect is an annual migrant during the growing season but does not overwinter successfully (Felt, 1925; Hardwick, 1965) (Fig. 10.2). Winter temperature is an overarching regulating condition that likely interacts with other abiotic factors, including soil moisture and soil type (Roach and Hopkins, 1979; Dillard *et al.*, 2023a, b), to drive high levels of overwintering mortality, but these relationships have not been studied in a geospatial framework.

After pupal eclosion in the spring, the overwintered adults disperse to oviposit on suitable non-crop weeds, cover crops, and small grains. Through a sequence of hosts, the spring population amplifies across several generations prior to long-distance dispersal (Stinner, 1979; Stinner *et al.*, 1982). Northward migration from southern overwintering areas typically occurs mid-summer and can result in significant larval infestations in sensitive host crops in northern latitudes. The overwintering temperature limits are an important distinguishing factor that differentiates overwintering and migratory regions and downstream management approaches. Annual northward *H. zea* migration and potential return flight to southerly cropping systems are the key connection between these regions (Pair *et al.*, 1987; Beerwinkle *et al.*, 1994; Gould *et al.*, 2002; Paula-Moraes *et al.*, 2024). The most important early validation of migration documented *H. zea* northward dispersal from southern Texas using presence of citrus pollen on adult moths as an indicator of long-distance migration from the subtropical region of the Rio Grande Valley into areas where citrus was not present (Westbrook *et al.*, 1997; Westbrook, 2008; Westbrook and López, 2010). Temporal trap catches of pollen-infested *H. zea* were related to synoptic weather data, which allowed researchers to estimate that moths could disperse from citrus producing source regions in southern Texas to central Texas, more than 600 km away, in a period of 1–2 nights (Westbrook *et al.*, 1997). In a recent reanalysis of this observational *H. zea* dispersal data, Wu *et al.* (2018) used two spatially explicit particle movement models, Hybrid Single-Particle Lagrangian Integrated Trajectory (HYSPLIT) and numerical trajectory, to simulate migration routes of moths and test predicted trajectory outcomes against the observed catch location for pollen-positive adults. The integration of high-resolution weather data predicted similar dispersal outcomes for pollen-infested moths caught in traps, confirming that spatially explicit particle trajectory models can accurately estimate the outcome of migration events (Wu *et al.*, 2018). Although moderate distance dispersal events can be predicted, researchers have known for decades that these sequential movements from southern overwintering zones to the north occur and remain one of the major drivers of *H. zea* issues in northern latitudes. Importantly, population movement from northerly to southerly latitudes in the autumn (return migration) may have important outcomes for population genetics, resistance management, and interregional IPM practices. Return flights have been documented by a small number of studies with limited geographic extents (Pair *et al.*, 1987; Beerwinkle *et al.*, 1994; Gould *et al.*, 2002), which limits broad generalization about southward autumn dispersal to suitable overwintering habitats in lower latitudes. Knowledge gaps about return migration currently limits our ability to infer these relationships in the context of applied production systems.

Connections between the southern source region and northern zones have been instrumental to understand the timing of *H. zea* migration events and predict infestation intensity for decades. To do this, entomologists use blacklight and pheromone trapping networks maintained by state cooperative extension programs to provide early warning of northward dispersal. They also often track the activity of more southerly networks as an indicator of potential *H. zea* movement. Connectivity was especially prominent among trap *H. zea* trap networks in the eastern US where a high density of traps has been maintained for decades (Fig. 10.5). To test this geographic connectivity concept, Lawton *et al.* (2022) collated a large *H. zea* activity data set to model population relationships between overwintering and migratory zones. Four decades of blacklight and pheromone *H. zea* trap data were compiled from monitoring networks located in 37 states and four Canadian provinces. The authors partitioned annual trap locations using winter temperatures into an overwintering zone, a transitional zone where overwintering was uncertain, and a lethal zone where pupae should not survive. Using this information, both within- and among-year activity was modeled for each zone (Lawton *et al.*, 2022). Results revealed similarity among zones during growing seasons that confirms the utility of southern regions for predicting large migratory dispersal events. The authors applied criteria extracting future temperature data from climate change ensemble models to estimate where *H. zea* may overwinter at higher latitudes in North America.

The key outcome of this large spatial analysis was that winter conditions in northerly regions are likely to become more favorable for *H. zea* overwintering in the future. Understanding survival limiting factors and meteorological events that favor long-distance *H. zea* dispersal may

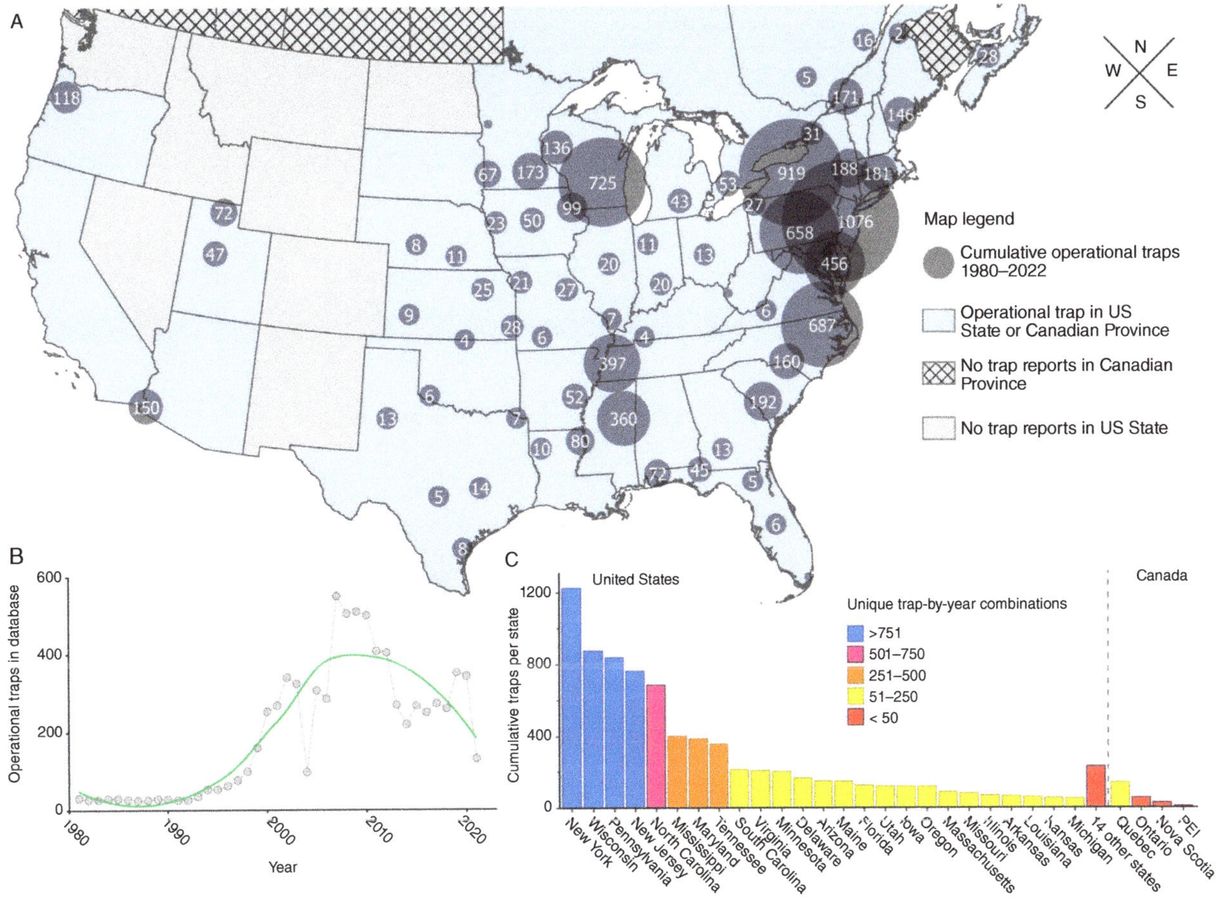

Fig. 10.5. (A–C) Distribution of *Helicoverpa zea* trapping in space and time across the contiguous United States (US) and eastern Canada (Lawton *et al.*, 2022). Trap data span many states and ecoregions where this insect is an economically important pest of vegetables and row crops. Increasing availability of personal computing explains the rapid increase in digital records after 2000. PEI, Prince Edward Island. (Original artwork by A.S. Huseth.)

also benefit future projections for potentially invasive species that have similar ecology and dispersal potentials. The ongoing risk of Old World bollworm, *Helicoverpa armigera* (Hübner), establishment in the US has motivated several authors to compare life histories, host plant use, and migratory potential of *H. zea* and *H. armigera* to understand invasion threat in the US (Kriticos *et al.*, 2015; Leite *et al.*, 2017; Haile *et al.*, 2021). In the future, approaches that leverage spatial modeling of *H. zea* may provide more quantitative risk assessment approaches to enable more effective *H. armigera* mitigation practices in North America.

Management Challenges for Northern Production Regions

As previously mentioned, sequential polyphagy, the propensity to evolve insecticide resistance, and long-distance migration are three key factors that annually contribute to the ongoing pest status of *H. zea* in both the southern and northern states (Fig. 10.2) (Kennedy and Storer, 2000; Hutchison *et al.*, 2007; Wu *et al.*, 2018; Lawton *et al.*, 2022). As *H. zea* populations cycle through multiple generations each year on southern crops (e.g. corn, cotton, soybean, vegetables), the pest is exposed to multiple insecticidal sprays as well as several Bt toxins expressed in both Bt cotton and Bt corn (Hutchison *et al.*, 2007; Reisig *et al.*, 2023). Because of the multi-generational exposure of the pest to several insecticides, selection pressure over several decades has resulted in significant levels of resistance to both synthetic pyrethroids and Bt in both the northern and southern states (Pietrantonio *et al.*, 2007; Hutchison *et al.*, 2007; Jacobson *et al.*, 2009; Dively *et al.*, 2023; Gassmann and Reisig, 2023). The phenomenon of *H. zea* resistance in the northern tier states can be attributed to two aspects of *H. zea* biology: (i) the pest is only known to overwinter south of the 40th latitude (Morey *et al.*, 2012) and (ii) evidence for northward migration of *H. zea* moths from the southern US via large-scale synoptic weather systems. These systems typically result when low-pressure fronts from the south that move west to east converge with high-pressure fronts from Canada and northern states and result in strong southerly wind flow (Lingren *et al.*, 1994; Sandstrom *et al.*, 2007; Westbrook and López, 2010; Wu *et al.*, 2018).

Consequently, indirect evidence of resistance genes within southern *H. zea* "source" populations moving north each summer (late July to August) has been reported for the northern states. One example of interregional connectivity of *H. zea* populations is the ongoing decline in pyrethroid sensitivity documented by specialty crop entomologists in northern states. Historically, pyrethroid insecticides have been used for managing *H. zea* on northern specialty crops, and several pyrethroid active ingredients used during the 1980s and 1990s provided excellent control in the Midwest region. Control in annual sweet corn field trials typically averaged or exceeded 95%. However, increasing bifenthrin and lambda-cyhalothrin tolerance (reduced mortality) began to emerge in *H. zea* populations in the early 2000s (Hutchison *et al.*, 2007). For example, in Minnesota and Wisconsin field trials, efficacy against *H. zea* larvae declined to only 40–50%. To quantify the extent of pyrethroid resistance in *H. zea* populations in the Midwest region, a three-year study was conducted using adult vial test pyrethroid bioassays across four midwestern states (Minnesota, Wisconsin, Illinois, and Indiana). In contrast to earlier pyrethroid resistance studies, adult vial test assays in midwestern states resulted in moth survival rates of 25–60%, depending on the cypermethrin dose. This result suggested field-evolved pyrethroid resistance had occurred in *H. zea* populations across a broad geography (Hutchison *et al.*, 2007). In a similar study focused on field collections from Indiana and Illinois, researchers tested lambda-cyhalothrin, one of the most common commercial pyrethroids, and documented comparable results to cypermethrin (Jacobson *et al.*, 2009). Importantly, adult vial test results for cypermethrin in the Midwest states during this period were very similar to assays of *H. zea* moths collected throughout Texas, ranging from 20% to 45% survival (Pietrantonio *et al.*, 2007). For 2004–2006, the Midwest results were quite similar to those from Louisiana, averaging around 45% survival (as summarized in Hutchison *et al.*, 2007).

Although the widespread use of Bt field corn in northern states has been correlated with an areawide suppression of *H. zea* in the Atlantic states (Dively *et al.*, 2018), the impact of pyrethroid resistance in many northern states has also resulted in continued feeding damage to several crops, including sweet corn, green (snap) beans, and peppers. These high value crops have ultra-low tolerances for feeding damage and/or insect contaminants that reduce marketable yields (Flood and Rabaey, 2007). One exception to this trend for Bt sweet corn is the use of hybrids expressing the vegetative Bt, Vip3A (e.g. Burkness *et al.*, 2010), where a multi-state network of Bt sentinel plot studies continue to show a high level of efficacy against *H. zea* (Dively *et al.*, 2023). In response to eroding pyrethroid susceptibility, in the mid-2010s sweet corn growers in northern states gradually shifted to alternative insecticides with a different mode of action, primarily diamides. It is important to note that diamides are now a common component of *H. zea* management in southern cotton and soybeans. The result of this regional deployment of diamides in southern states has greatly increased the likelihood of *H. zea* resistance selection that, in turn, will have far-reaching impacts on both southern and northern production systems due to clear migratory linkages and similarities in insecticidal tools.

Regarding the use of Bt sweet corn trials for monitoring Bt resistance in the Midwest states, baseline data were collected in Minnesota in 1995, prior to commercial release of Bt field corn in 1996 (Burkness *et al.*, 2001), with additional studies conducted in Minnesota, Illinois, and Wisconsin from 1997 to 1998 (Burkness *et al.*, 2002). In most of these trials, all using Cry1Ab only, it was common to see first instars surviving on the silk tissue or ear tips of Bt sweet corn. Although the overall efficacy was very high against later *H. zea* instars, providing over 95% marketable ears, the data suggested that Cry1Ab did not meet the high-dose criteria (i.e. killing greater than 99% of susceptible insects in the field). In contrast, mortality of European corn borer, *Ostrinia nubilalis* (Hübner) (Lepidoptera: Crambidae), larvae was 99–100% in early trials, demonstrating the high-dose effect. In Minnesota, the efficacy of Cry1Ab sweet corn began to significantly decline in 2014–2015 (W.D. Hutchison, unpublished data); however, the trend was not fully documented until 2018–2020 (Dively *et al.*, 2021). During this period, it became evident that Cry1Ab resistance was common across most Midwest and Atlantic coast states. These clear changes in Bt efficacy were mirrored by eroding susceptibility to similar toxins expressed in Bt cotton and field corn grown in southern landscapes (see following section). Similar declines in pyrethroid and Bt susceptibility across multiple latitudes further reinforces the hypothesis that management practices targeting *H. zea* cannot be considered in isolation and highlights the need for more coordinated management of this pest among multiple stakeholders across a broad geography.

In summary, trends in field-evolved pyrethroid and Bt resistance reflect indirect evidence for long-distance movement of insecticide and Bt resistance genes from southern to northern states, particularly for Minnesota and Wisconsin, where *H. zea* are not known to currently overwinter (Morey *et al.*, 2012; Lawton *et al.*, 2022). Future research should be conducted to connect *H. zea* populations from multiple southern and northern states and elucidate genetic markers that could be used to accurately track migration-enhanced resistance. This may enable the development of more proactive IPM programs that harmonize IRM recommendations between southern and northern tier states.

Resistance Management: Southern Row Crop Perspective

The evolution of *H. zea* resistance to genetically engineered cotton and corn that express lepidopteran active Bt toxins has been a significant issue for entomologists across the southern US (Reisig *et al.*, 2018, 2023; Niu *et al.*, 2021; Tabashnik *et al.*, 2023). One major challenge has been to understand where and when resistance evolved to different Bt toxins, which is one of the contributing drivers of geospatial research in this area (Carrière and Onstad, 2023). Prior to widespread Bt crop deployment, spatially explicit computational studies were conducted to describe which agroecosystem components were the most influential resistance selection factors in simulated landscapes. To do this, Storer *et al.* (2003) included varying abundance of cotton and corn with spatiotemporal population dynamics to understand selection processes. Their results highlighted that local deployment of Bt crops was important despite long-distance dispersal capability of this pest. Moreover, their results also showed that intensification of Bt crop production on clusters of agricultural fields could strongly influence the rate of *H. zea* resistance evolution. Following the emergence of widespread field-evolved Bt resistance (Tabashnik and Carrière, 2017), Caprio *et al.* (2019) used a similar simulation modeling approach to estimate the durability of multiple Bt toxins expressed in cotton and corn. Importantly, the spatially explicit simulation model included key life table components of *H. zea* on different host crops. The models also included phenological differences in *H. zea* crop suitability and the affect these differences had on the trajectory of toxin durability. Including phenology in a multi-crop simulation model approach adds a realistic element to estimated durability outcomes and, in turn, more closely represents realistic production systems. Building on this approach, the spatial research community can test these general resistance selection models using an unprecedented amount of remotely sensed crop information and data-driven statistical models.

With the availability of reasonably accurate and spatially explicit CDL data, researchers have extended these landscape-scale analyses to estimate how the abundance of both corn and cotton in the landscape relates to the Bt crop injury or *H. zea* resistance levels measured in the field. Because Bt toxins expressed in genetically engineered corn and cotton have been rapidly adopted across the region, researchers have used the near 100% incidence of Bt in these two crops as a convenient proxy for recent resistance selection pressure. These place-based relationships are important because the response to resistance is often increased pesticide use, which leads to unequal input intensity depending on the incidence of the crop. For example, the evolution of *H. zea* Cry1Ac Bt resistance triggered major changes to insect management in cotton, ultimately driving an increase in sprays targeting the pest and economic losses for growers (Reisig *et al.*, 2018).

To understand where resistant populations occur, a combination of long-term on-farm surveys for *H. zea*-damaged cotton bolls and replicated on-farm variety trials were used to document both the spatial structure and landscape drivers of *H. zea* injury to Bt cotton (Dorman *et al.*, 2021). The authors reported a significant increase in observed boll injury after 2012, with a positive relationship between the frequency of Bt cotton boll damage and the intensity of Bt corn and cotton production in the surrounding landscape. This result corroborated Storer *et al.* (2003) simulations of Bt crop intensification and resistance. This geospatial analysis provided useful information about the places where greater resistance selection and damage are most likely occurring. Although the area of Bt crops grown in these landscapes may not vary appreciably, this landscape perspective does help target places to increase IRM messaging and adoption of mitigation. Specifically, changing grower attitudes toward adopting refuges in Bt corn in intensive corn–cotton production areas has been a major initiative to address ongoing *H. zea* resistance evolution to similar toxins expressed in both crops (Reisig, 2017; Reisig and Kurtz, 2018).

Using a similar landscape analysis approach, Arends *et al.* (2021) showed that the relative abundance of Bt corn, cotton, and non-Bt soybean in the previous year described significant variation measured in Cry1Ac bioassay screens of *H. zea* populations collected from corn. Although highly significant, these mixed relationships are challenging to understand because crop interactions included in the model require complex ecological assumptions that include sequences of crop–pest interactions spanning multiple generations and growing seasons. Corn kernel consumption by *H. zea* was also estimated on non-Bt and Bt corn expressing Cry1A.105 + Cry2Ab2 as an alternative field-level measure of sensitivity among many on-farm locations. Results showed that increasing abundance of Bt corn in the current season was negatively related to corn kernel injury, suggesting that a combination of pest dilution in corn and variable resistance status likely contributed to reduced injury as corn abundance increased (Arends *et al.*, 2022). Feeding damage assessments in vegetable crops have also been used to creatively quantify declining *H. zea* populations because of Bt suppression in the Mid-Atlantic region of the US. Dively *et al.* (2018) showed that increasing abundance of Bt crops was associated with decreasing damage and insecticide sprays in vegetables. This areawide suppression of *H. zea* populations had an unanticipated benefit for specialty crop growers that translated into important sustainability improvements through the reduction in crop loss and pesticide inputs. These results complement previous research focused on *O. nubilalis* suppression by Bt corn in the midwestern US (Hutchison *et al.*, 2010).

Connecting *H. zea* population dynamics to Bt crop composition has been another theme in the southern US where these crops dominate agroecosystems. These studies use *H. zea* abundance from traps to estimate host relationships in the landscape. Dorman *et al.* (2021) used a spatiotemporal Bayesian hierarchical model to estimate the relative importance of host patch abundance for *H. zea*. Results showed that the abundance of a key non-Bt refuge crop, soybean, was the most important positive predictor of *H. zea* trap catch in North Carolina. Locations with greater soybean area predicted around 40% higher *H. zea* than sites with fewer soybeans (Dorman *et al.*, 2021). Importantly, researchers extended model outcomes to visualize the relationship between non-Bt soybean and Bt crops using a simple ratio of abundance from randomly selected points distributed across the landscape (Dorman *et al.*, 2021). Interpolated ratios highlighted places where the abundance of refuge soybean was lacking due to dominance of Bt crops on available agricultural land. One limitation of these modeling approaches is the lack of temporal suitability when connecting moth abundance to crop patches in the landscape, particularly soybean, as discussed above.

It is important to consider the value of these geospatial crop history data sets to help guide resistance management decisions in actual cropping landscapes. Undoubtedly, selection for resistance is a process that occurs over a long period, and this time component is critical for landscape-level resistance modeling efforts (Carrière and Onstad, 2023). In these cases, the integration of multiple years of cropping data may improve inference and describe the frequency of resistant phenotypes in populations. The recent release of georeferenced crop frequency data will streamline analyses that include spatially explicit longitudinal crop data in arthropod-landscape studies (USDA-NASS, 2022, 2024). While more traditional deterministic and stochastic resistance models do provide valuable forecasts of resistance outcomes, the integration of actual crop production data into visualizations and maps can help communicate potential resistance issues. Using these visualizations, stakeholders could more effectively implement proactive resistance mitigation informed by actual landscape composition in space and time. Similar approaches leveraging cropping patterns have been useful tools to help describe patterns of pest resistance and communicate the importance of proactive IRM tactics to alleviate reliance on additional pesticides to manage resistant populations (Huseth *et al.*, 2015, 2018; Gassmann, 2021; Carrière and Onstad, 2023). This element of realistic cropping systems can strengthen credibility of applied entomologists working toward convincing producers of the utility of changing practices or adhering to refuge requirements targeting a problematic pest like *H. zea*.

Knowledge Gaps Limiting Prediction in Landscapes

One major limitation highlighted above is the temporal crop host suitability knowledge gap in contemporary agroecosystems. Closing this gap with host phenology and spatial abundance data would enable more robust prediction of *H. zea* infestation and damage across the landscape. A key component of this effort would be improving knowledge of sequential non-crop host use both during the growing season and periods outside typical cropping cycles when this insect is active and reproducing. In one longitudinal study across the growing season, *H. zea* larvae were abundant on weedy hosts through May, but became less numerous in June, July, and early August (Roach, 1975). During late August through October, *H. zea* larvae were once again captured on a number of weedy plants, suggesting their importance as late-season hosts when crops were no longer suitable (Roach, 1975). Two other studies recognized the importance of early-season hosts and suggested that shifts in early-season weed hosts in combination with *H. zea* larval parasites associated with these weed species are a determinant of population buildups during the growing season (Roach, 1975; Mueller and Phillips, 1983). The impact of these interactions, particularly in the spring, are difficult to quantify and integrate into existing landscape modeling frameworks due to an absence of fine-scale host data. Accounting for these effects could improve the accuracy and precision of time-series models forecasting seasonal *H. zea* population development, particularly in southern US systems where weeds are a key host for the first larval generation.

Changing tillage and agronomic practices over the past several decades are a second significant change to areawide production practices that may affect *H. zea*, but these are difficult to incorporate into spatial models. Broad adoption of no-till and conservation tillage (>30% of soil surface undisturbed) has occurred in the southeastern US, especially in the highly eroded Piedmont physiographic region (Cassel *et al.*, 1995). In other production systems, tillage operations have been widely adopted as a cultural control tactic for *Helicoverpa* spp. Tillage tactics, commonly called pupal busting, have been used as a component of Bt resistance management plans in Australia to limit interseasonal carryover of resistance genes (Whitehouse *et al.*, 2023). Pupal busting was widely adopted in Australian cotton systems for years, but this practice may decline as soil and water conservation goals incentivize less intensive tillage (Whitehouse *et al.*, 2023). While these tillage management approaches targeting *Helicoverpa* spp. may be in flux, the Australian case study represents a clear example of the connection between soil disturbance and pupal survival. As a result, it is logical to assume that declining tillage intensity in the US may also have long-term effects on overwintering *H. zea*. In the US, the area under intensive tillage has decreased by 24% from 2012 to 2017 (USDA-NASS, 2017). Concurrent gains in reduced tillage and no-till practices indicate that the level of soil disturbance has reduced markedly across millions of acres of US farmland (USDA-NASS, 2017). Although we do not know how these soil conservation practices will affect *H. zea* populations, changing tillage intensity across broad spatial extents represents a reasonable area for future spatial ecology studies focused on this pest.

Concurrently, cover crops have grown in popularity across the US, increasing by approximately 2 million ha from 2012 to 2017 (USDA-NASS, 2017). Many of the most significant gains have been recorded in the eastern US where *H. zea* is a persistent pest. Despite this, few studies have evaluated the relationship between cover crops and *H. zea*. In one on-farm study, researchers evaluated Heliothinae (*H. zea* and *C. virescens*) populations and beneficial insect populations in cotton fields planted behind cereal rye, a legume mix, legume + rye (*Secale cereale*), crimson clover (*Trifolium incarnatum*), and conventional till treatments (Tillman *et al.*, 2004). On average, cotton plots that followed rye cover crop exceeded the economic threshold for Heliothinae larvae significantly fewer times than control plots (mean 0.3 versus 2.0 times exceeding economic threshold, respectively). The authors hypothesized that beneficial insects could be one explanation for carryover differences between cover crops and the subsequent cotton crop, but these were not consistent among cover crops. The authors also suggested that differences in Heliothinae larval densities were not due to adult ovipositional choice because they observed similar Heliothinae egg densities among treatments. However, they did not address the effect of plant quality on larval development. For example, in a similar study, cotton that followed legumes in a cover crop was shorter and yellower than other treatments which could affect *H. zea* host preference (Gaylor *et al.*, 1984). In addition to influencing the cash crop behind the cover crop, cover crop species can have a direct influence on *H. zea* populations if the cover crop that is planted is a reproductive host. One such popular cover crop species, crimson clover, is an excellent reproductive host for *H. zea*, especially when it flowers (Ellsbury *et al.*, 1989). Therefore, local selection of specific cover crops may influence early-season *H. zea* populations in some landscapes. These shifts in production practices may provide additional insight to explain unexplained variation detected in statistical approaches designed to predict *H. zea* abundance during the growing season.

Summary and Conclusions

Linking landscapes to pests is a promising area of research with a goal of improving infestation prediction and efficacy of mitigation strategies. Spatial ecology of *H. zea* is an excellent example of applied geospatial science used to understand the interaction of pests with the surrounding agroecosystem, with researchers using a continuum of approaches from simple cartographic representations of life tables to sophisticated spatial regression models. The diversity of approaches has improved our understanding of this pest, but they have not displaced the need to continue research into efficient scouting and economic thresholds. As farm sizes continue to grow, it will be important to integrate these spatial approaches into the IPM portfolio as another tool to improve the accuracy and precision of scouting activity while minimizing unnecessary effort and cost. One key limitation to this effort is the lack of near real-time crop production data at spatial scales that match the dispersal potential of this pest. Future efforts to improve the temporal resolution of spatially explicit data products will open new doors to explore key relationships between *H. zea* and landscapes in space and time.

Acknowledgments

We would like to thank applied entomologists and extension agents for decades of effort maintaining *H. zea* trap networks across North America. Parts of the research presented here was supported by Crop Protection and Pest Management competitive grant 2017-70006-27205 and Biotechnology Risk Assessment Grant 2020-33522-32272 from the USDA. We also acknowledge funding from USDA-NIFA Hatch Projects NC2884 and NC2778.

References

Allen, K.C. and Luttrell, R.G. (2009) Spatial and temporal distribution of heliothines and tarnished plant bugs across the landscape of an Arkansas farm. *Crop Protection* 28, 722–727. DOI: 10.1016/j.cropro.2009.04.007

Allen, K.C. and Luttrell, R.G. (2011) Temporal and spatial distribution of *Helicoverpa zea* and *Heliothis virescens* (Lepidoptera: Noctuidae) moths in pheromone traps across agricultural landscapes in Arkansas. *Journal of Entomological Science* 46, 269–283. DOI: 10.18474/0749-8004-46.4.269

Arends, B., Reisig, D.D., Gundry, S., Huseth, A.S., Reay-Jones, F. *et al.* (2021) Effectiveness of the natural resistance management refuge for *Bt*-cotton is dominated by local abundance of soybean and maize. *Scientific Reports* 11, 17601. DOI: 10.1038/s41598-021-97123-8

Arends, B.R., Reisig, D.D., Gundry, S., Greene, J.K., Kennedy, G.G. *et al.* (2022) *Helicoverpa zea* (Lepidoptera: Noctuidae) feeding incidence and survival on Bt maize in relation to maize in the landscape. *Pest Management Science* 78, 2309–2315. DOI: 10.1002/ps.6855

Beerwinkle, K.R., Lopez, J.D., Witz, J.A., Schleider, P.G., Eyster, R.S. *et al.* (1994) Seasonal radar and meteorological observations associated with nocturnal insect flight at altitudes to 900 meters. *Environmental Entomology* 23, 676–683. DOI: 10.1093/ee/23.3.676

Blanco, C.A., Terán-Vargas, A.P., López, J.D., Jr, Kauffman, J.V. and Wei, X. (2007) Densities of *Heliothis virescens* and *Helicoverpa zea* (Lepidoptera: Noctuidae) in three plant hosts. *Florida Entomologist* 90, 742–750.

Bradley, J.R. (1996) Major developments of insect and mite pests in cotton. In: King, E.G., Phillips, J.R. and Coleman, R.J. (eds) *Cotton Insects and Mites: Characterization and Management*. Memphis, TN: The Cotton Foundation, pp. 1–13.

Burkness, E.C., Hutchison, W.D., Bolin, P.C., Bartels, D.W., Warnock, D. *et al.* (2001) Field efficacy of Bt sweet corn against European corn borer (Lepidoptera: Crambidae) and corn earworm (Lepidoptera: Noctuidae). *Journal of Economic Entomology* 94, 197–203. DOI: 10.1603/0022-0493-94.1.197

Burkness, E.C., Hutchison, W.D., Weinzierl, R.A., Wedberg, J.L., Wold, S. and Shaw, J.T. (2002) Efficacy and risk efficiency of sweet corn hybrids expressing a *Bacillus thuringiensis* toxin for Lepidopteran pest management in the Midwestern US. *Crop Protection* 21, 157–169. DOI: 10.1016/S0261-2194(01)00080-1

Burkness, E.C., Dively, G.P., Patton, T., Morey, A.C. and Hutchison, W.D. (2010) Novel Vip3A *Bacillus thuringiensis* (Bt) maize approaches high-dose efficacy against *Helicoverpa zea* (Lepidoptera: Noctuidae) under field conditions: implications for resistance management. *GM Crops* 1, 337–343. DOI: 10.4161/gmcr.1.5.14765

Caprio, M.A., Kurtz, R., Catchot, A., Kerns, D., Reisig, D. *et al.* (2019) The corn–cotton agroecosystem in the Mid-Southern United States: what insecticidal event pyramids should be used in each crop to extend Vip3A durability. *Journal of Economic Entomology* 112, 2894–2906. DOI: 10.1093/jee/toz208

Caron, R.E., Bradley, J.R., Pleasants, R.H., Rabb, R.L. and Stinner, R.E. (1978) Overwinter survival of *Heliothis zea* produced on late-planted field corn in North Carolina. *Environmental Entomology* 7, 193–196. DOI: 10.1093/ee/7.2.193

Carrière, Y. and Onstad, D.W. (2023) The role of landscapes in insect resistance management. In: Onstad, D.W. and Knolhoff, L.M. (eds) *Insect Resistance Management*, 3rd edn. Cambridge, MA: Academic Press, pp. 329–379. DOI: 10.1016/B978-0-12-823787-8.00006-4

Carrière, Y., Goodell, P.B., Ellers-Kirk, C., Larocque, G., Dutilleul, P. *et al.* (2012) Effects of local and landscape factors on population dynamics of a cotton pest. *PLoS ONE* 7, e39862. DOI: 10.1371/journal.pone.0039862

Cassel, D.K., Raczkowski, C.W. and Denton, H.P. (1995) Tillage effects on corn production and soil physical conditions. *Soil Science Society of America Journal* 59, 1436–1443. DOI: 10.2136/sssaj1995.03615995005900050033x

Cunningham, J.P. and Zalucki, M.P. (2014) Understanding heliothine (Lepidoptera: Heliothinae) pests: what is a host plant? *Journal of Economic Entomology* 107, 881–896. DOI: 10.1603/EC14036

Del Pozo-Valdivia, A.I., Reisig, D.D., Braswell, L., Greene, J., Roberts, P. and Taylor, S.V. (2021) Economic injury levels for *Bt*-resistant *Helicoverpa zea*. *Journal of Economic Entomology* 114, 747–756. DOI: 10.1093/jee/toab012

Diffenbaugh, N.S., Krupke, C.H., White, M.A. and Alexander, C.E. (2008) Global warming presents new challenges for maize pest management. *Environmental Research Letters* 3, 044007. DOI: 10.1088/1748-9326/3/4/044007

Dillard, D., Reisig, D.D. and Reay-Jones, F.P. (2023a) *Helicoverpa zea* (Lepidoptera: Noctuidae) in-season and overwintering pupation response to soil type. *Environmental Entomology* 52, 67–73. DOI: 10.1093/ee/nvac106

Dillard, D., Reisig, D.D., Schug, H.T. and Burrack, H.J. (2023b) Moisture and soil type are primary drivers of *Helicoverpa zea* (Lepidoptera: Noctuidae) pupation. *Environmental Entomology* 52, 847–852. DOI: 10.1093/ee/nvad074

Dively, G.P., Venugopal, P.D., Bean, D., Whalen, J., Holmstrom, K. *et al.* (2018) Regional pest suppression associated with widespread Bt maize adoption benefits vegetable growers. *Proceedings of the National Academy of Sciences* 115, 3320–3325. DOI: 10.1073/pnas.1720692115

Dively, G.P., Kuhar, T.P., Taylor, S., Doughty, H.B., Holmstrom, K. *et al.* (2021) Sweet corn sentinel monitoring for lepidopteran field-evolved resistance to Bt toxins. *Journal of Economic Entomology* 114, 307–319. DOI: 10.1093/jee/toaa264

Dively, G.P., Kuhar, T.P., Taylor, S., Doughty, H.B., Holmstrom, K. *et al.* (2023) Extended sentinel monitoring of *Helicoverpa zea* resistance to Cry and Vip toxins in Bt sweet corn: assessing phenotypic and allele frequencies of resistance. *Insects* 14, 577. DOI: 10.3390/insects14070577

Dorman, S.J., Schürch, R., Huseth, A.S. and Taylor, S.V. (2020) Landscape and climatic effects driving spatiotemporal abundance of *Lygus lineolaris* (Hemiptera: Miridae) in cotton agroecosystems. *Agriculture, Ecosystems & Environment* 295, 106910. DOI: 10.1016/j.agee.2020.106910

Dorman, S.J., Hopperstad, K.A., Reich, B.J., Kennedy, G. and Huseth, A.S. (2021) Soybeans as a non-*Bt* refuge for *Helicoverpa zea* in maize-cotton agroecosystems. *Agriculture, Ecosystems & Environment* 322, 107642. DOI: 10.1016/j.agee.2021.107642

Ellsbury, M.M., Burkett, G.A. and Davis, F.M. (1989) Development and feeding behavior of *Heliothis zea* (Lepidoptera: Noctuidae) on leaves and flowers of crimson clover. *Environmental Entomology* 18, 323–327. DOI: 10.1093/ee/18.2.323

Fehr, W.R. and Caviness, C.E. (1977) *Stages of Soybean Development*. Special Report 80. Ames, IA: Iowa State University Cooperative Extension, pp. 3–11. Available at: http://lib.dr.iastate.edu/specialreports/87 (accessed April 22, 2024).

Felt, E.P. (1925) The dissemination of insects by air currents. *Journal of Economic Entomology* 18, 152–158. DOI: 10.1093/jee/18.1.152

Feng, H., Gould, F., Huang, Y., Jiang, Y. and Wu, K. (2010) Modeling the population dynamics of cotton bollworm *Helicoverpa armigera* (Hübner) (Lepidoptera: Noctuidae) over a wide area in northern China. *Ecological Modelling* 221, 1819–1830. DOI: 10.1016/j.ecolmodel.2010.04.003

Fitt, G.P. (1989) The ecology of *Heliothis* species in relation to agroecosystems. *Annual Review of Entomology* 34, 17–53. DOI: 10.1146/annurev.en.34.010189.000313

Fitt, G.P., Dillon, M.L. and Hamilton, J.G. (1995) Spatial dynamics of *Helicoverpa* populations in Australia: simulation modelling and empirical studies of adult movement. *Computers and Electronics in Agriculture* 13, 177–192. DOI: 10.1016/0168-1699(95)00024-X

Flood, B.R. and Rabaey, T.L. (2007) Potential impact of pyrethroid resistance in *Helicoverpa zea* to the Midwest processing industry: sweet corn and snap beans. *Plant Health Progress* 8, 61. DOI: 10.1094/PHP-2007-0719-06-RV

Gagic, V., Holding, M., Venables, W.N., Hulthen, A.D. and Schellhorn, N.A. (2021) Better outcomes for pest pressure, insecticide use, and yield in less intensive agricultural landscapes. *Proceedings of the National Academy of Sciences* 118, e2018100118. DOI: 10.1073/pnas.2018100118

Gassmann, A.J. (2021) Resistance to Bt maize by western corn rootworm: effects of pest biology, the pest–crop interaction and the agricultural landscape on resistance. *Insects* 12, 136. DOI: 10.3390/insects 12020136

Gassmann, A.J. and Reisig, D.D. (2023) Management of insect pests with Bt crops in the United States. *Annual Review of Entomology* 68, 31–49. DOI: 10.1146/annurev-ento-120220-105502

Gaylor, M.J., Fleischer, S.J., Muehleisen, D.P. and Edelson, J.V. (1984) Insect populations in cotton produced under conservation tillage. *Journal of Soil and Water Conservation* 39, 61–64.

Getz, W.M. and Gutierrez, A.P. (1982) A perspective on systems analysis in crop production and insect pest management. *Annual Review of Entomology* 27, 447–466. DOI: 10.1146/annurev.en.27.010182.002311

Gould, F.R., Blair, N.E., Reid, M., Rennie, T.L., Lopez, J. and Micinski, S. (2002) *Bacillus thuringiensis*-toxin resistance management: stable isotope assessment of alternate host use by *Helicoverpa zea*. *Proceedings of the National Academy of Sciences* 99, 16581–16586. DOI: 10.1073/pnas.242382499

Graham, H.M., Hernandez, N.S., Jr and Llanes, J.R. (1972) The role of host plants in the dynamics of populations of *Heliothis* spp. *Environmental Entomology* 1, 424–431. DOI: 10.1093/ee/1.4.424

Gutierrez, A.P., Falcon, L.A., Loew, W., Leipzig, P.A. and van den Bosch, R. (1975) An analysis of cotton production in California: a model for Acala cotton and the effects of defoliators on its yields. *Environmental Entomology* 4, 125–136. DOI: 10.1093/ee/4.1.125

Haile, F., Nowatzki, T. and Storer, N. (2021) Overview of pest status, potential risk, and management considerations of *Helicoverpa armigera* (Lepidoptera: Noctuidae) for US soybean production. *Journal of Integrated Pest Management* 12, 3. DOI: 10.1093/jipm/pmaa030

Hardwick, D.F. (1965) The corn earworm complex. *The Memoirs of the Entomological Society of Canada* 97(S40), 5–247. DOI: 10.4039/entm9740fv

Hardwick, D.F. (1996) *A Monograph to the North American Heliothentinae (Lepidoptera: Noctuidae)*. Almonte, Canada: Agriculture Canada, Centre for Land and Biological Resources Research.

Hargis, C.D., Bissonette, J.A. and David, J.L. (1998) The behavior of landscape metrics commonly used in the study of habitat fragmentation. *Landscape Ecology* 13, 167–186. DOI: 10.1023/A:1007965018633

Head, G., Jackson, R.E., Adamczyk, J., Bradley, J.R., Van Duyn, J. *et al.* (2010) Spatial and temporal variability in host use by *Helicoverpa zea* as measured by analyses of stable carbon isotope ratios and gossypol residues. *Journal of Applied Ecology* 47, 538–592. DOI: 10.1111/j.1365-2664.2010.01796.x

Huseth, A.S., Petersen, J.D., Poveda, K., Szendrei, Z., Nault, B.A. *et al.* (2015) Spatial and temporal potato intensification drives insecticide resistance in the specialist herbivore, *Leptinotarsa decemlineata*. *PLoS ONE* 10, e0127576. DOI: 10.1371/journal.pone.0127576

Huseth, A.S., Chappell, T.M., Chitturi, A., Jacobson, A.L. and Kennedy, G.G. (2018) Insecticide resistance signals negative consequences of widespread neonicotinoid use on multiple field crops in the US cotton belt. *Environmental Science & Technology* 52, 2314–2322. DOI: 10.1021/acs.est.7b06015

Hutchison, W.D., Burkness, E.C., Jensen, B., Leonard, B.R., Temple, J. *et al.* (2007) Evidence for decreasing *Helicoverpa zea* susceptibility to pyrethroid insecticides in the Midwestern U.S. *Plant Health Progress* 8(1). DOI: 10.1094/PHP-2007-0719-02-RV

Hutchison, W.D., Burkness, E.C., Mitchell, P.D., Moon, R.D., Leslie, T.W. *et al.* (2010) Areawide suppression of European corn borer with Bt maize reaps savings to non-*Bt* maize growers. *Science* 330, 222–225. DOI: 10.1126/science.1190242

Jackson, R.E., Bradley, J.R., Van Duyn, J., Leonard, B.R., Allen, K.C. *et al.* (2008) Regional assessment of *Helicoverpa zea* populations on cotton and non-cotton crop hosts. *Entomologia Experimentalis et Applicata* 126, 89–106. DOI: 10.1111/j.1570-7458.2007.00653.x

Jacobson, A., Foster, R., Krupke, C., Hutchison, W., Pittendrigh, B. *et al.* (2009) Resistance to pyrethroid insecticides in *Helicoverpa zea* (Lepidoptera: Noctuidae) in Indiana and Illinois. *Journal of Economic Entomology* 102, 2289–2295. DOI: 10.1603/029.102.0634

Johnson, M.W., Stinner, R.E. and Rabb, R.L. (1975) Ovipositional response of *Heliothis zea* (Boddie) to its major hosts in North Carolina. *Environmental Entomology* 4, 291–297. DOI: 10.1093/ee/4.2.291

Karp, D.S., Chaplin-Kramer, R., Meehan, T.D., Martin, E.A., DeClerck, F. *et al.* (2018) Crop pests and predators exhibit inconsistent responses to surrounding landscape composition. *Proceedings of the National Academy of Sciences* 115, E7863–E7870. DOI: 10.1073/pnas.1800042115

Kennedy, G.G. and Margolies, D.C. (1985) Mobile arthropod pests: management in diversified agroecosystems. *Bulletin of the Entomological Society of America* 31(3), 21–35. DOI: 10.1093/besa/31.3.21

Kennedy, G.G. and Storer, N.P. (2000) Life systems of polyphagous arthropod pests in temporally unstable cropping systems. *Annual Review of Entomology* 45, 467–493. DOI: 10.1146/annurev.ento.45.1.467

Kogan, M., Helm, C.G., Kogan, J. and Brewer, E. (1989) Distribution and economic importance of *Heliothis virescens* and *Heliothis zea* in North, Central and South America and of their natural enemies and host plants. In: Kingand, E.G. and Jackson, R.D. (eds) *Proceedings of the Workshop on Biological Control of Heliothis: Increasing the Effectiveness of Natural Enemies*. New Delhi, India: US Department of Agriculture, Far Eastern Regional Research Office, pp. 241–298. DOI: 10.1017/S0007485300051312

Kriticos, D.J., Ota, N., Hutchison, W.D., Beddow, J., Walsh, T. *et al.* (2015) The potential distribution of invading *Helicoverpa armigera* in North America: is it just a matter of time? *PLoS ONE* 10, e0119618. DOI: 10.1371/journal.pone.0119618

Lawton, D., Huseth, A.S., Kennedy, G.G., Morey, A.C., Hutchison, W.D. *et al.* (2022) Pest population dynamics are related to a continental overwintering gradient. *Proceedings of the National Academy of Sciences* 119, e2203230119. DOI: 10.1073/pnas.2203230119

Leite, N.A., Correa, A.S., Michel, A.P., Alves-Pereira, A., Pavinato, V.A.C. *et al.* (2017) Pan-American similarities in genetic structures of *Helicoverpa armigera* and *Helicoverpa zea* (Lepidoptera: Noctuidae) with implications for hybridization. *Environmental Entomology* 46, 1024–1034. DOI: 10.1093/ee/nvx088

Lingren, P.D., Westbrook, J.K., Bryant, V.M., Raulston, J.R., Esquivel, J.F. and Jones, G.D. (1994) Origin of corn earworm (Lepidoptera: Noctuidae) migrants as determined by citrus pollen markers and synoptic weather systems. *Environmental Entomology* 23, 562–570. DOI: 10.1093/ee/23.3.562

Logan, J.A., Stinner, R.E., Rabb, R.L. and Bacheler, J.S. (1979) A descriptive model for predicting spring emergence of *Heliothis zea* populations in North Carolina. *Environmental Entomology* 8, 141–146. DOI: 10.1093/ee/8.1.141

McGarigal, K. (1995) *FRAGSTATS: Spatial Pattern Analysis Program for Quantifying Landscape Structure*. General Technical Report PNW-GTR-351. Portland, OR: US Department of Agriculture, Forest Service, Pacific Northwest Research Station. Available at: https://www.fs.usda.gov/pnw/pubs/pnw_gtr351.pdf (accessed April 22, 2024).

Meisner, M.H., Zaviezo, T. and Rosenheim, J.A. (2017) Landscape crop composition effects on cotton yield, *Lygus hesperus* densities and pesticide use. *Pest Management Science* 73, 232–239. DOI: 10.1002/ps.4290

Morey, A.C., Hutchison, W.D., Venette, R.C. and Burkness, E.C. (2012) Cold hardiness of *Helicoverpa zea* (Lepidoptera: Noctuidae) pupae. *Environmental Entomology* 41, 172–179. DOI: 10.1603/EN11026

Mueller, T.F. and Phillips, J.R. (1983) Population dynamics of *Heliothis* spp. in spring weed hosts in southeastern Arkansas: survivorship and stage-specific parasitism. *Environmental Entomology* 12, 1846–1850. DOI: 10.1093/ee/12.6.1846

Niu, Y., Oyediran, I., Yu, W., Lin, S., Dimase, M. *et al.* (2021) Populations of *Helicoverpa zea* (Boddie) in the southeastern United States are commonly resistant to Cry1Ab, but still susceptible to Vip3Aa20 expressed in MIR 162 corn. *Toxins* 13, 63. DOI: 10.3390/toxins13010063

Pair, S.D., Raulston, J.R., Rummel, D.R., Westbrook, J.K., Wolf, W.W. *et al.* (1987) Development and production of corn earworm and fall armyworm in the Texas High Plains: evidence for reverse fall migration. *Southwestern Entomologist* 12, 89–99.

Paula-Moraes, S.V., Calixto, E.S., Santos, A.A., Reay-Jones, F.P., Reisig, D.D. *et. al.* (2024) Continental-scale migration patterns and origin of *Helicoverpa zea* (Lepidoptera: Noctuidae) based on a biogeochemical marker. *Environmental Entomology* 53, 487–497. DOI: 10.1093/ee/nvae034

Peng, J., Wang, Y., Zhang, Y., Wu, J., Li, W. *et al.* (2010) Evaluating the effectiveness of landscape metrics in quantifying spatial patterns. *Ecological Indicators* 10, 217–223. DOI: 10.1016/j.ecolind.2009.04.017

Phillips, W.J. and Barber, G.W. (1929) *A Study of Hibernation of the Corn Earworm*. Technical Bulletin 40. Blacksburg, VA: Virginia Agricultural Experiment Station.

Pietrantonio, P.V., Junek, T.A., Parker, R., Mott, D., Siders, K. *et al.* (2007) Detection and evolution of resistance to the pyrethroid cypermethrin in *Helicoverpa zea* (Lepidoptera: Noctuidae) populations in Texas. *Environmental Entomology* 36, 1174–1188. DOI: 10.1603/0046-225X(2007)36[1174:DAEORT]2.0.CO;2

Quaintance, A.L. and Brues, C.T. (1905) *The Cotton Bollworm*. Bulletin 50. Washington, DC: US Government Printing Office. DOI: 10.5962/bhl.title.109927

Rabb, R.L. (1978) A sharp focus on insect populations and pest management from a wide-area view. *Bulletin of the Entomological Society of America* 24, 55–61. DOI: 10.1093/besa/24.1.55

Reisig, D. and Goldsworthy, E. (2023) Yield and efficacy of selected insecticides on bollworm in cotton, 2022. *Arthropod Management Tests* 48, tsad082. DOI: 10.1093/amt/tsad082

Reisig, D., Buntin, G.D., Greene, J.K., Paula-Moraes, S.V., Reay-Jones, F. *et al.* (2023) Magnitude and extent of *Helicoverpa zea* resistance levels to Cry1Ac and Cry2Ab2 across the southeastern USA. *Insects* 14, 262. DOI: 10.3390/insects14030262

Reisig, D.D. (2017) Factors associated with willingness to plant non-Bt maize refuge and suggestions for increasing refuge compliance. *Journal of Integrated Pest Management* 8, 9. DOI: 10.1093/jipm/pmx002

Reisig, D.D. and Kurtz, R. (2018) Bt resistance implications for *Helicoverpa zea* (Lepidoptera: Noctuidae) insecticide resistance management in the United States. *Environmental Entomology* 47, 1357–1364. DOI: 10.1093/ee/nvy142

Reisig, D.D., Huseth, A.S., Bacheler, J.S., Aghaee, M.A., Braswell, L. *et al.* (2018) Long-term empirical and observational evidence of practical *Helicoverpa zea* resistance to cotton with pyramided Bt toxins. *Journal of Economic Entomology* 111, 1824–1833. DOI: 10.1093/jee/toy106

Roach, S.H. (1975) *Heliothis* spp.: larvae and associated parasites and diseases on wild host plants in the Pee Dee area of South Carolina. *Environmental Entomology* 4, 725–728. DOI: 10.1093/ee/4.5.725

Roach, S.H. and Hopkins, A.R. (1979) *Heliothis* spp.: behavior of prepupae and emergence of adults from different soils at different moisture levels. *Environmental Entomology* 8, 388–391. DOI: 10.1093/ee/8.3.388

Rosenheim, J.A., Parsa, S., Forbes, A.A., Krimmel, W.A., Hua Law, Y. *et al.* (2011) Ecoinformatics for integrated pest management: expanding the applied insect ecologist's tool-kit. *Journal of Economic Entomology* 104, 331–342. DOI: 10.1603/EC10380

Rosenheim, J.A., Cluff, E., Lippey, M.K., Cass, B.N., Paredes, D. *et al.* (2022) Increasing crop field size does not consistently exacerbate insect pest problems. *Proceedings of the National Academy of Sciences* 119, e2208813119. DOI: 10.1073/pnas.2208813119

Sandstrom, M.A., Changnon, D. and Flood, B.R. (2007) Improving our understanding of *Helicoverpa zea* migration in the Midwest: assessment of source populations. *Plant Health Progress* 8(1). DOI: 10.1094/PHP-2007-0719-08-RV

Santos, I.B.D., Paula-Moraes, S.V., Beuzelin, J.M., Hahn, D.A., Perera, O.P. *et al.* (2023) Factors affecting population dynamics of *Helicoverpa zea* (Lepidoptera: Noctuidae) in a mixed landscape with Bt cotton and peanut. *Insects* 14, 395. DOI: 10.3390/insects14040395

Sivakoff, F.S., Rosenheim, J.A., Dutilleul, P. and Carrière, Y. (2013) Influence of the surrounding landscape on crop colonization by a polyphagous insect pest. *Entomologia Experimentalis et Applicata* 149, 11–21. DOI: 10.1111/eea.12101

Stadelbacher, E.A. (1981) Role of early-season wild and naturalized host plants in the buildup of the F_1 generation of *Heliothis zea* and *H. virescens* in the Delta of Mississippi. *Environmental Entomology* 10, 766–770. DOI: 10.1093/ee/10.5.766

Stadelbacher, E.A., Graham, H.M., Harris, V.E., Lopez, J.D., Phillips, J.R. and Roach, S.H. (1986) *Heliothis* populations and wild host plants in the Southern US. In: Johnson, S.J., King, E.G. and Bradley, J.R., Jr (eds) *Theory and Tactics of* Heliothis *Population Management: Cultural and Biological Control*, Vol. 1. Southern Cooperative Series Bulletin 163. Stillwater, OK: Oklahoma State University, Agricultural Experiment Station.

Stinner, R.E. (1979) Biological monitoring essentials in studying wide-area moth movement. In: Rabb, R.L. and Kennedy, G.G. (eds) *Movement of Highly Mobile Insects: Concepts and Methodology in Research*. Raleigh, NC: North Carolina State University, pp. 199–211.

Stinner, R.E., Rabb, R.L. and Bradley, J.R. (1974) Population dynamics of *Heliothis zea* (Boddie) and *H. virescens* (F.) in North Carolina: a simulation model. *Environmental Entomology* 3, 163–168. DOI: 10.1093/ee/3.1.163

Stinner, R.E., Butler, G.D., Bacheler, J.S. and Tuttle, C. (1975) Simulation of temperature-dependent development in population dynamics models. *The Canadian Entomologist* 107, 1167–1174. DOI: 10.4039/Ent1071167-11

Stinner, R.E., Regniere, J. and Wilson, K. (1982) Differential effects of agroecosystem structure on dynamics of three soybean herbivores. *Environmental Entomology* 11, 538–543. DOI: 10.1093/ee/11.3.538

Storer, N.P., Peck, S.L., Gould, F., Van Duyn, J.W. and Kennedy, G.G. (2003) Spatial processes in the evolution of resistance in *Helicoverpa zea* (Lepidoptera: Noctuidae) to Bt transgenic corn and cotton in a mixed agroecosystem: a biology-rich stochastic simulation model. *Journal of Economic Entomology* 96, 156–172. DOI: 10.1093/jee/96.1.156

Tabashnik, B. and Carrière, Y. (2017) Surge in insect resistance to transgenic crops and prospects for sustainability. *Nature Biotechnology* 35, 926–935. DOI: 10.1038/nbt.3974

Tabashnik, B.E., Fabrick, J.A. and Carrière, Y. (2023) Global patterns of insect resistance to transgenic Bt crops: the first 25 years. *Journal of Economic Entomology* 116, 297–309. DOI: 10.1093/jee/toac183

Tillman, G., Schomberg, H., Phatak, S., Mullinix, B., Lachnicht, S. *et al.* (2004) Influence of cover crops on insect pests and predators in conservation tillage cotton. *Journal of Economic Entomology* 97, 1217–1232. DOI: 10.1093/jee/97.4.1217

Turner, M.G. and Gardner, R.H. (2015) Landscape metrics. In: Turner, M.G. and Gardner, R.H. (eds) *Landscape Ecology in Theory and Practice: Pattern and Process*, 2nd edn. New York: Springer, pp. 97–142. DOI: 10.1007/978-1-4939-2794-4

USDA-NASS (2017) *Land Use Practices*. Washington, DC: US Department of Agriculture, National Agricultural Statistics Service. Available at: https://www.nass.usda.gov/Publications/Highlights/2020/census-land-use-practices.pdf (accessed December 19, 2023).

USDA-NASS (2022) *CropScape – Cropland Data Layer*. Washington, DC: US Department of Agriculture, National Agricultural Statistics Service. Available at: https://nassgeodata.gmu.edu/CropScape/ (accessed August 10, 2023).

USDA-NASS (2024) *National Crop Frequency Layer 2008–2023*. Washington, DC: US Department of Agriculture, National Agricultural Statistics Service. Available at: https://www.nass.usda.gov/Research_and_Science/Cropland/Release/ (accessed May 4, 2024).

Vann, R.A., Drake-Stowe, K., Buol, G.S. and Dunphy, E.J. (2021) Production practices that maximize soybean yield: what we have learned from the North Carolina soybean yield contest. *Agronomy Journal* 113, 3399–3417. DOI: 10.1002/agj2.20728

Whitehouse, M.E., Tann, C.R. and Braunack, M.V. (2023) Is 'pupae busting' or destroying overwintering pupae of *Helicoverpa* spp. (Lepidoptera: Noctuidae) still relevant today in Australian Bt cotton? *Austral Entomology* 22, 392–409. DOI: 10.1111/aen.12669

Westbrook, J.K. (2008) Noctuid migration in Texas within the nocturnal aeroecological boundary layer. *Integrative and Comparative Biology* 48, 99–106. DOI: 10.1093/icb/icn040

Westbrook, J.K. and López, J.D., Jr (2010) Long-distance migration in *Helicoverpa zea*: what we know and need to know. *Southwestern Entomologist* 35, 355–360. DOI: 10.3958/059.035.0315

Westbrook, J.K., Wolf, W.W., Lingren, P.D., Raulston, J.R., Lopez, J.D., Jr *et al.* (1997) Early-season migratory flights of corn earworm (Lepidoptera: Noctuidae). *Environmental Entomology* 26, 12–20. DOI: 10.1093/ee/26.1.12

Wu, Q.L., Hu, G., Westbrook, J.K., Sword, G.A. and Zhai, B.P. (2018) An advanced numerical trajectory model tracks a corn earworm moth migration event in Texas, USA. *Insects* 9, 115. DOI: 10.3390/insects9030115

11 Nature's Workforce: Understanding the Role of Wild Bees in Agroecosystems

Isaac L. Esquivel[1,*] **and Katherine A. Parys**[2]
[1]*Department of Entomology & Nematology, University of Florida, North Florida Research and Education Center, Quincy, FL, USA;* [2]*US Department of Agriculture, Agricultural Research Service, Pollinator Health in Southern Crop Ecosystems Research Unit, Stoneville, MS, USA*

Abstract
Large-scale agroecosystems are often a mosaic of different annual and perennial crops, semi-natural habitats, and natural habitats within various land-use systems. In these systems, pest and beneficial insect population patterns increase and decline at varying degrees within and across crops. Different biotic and abiotic interactions within these systems often influence these patterns, such as landscape configurations, climatic conditions, and on-farm practices. Pollination by insects is a critical ecosystem service necessary to produce most crops, including those providing essential nutrients for food security. When managed appropriately, agricultural landscapes can provide a habitat for many insect species. However, we are amid an unprecedented decline in managed and wild pollinator populations, diversity, and health worldwide. Large-scale agroecosystems dominate North American, European, and selected other rural landscapes worldwide, covering roughly 137 million ha in the United States alone. This anthropogenic intensification reduces natural or semi-natural land harboring wild pollinators and diverse floral resources. While there are many forms of pollinators, most pollination ecologists would agree that bees (Hymenoptera: Anthophila) are recognized as the most predominant pollinators for most plants. This chapter will dive into the relationship between pollinators and the crops they pollinate in large-scale agroecosystems (including honey bees, wild bees, and non-bee pollinators), benefits to pollinators from crops and benefits to crops from bees, and integrated pest and pollinator management (IPPM) systems. Implementing landscape-scale conservation strategies, such as promoting habitat restoration, reducing pesticide use, creating floral resources, and enhancing connectivity, can help support diverse and resilient pollinator populations.

Introduction

Pollinators are responsible for pollinating 87% of the world's wild and cultivated plant species and are essential for maintaining healthy ecosystems (Ollerton *et al.*, 2011). They contribute to the pollination of wild plants, promoting biodiversity and the functioning of natural habitats (Osterman *et al.*, 2021). Moreover, bees play a role in the reproduction of trees and flowering plants that provide habitats and food sources for other animals (Winfree *et al.*, 2011). Many agricultural industries heavily depend on pollination services provided by managed and wild bees (Cumming *et al.*, 2014). Bees and other pollinators benefit farmers and crop producers and contribute to the food supply chain and local economies (Calderone, 2012). The global economic value

*Corresponding author: Isaac.Esquivel@ufl.edu

DOI: 10.1079/9781800622777.0011

of pollination services provided by managed and wild bees, flies, butterflies, and others in agriculture is estimated at anywhere between US$235 and 577 billion annually. In the United States (US), the benefits of pollinators for commercial crops range from US$12 to 74 billion, with native pollinators accounting for US$3 billion of the total (Calderone, 2012). Despite their importance, bees and other pollinators face numerous challenges, including habitat loss, pesticide exposure, climate change, and diseases (Kremen *et al.*, 2002; Gérard *et al.*, 2020; Senapathi *et al.*, 2021). These factors have contributed to the worldwide bee population decline, with consequences for both commercial agriculture and natural ecosystems. The efficiency and amount of pollination services are expected to decrease with this decline, while the need for these services is not (Aizen *et al.*, 2019).

Farmland covers more than 35% of the earth's terrestrial surface areas and continues to expand and intensify to meet the growing food and fiber demands of human populations (Cumming *et al.*, 2014). Agroecosystems are a dominant feature in rural landscapes across Europe, North America, and selected other locations, covering roughly 137 million ha in the US alone (Bigelow and Borchers, 2017). This anthropogenic intensification reduces natural or semi-natural land harboring wild non-*Apis* bees, other pollinators, and diverse floral resources (Deguines *et al.*, 2014). While there are many forms of pollinators, most pollination ecologists would agree that bees (Hymenoptera: Anthophila) are recognized as predominant pollinators for most plants (Winfree *et al.*, 2011). This predominance is mainly attributable to around 20,000 species of bees being obligate florivores, and both larval and adult life stages feed directly or indirectly on floral products (Michener, 2007). The recent population declines in bees and other insects could have significant ecological consequences.

Understanding the impacts of large-scale agricultural production on species groups that provide critical ecosystem services, such as pollination in agroecosystems, is crucial for food security and conservation (Aizen *et al.*, 2019). Protecting and promoting bee populations through sustainable practices, habitat conservation, and reducing pesticide use are crucial steps in ensuring their continued role in commercial agriculture. In this chapter, we broadly address landscape effects on pollinators in agricultural systems, differences in pollinator-dependent and pollinator-independent crops, benefits to crops and their potential benefit to bees (managed honey bees and wild bees), non-bee pollinators, and integrated pest and pollinator management (IPPM) systems.

Managed Bees

Significant differences exist between the traits and required resources of managed and wild pollinators and their role in crop pollination. Managed pollinators consist primarily of social bees that often exhibit more significant levels of foraging due to high colony-level resource demands, though a few solitary bee species are exceptions, including the Megachilids *Osmia cornuta* (Latr.), *Osmia lignaria* (Say), and *Megachile rotundata* (F.) (Pitts-Singer and Cane, 2011). Managed honey bees, inclusive of the European honey bee, *Apis mellifera* (L.), and other honey bee species in the genus *Apis* (Hymenoptera: Apidae), and wild bees are two distinct groups of bees that differ not only in their geographic origin but have different behaviors, habitat preferences, and ecological roles within the larger agroecosystem (Papa *et al.*, 2022). Honey bees are native to parts of Asia but have been widely introduced and managed worldwide for honey production and pollination services (Steffan-Dewenter and Kuhn, 2003). In contrast, with the exception of a few species (some mentioned above), wild bees are unmanaged by humans and often naturally occurring species that have evolved and adapted to specific ecosystems over time.

Honey bees (*Apis* spp.) are eusocial insects, living in large colonies with a queen, workers, and drones. They have complex hierarchical organization and division of labor within the colony. Honey bees build and maintain socially complex hives, usually in cavities such as tree hollows or artificial structures like beehives (Seeley and Morse, 1976). They construct hexagonal wax cells to store honey, pollen, and raise brood. Honey bee colonies can grow to large numbers, with tens of thousands of individuals. They can build and maintain substantial populations due to their eusocial nature and a large workforce.

Wild (also referred to as native bees) on the other hand, exhibit a wide range of social structures ranging from solitary, to semi-social, social, and parasitic (Michener, 2007). These bees exhibit diverse nesting habits, including ground-nesting, stem nesting, and using pre-existing cavities (Linsley *et al.*, 1952). Native bees, being mostly solitary or semi-social, typically have smaller populations. However, native bee communities can be highly diverse, with numerous species contributing pollination services in an area (Gill and O'Neal, 2015; Landaverde-González *et al.*, 2017; Parys *et al.*, 2020).

Honey bees are renowned for their broad polylectic diet and pollination efficiency across a wide variety of crops and other plants (Carpes *et al.*, 2009; Lau *et al.*, 2018; Layek *et al.*, 2020; Stoner *et al.*, 2022). They have specialized morphological adaptations, such as scopa on their hind legs, which allow them to pack and carry significant amounts of pollen back to the colony (Michener *et al.*, 1978). Native bees also contribute to pollination, and some species exhibit specific relationships with certain plant taxa that result in specialized pollination strategies (Fowler, 2016). Bumble bees (*Bombus* spp.), along with other groups of bees, have the foraging behavior and ability to perform "buzz pollination" (vibrating their bodies at specific frequencies to release pollen from flowers), making them effective pollinators for many crops in the plant family Solanaceae (e.g. tomatoes, eggplants) (De Luca and Vallejo-Marín, 2013).

Some wild bees are also managed. Eight solitary bees, in particular, leafcutter and mason bee species (Megachilidae: *Megachile* spp. and *Osmia* spp., respectively) but also the alkali bee (Halictidae: *Nomia melanderi*), are currently managed for crop pollination (Osterman *et al.*, 2021). The alfalfa leafcutting bee, *M. rotundata*, was initially introduced to the US before the 1940s and has become the most managed wild solitary bee worldwide (Pitts-Singer and Cane, 2011). Around 800 million alfalfa leafcutter bees are traded commercially per year in North America, and an additional 1.6 million are promoted in and around alfalfa fields in the US (Osterman *et al.*, 2021). Bumbles bees, specifically *Bombus impatiens* (Cresson) and *Bombus terrestris* (L.), which are social bees, are another group of bees that are commercially managed for crop pollination (Osterman *et al.*, 2021). Since the initial commercialization of *B. terrestris* in the 1980s, the volume of managed colonies of this species traded annually has seen a significant increase, reaching 1 million colonies by 2006 (Osterman *et al.*, 2021). Recently, stingless bees have been studied in Brazil for pollination services, but their use in pollination management is not yet an established practice (Osterman *et al.*, 2021). Osterman *et al.* (2021) report an additional 15 species that have been or are under consideration as managed pollinators, primarily for crop pollination (e.g. strawberry, cucumber, tomatoes, habanero, sweet pepper) in enclosures.

All bees, including honey bees, other managed bees, and wild/native bees, face challenges and conservation concerns. Habitat loss, pesticides, diseases, and parasites have impacted honey bees (Hristov *et al.*, 2020). Native bees, particularly those with specialized habitat or foraging requirements, are vulnerable to habitat destruction, fragmentation, pesticide exposure, and climate change (Kremen *et al.*, 2002; Barbosa *et al.*, 2015; Freitas *et al.*, 2009; Deutsch *et al.*, 2023). Conservation efforts promote habitat restoration, reduce pesticide use, and raise awareness about the importance of honey bees and native bees (Winfree, 2010; Wilson *et al.*, 2017; Penn *et al.*, 2019; van Vierssen Trip *et al.*, 2020). It is important to note that groupings of bees like managed versus unmanaged or native versus wild are not mutually exclusive in their roles as pollinators. Each of these groupings or classifications contribute significantly to pollination services and have complementary ecological functions.

Non-bee Pollinators

While bees are the most well-known and essential group of crop pollinators, several other non-bee pollinators also play a role in pollination, including various additional insects, birds, bats, and even small mammals (Rader *et al.*, 2016). Butterflies and moths (Insecta: Lepidoptera) are important pollinators for a wide variety of plant families around the world, including certain crops, particularly those with large, colorful flowers (Hahn and Brühl, 2016; Buxton *et al.*, 2018). Adults have an elongated flexible proboscis used to access nectar from flowers, and as they feed on nectar, they unintentionally pick up and transfer pollen (Macgregor and Scott-Brown, 2020). Some crops reliant on pollination by Lepidoptera include a variety of smaller and often regional commodities, including macadamia, papaya, and some night-blooming cucurbits (Martins and Johnson, 2009; Chen *et al.*, 2016; Lu *et al.*, 2021; Santos *et al.*, 2020). It is important to note that although lepidopterans may be pollinators in their adult form, larvae of these insects are herbivorous and, in some cases, destructive crop pests. While some lepidopteran larvae are specific to a small group of plants, some, like the tobacco hornworm, *Manduca sexta* (L.) (Sphingidae), are highly polyphagous and pests of many solanaceous vegetable crops including tomatoes and tobacco (Cranshaw, 2004).

Flies (Insecta: Diptera), including hoverflies (Sryphidae) and beeflies (Bombyliidae), are significant crop pollinators in many agricultural landscapes and are known to pollinate over 100 cultivated crops (Larson *et al.*, 2001; Ssymank *et al.*, 2008). They are often attracted to flowers with strong odors or those that mimic decaying organic matter. Flies are known for their ability to visit many flowers quickly, making them efficient pollinators (Cook *et al.*, 2020). Crops, including rape, apple, mango, and strawberries, often benefit from fly pollination (Jauker and Wolters, 2008; Rader *et al.*, 2013; Saeed *et al.*, 2016; Stavert *et al.*, 2018; Finch *et al.*, 2023). In addition to providing pollination services to agroecosystems, sryphid flies also contribute to biological control services as they are predators of aphids and other piercing-sucking insects like psyllids, whitefly, and scales (Dunn *et al.*, 2020). Sryphids are known to cover large-sized fields when aphid prey are present, but less is known of their range into crops for floral resources in the absence of aphids (Dunn *et al.*, 2020). This lack of knowledge often makes sryphids' dual pollination and pest control roles overlooked.

Beetles (Insecta: Coleoptera) were among the first insects to pollinate flowering plants in the evolutionary record. While they are often not as efficient as bees in pollen transfer, they play a role in pollinating certain crops. Generally, beetle-pollinated flowers tend to be large, bowl-shaped, or flat and often emit strong fragrances. In Brazil, species in the genus *Elaeidobius* (Coleoptera: Curculionidae), are known to pollinate araticum, atemoya, cupuassu, African oil palm, soursop, and tucumã palm (Giannini *et al.*, 2015). The genus *Elaeidobius* are important worldwide as pollinators of oil palm and are effective pollinators in palm plantations (Yue *et al.*, 2015; Auffray *et al.*, 2017; Yousefi *et al.*, 2020). Additional examples of crops that can be beetle-pollinated include certain fruits like custard apples and other tropical fruit species (Blanche and Cunningham, 2005; Paz *et al.*, 2021).

Certain bird species, especially hummingbirds, are also important pollinators for specific crops, particularly those with long, tubular flowers (Cronk and Ojeda, 2008; Regan *et al.*, 2015). Hummingbirds have long beaks and specialized tongues that allow them to reach nectar deep within flowers, and they inadvertently transfer pollen as they move from flower to flower (Claudino *et al.*, 2022; Leimberger *et al.*, 2022). Crops such as certain types of bananas and papaya utilize bird pollination (Nun, 1976; Kiew, 1987). Other commercial tropical crops like nutmeg rely on frugivorous birds for seed dispersal (Sharma and Armstrong, 2013).

Bats are also pollinators, particularly in tropical and desert regions. They are often attracted to night-blooming flowers that are usually pale or white and emit strong fragrances. Bats have long tongues that can reach deep into flowers to access nectar, and transfer pollen while moving from plant to plant (Diniz *et al.*, 2019). Some crops that rely on bat pollination include agaves, pitayas, bananas, and certain types of fruits like durian and mango (Ayensu, 1974; Nun, 1976; Molina-Freaner and Eguiarte, 2003; Bumrungsri *et al.*, 2009; Roberto-Emiliano *et al.*, 2016; Tremlett *et al.*, 2020; Baqi *et al.*, 2022). Pollination services provided by bats to regional pitaya production in Mexico are worth around US$2500 per hectare through increasing fruit yield and size (Tremlett *et al.*, 2021).

With the increasing demand for pollination services in crop production, the impact of wild pollinators and their functional diversity in sustaining pollination services will become increasingly important, considering environmental change over time. While honey bees and wild bees remain the most significant and efficient crop pollinators, these other non-bee pollinators contribute to pollination systems' overall diversity and resilience (Requier *et al.*, 2023). Although commonly less efficient than bees, they may become more important in years when they are highly abundant relative to wild bees. Protecting and conserving these diverse pollinator species and their habitats is crucial for maintaining healthy ecosystems and ensuring the pollination services for a wide range of crops in both small and large-scale agroecosystems.

Pollinator-dependent and Pollinator-independent Crops

Pollination by insects is a critical ecosystem service necessary to produce most crops, including those providing essential nutrients for food security (Reilly *et al.*, 2020). However, in terms of pollination dependence, not all crops are the same. Some crops rely on pollination for fruit and seed production (pollinator-dependent), while others are self-compatible, wind-pollinated, or parthenocarpic (pollinator-independent). In a hallmark review, Klein *et al.* (2007) indicated that out of 252 crops examined by the United Nations Food and Agriculture Organization, 87 crops worldwide rely on animal pollination to produce yield, while only 28 crops do not depend on pollinators. However, it was noted that 60% of global agricultural production comes from crops that do not require pollination, while an additional 35% relies on pollinators, and 5% has an unknown significance. After narrowing down the crops used for human consumption and producing at least 4 million tonnes annually, 124 crops were considered. Of these, 24 self-pollinated, wind-pollinated, or parthenocarpic crops were excluded (pollinator-independent). The study found that 13 crops were strictly pollinator-dependent, 30 crops were highly pollinator-dependent, 27 crops were moderately pollinator-dependent, and 21 crops were slightly dependent on pollinators. In comparison, pollination was unimportant for seven crops. The remaining nine crops had unknown significance regarding pollination (Table 11.1).

The decline in pollinator abundance and diversity is expected to reduce the efficiency and quantity of pollination services in these crops (Aizen *et al.*, 2019). However, the demand for these services has not decreased as the proportion of area cultivated with pollinator-dependent crops planted increased by approximately 70% from 1961 to 2016 compared to a 17.3% increase in pollinator-independent crops (Aizen *et al.*, 2019). However, this decline will only affect crop yield if a lack of pollination truly limits the crop. When it comes to pollinator-dependent crops, there is typically an expected positive correlation between pollination and yield until a plateau is reached (Reilly *et al.*, 2020). This means that as pollination increases, so does the crop yield until a point is reached where additional pollinators no longer make a difference. A positive relationship between pollination and yield is expected if a crop is pollination-limited, meaning that more pollination will result in higher yields. However, there will be no relationship between pollination and yield if pollination is not limited (Ashman *et al.*, 2004; Reilly *et al.*, 2020).

In large-scale production agriculture, where crops often differ in pollination requirements and visitation rates, there is an expectation that crops with lower pollinator visitation have lower yields. The relationship between visitation and yield may not be apparent in fields with high pollinator visitation rates due to other yield-limiting abiotic and biotic factors. In optimal growing conditions, crop yields may not be limited if the pollination threshold is met, regardless of dependence. This means that pollination will not increase yields if other factors, such as fertility, water stress, pests, and disease pressure, are more restrictive (Ashman *et al.*, 2004). In almonds, pollination and irrigation have a significant interaction with yield. Klein *et al.* (2015) found a significant interaction between pollination and yield from decreased yield in hand and open-pollinated treatments under water reduction but no water reduction effect on yield in pollinator exclusion treatments. This suggests a threshold of pollination is needed before the negative relationship between pollination and reduced water is apparent.

Benefits to Crops versus Benefits to Bees

In pollinator-dependent crops, a vast literature base supports the benefit of pollination services that pollinators provide. Bee pollination can enhance the yield and quality of crops by ensuring better fruit set, uniformity, and size (Table 11.1). However, even in these crops, there is evidence of pollen limitation in the presence of both managed honey bees and wild bees on small- and large-scale commercial systems. For example, Reilly *et al.* (2020) found evidence of pollen limitation in sweet cherry, tart cherry, blueberry,

Table 11.1. Select common crops and their pollinator-dependence status, modes of pollination without floral visitors (i.e. passive self-pollination or wind pollination), potential benefits from pollination, main groups of bees (honey bees, bumble bees, solitary bees), and other non-bee pollinators and select references of yield increases from pollination.

Commodity name	Crop species	Pollination dependence	Pollination without visitors	Benefits from pollination	Pollinators	References
Cotton	*Gossypium hirustum*	Pollinator-independent	Passive self-pollination	Yield increase: none-modest	Honey bees, bumble bees, solitary bees	Pires *et al.*, 2014; Cusser *et al.*, 2016; Esquivel *et al.*, 2020
Soybean	*Glycine max*	Pollinator-independent	Passive self-pollination	Yield increase: none-modest	Honey bees, bumble bees, solitary bees	de O. Milfont *et al.*, 2013; Cunningham-Minnick *et al.*, 2019; Blanco *et al.*, 2020
Canola	*Brassica napus*	Pollinator-independent	Passive self-pollination, wind pollination	Yield increase: modest	Honey bees, bumble bees, solitary bees	Witter *et al.*, 2015; Adamidis *et al.*, 2019; Amro, 2021
Watermelon	*Citrullus lanatus*	Pollinator-dependent	None	Required for yield	Honey bees, bumble bees, solitary bees	Rader *et al.*, 2013; Campbell *et al.*, 2019
Cantaloupe	*Cucumis melo*	Pollinator-dependent	Self-pollination in some varieties	Required for yield	Honey bees, bumble bees, solitary bees	Eischen *et al.*, 1994; Mussen and Thorp, 1997
Fava bean	*Vinica fava*	Pollinator-independent	Passive self-pollination	Yield increase: none-modest	Honey bees, bumble bees, solitary bees	Cunningham and Le Feuvre, 2013; Bishop *et al.*, 2016; Beyer *et al.*, 2022
Coffee	*Coffea arabica*	Pollinator-independent	Passive self-pollination	Yield increase: modest	Honey bees, stingless bees, solitary bees	Ricketts *et al.*, 2004; De Marco and Coelho, 2004
Almond	*Amygdalus communis*	Pollinator-dependent	Limited self-pollination in some varieties, most self-incompatible	Yield increase: significant-required	Honey bees, bumble bees, solitary bees, hover flies	Bosch, 1994; Pitts-Singer *et al.*, 2018; Sáez *et al.*, 2020
Blueberries	*Vaccinium* spp.	Pollinator-dependent	Passive self-pollination	Yield increase: significant	Honey bees, bumble bees, solitary bees	Isaacs and Kirk, 2010; Gibbs *et al.*, 2016
Strawberry	*Fragaria* spp.	Pollinator-independent	Passive self-pollination	Yield increase: modest	Honey bees, bumble bees, solitary bees, hover flies	Horth and Campbell, 2018; Martin *et al.*, 2019
Tomato	*Lycopersicon esculentum*	Pollinator-dependent	Wind or buzz pollination required	Yield increase: slight	Honey bees, bumble bees	Santos *et al.*, 2014; Cooley and Vallejo-Marín, 2021; Toni *et al.*, 2021
Apple	*Malus domestica*	Pollinator-dependent	Some self-pollination	Required for yield	Honey bees, bumble bees, solitary bees, hover flies	Blitzer *et al.*, 2016; Geslin *et al.*, 2017
Peanut	*Arachis hypogea*	Pollinator-independent	Passive self-pollination	Unknown, some cases with slight increase	Honey bees, bumble bees, solitary bees	Blanche *et al.*, 2006

watermelon, and pumpkin across major producing regions across the US. While managed honey bees are often used to provide and enhance necessary pollination in these crops, they are often not the most efficient. In their review, Garibaldi *et al.* (2013) found that wild bees often pollinate crops more effectively: wild bee visitation enhanced fruit set of certain crops, including almonds, blueberries, and others by twice as much as the same visitation by honey bees. In floral visits by wild bees and honey bees, pollination by managed honey bees supplemented, rather than substituted for, pollination by wild bees.

When it comes to pollinator-independent crops, bees (both managed and wild species) regularly visit and utilize them despite not being required for crop production, including cotton, corn, and soybeans (Gill and O'Neal, 2015; Wheelock and O'Neal, 2016; Wheelock *et al.*, 2016; Parys *et al.*, 2020; Esquivel *et al.*, 2021a) (Fig. 11.1). Past literature on bees' effects on these crops is often limited, and reports are mixed but primarily show positive impacts on yield. Recent reviews highlight the growing research base in pollinator-independent crops, and some, like cotton and soybean that are typically planted in large-scale systems, benefit from pollinator visitation (Kral-O'Brien *et al.*, 2022; Esquivel *et al.*, 2021a).

Cotton is a highly managed large-scale crop that holds significant economic and cultural importance as a fiber crop worldwide. Early studies on bees in cotton were conducted to determine the most suitable candidates for hybrid cottonseed production (Moffett *et al.*, 1976a, b; Berger *et al.*, 1985, 1988). Various types of bees across the genera *Bombus*, *Diadasia*, *Melissodes*, *Agapostemon*, and *Perdita* were seen visiting flowers alongside honey bees in cotton fields (Berger *et al.*, 1985; Waller *et al.*, 1985; Berger *et al.*, 1988). Despite the diverse range of wild bees, only three species, including two *Melissodes* species and managed *A. mellifera*, were potentially abundant enough to aid in hybrid cotton seed production (Moffett *et al.*, 1976a, b). Wild bee diversity in cotton has become more apparent within the last 10 years. Wild bee communities in cotton production areas range from 33 to 41 species in Mississippi and Texas (Parys *et al.*, 2020), 32 in Burkina Faso (Stein *et al.*, 2017), 20 in Zambia (Mayes and Petrillo, 2017), and 29 in Brazil (Pires *et al.*, 2014).

In addition to cotton supporting a diverse wild bee community, several studies have demonstrated that a diverse population of native bees in cotton-growing areas worldwide is associated with higher cotton yields. For instance, Pires *et al.* (2014) identified a link between an increase in wild bee diversity and a 12% rise in cotton seed weight, alongside a 17% increase in seed quantity in Brazil. Similarly, using pollinator exclusion and hand crosses, Cusser *et al.* (2016) discovered an up to 18% rise in seed cotton weights during a pollinator manipulation experiment conducted with wild bee populations in large-scale cotton systems in Texas. In more manipulative experiments, Esquivel *et al.* (2020) found a direct link between pollination by the wild bee *Melissodes tepaneca* (Cresson) (Hymenoptera: Apidae) and increased cotton yield. Flowers exposed to pollination by *M. tepaneca* over 2 weeks produced a yield gain of 31% per boll and a 12–15% on a whole-plant basis compared to flowers excluded from *M. tepaneca*.

Although soybean as a crop is essentially pollinator-independent, it is also cross-pollinated mainly by flower-visiting insects. Honey bees are the most abundant and frequent flower-visiting insects of soybean, although there has been a recent increase in the literature base that documents a diverse wild bee community in soybean worldwide (de O. Milfont *et al.*, 2013). These studies, however, range in survey methods and taxonomic resolution. In the US, 32–55 species of wild bees were collected across multiple soybean studies (Gill and O'Neal, 2015; Cunningham-Minnick *et al.*, 2019), while 14 were found in Brazil (de O. Milfont *et al.*, 2013), and 33 in Argentina (Le Féon *et al.*, 2016). In terms of yield benefits from bee pollinators, recent studies are inconclusive at best, with some supporting an increase in yield while others show no benefit. Increased soybean yield was observed in 15 of 23 different studies comparing treatments of soybeans exposed to wild bees, supplemented with *A. mellifera*, caged with bees, and supplemental habitat provided for wild bees compared to soybeans excluded from pollinators (Garibaldi *et al.*, 2021). In terms of yield benefits across various treatments, the studies ranged greatly in yield (0–126%), pod set (0–91%), and seed set (0–249%), with an average increase of 21% in total yield (Garibaldi *et al.*, 2021).

Fig. 11.1. Bees (Apidae) foraging on crop and wild plant flowers. (A) *Xylocopa virginica* (L.) foraging in a cotton bloom. (B) *Peponapis prunosa* (Say) foraging in a squash bloom. (C) *Bombus griseocolis* (De Geer) foraging on a sunflower bloom. (D) *X. virginica* foraging on a red clover bloom. (E) *Bombus impatens* (Cresson) foraging on a blueberry bloom. (F) *Apis mellifera* (L.) foraging on a clover bloom. (Photo credit: Katherine A. Parys.)

Other crops like canola, coffee, and fava bean have abundant literature on the yield benefits from the diversity and abundance of wild bees. Bees in the genera *Andrena*, *Agapostemon*, *Bombus*, *Lasioglossum*, *Halictus*, and *Melissodes* have been observed visiting canola (Morandin and Winston, 2005). In addition to wild bees being attracted to canola flowers, evidence suggests insect visits increase canola yields. For example, Bommarco *et al.* (2012) found that bees increased seed weight per plant by 18% and market value by 20% in canola fields. Further, insect pollination enhanced seed quality, producing heavier seeds, higher oil content, and lower chlorophyll contents.

Coffee, both lowland and highland, are pollinator-independent and mainly wind-pollinated. However, both can benefit from pollination from bees for increased out-crossing and fruit set, with a yield increase of up to 15.8% in lowland and 12.3% in highland coffee (Klein *et al.*, 2003). Fava bean is another crop with mixed results on the benefits of bee pollination and increased yields. Insect pollination increases the seed weight per plant of fava beans by 16–40% and the average yield by a third (Cunningham and Le Feuvre, 2013; Bishop *et al.*, 2016; Rader *et al.*, 2016; Bishop *et al.*, 2020). However, certain cultivars are more responsive to pollination than others and interact with other environmental factors.

All these crops share the trait of being pollinator-independent, but they are considered mass-flowering crops that can produce large floral resources. For example, canola fields can provide from 350,000 to 700,000 plants per hectare, with each plant capable of producing more than 100 flowers over 4 weeks. Soybean can produce up to 800 florets per plant and as many as a half-million florets per acre, making them potentially incredible nectar and pollen resources for bees (Westphal *et al.*, 2003; Gill and O'Neal, 2015). Studies have demonstrated that soybeans are an important nectareous resource for honey production by honey bees, even resulting in monofloral honey from soybean in some cases. Moreover, soybeans have been reported as the most important source of pollen for honey bee colonies during peak flowering (Blettler *et al.*, 2018).

Traditionally, pollinator-independent crops are not considered to be resources for bees. However, after analyzing honey bee waggle dances, Silliman *et al.* (2022) found that up to half of their communicated locations were to row crops coinciding with peak bloom. Most visitations were to peanut, followed by corn, cotton, and soybean in order of visitation. Pollinator-independent crops can provide foraging opportunities to honey bees in these large-scale agroecosystems, but despite their presence, little knowledge is known of these relationships for wild bees. It is unclear if these crops are used for pollen and nectar resources for wild bees or if they can satisfy the nutritional requirements for bee health (Lau *et al.*, 2023).

Previous research has shown that bees exist and sometimes thrive in pollinator-independent row-crop agroecosystems and highlights the benefits of bees to these crops in the form of yield increases (Garibaldi *et al.*, 2013 Esquivel *et al.*, 2021b; Kral-O'Brien *et al.*, 2022). It is currently uncertain whether these crops serve as sources of pollen rather than just nectar, and whether pollen alone can provide the necessary protein and amino acids for development (Esquivel *et al.*, 2021a; Kral-O'Brien *et al.*, 2022). Pollen trap data from honey bee hives suggests bees use pollen from pollinator-independent crops including corn, canola, and soybean (Kral-O'Brien *et al.*, 2022). Honey bees carried larger pollen loads during peak pollinator-independent crop blooming times flowering in a large-scale agricultural landscape, suggesting substantial usage of pollen from these crops. Kral-O'Brien *et al.*'s review (2022) also highlighted pollen protein and amino acid nutrition, an important consideration when studying bee nutritional ecology.

While landscapes with pollinator-independent crops can provide essential resources for bees, it is important to note that they cannot offer stable resources over time or nesting habitats because of the ephemeral flowering nature of large-scale crops and how they are typically managed. Growers planting such crops can still aid in the conservation efforts for bees and other beneficial insects in large-scale systems. Examples of conservation efforts include reducing or changing pesticide application regimens, increasing overall crop diversity or crop rotation, and planting crops or crop varieties more attractive to pollinators (O'Brien *et al.*, 2021). Pollinator-independent crops can provide pulses of essential resources for wild and managed bees. Still, additional floral resources and management of semi-natural habitats can increase diet mixing and contribute to sustainable bee reproduction in large-scale agricultural landscapes (McAulay and Forrest, 2019).

Landscape Effects on Pollinator Communities

Large-scale agricultural systems are often a mosaic of different annual and perennial crops, semi-natural habitats, and natural habitats. The configuration of landscapes in agroecosystems (size, shape, and spatial arrangement of land use patches), in addition to their composition (proportion of land use types), is increasingly suggested as a key factor in determining biodiversity and associated ecosystem services in agricultural landscapes (Fahrig *et al.*, 2011). Studies on the impact of landscapes on pollinators have primarily centered on the significance of natural and semi-natural areas adjacent to farms. These areas offer crucial habitats, floral resources, and nesting sites that promote the survival of wild bees (Williams and Kremen, 2007). Based on data analyses across various taxa, crop species, and biomes, it has been established that bee visitation, richness, and stability progressively improve as the distance from these diverse habitats decreases (Ricketts *et al.*, 2008; Garibaldi *et al.*, 2011). In turn, semi-natural habitats surrounding agricultural fields are often used as a proxy for nesting habitat and a predictor of bee abundance and diversity (Kruess and Tscharntke, 2002). All bees collect pollen and nectar for their developing larvae. However, landscape changes have altered the resources naturally available for pollinators. Farmland and cropping systems typically contain few crops that dominate the landscape, and few natural corridors with non-crop plants. This creates a unique nutritional landscape that bees temporally can access.

Wild bees need two essential resources to survive in a landscape: pollen and nectar for food, with flowering plants as the most prominent source, and soil, woody vegetation, and debris for nesting habitat. Bees are thought to track floral resources within the landscape (Ghazoul, 2006). They often exhibit increased visitation in patches of high floral density (Hegland and Boeke, 2006). Aside from the availability of floral resources, wild bee diversity and abundance often depend on landscape components that drive the distribution of important nesting resources (Fig.11.2). Bees exhibit a distinct foraging behavior and actively seek out floral resources within their flight range. This range, however, is often limited, but even the smallest bees tend to forage within a few hundred meters of their nesting site (Greenleaf *et al.*, 2007). This behavior potentially influences their response to changes in land cover. Further, certain bee species have specific nesting requirements (Michener, 2007), which indicates that the availability of nesting resources could significantly influence their response to changes in landcover (Fig. 11.3). Previous research has demonstrated that when natural habitats are converted to agricultural land and nesting substrates are eliminated, bee populations may decrease or become extinct in certain areas (Potts *et al.*, 2005).

Fig. 11.2. (A) Female and (B) male *Epimelissodes aegis* LaBerge (Hymenoptera: Apidae) emerge from a ground nest between rows in a cotton field. (Photo credit: Isaac L. Esquivel.)

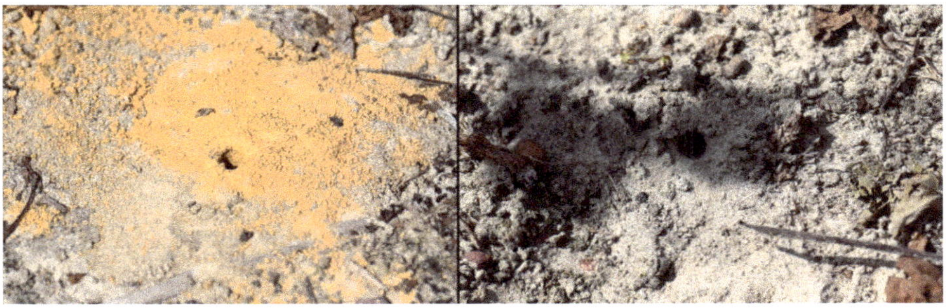

Fig. 11.3. Nest entrances of *Epimelissodes aegis* LaBerge (Hymenoptera: Apidae) between cotton rows. An aggregation of nests was found within a cotton field in Quincy, FL. (Photo credit: Isaac L. Esquivel.)

Landscape composition, local-scale crops, and semi-natural habitat configuration can influence pollinator communities, foraging behavior, and pollination services in large-scale agroecosystems. Large, continuous areas of natural and semi-natural habitats are generally more beneficial for supporting diverse pollinator communities (Grab *et al.*, 2019). Fragmentation can lead to the loss of suitable nesting sites, foraging resources, and lack of connectivity between patches, negatively impacting pollinator populations (Rathcke and Jules, 1993; Winfree *et al.*, 2009; Williams *et al.*, 2010). However, this is not always the case. For example, Esquivel *et al.* (2021b) found that different crop interfaces significantly affected the native bee communities in large-scale agroecosystems. Crop–crop interfaces tended to have more consistent species richness and abundance of native bees compared to interfaces of crop–semi-natural habitats. Further, Westphal *et al.* (2003) demonstrated that densities of bumblebees were not related to the proportion of semi-natural habitat within an agricultural landscape but were driven by the availability of mass-flowering crops within the landscape that were highly rewarding to foragers.

Monoculture agricultural landscapes are dominated by a single crop and often largely devoid of natural and semi-natural habitats. These landscapes tend to support fewer pollinator species than diverse landscapes (Wang *et al.*, 2019). Including elements like hedgerows, field margins, wildflower strips, and other natural or semi-natural habitats in agricultural areas can provide additional foraging and nesting opportunities for pollinators (Zhang *et al.*, 2023). Placing strips of native and perennial vegetation at an edge or along contours within a crop field increases water-holding capacity and reduces soil and nutrient losses from cropland while increasing biodiversity across several taxa (Zhang *et al.*, 2023). Prairie strips increase the abundance and diversity of pollinators, as the associated flowering native plants are attractive to pollinators (Schulte *et al.*, 2017). The presence of connected habitats and corridors can facilitate the movement and dispersal of pollinators across the landscape. Connectivity allows pollinators to access various resources and increases gene flow between populations (Sutter *et al.*, 2018). Creating and maintaining habitat corridors, such as native plant hedgerows or linear wildflower strips, can enhance pollinators' landscape connectivity (Sutter *et al.*, 2018; see also Chapter 12, this volume).

The availability of flowers within a specific habitat and the spatial distribution of foraging habitats at the landscape scale affect pollinators' foraging decisions, which in turn, drives the distribution of pollinators in the landscape (Holzschuh *et al.*, 2011; Westphal *et al.*, 2006). This can positively affect crop yields, especially in landscapes with greater biodiversity. Beyer *et al.* (2022) found that increasing covers of fava bean and semi-natural habitats positively influenced bumblebee densities, while at the same time, high oilseed rape covers negatively affected short-tongued bumblebee densities in fava fields. Early mass-flowering crops can change the community composition of pollinators. Diekötter *et al.* (2010) showed that increasing the amount of early-flowering oilseed rape in the landscape decreased the densities of long-tongued bumblebees on late-flowering red clover.

Different species of bees have specific environmental requirements and temperature tolerances (Michener, 2007). Climate change can disrupt pollinators' natural balance and their limited natural floral resources, affecting the suitability and availability of habitats in large-scale agroecosystems (Gérard *et al.*, 2020). Changes in the distribution and abundance of pollinators can have cascading effects on plant–pollinator interactions and ecosystem dynamics. Mismatches in plant–pollinator interactions can occur through reduced co-occurrence between bees and plants in a shared habitat (Menzel *et al.*, 2006). This reduction can be spatial and temporal, driven by a change in the plant's flowering period and the pollinator's phenology (Menzel *et al.*, 2006). In addition to these mismatches, climate change can affect the pollinator's foraging behavior, the plant's attractiveness, and the quality and quantity of plant resources (Gérard *et al.*, 2020). For example, heat stress can decrease the concentration of soluble carbohydrates and proline and alter pollen's phospholipid content (Pressman *et al.*, 2002; Santiago and Sharkey, 2019).

Large-scale agricultural systems often rely on substantial amounts of herbicides and insecticides, which can also affect pollinators. Studies have shown that pesticide exposure and poor nutrition can drastically alter individual bee physiology, behavior, overall colony health, and honey bee survival (Tosi *et al.*, 2017; Schmolke *et al.*, 2021). While many studies have examined honey bee health, research on wild bees is more limited. However, some recent research also focuses on understanding the implications of such interactive effects on wild bees (Hladik *et al.*, 2016). For example, pesticide-exposed blue orchard bees, *Osmia lignaria* (Say) (Hymenoptera: Megachilidae), one of the few wild bees that are managed, showed a dramatically delayed onset of nesting, earlier cessation, and a substantially lower rate of offspring provisioning, resulting in reduced production. Further, females exposed to imidacloprid began nesting 3.6 days later and spent 5.2 fewer days nesting than unexposed control bees, eliminating nearly a week of potential nesting days (Stuligross and Williams, 2020). Other studies indicate that results with wild bees vary widely depending on the pesticide used, the mode of action, the type of exposure (oral or topical), and the species of bee studied (Piovesan *et al.*, 2020).

Management practices in large-scale agricultural systems aimed at mitigating the negative effect of pest species negatively affecting yield can indirectly affect crop yields. Pesticides, in general, are frequently used to control the activities of insect pests and plant pathogens that are detrimental to yield. However, these compounds can also indirectly affect beneficial insects, including natural enemies and pollinators. Pests often exhibit life-history strategies that allow their populations to recover faster than those of beneficial insects (Chagnon *et al.*, 2015). Pesticides are associated with declines of insectivorous birds, pollinators, and natural enemies, ultimately affecting the valuable ecosystem services they provide, like natural pest control and pollination in the long term (Chagnon *et al.*, 2015). As noted above, large-scale agricultural systems possess many interactions and challenges for pollinators, yet some species persist and thrive in such systems. However, large-scale systems like cotton production in Texas may represent a win-win for pollinator stewardship and crop productivity, and such systems may be a valuable learning arena to explore the underlying mechanisms that benefit pollinators and bolster ecosystem services in the form of pollination (Esquivel *et al.*, 2020).

IPPM Systems

IPPM systems are holistic approaches to pest management that aim to minimize the impact of pesticides on pollinators while simultaneously effectively controlling pests (Egan *et al.*, 2020; Lundin *et al.*, 2021). These systems recognize the importance of both pest control using integrated pest management strategies alongside pollinator conservation, seeking to strike a balance between the two (Kogan, 1998; Biddinger and Rajotte, 2015). Recent work has suggested that additional research is needed to fully understand suggested mitigation measures to protect pollinators (Straw and Stanley, 2023).

Regular monitoring and accurate identification of pests are essential for IPPM (Egan *et al.*, 2020). Farmers can make informed decisions about the appropriate timing and methods utilized for pest control interventions by understanding pest populations and their life cycles (Shackelford *et al.*, 2013). IPPM in some crops emphasizes cultural practices that promote healthy plants and reduce pest pressure (Egan *et al.*, 2020). This includes selecting pest-resistant crop varieties, practicing crop rotation, optimizing irrigation and fertilization, and managing plant nutrition (Merle *et al.*, 2022). Biological control is also integral to IPPM and utilizes beneficial organisms, such as predatory insects, parasitoids, and microbial agents, to suppress pest populations (Lundin *et al.*, 2021). By conserving and enhancing natural enemies of pests, farmers can reduce reliance on chemical pesticides (Holland *et al.*, 2017).

When pesticide application is necessary, IPPM focuses on using pesticides selectively, with minimal impacts on pollinators (Leach *et al.*, 2022). This involves choosing pesticides with lower toxicity to non-target organisms, using the least toxic formulations, applying them at the right times to avoid negative impacts, and in appropriate quantities (Phan *et al.*, 2023). Precision application techniques using geospatial technology can also help minimize pesticide exposure to non-target areas (Adan *et al.*, 2021).

IPPM programs also can recognize the importance of providing suitable habitats for pollinators (Penn *et al.*, 2021). Enhancing biodiversity in and around agricultural landscapes can support healthy pollinator populations (Lichtenberg *et al.*, 2017). This often involves creating and preserving natural or semi-natural habitats, such as hedgerows, wildflower strips, or flowering cover crops, that provide food and nesting resources for pollinators and other beneficial biological control organisms (Dolezal *et al.*, 2019; Belien *et al.*, 2021; Garibaldi *et al.*, 2021; Zhang *et al.*, 2023). For example, hedgerows promote pollinator spillover on adjacent strawberry plants, leading to higher pollinator abundance in the field and increased fruit weight and market value (Castle *et al.*, 2019). Further, when strips of native prairie vegetation were planted within crop fields, it increased the diversity and abundance of wild pollinators (Kordbacheh *et al.*, 2020).

Programs focusing on IPPM encourage collaboration among farmers, researchers, extension agents, and other relevant stakeholders—emphasizing the importance of education and training for farmers, agricultural advisors, and other stakeholders (Khanal *et al.*, 2022; Whitehair *et al.*, 2022). Individuals can make informed decisions and implement appropriate practices by increasing awareness and knowledge about pollinator-friendly pest management strategies (Singh, 2012; Schreinemachers *et al.*, 2020). Sharing experiences, best practices, and research findings can improve the implementation and effectiveness of IPPM systems, while monitoring and evaluating the outcomes allow for continuous improvement and adaptation of strategies. These IPPM systems provide a comprehensive framework for sustainable pest management: safeguarding pollinators, promoting biodiversity, and allowing producers to weigh costs and benefits (Flöhr *et al.*, 2020). By adopting these integrated approaches, farmers can minimize pesticide risks, reduce negative impacts on pollinators, enhance ecological resilience, increase economic development, and ensure agricultural systems' long-term productivity and sustainability (Wangithi *et al.*, 2022).

Summary and Conclusions

It is undeniable that pollinators play a crucial role in the production of food and fiber crops that is necessary for human well-being and food security. Pollinators often exist, and some thrive in large and small-scale agricultural systems. However, many biotic and abiotic interactions in agricultural systems across different crops, landscape configurations, climatic conditions, and farm practices affect pollinators with variable outcomes. There are still gaps in the literature, particularly on pollinators in pollinator-independent crops, in terms of yield benefits and nutrition of those crops' resources to bees. Understanding the landscape effects on pollinator communities is crucial for their conservation and management. Implementing landscape-scale conservation strategies, such as promoting habitat restoration, reducing pesticide use, creating floral resources, and enhancing connectivity, can help support diverse and resilient pollinator populations. These efforts are essential not only for the well-being of pollinators but also for maintaining the critical ecosystem services they provide, including pollination of wild plants and agricultural crops.

This chapter touched broadly on landscape effects on pollinators in agricultural systems, differences in pollinator-dependent and pollinator-independent crops, benefits to crops and their potential benefit to bees (managed honey bees and wild bees), non-bee pollinators, and IPPM systems. There is a vast literature base on each topic covered in this chapter. We provide examples of literature that dives into more details for certain topics: review of pollination in pollinator-independent crops (e.g. Esquivel *et al.*, 2021a), review of the use of pollinator-independent crops by bees (e.g. Kral-O'Brien *et al.*, 2022), importance of pollinators in changing landscapes (e.g. Klein *et al.*, 2007), and review of worldwide managed pollinators (Osterman *et al.*, 2021).

Acknowledgements

The findings and conclusions in this publication are those of the author(s) and should not be construed to represent any official USDA or US Government determination or policy. Mention of trade names or commercial products in this publication is solely for the purpose of providing specific information. It does not imply a recommendation or endorsement by the USDA.

References

Adan, M., Abdel-Rahman, E.M., Gachoki, S., Muriithi, B.W., Lattorff, H.M.G. *et al.* (2021) Use of earth observation data to guide the implementation of integrated pest and pollinator management (IPPM) technologies in an avocado production system. *Remote Sensing Applications: Society and Environment* 23, 100566. DOI: 10.1016/j.rsase.2021.100566

Adamidis, G.C., Cartar, R.V., Melathopoulos, A.P., Pernal, S.F. and Hoover, S.E. (2019) Pollinators enhance crop yield and shorten the growing season by modulating plant functional characteristics: a comparison of 23 canola varieties. *Scientific Reports* 9, 14208. DOI: 10.1038/s41598-019-50811-y

Aizen, M.A., Aguiar, S., Biesmeijer, J.C., Garibaldi, L.A., Inouye, D.W. *et al.* (2019) Global agricultural productivity is threatened by increasing pollinator dependence without a parallel increase in crop diversification. *Global Change Biology* 25, 3516–3527. DOI: 10.1111/gcb.14736

Ashman, T., Knight, T.M., Steets, J.A., Amarasekare, P., Burd, M. *et al.* (2004) Pollen limitation of plant reproduction: ecological and evolutionary causes and consequences. *Ecology* 85, 2408–2421. DOI: 10.1890/03-8024

Amro, A.M. (2021) Pollinators and pollination effects on three canola (*Brassica napus* L.) cultivars: a case study in Upper Egypt. *Journal of King Saud University Science* 33, 101240. DOI: 10.1016/j.jksus.2020.101240

Auffray, T., Frérot, B., Poveda, R., Louise, C. and Beaudoin-Ollivier, L. (2017) Diel patterns of activity for insect pollinators of two oil palm species (Arecales: Arecaceae). *Journal of Insect Science* 17, 45. DOI: 10.1093/jisesa/iex018

Ayensu, E.S. (1974) Plant and bat interactions in West Africa. *Annals of the Missouri Botanical Garden* 61, 702–727. DOI: 10.2307/2395025

Baqi, A., Lim, V.-C., Yazid, H., Khan, F.A.A., Lian, C.J. *et al.* (2022) A review of durian plant-bat pollinator interactions. *Journal of Plant Interactions* 17, 105–126. DOI: 10.1080/17429145.2021.2015466

Barbosa, W.F., Smagghe, G. and Guedes, R.N.C. (2015) Pesticides and reduced-risk insecticides, native bees and pantropical stingless bees: pitfalls and perspectives. *Pest Management Science* 71, 1049–1053. DOI: 10.1002/ps.4025

Belien, T., Raymaekers, S., Eeraerts, M., Mommaerts, V., Claus, G. *et al.* (2021) Towards integrated pest and pollinator management in intensive pear cultivation: a case study from Belgium. *Insects* 12, 901. DOI: 10.3390/insects12100901

Berger, L.A., Moffet, J.O. and Rummel, D.R. (1985) Seasonal cycles of *Agapostemon angelicus* Cockerell relative to hybrid cottonseed production in Texas (Hymenoptera: Halictidae). *Journal of the Kansas Entomological Society* 58, 1–8.

Berger, L.A., Vaissiére, B., Moffett, J.O. and Merritt, S.J. (1988) *Bombus* spp. (Hymenoptera: Apidae) as pollinators of male-sterile Upland cotton on the Texas High Plains. *Environmental Entomology* 17, 789–794. DOI: 10.1093/ee/17.5.789

Beyer, N., Gabriel, D. and Westphal, C. (2022) Landscape composition modifies pollinator densities, foraging behavior and yield formation in faba beans. *Basic and Applied Ecology* 61, 30–40. DOI: 10.1016/j.baae.2022.03.002

Biddinger, D.J. and Rajotte, E.G. (2015) Integrated pest and pollinator management – adding a new dimension to an accepted paradigm. *Current Opinion in Insect Science* 10, 204–209. DOI: 10.1016/j.cois.2015.05.012

Bigelow, D.P. and Borchers, A. (2017) *Major Uses of Land in the United States, 2012*. EIB-178, Washington, DC: US Department of Agriculture, Economic Research Service.

Bishop, J., Jones, H.E., Lukac, M. and Potts, S.G. (2016) Insect pollination reduces yield loss following heat stress in faba bean (*Vicia faba* L.). *Agriculture, Ecosystems & Environment* 220, 89–96. DOI: 10.1016/j.agee.2015.12.007

Bishop, J., Garratt, M.P.D. and Breeze, T.D. (2020) Yield benefits of additional pollination to faba bean vary with cultivar, scale, yield parameter and experimental method. *Scientific Reports* 10, 2102. DOI: 10.1038/s41598-020-58518-1

Blettler, D.C., Fagúndez, G.A. and Caviglia, O.P. (2018) Contribution of honeybees to soybean yield. *Apidologie* 49, 101–111. DOI: 10.1007/s13592-017-0532-4

Blanche, R. and Cunningham, S.A. (2005) Rain forest provides pollinating beetles for atemoya crops. *Journal of Economic Entomology* 98, 1193–1201. DOI: 10.1603/0022-0493-98.4.1193

Blanche, K.R., Hughes, M., Ludwig, J.A. and Cunningham, S.A. (2006) Do flower tripping bees enhance yields in peanut varieties grown in north Queensland? *Australian Journal of Experimental Agriculture* 46, 15291534. DOI: 10.1071/EA05190

Blanco, C., Hernandez, G., McGee, J., Corona, M. and Conover, K. (2020) Excluding flying insects during flowering does not affect soybean yield. *Southwestern Entomologist* 45, 381. DOI: 10.3958/059.045.0207

Blitzer, E., Gibbs, J., Park, M. and Danforth, B. (2016) Pollination services for apple are dependent on diverse wild bee communities. *Agriculture Ecosystems & Environment* 221, 1–7. DOI: 10.1016/j.agee.2016.01.004

Bommarco, R., Marini, L. and Vaissière, B.E. (2012) Insect pollination enhances seed yield, quality, and market value in oilseed rape. *Oecologia* 169, 1025–1032. DOI: 10.1007/s00442-012-2271-6

Bosch, J. (1994) *Osmia cornuta* Latr. (Hym., Megachilidae) as a potential pollinator in almond orchards. *Journal of Applied Entomology* 117, 151–157. DOI: 10.1111/j.1439-0418.1994.tb00720.x

Bumrungsri, S., Sripaoraya, E., Chongsiri, T., Sridith, K. and Racey, P. (2009) The pollination ecology of durian (*Durio zibethinus*, Bombacaceae) in southern Thailand. *Journal of Tropical Ecology* 25, 85–92. DOI: 10.1017/S0266467408005531

Buxton, M.N., Anderson, B.J. and Lord, J.M. (2018) The secret service – analysis of the available knowledge on moths as pollinators in New Zealand. *New Zealand Journal of Ecology* 42, 1–9. DOI: 10.20417/nzjecol.42.11

Calderone, N.W. (2012) Insect pollinated crops, insect pollinators and US agriculture: trend analysis of aggregate data for the period 1992–2009. *PLoS ONE* 7, e37235. DOI: 10.1371/journal.pone.0037235

Campbell, J., Stanley-Stahr, C., Bammer, M., Daniels, J. and Ellis, J. (2019) Contribution of bees and other pollinators to watermelon (*Citrullus lanatus* Thunb.) pollination. *Journal of Apicultural Research* 58, 597–603. DOI: 10.1080/00218839.2019.1614271

Carpes, S.T., Cabral, I.S.T., Luz, C.F.P., Capeletti, J.P., Alencar, S.M. *et al.* (2009) Palynological and physico-chemical characterization of *Apis mellifera* L. bee pollen in the Southern region of Brazil. *Journal of Food, Agriculture & Environment* 7, 667–673. DOI: 10.1234/4.2009.2739 (accessed December 1, 2023).

Castle, D., Grass, I. and Westphal, C. (2019) Fruit quantity and quality of strawberries benefit from enhanced pollinator abundance at hedgerows in agricultural landscapes. *Agriculture, Ecosystems & Environment* 275, 14–22. DOI: 10.1016/j.agee.2019.01.003

Chagnon, M., Kreutzweiser, D., Mitchell, E.A., Morrissey, C.A., Noome, D.A. *et al.* (2015) Risks of large-scale use of systemic insecticides to ecosystem functioning and services. *Environmental Science and Pollution Research* 22, 119–134. DOI: 10.1007/s11356-014-3277-x

Chen, G., Liu, C.-Q. and Sun, W. (2016) Pollination and seed dispersal of *Aquilaria sinensis* (Lour.) Gilg (Thymelaeaceae): an economic plant species with extremely small populations in China. *Plant Diversity* 38, 227–232. DOI: 10.1016/j.pld.2016.09.006

Claudino, R.M., Antonini, Y., Martins, C., Beirão, M.V., Braga, É.M. and de Azevedo, C.S. (2022) Is bigger always better? Neither body size nor aggressive behavior predicts specialization of hummingbirds in a rocky outcrop. *Journal of Zoology* 317, 68–76. DOI: 10.1111/jzo.12961

Cooley, H. and Vallejo-Marín, M. (2021) Buzz-pollinated crops: a global review and meta-analysis of the effects of supplemental bee pollination in tomato. *Journal of Economic Entomology* 114, 505–519. DOI: 10.1093/jee/toab009

Cook, D.F., Voss, S.C., Finch, J.T.D., Rader, R.C., Cook, J.M. and Spurr, C.J. (2020) The role of flies as pollinators of horticultural crops: an Australian case study with worldwide relevance. *Insects* 11, 341. DOI: 10.3390/insects11060341

Cranshaw, W. (2004) Hornworms/sphinx moths. In: *Garden Insects of North America: The Ultimate Guide to Backyard Bugs*. Princeton, NJ: Princeton University Press, pp. 146–149.

Cronk, Q. and Ojeda, I. (2008) Bird-pollinated flowers in an evolutionary and molecular context. *Journal of Experimental Botany* 59, 715–727. DOI: 10.1093/jxb/ern009

Cumming, G.S., Buerkert, A., Hoffmann, E.M., Schlecht, E., von Cramon-Taubadel, S. *et al.* (2014) Implications of agricultural transitions and urbanization for ecosystem services. *Nature* 515, 50–57. DOI: 10.1038/nature13945

Cunningham, S.A. and Le Feuvre, D. (2013) Significant yield benefits from honeybee pollination of faba bean (*Vicia faba*) assessed at field scale, *Field Crops Research* 149, 269–275. DOI: 10.1016/j.fcr.2013.05.019

Cunningham-Minnick, M.J., Peters, V.E. and Crist, T.O. (2019) Nesting habitat enhancement for wild bees within soybean fields increases crop production. *Apidologie* 50, 833–844. DOI: 10.1007/s13592-019-00691-y

Cusser, S., Neff, J.L. and Jha, S. (2016) Natural land cover drives pollinator abundance and richness, leading to reductions in pollen limitation in cotton agroecosystems. *Agriculture, Ecosystems & Environment* 226, 33–42. DOI: 10.1016/j.agee.2016.04.020

Deguines, N., Jono, C., Baude, M., Henry, M., Julliard, R. and Fontaine, C. (2014) Large-scale trade-off between agricultural intensification and crop pollination services. *Frontiers in Ecology and the Environment* 12, 212–217. DOI: 10.1890/130054

De Luca, P.A. and Vallejo-Marín, M. (2013) What's the 'buzz' about? The ecology and evolutionary significance of buzz-pollination. *Current Opinion in Plant Science* 16, 429–435. DOI: 10.1016/j.pbi.2013.05.002

De Marco, P. and Coelho, F.M. (2004) Services performed by the ecosystem: forest remnants influence agricultural cultures' pollination and production. *Biodiversity Conservation* 13 1245–1255. DOI: 10.1023/B:BIOC.0000019402.51193.e8

de O. Milfont, M., Rocha, E.E.M., Lima, A.O.N. and Freitas, B.M. (2013) Higher soybean production using honeybee and wild pollinators, a sustainable alternative to pesticides and autopollination. *Environmental Chemistry Letters* 11, 335–341. DOI: 10.1007/s10311-013-0412-8

Deutsch, K.R., Graham, J.R., Boncristiana, H.F., Bustamente, T., Mortensen, A.N. *et al.* (2023) Widespread distribution of honey bee-associated pathogens in native bees and wasps: trends in pathogen prevalence and co-occurrence. *Journal of Invertebrate Pathology* 200, 107973. DOI: 10.1016/j.jip.2023.107973

Diekötter, T., Kadoya, T., Wolters, V. and Jauker, F. (2010) Oilseed rape crops distort plant–pollinator interactions. *Journal of Applied Ecology* 47, 209–214. DOI: 10.1111/j.1365-2664.2009.01759.x

Diniz, U.M., Lima, S.A. and Machado, I.C.S. (2019) Short-distance pollen dispersal by bats in an urban setting: monitoring the movement of a vertebrate pollinator through fluorescent dyes. *Urban Ecosystems 22*, 281–291. DOI: 10.1007/s11252-019-0825-7

Dolezal, A.G., St. Clair, A.L., Zhang, G., Toth, A.L. and O'Neal, M.E. (2019) Native habitat mitigates feast–famine conditions faced by honey bees in an agricultural landscape. *Proceedings of the National Academy of Sciences* 116, 25147–25155. DOI: 10.1073/pnas.1912801116

Dunn, L., Lequerica, M., Reid, C.R. and Latty, T. (2020) Dual ecosystem services of syrphid flies (Diptera: Syrphidae): pollinators and biological control agents. *Pest Management Science* 76, 1973–1979. DOI: 10.1002/ps.5807

Egan, P.A., Dicks, L.V., Hokkanen, H.M. and Stenberg, J.A. (2020) Delivering integrated pest and pollinator management (IPPM). *Trends in Plant Science* 25, 577–589. DOI: 10.1016/j.tplants.2020.01.006

Eischen, F.A., Underwood, B.A. and Collins, A.M. (1994) The effect of delaying pollination on cantaloupe production. *Journal of Apicultural Research* 33, 180–184. DOI: 10.1080/00218839.1994.11100868

Esquivel, I.L., Coulson, R.N. and Brewer, M.J. (2020) A native bee, *Melissodes tepaneca* (Hymenoptera: Apidae) benefits cotton production. *Insects* 11, 487. DOI: 10.3390/insects11080487

Esquivel, I.L., Parys, K.A. and Brewer, M.J. (2021a) Pollination by non-*Apis* bees and potential benefits in self-pollinating crops. *Annals of the Entomological Society of America* 114, 257–266. DOI: 10.1093/aesa/saaa059

Esquivel, I.L., Parys, K.A., Wright, K.W., Eubanks, M.D., Oswald, J.D. *et al.* (2021b) Crop and semi-natural habitat configuration affects diversity and abundance of native bees (Hymenoptera: Anthophila) in a large-field cotton agroecosystem. *Insects* 12, 601. DOI: 10.3390/insects12070601

Fahrig, L., Baudry, J., Brotons, L., Burel, F.G., Crist, T.O. *et al.* (2011) Functional landscape heterogeneity and animal biodiversity in agricultural landscapes. *Ecology Letters* 14, 101–112. DOI: 10.1111/j.1461-0248.2010.01559.x

Finch, J., Gilpin, A.-M. and Cook, J. (2023) Fishing for flies: testing the efficacy of "stink stations" for promoting blow flies as pollinators in mango orchards. *Journal of Pollination Ecology* 32, 79–100. DOI: 10.26786/1920-7603(2023)711

Flöhr, A., Stenberg, J.A. and Egan, P.A. (2020) The joint economic impact level (jEIL): a decision metric for integrated pest and pollinator management. In: Gao, Y., Hokkanen, H. and Menzler-Hokkanen, I. (eds) *Integrative Biological Control (Progress in Biological Control 20)*. Cham, Switzerland: Springer, pp. 17–38. DOI: 10.1007/978-3-030-44838-7_2

Fowler, J. (2016) Specialist bees of the Northeast: host plants and habitat conservation. *Northeastern Naturalist* 23, 305–320. DOI: 10.1656/045.023.0210

Freitas, B.M., Imperatriz-Fonseca, V.L., Medina, L.M., Klienert, A.M.P., Galetto, L. *et al.* (2009) Diversity, threats and conservation of native bees in the neotropics. *Apidologie* 40, 332–3476. DOI: 10.1051/apido/2009012

Garibaldi, L.A., Steffan-Dewenter, I., Kremen, C., Morales, J.M., Bommarco, R. *et al.* (2011) Stability of pollination services decreases with isolation from natural areas despite honey bee visits. *Ecology Letters* 14, 1062–1072. DOI: 10.1111/j.1461-0248.2011.01669.x

Garibaldi, L.A., Steffan-Dewenter, I., Winfree, R., Aizen, M.A., Bommarco, R. *et al.* (2013) Wild pollinators enhance fruit set of crops regardless of honey bee abundance. *Science* 339, 1608–1611. DOI: 10.1126/science.1230200

Garibaldi, L.A., Schulte, L.A., Jodar, D.N.N., Carella, D.S.G. and Kremen, C. (2021) Time to integrate pollinator science into soybean production. *Trends in Ecology & Evolution* 36, 573–575. DOI: 10.1016/j.tree.2021.03.013

Ghazoul, J. (2006) Floral diversity and the facilitation of pollination. *Journal of Ecology* 94, 295–304. DOI: 10.1111/j.1365-2745.2006.01098.x

Gérard, M., Vanderplanck, M., Wood, T. and Michez, D. (2020) Global warming and plant–pollinator mismatches. *Emerging Topics in Life Sciences* 4, 77–86. DOI: 10.1042/ETLS20190139

Geslin, B., Aizen, M., Garcia, N., Pereira, A., Vaissiere, B. and Garibaldi, L. (2017) The impact of honey bee colony quality on crop yield and farmers' profit in apples and pears. *Agriculture, Ecosystems & Environment* 248, 153–161. DOI: 10.1016/j.agee.2017.07.035

Giannini, T.C., Boff, S., Cordeiro, G.D., Cartolano, E.A., Veiga, A.K. *et al.* (2015) Crop pollinators in Brazil: a review of reported interactions. *Apidologie* 46, 209–223. DOI: 10.1007/s13592-014-0316-z

Gibbs, J., Elle, E., Bobiwash, K., Haapalainen, T. and Isaacs, R. (2016) Contrasting pollinators and pollination in native and non-native regions of highbush blueberry production. *PLoS ONE* 11, e0158937. DOI: 10.1371/journal.pone.0158937

Gill, K.A. and O'Neal, M.E. (2015) Survey of soybean insect pollinators: community identification and sampling method analysis. *Environmental Entomology* 44, 488–498. DOI: 10.1093/ee/nvv001

Grab, H., Branstetter, M., Amon, N., Urban-Mead, K.R., Park, M.G. *et al.* (2019) Agriculturally dominated landscapes reduce bee phylogenetic diversity and pollination services. *Science* 363, 282–284. DOI: 10.1126/science.aat6016

Greenleaf, S.S., Williams, N.M., Winfree, R. and Kremen, C. (2007) Bee foraging ranges and their relationship to body size. *Oecologia* 153, 589–596. DOI: 10.1007/s00442-007-0752-9

Hahn, M. and Brühl, C.A. (2016) The secret pollinators: an overview of moth pollination with a focus on Europe and North America. *Arthropod-Plant Interactions* 10, 21–28. DOI: 10.1007/s11829-016-9414-3

Hegland, S.J. and Boeke, L. (2006) Relationships between the density and diversity of floral resources and flower visitor activity in a temperate grassland community. *Ecological Entomology* 31, 532–538. DOI: 10.1111/j.1365-2311.2006.00812.x

Hladik, M.L., Vandever, M. and Smalling, K.L. (2016) Exposure of native bees foraging in an agricultural landscape to current-use pesticides. *Science of The Total Environment* 542, 469–477. DOI: 10.1016/j.scitotenv.2015.10.077

Holland, J.M., Douma, J.C., Crowley, L., James, L., Kor, L. *et al.* (2017) Semi-natural habitats support biological control, pollination and soil conservation in Europe. A review. *Agronomy for Sustainable Development* 37, 31. DOI: 10.1007/s13593-017-0434-x

Holzschuh, A., Dormann, C.F., Tscharntke, T. and Steffan-Dewenter, I. (2011) Expansion of mass-flowering crops leads to transient pollinator dilution and reduced wild plant pollination. *Proceedings of the Royal Society B: Biological Sciences* 278, 3444–3451. DOI: 10.1098/rspb.2011.0268

Horth, L. and Campbell, L.A. (2018) Supplementing small farms with native mason bees increases strawberry size and growth rate. *Journal of Applied Ecology* 55, 591–599. DOI: 10.1111/1365-2664.12988

Hristov, P., Shumkova, R., Palova, N. and Neov, B. (2020) Factors associated with honey bee colony losses: a mini-review. *Veterinary Sciences* 7, 166. DOI: 10.3390/vetsci7040166

Isaacs, R. and Kirk, A. (2010) Pollination services provided to small and large highbush blueberry fields by wild and managed bees. *Journal of Applied Ecology* 47, 841–849. DOI: 10.1111/j.1365-2664.2010.01823.x

Jauker, F. and Wolters, V. (2008) Hover flies are efficient pollinators of oilseed rape *Oecologia* 156, 819–823. DOI: 10.1007/s00442-008-1034-x

Kiew, R. (1987) Notes on the natural history of the Johore Banana, *Musa gracilis* Holttum. *Malayan Nature Journal* 41, 239–248.

Khanal, B., Schoengold, K., Mieno, T. and Schulte Moore, L. (2022) The impact of policy design on willingness to pay for ecosystem services from prairie strips. *Journal of the Agricultural and Applied Economics Association* 1, 352–369. DOI: 10.1002/jaa2.33

Klein, A.M., Steffan-Dewenter, I. and Tscharntke, T. (2003) Bee pollination and fruit set of *Coffea arabica* and *C. canephora* (Rubiaceae). *American Journal of Botany* 90, 153–157. DOI: 10.3732/ajb.90.1.153

Klein, A.M., Vaissière, B.E., Cane, J.H., Steffan-Dewenter, I., Cunningham, S.A. *et al.* (2007) Importance of pollinators in changing landscapes for world crops. *Proceedings of the Royal Society B: Biological Sciences* 274, 303–313. DOI: 10.1098/rspb.2006.3721

Klein, A.M., Hendrix, S.D., Clough, Y., Scofield, A. and Kremen, C. (2015) Interacting effects of pollination, water and nutrients on fruit tree performance. *Plant Biology* 17, 201–208. DOI: 10.1111/plb.12180

Kogan, M. (1998) Integrated pest management: historical perspectives and contemporary developments. *Annual Review of Entomology* 43, 243–270. DOI: 10.1146/annurev.ento.43.1.243

Kordbacheh, F., Liebman, M. and Harris, M. (2020) Strips of prairie vegetation placed within row crops can sustain native bee communities. *PLoS ONE* 15, e0240354. DOI: 10.1371/journal.pone.0240354

Kral-O'Brien, K.C., Hovick, T.J. and Harmon, J.P. (2022) Quid pro quo? A review on bee utilization of pollinator-independent crops. *Annals of the Entomological Society of America* 115, 1–9. DOI: 10.1093/aesa/saab029

Kremen, C., Williams, N.M. and Thorp, R.W. (2002) Crop pollination from native bees at risk from agricultural intensification. *Proceedings of the National Academy of Sciences* 99, 16812–16816. DOI: 10.1073/pnas.262413599

Kruess, A. and Tscharntke, T. (2002) Contrasting responses of plant and insect diversity to variation in grazing intensity. *Biological Conservation* 106, 293–302. DOI: 10.1016/S0006-3207(01)00255-5

Landaverde-González, P., Quezada-Euán, J.J.G., Theodorou, P., Murray, T.E., Husemann, M. *et al.* (2017) Sweat bees on hot chillies: provision of pollination services by native bees in traditional slash-and-burn agriculture in the Yucatán Peninsula of tropical Mexico. *Journal of Applied Ecology* 54, 1814–1824. DOI: 10.1111/1365-2664.12860

Larson, B.M.H., Kevan, P.G. and Inouye, D.W. (2001) Flies and flowers: taxonomic diversity of anthophiles and pollinators. *Canadian Entomologist* 133, 439–465. DOI: 10.4039/Ent133439-4

Lau, P., Bryant, V. and Rangel, J. (2018) Determining the minimum number of pollen grains needed for accurate honey bee (*Apis mellifera*) colony pollen pellet analysis. *Palynology* 42, 36–42. DOI: 10.1080/01916122.2017.1306810

Lau, P.W., Esquivel, I.L., Parys, K.A., Hung, K.-L.J. and Chakrabarti, P.B. (2023) The nutritional landscape in agroecosystems: a review on how resources and management practices can shape pollinator health in agricultural environments. *Annals of the Entomological Society of America* 116, 261–275. DOI: 10.1093/aesa/saad023

Layek, U., Manna, S.S. and Karmaker, P. (2020) Pollen foraging behaviour of honey bee (*Apis mellifera* L.) in southern West Bengal, India. *Palynology* 44, 114–126. DOI: 10.1080/01916122.2018.1533898

Le Féon, V., Poggio, S.L., Torretta, J.P., Bertrand, C., Molina, G.A. *et al.* (2016) Diversity and life-history traits of wild bees (Insecta: Hymenoptera) in intensive agricultural landscapes in the Rolling Pampa, Argentina. *Journal of Natural History* 50, 1175–1196. DOI: 10.1080/00222933.2015.1113315

Leach, A., Pecenka, J. and Kaplan, I. (2022) Does IPPM bear fruit? Evaluating reduced-risk insecticide programmes on pests, pollinators and marketable yield. *Journal of Applied Ecology* 59, 2993–3002. DOI: 10.1111/1365-2664.14294

Leimberger, K.G., Dalsgaard, B., Tobias, J.A., Wolf, C. and Betts, M.G. (2022) The evolution, ecology, and conservation of hummingbirds and their interactions with flowering plants. *Biological Reviews* 97, 923–959. DOI: 10.1111/brv.12828

Lichtenberg, E.M., Kennedy, C.M., Kremen, C., Batáry, P., Berendse, F. *et al.* (2017) A global synthesis of the effects of diversified farming systems on arthropod diversity within fields and across agricultural landscapes. *Global Change Biology* 23, 4946–4957. DOI: 10.1111/gcb.13714

Linsley, E.G., MacSwain, J.W. and Smith, R.F. (1952) Outline for ecological life histories of solitary and semi-social bees. *Ecology* 33, 558–567. DOI: 10.2307/1931531

Lu, Q.-B., Liu, C.-Q. and Huang, S.-X. (2021) Moths pollinate four crops of Cucurbitaceae in Asia. *Journal of Applied Entomology* 145, 499–507. DOI: 10.1111/jen.12871

Lundin, O., Rundlöf, M., Jonsson, M., Bonnarco, R. and Williams, N.M. (2021) Integrated pest and pollinator management – expanding the concept. *Frontiers in Ecology and the Environment* 19, 283–291. DOI: 10.1002/fee.2325

Macgregor, C.J. and Scott-Brown, A.S. (2020) Nocturnal pollination: an overlooked ecosystem service vulnerable to environmental change. *Emerging Topics in Life Sciences* 4, 19. DOI: 10.1042/ETLS20190134

Martin, C.D., Fountain, M.T. and Brown, M.J.F. (2019) Varietal and seasonal differences in the effects of commercial bumblebees on fruit quality in strawberry crops. *Agriculture, Ecosystems & Environment* 281, 124–133. DOI: 10.1016/j.agee.2019.04.007

Martins, D.J. and Johnson, S.D. (2009) Distance and quality of natural habitat influence hawkmoth pollination of cultivated papaya. *International Journal of Tropical Insect Science* 29, 114–123. DOI: 10.1017/S1742758409990208

Mayes, D.M. and Petrillo, H. (2017) Cotton flower-visiting insects in small-scale farm fields in Mwachisompola, Zambia. *Journal of the Kansas Entomological Society* 90, 122–130. DOI: 10.2317/JKES1703.1

McAulay, M.K. and Forrest, J.R.K. (2019) How do sunflower pollen mixtures affect survival of queenless microcolonies of bumblebees (*Bombus impatiens*)? *Arthropod-Plant Interactions* 13, 517–529. DOI: 10.1007/s11829-018-9664-3

Menzel, A., Sparks, T.H., Estrella, N., Koch, E., Aasa, A. *et al.* (2006) European phenological response to climate change matches the warming pattern. *Global Change Biology* 12, 1969–1976. DOI: 10.1111/j.1365-2486.2006.01193.x

Merle, I., Hipólito, J. and Requier, F. (2022) Towards integrated pest and pollinator management in tropical crops. *Current Opinions in Insect Science* 50, 100866. DOI: 10.1016/j.cois.2021.12.006

Michener, C.D. (2007) *The Bees of the World*. Baltimore, MD: John Hopkins University Press.

Michener, C.D., Winston, M.L. and Jander, R. (1978) Pollen manipulation and related activities and structures in bees of the family Apidae. *University of Kansas Science Bulletin* 51, 575–601. DOI: 10.5962/bhl.part.17249

Moffett, J.O., Stith, L.S., Burkhardt, C.C. and Shipman, C.W. (1976a) Fluctuation of wild bee and wasp visits to cotton flowers. *Arizona Academy of Science* 11, 64–68.

Moffett, J.O., Stith, L.S., Burkhardt, C.C. and Shipman, C.W. (1976b) Insect visitors to cotton flowers. *Arizona Academy of Science* 11, 47–48.

Morandin, L.A. and Winston, M.L. (2005) Wild bee abundance and seed production in conventional, organic, and genetically modified canola. *Ecological Applications* 15, 871–881. DOI: 10.1890/03-5271

Molina-Freaner, F. and Eguiarte, L.E. (2003) The pollination biology of two paniculate agaves (Agavaceae) from northwestern Mexico: contrasting roles of bats as pollinators. *American Journal of Botany* 90, 1016–1024. DOI: 10.3732/ajb.90.7.1016

Mussen, E.C. and Thorp, R.W. (1997) *Honey Bee Pollination of Cantaloupe, Cucumber, & Watermelon*. Publication 7224. Berkeley, CA: University of California Agriculture and Natural Resources. DOI: 10.3733/ucanr.7224

Nun, N. (1976) Studies on pollination in Musaceae. *Annals of Botany* 40, 167–177. DOI: 10.1093/oxfordjournals.aob.a085120

O'Brien, P., Kral-O'Brien, K. and Hatfield, J.L. (2021) Agronomic approach to understanding climate change and food security. *Agronomy Journal* 113, 4616–4626. DOI: 10.1002/agj2.20693

Ollerton, J., Winfree, R. and Tarrant, S. (2011) How many flowering plants are pollinated by animals? *Oikos* 120, 321–326. DOI: 10.1111/j.1600-0706.2010.18644.x

Osterman, J., Aizen, M.A., Biesmeijer, J.C., Bosch, J., Howlett, B.G. *et al.* (2021) Global trends in the number and diversity of managed pollinator species. *Agriculture, Ecosystems & Environment* 322, 107653. DOI: 10.1016/j.agee.2021.107653

Papa, G., Maier, R., Durazzo, A., Lucarini, M., Karabagias, I.K. *et al.* (2022) The honey bee *Apis mellifera*: an insect at the interface between human and ecosystem health. *Biology* 11, 233. DOI: 10.3390/biology11020233

Parys, K.A., Esquivel, I.L., Wright, K.W., Griswold, T. and Brewer, M.J. (2020) Native pollinators (Hymenoptera: Anthophila) in cotton grown in the Gulf South, United States. *Agronomy* 10, 698. DOI: 10.3390/agronomy10050698

Paz, F.S., Pinto, C.E., de Brito, R.M., Imperatriz-Fonseca, V.L. and Giannini, T.C. (2021) Edible fruit plant species in the Amazon forest rely mostly on bees and beetles as pollinators. *Journal of Economic Entomology* 114, 710–722. DOI: 10.1093/jee/toaa284

Penn, H.J., Penn, J., Cunningham-Minnick, M. and Hu, W. (2021) Kentucky soybean farmers' supportiveness of two integrated pest and pollinator management tactics. *Journal of Integrated Pest Management* 12, 12. DOI: 10.1093/jipm/pmab005

Penn, J., Hu, W. and Penn, H.J. (2019) Support for solitary bee conservation among the public versus beekeepers. *American Journal of Agricultural Economics* 101, 1386–1400. DOI: 10.1093/ajae/aaz050

Phan, N.T., Rajotte, E.G., Smagghe, G., Ren, X.-Z., Biddinger, D.B. and Joshi, N.K. (2023) Agricultural pesticide regulatory environment for pollinator protection across geographical regions. *Frontiers in Sustainable Food Systems* 7, 1241601. DOI: 10.3389/fsufs.2023.1241601

Piovesan, B., Padilha, A.C., Morais, M.C., Botton, M., Grutzmacher, A.D. and Zotti, M.J. (2020) Effects of insecticides used in strawberries on stingless bees *Melipona quadrifasciata* and *Tetragonisca fiebrigi* (Hymenoptera: Apidae). *Environmental Science and Pollution Research* 27, 42472–42480. DOI: 10.1007/s11356-020-10191-7

Pires, V.C., Silveira, F.A., Sujii, E.R., Torezani, K.R., Rodrigues, W.A. *et al.* (2014) Importance of bee pollination for cotton production in conventional and organic farms in Brazil. *Journal of Pollination Ecology* 13, 151–160. DOI: 10.26786/1920-7603(2014)20

Pitts-Singer, T.L. and Cane, J.H. (2011) The alfalfa leafcutting bee, *Megachile rotundata*: the world's most intensively managed solitary bee. *Annual Review of Entomology* 56, 221–237. DOI: 10.1146/annurev-ento-120709-144836

Pitts-Singer, T.L., Artz, D.R., Peterson, S.S., Boyle, N.K. and Wardell, G.I. (2018) Examination of a managed pollinator strategy for almond production using *Apis mellifera* (Hymenoptera: Apidae) and *Osmia lignaria* (Hymenoptera: Megachilidae). *Environmental Entomology* 47, 364–377. DOI: 10.1093/ee/nvy009

Potts, S.G., Vulliamy, B., Roberts, S., O'Toole, C., Dafni, A. *et al.* (2005) Role of nesting resources in organising diverse bee communities in a Mediterranean landscape. *Ecological Entomology* 30, 78–85. DOI: 10.1111/j.0307-6946.2005.00662.x

Pressman, E., Peet, M.M. and Pharr, D.M. (2002) The effect of heat stress on tomato pollen characteristics is associated with changes in carbohydrate concentration in the developing anthers. *Annals of Botany* 90, 631–636. DOI: 10.1093/aob/mcf240

Rader, R., Edwards, W., Westcott, D.A., Cunningham, S.A. and Howlett, B.G. (2013) Diurnal effectiveness of pollination by bees and flies in agricultural *Brassica rapa*: implications for ecosystem resilience. *Basic and Applied Ecology* 14, 20–27. DOI: 10.1016/j.baae.2012.10.011

Rader, R., Reilly, J., Bartomeus, I. and Winfree, R. (2013) Native bees buffer the negative impact of climate warming on honey bee pollination of watermelon crops. *Global Change Biology* 19, 3103–3110. DOI: 10.1111/gcb.12264

Rader, R., Bartomeus, I., Garibaldi, L.A., Garratt, M.P., Howlett, B.G. *et al.* (2016) Non-bee insects are important contributors to global crop pollination. *Proceedings of the National Academy of Sciences* 113, 146–151. DOI: 10.1073/pnas.1517092112

Rathcke, B.J. and Jules, E.S. (1993) Habitat fragmentation and plant–pollinator interactions. *Current Science* 65, 273–277.

Regan, E.C., Santini, L., Ingwall-King, L., Hoffmann, M., Rondinini, C. *et al.* (2015) Global trends in the status of bird and mammal pollinators. *Conservation Letters* 8, 397–403. DOI: 10.1111/conl.12162

Reilly, J.R., Artz, D.R., Biddinger, D., Bobiwash, K., Boyle *et al.* (2020) Crop production in the USA is frequently limited by a lack of pollinators. *Proceedings of the Royal Society B: Biological Sciences* 287, 20200922. DOI: 10.1098/rspb.2020.0922

Requier, F., Pérez-Méndez, N., Andersson, G.K., Blareau, E., Merle, I. *et al.* (2023) Bee and non-bee pollinator importance for local food security. *Trends in Ecology & Evolution* 38, 196–205. DOI: 10.1016/j.tree.2022.10.006

Ricketts, T.H., Daily, G.C., Ehrlich, P.R. and Michener, C.D. (2004) Economic value of tropical forest to coffee production. *Proceedings of the National Academy of Sciences* 101, 12579–12582. DOI: 10.1073/pnas.0405147101

Ricketts, T.H., Regetz, J., Steffan-Dewenter, I., Cunningham, S.A., Kremen, C. *et al.* (2008) Landscape effects on crop pollination services: are there general patterns? *Ecology Letters* 11, 499–515. DOI: 10.1111/j.1461-0248.2008.01157.x

Roberto-Emiliano, T.-S., Luis, E.E., David, S.-P. and Rodrigo, A.M. (2016) Save our bats, save our tequila: industry and science join forces to help bats and *agaves*. *Natural Areas Journal* 36, 523–530. DOI: 10.3375/043.036.0417

Saeed, S., Naqqash, M.N., Jaleel, W., Saeed, Q. and Ghouri, F. (2016) The effect of blow flies (Diptera: Calliphoridae) on the size and weight of mangos (*Mangifera indica* L.). *PeerJ* 4, e2076. DOI: 10.7717/peerj.2076

Sáez, A., Aizen, M.A., Medici, S., Viel, M., Villalobos, E. and Negri, P. (2020) Bees increase crop yield in an alleged pollinator-independent almond variety. *Scientific Reports* 10, 3177. DOI: 10.1038/s41598-020-59995-0

Santiago, J.P. and Sharkey, T.D. (2019) Pollen development at high temperature and role of carbon and nitrogen metabolites. *Plant, Cell & Environment* 42, 2759–2775. DOI: 10.1111/pce.13576

Santos, A.O.R., Bartelli, B.F. and Nogueira-Ferreira, F.H. (2014) Potential pollinators of tomato, *Lycopersicon esculentum* (Solanaceae) in open crops and the effect of a solitary bee in fruit set and quality. *Journal of Economic Entomology* 107, 987–994. DOI: 10.1603/EC13378

Santos, R.D.S., de O. Milfont, M., Silva, M.M., Carneiro, L.T. and Castro, C.C. (2020) Butterflies provide pollination services to macadamia in northeastern Brazil. *Scientia Horticulturae* 259, 108818. DOI: 10.1016/j.scienta.2019.108818

Schmolke, A., Galic, N., Feken, M., Thompson, H., Sgolastra, F. *et al.* (2021) Assessment of the vulnerability to pesticide exposures across bee species. *Environmental Toxicology and Chemistry* 40, 2640–2651. DOI: 10.1002/etc.5150

Schreinemachers, P., Grovermann, C., Praneetvatakul, S., Heng, P., Nguyen, T.T.L. *et al.* (2020) How much is too much? Quantifying pesticide overuse in vegetable production in Southeast Asia. *Journal of Cleaner Production* 244, 118738. DOI: 10.1016/J.JCLEPRO.2019.118738

Schulte, L.A., Niemi, J., Helmers, M.J., Liebman, M., Arbuckle, J.G. *et al.* (2017) Prairie strips improve biodiversity and the delivery of multiple ecosystem services from corn–soybean croplands. *Proceedings of the National Academy of Sciences* 114, 11247–11252. DOI: 10.1073/pnas.1620229114

Seeley, T.D. and Morse, R.A. (1976) The nest of the honey bee (*Apis mellifera* L.). *Insects Sociaux* 23, 495–512. DOI: 10.1007/BF02223477

Senapathi, D., Fründ, J., Albrecht, M., Garratt, M.P., Kleijn, D. *et al.* (2021) Wild insect diversity increases inter-annual stability in global crop pollinator communities. *Proceedings of the Royal Society B: Biological Sciences* 288, 20210212. DOI: 10.1098/rspb.2021.0212

Shackelford, G., Steward, P.R., Benton, T.G., Kunin, W.E., Potts, S.G. *et al.* (2013) Comparison of pollinators and natural enemies: a meta-analysis of landscape and local effects on abundance and richness in crops. *Biological Reviews* 88, 1002–1021. DOI: 10.1111/brv.12040

Sharma, M.V. and Armstrong, J.E. (2013) Pollination of *Myristica* and other nutmegs in natural populations. *Tropical Conservation Science* 6, 595–607. DOI: 10.1177/194008291300600502

Silliman, M.R., Schürch, R., Malone, S., Taylor, S.V. and Couvillon, M.J. (2022) Row crop fields provide mid-summer forage for honey bees. *Ecology and Evolution* 12, e8979. DOI: 10.1002/ece3.8979

Singh, K.M. (2012) Dangers of pesticide misuse: challenges and strategies. *SSRN* 1989829. DOI: 10.2139/SSRN.1989829

Ssymank, A., Kearns, C.A., Pape, T. and Thompson, F.C. (2008) Pollinating flies (Diptera): a major contribution to plant diversity and agricultural production, *Biodiversity* 9, 86–89. DOI: 10.1080/14888386.2008.9712892

Stavert, J.R., Pattemore, D.E., Bartomeus, I., Gaskett, A.C. and Beggs, J.R. (2018) Exotic flies maintain pollination services as native pollinators decline with agricultural expansion. *Journal of Applied Ecology* 55, 1737–1746. DOI: 10.1111/1365-2664.13103

Steffan-Dewenter, I. and Kuhn, A. (2003) Honeybee foraging in differentially structured landscapes. *Proceedings of the Royal Society B: Biological Sciences* 270, 569–575. DOI: 10.1098/rspb.2002.2292

Stein, K., Coulibaly, D., Stenchly, K., Goetze, D., Porembski, S. *et al.* (2017) Bee pollination increases yield quantity and quality of cash crops in Burkina Faso, West Africa. *Scientific Reports* 7, 17691. DOI: 10.1038/s41598-017-17970-2

Stoner, K.A., Nurse, A., Koethe, R.W., Hatala, M.S. and Lehmann, D.M. (2022) Where does honey bee (*Apis mellifera* L.) pollen come from? A study of pollen collected from colonies at ornamental plant nurseries. *Insects* 13, 744. DOI: 10.3390/insects13080744

Straw, E.A. and Stanley, D.A. (2023) Weak evidence base for bee protective pesticide mitigation measures, *Journal of Economic Entomology* 116, 1604–1612. DOI: 10.1093/jee/toad118

Stuligross, C. and Williams, N.M. (2020) Pesticide and resource stressors additively impair wild bee reproduction. *Proceedings of the Royal Society B: Biological Sciences* 287, 20201390. DOI: 10.1098/rspb.2020.1390

Sutter, L., Albrecht, M. and Jeanneret, P. (2018) Landscape greening and local creation of wildflower strips and hedgerows promote multiple ecosystem services. *Journal of Applied Ecology* 55, 612–620. DOI: 10.1111/1365-2664.12977

Tosi, S., Burgio, G. and Nieh, J.C. (2017) A common neonicotinoid pesticide, thiamethoxam, impairs honey bee flight ability. *Scientific Reports* 7, 1201. DOI: 10.1038/s41598-017-01361-8

Tremlett, C.J., Peh, K.S.H., Zamora-Gutierrez, V. and Schaafsma, M. (2021) Value and benefit distribution of pollination services provided by bats in the production of cactus fruits in central Mexico. *Ecosystem Services* 47, 101197. DOI: 10.1016/j.ecoser.2020.101197

Tremlett, C.J., Moore, M., Chapman, M.A., Zamora-Gutierrez, V. and Peh, K.S.H. (2020) Pollination by bats enhances both quality and yield of a major cash crop in Mexico. *Journal of Applied Ecology* 57, 450–459. DOI: 10.1111/1365-2664.13545

Toni, H.C., Djossa, B.A., Ayenan, M.A.T. and Teka, O. (2021) Tomato (*Solanum lycopersicum*) pollinators and their effect on fruit set and quality. *Journal of Horticultural Science and Biotechnology* 96, 1–13. DOI: 10.1080/14620316.2020.1773937 van Vierssen Trip, N., MacPhail, V.J., Colla, S.R. and Olivastri, B. (2020) Examining the public's awareness of bee (Hymenoptera: Apoidae: Anthophila) conservation in Canada. *Conservation Science and Practice* 2, e293. DOI: 10.1111/csp2.293

Waller, G.D., Vaissiere, B.E., Moffett, J.O. and Martin, J.H. (1985) Comparison of carpenter bees (*Xylocopa varipuncta* Patton) (Hymenoptera: Anthophoridae) and honey bees (*Apis mellifera* L.) (Hymenoptera: Apidae) as pollinators of male-sterile cotton in cages. *Journal of Economic Entomology* 78, 558–561. DOI: 10.1093/jee/78.3.558

Wang, X., Hua, F., Wang, L., Wilcove, D.S. and Yu, D.W. (2019) Thie biodiversity benefit of native forests and mixed-species plantations over monoculture plantations. *Diversity and Distributions* 25, 1721–1735. DOI: 10.1111/ddi.12972

Wangithi, C., Muriithi, B.W., Diiro, G., Dubois, T., Mohamed, S. *et al.* (2022) Synergies of integrated pest and pollinator management in avocado farming in East Africa: an ex-ante economic analysis. *PLoS ONE* 17, e0271241. DOI: 10.1371/journal.pone.0271241

Westphal, C., Steffan-Dewenter, I. and Tscharntke, T. (2003) Mass flowering crops enhance pollinator densities at a landscape scale. *Ecology Letters* 6, 961–965. DOI: 10.1046/j.1461-0248.2003.00523.x

Westphal, C., Steffan-Dewenter, I. and Tscharntke, T (2006) Bumblebees experience landscapes at different spatial scales: possible implications for coexistence. *Oecologia* 149, 289–300. DOI: 10.1007/s00442-006-0448-6

Wheelock, M.J. and O'Neal, M.E. (2016) Insect pollinators in Iowa cornfields: community identification and trapping method analysis. *PLoS ONE* 11, e0143479. DOI: 10.1371/journal.pone.0143479

Wheelock, M.J., Rey, K.P. and O'Neal, M.E. (2016) Defining the insect pollinator community found in Iowa corn and soybean fields: implications for pollinator conservation. *Environmental Entomology* 45, 1099–1106. DOI: 10.1093/ee/nvw087

Whitehair, R., Grudens-Schuck, N. and Schulte, L.A. (2022) Program evaluation of a workshop on prairie strips for farm advisors: framing the co-occurring outcomes of low knowledge acquisition and high confidence. *Horticulturae* 8, 8121215. DOI: 10.3390/horticulturae8121215

Williams, N.M. and Kremen, C. (2007) Resource distributions among habitats determine solitary bee offspring production in a mosaic landscape, *Ecological Applications* 17, 910–921. DOI: 10.1890/06-0269

Williams, N.M., Crone, E.E., Roulston, T.H., Minckley, R.L., Packer, L. *et al.* (2010) Ecological and life-history traits predict bee species responses to environmental disturbances. *Biological Conservation* 143, 2280–2291. DOI: 10.1016/j.biocon.2010.03.024

Wilson, J.S., Forister, M.L. and Carril, O.M. (2017) Interest exceeds understanding of public support of bee conservation. *Frontiers in Ecology and the Environment* 15, 460–466. DOI: 10.1002/fee.1531

Winfree, R. (2010) The conservation and restoration of wild bees. *Annals of the New York Academy of Sciences*, 1195, 169–197. DOI: 10.1111/j.1749-6632.2010.05449.x

Winfree, R., Aguilar, R., Vázquez, D.P., LeBuhn, G. and Aizen, M.A. (2009) A meta-analysis of bees' responses to anthropogenic disturbance. *Ecology* 90, 2068–2076. DOI: 10.1890/08-1245.1

Winfree, R., Bartomeus, I. and Cariveau, D.P. (2011) Native pollinators in anthropogenic habitats. *Annual Review of Ecology, Evolution, and Systematics* 42, 1–22. DOI: 10.1146/annurev-ecolsys-102710-145042

Witter, S., Nunes-Silva, P., Lisboa, B.B., Tirelli, F.P., Sattler, A. *et al.* (2015) Stingless bees as alternative pollinators of canola. *Journal of Economic Entomology* 108, 880–886. DOI: 10.1093/jee/tov096

Yue, J., Yan, Z., Bai, C., Chen, Z., Lin, W. *et al.* (2015) Pollination activity of *Elaeidobius kamerunicus* (Coleoptera: Curculionoidea) on oil palm on Hainan Island. *Florida Entomologist* 98, 499–505.

Yousefi, M., Rafie, A.S.M., Abd Aziz, S. and Azrad, S. (2020) Introduction of current pollination techniques and factors affecting pollination effectiveness by *Elaeidobius kamerunicus* in oil palm plantations on regional and global scale: a review. *South African Journal of Botany* 132, 171–179. DOI: 10.1016/j.sajb.2020.04.017

Zhang, G., Murray, C.J., St. Clair, A.L., Cass, R.P., Dolezal, A.G. *et al.* (2023) Native vegetation embedded in landscapes dominated by corn and soybean improves honey bee health and productivity. *Journal of Applied Ecology* 60, 1032–1043. DOI: 10.1111/1365-2664.14397

12 Honey Beekeeping in Agricultural Landscapes: Addressing the Stresses of Pesticide Exposure and Limited Forage

Matthew E. O'Neal* and Maura Hall
Department of Plant Pathology, Entomology and Microbiology, Iowa State University, Ames, IA, USA

Abstract

Honey beekeeping suffers from stressors encountered at a landscape scale. In this chapter, two of these, lack of forage and pesticide exposure, are reviewed in the context of the United States Corn Belt. Focus is placed on observations and research conducted in this region and specifically in the state of Iowa, that reveal the impact of these stressors as well as methods to mitigate their impacts. This includes a summary of experiments demonstrating the value of government-supported conservation practices implementation at the edges of crop fields. One practice discussed in detail, prairie strips, can improve honey bee productivity despite being embedded in a landscape committed to conventional corn and soybean production. Although this practice comes with a potential for exposure to insecticides, including neonicotinoids, evidence from commercial-scale implementation of this practice suggests the potential impact of this exposure is limited. Efforts to further protect honey bees from exposure to insecticides are summarized, revealing a potential to practice beekeeping more sustainably within landscapes committed to annual crop production.

Introduction

Honey beekeeping in North America, especially the United States (US), is a form of agriculture practiced in multiple landscapes to achieve two major goals: crop pollination and honey production. These goals can be achieved because of the broad host and foraging range of honey bees (Apidae: Hymenoptera) and the capacity to move colonies great distances to crops that either require pollination or to locations considered suitable for honey production. This movement exposes honey bees to a variety of stressors that can negatively affect productivity and health. The impact of these stressors has been observed in declining honey production and poor overwintering success. This is important as honey bees are the most valuable managed pollinator species, pollinating over 150 crops. In the US alone, pollination by honey bees has an annual value of around US$15 billion (Calderone, 2012). Although the global amount of managed honey bees, honey, and beeswax production has increased over a six-decade period

*Corresponding author: oneal@iastate.edu

DOI: 10.1079/9781800622777.0012

(1961–2017) (Phiri *et al.*, 2022), the demand for pollination services by honey bees is increasing faster than the capacity to manage honey bee colonies (Mashilingi *et al.*, 2022).

Honey production in the US has declined during the past 33-year period (1990–2023) (USDA-NASS, 2024) ranging from 28 kg (61.5 lb) per colony in 1990 to 21.4 kg (47 lb) per colony in 2020 (Fig. 12.1). In that same period, the number of honey bee colonies managed in the US has also fluctuated, from a peak of 3.2 million in 1991 to 2.3 million in 2008. Since that nadir, beekeepers have increased colony numbers to 2.66 million by 2022. Although the number of managed colonies has increased, significant year-to-year losses occur. The most recent winter survey (2022–2023) of honey beekeepers in the US estimated 37.4% of their colonies were lost after the winter (Steinhauer *et al.*, 2023). Combined with summer losses, an estimated 48.2% of all managed colonies in the US did not survive a 12-month period. This is remarkable as sustainably managed honey bee colonies could survive indefinitely, if their needs are met.

These losses are attributed to a combination of stressors that contribute to poor health and reduced productivity: pathogens and pests (e.g. viruses, fungi, and Varroa mite), pesticide exposure, and poor forage availability and quality. The latter two stressors are experienced by honey bees across landscapes committed to agriculture. Through observations and research conducted in the US, we explore how these stressors have increased during the past three decades and how they could be mitigated into the future. This chapter will focus primarily on the US Corn Belt, a geographical region that includes 13 states, including the top four producers of corn (Iowa, Illinois, Nebraska, and Minnesota) and soybean (Illinois, Iowa, Minnesota, and Indiana). As of 2022, Corn Belt states were home to 41% of the 2.6 million colonies kept in the US, including three states in the top ten (North Dakota, South Dakota, and Minnesota).

Stressors Impacting Honey Bee Health and Survival

Although honey bees are kept in the Corn Belt, the primary crops grown in this region (corn and soybean) do not require insect pollination to achieve optimal yield. Despite being self-pollinated,

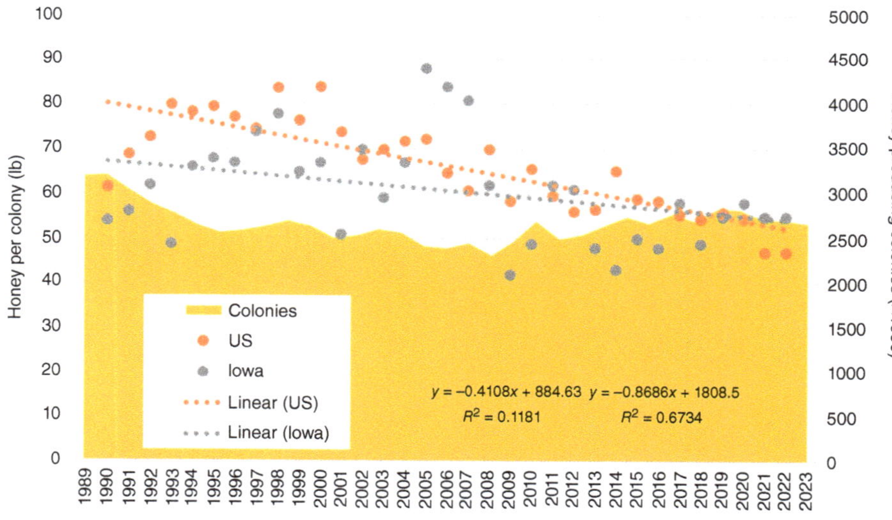

Fig. 12.1. Summary of US Department of Agriculture data of honey production (left *y*-axis) for the entire United States (US) and Iowa, and honey-producing colonies (right *y*-axis) for the entire US over a 33-year period (1990–2023). Linear regression analysis of honey production was conducted for the US (R^2 = 0.67) and Iowa (R^2 = 0.12) (USDA-NASS, 2024). (Original artwork by M.E. O'Neal.)

there is growing evidence that soybean yield is positively affected, with an average yield gain of 21%, when flowers are visited by bees (Garibaldi *et al.*, 2021; see Chapter 11, this volume). As promising as this evidence is, improving soybean production by increasing bee pollination is not a component of soybean production in the Corn Belt. From the perspective of a beekeeper, many parts of the Corn Belt lack sufficient non-cropped areas to provide significant forage when crops are no longer in anthesis (e.g. soybean, clover). A lack of forage is considered a major driver for declining bee diversity in this region (Koh *et al.*, 2016) and as we explore below, contributes to reduced honey bee productivity (Otto *et al.*, 2016).

Insecticides have been documented to cause both acute and chronic adverse effects on honey bees including mortality, foraging impairment, potential increased susceptibility to diseases and viruses, and as well as overall decrease in colony fitness (Pisa *et al.*, 2015). However, the extent to which these effects occur in real-world scenarios is dependent on honey bee direct and indirect insecticide exposure. Insecticides are commonly used in the production of corn and soybeans globally, but their use has decreased in corn in the US with the adoption of genetically modified crops conferring resistance to insects. However, this trend did not occur for soybeans. Most cultivated soybeans in the US are genetically modified to be resistant to herbicides but not resistant to insects. Insecticide use increased on soybeans after the establishment of the invasive soybean aphid, *Aphis glycines* (LeConte) (Aphididae: Hemiptera). Furthermore, seed-applied insecticide use increased in both crops. Below, we explore the response of honey bees to these changes in the US and if the impact of this stressor can be mitigated.

This review will not focus on pathogens and pests that attack honey bees. This topic has recently been reviewed (Ullah *et al.*, 2021; Nekoei *et al.*, 2023), highlighting the various pathogens that negatively impact honey beekeeping. Typically, Varroa mites are included in this category, and considered by beekeepers as the most significant contributor to honey bee health (Cass *et al.*, 2022), despite recommendations for maintaining low mite loads in colonies (HBHC, 2021a). The impact of pathogens and pests is not limited to the health of individual honey bees. For example, Varroa mite infestations can interfere with the capacity of honey bees to use forage at a landscape scale, preventing them from taking advantage of forage that is readily available in a surrounding landscape with more uncultivated features (Dolezal *et al.*, 2016a). However, we are unaware of recommendations for managing Varroa mites depending upon variation in the landscape in which apiaries are kept. Pathogens also interact additively to synergistically with poor nutrition and pesticides (Siviter *et al.*, 2021). For example, mortality from viral infections is reduced when honey bees have access to a higher quality diet (Dolezal *et al.*, 2019a; Zhang *et al.*, 2020). A general hypothesis explaining this interaction notes that a diet comprised of a greater diversity of plants, with greater nutritional quality, increases the immune response of individual honey bees (Dolezal and Toth, 2018). A recent review of the impact of pesticides and viruses on honey bee health indicates a lack of studies that address this complex interaction (Harwood and Dolezal, 2020). For this chapter, we focus on the two main stressors—forage and pesticides—for which a specific link to the surrounding landscape has been documented and recommendations exist for managing them.

Complicating the impact of these stressors on honey bees is the growing evidence that extreme weather events affect honey bee survival. This is a relatively recent area of study, but highly relevant given the growing body of evidence of global climate change on invertebrates, plants, and agriculture in general. Summer temperatures and precipitation have both been observed to explain variation in honey bee overwintering success and colony size during the growing season (Calovi *et al.*, 2021; Insolia *et al.*, 2022; Quinlan *et al.*, 2023). These facets of weather are anticipated to become more extreme in the future as the global climate changes. To what extent the combined impact of land and pesticide use on honey bees will be affected by a changing climate is unclear, but it is likely to be harmful in aggregate.

In regions like the Corn Belt, where commodity-scale agriculture is practiced, a lack of forage and pesticide exposure are experienced by honey bees at a landscape scale. Recent examples reveal how pesticide use and forage availability affect honey bees in the state of Iowa where corn and soybean comprise more than 70% of the landscape. These studies reveal opportunities for more sustainable beekeeping in landscapes committed to conventional production of annual crops. A basic understanding of honey bee biology and foraging behavior is required to appreciate how an agricultural-dominated landscape impacts honey bee health and productivity.

Biology and Beekeeping Impacting Response to Landscape Variation

Honey bees are a social insect that live in colonies comprised of a single reproductive queen responsible for all bees within the colony (Winston, 1987). During a growing season, a colony has thousands of females (workers) with limited reproductive capacity and responsibilities for all labor within the colony. A limited number of males (drones) that mate with queens from other colonies during a brief period of a growing season may also be present. The workers are responsible for care of the immature bees reared within the colony and bringing resources to the colony by foraging across several kilometers around the colony. The early period of a worker's adulthood includes caring for larval honey bees, nursing them by feeding them food derived from nectar and pollen collected by older foraging workers. Eventually workers die from their labor, sometimes unable to return to the hive due to worn wings that prevent them from returning from foraging. The lifespan of an adult worker bee can be several weeks, but the queen's lifespan can extend for up to 3 years. If a queen fails to produce workers, she will be replaced either by the colony or a beekeeper. In this way, a colony can persist indefinitely. During the growing season, as resources diminish and seasonal changes occur, the colony transitions with the queen ceasing to lay eggs and produce workers. The remaining workers then cluster around the queen to feed her and maintain a viable temperature. If stored honey and pollen are sufficient and no outbreaks of a pathogen or pest occur, the colony can survive until the next growing season.

Beekeeping as a form of agriculture has developed dramatically since honey bees were introduced to North America with European settlers and the invention of the removable frame hive by Reverend Langstroth in 1851. In North America, beekeepers are classified into three categories based on the number of colonies they maintain (Kulhanek *et al.*, 2017): hobbyists (≤50 colonies), sideliners (51–500), and commercial (>500). In the US, sideliners and most commercial beekeepers rent colonies to multiple farmers growing crops requiring bee pollination (Bond *et al.*, 2021), moving colonies by sealing the boxes and travelling often hundreds of kilometers to a farm. This includes transportation of colonies in the spring to California and Texas with subsequent movement to states in the Pacific Northwest and Southeast, respectively. An estimated 68% of all managed honey bees in the US are used for almond production in California where nearly 100% of US almonds are grown. The scale of almond production in California is remarkable, as it is the most valuable export crop for the state (CDFA, 2020). In 2020, this industry was serviced by an estimated 610,000 colonies brought into California, which were combined with 540,000 that are kept in the state. By the end of the pollination season many migratory colonies are returned to the Northern Great Plains region for summer honey production and overwintering (Bond *et al.*, 2021).

The number of colonies that can survive and the production of honey on a per colony basis varies based on the resources surrounding an apiary. Ecologically, the parameter that explains how many of a species that can reside in an environment is referred to as its carrying capacity. Estimating the impact of a landscape on the carrying capacity of honey bees is challenging for two reasons. First, honey bees are a social organism whose reproduction is limited to a single queen per colony. Ecologically, the carrying capacity for honey bees is not measured by the number of workers but at the colony level, as that is the reproductive unit. The worker population within a colony is a function of the resources available and can vary throughout a growing season. When resources are limited, the population of workers shrink as the queen ceases egg laying. Beekeepers maintain colonies during times of limited resources and even stimulate worker production by supplementally feeding colonies. Secondly, improving the ecological carrying capacity of honey bees is not always in the best interest of the beekeeper, as this will result in new queens being produced, not necessarily more honey. With the production of a new queen, a colony will produce a swarm that includes the old queen with a substantial number of workers to begin a new colony elsewhere. This is counter-productive for a beekeeper, as the new queen will have a reduced labor pool for crop pollination and honey production. Beekeepers optimize colony size by preventing swarming (e.g. killing queens in the larval stage) and providing space to prevent congestion which can stimulate swarming.

Beekeepers use the term carrying capacity to describe how much honey could be produced at a location, not how many colonies are supported. Beekeepers make predictions about how much honey could be produced in a landscape based on the most common flowering plant. Efforts to estimate this carrying capacity have been attempted (Al-Ghamdi *et al.*, 2016), but are site-specific and require extensive knowledge of how much honey can be produced from nectar collected from a single or limited number of plants. For example, *Ziziphus* spp. (Rhamnaceae: Rosales) flowering phenology and physiology are well studied (Alqarni, 2015), and based on the amount of this plant in a landscape, the amount of honey can be predicted (Adgaba *et al.*, 2017). Where this plant is very common (e.g. the coastal region of Saudi Arabia), it is the sole source of nectar, resulting in a monofloral honey. However, in landscapes with greater diversity of flowering plants, honey bees mix nectar from multiple sources gathered throughout a growing season.

Honey bees are super-generalists capable of foraging on a wide range of plants (McGregor, 1976), making them a useful source of pollination for multiple crop species, and capable of producing honey from multiple plant species. This polyphagy is challenging for estimating how much honey can be produced in landscapes with more floral diversity. Complicating this challenge is the foraging range of workers that can extend several kilometers from a colony. Decoded waggle dances of foraging workers reveal honey bees can travel from 0.5 to 5.5 km from their colonies in search of pollen and nectar (Visscher and Seeley, 1982; Beekman and Ratnieks, 2001, respectively). This range allows managed honey bee colonies to utilize multiple sources of flowers, including those found in crop fields and non-cropped areas. Field studies in the US indicate honey bees have a preference for non-native species (Urbanowicz *et al.*, 2020), but they will use several species that are native to North America (Carr-Markell *et al.*, 2020), especially later in the growing season when crops cease blooming (Zhang *et al.*, 2021).

An effort to determine locations most suitable for honey production that have multiple sources of nectar was attempted in North Dakota (Gallant *et al.*, 2014), part of the Corn Belt and a leading producer of honey in the US. Large-scale commercial beekeepers bring their colonies to this region during the summer for honey produced from multiple crops (e.g. alfalfa, canola, sunflowers) and non-cropped areas (e.g. wind-breaks, fence lines) that contain deciduous trees and shrubs managed under the US Department of Agriculture (USDA)'s Conservation Reserve Program (CRP). Mapping locations for apiaries in North Dakota was based on expert opinion regarding what amount and form of land use would produce sufficient honey for colonies kept at a commercial-scale density (i.e. 100 per apiary). This mapping effort revealed two challenges to estimating the value of a location for honey production (Gallant *et al.*, 2014). First, the quality of mapping data for multiple crops was limited, resulting in an underestimation of the actual number of apiaries that could be kept in North Dakota. Second, dramatic changes in land use occurred during an 8-year period (2002–2010). During this period, crops associated with honey production (e.g. sunflower, canola) and land that was not farmed and enrolled in CRP declined in acreage. This included over around 400,000 ha converted from sunflowers and CRP to corn and soybeans (Gallant *et al.*, 2014). Despite soybeans being a significant contributor to honey (Lin *et al.*, 2022), the authors suggest this change contributed to a decline in honey produced in the Dakotas.

A multi-year, on-farm study in Iowa focused on the relationship between the complexity of the surrounding landscape and honey production (Dolezal *et al.*, 2019b). Sentinel apiaries comprised of four colonies were deployed at the edge of conventional soybean fields surrounded primarily by either other soybean and corn fields (i.e. a simple landscape) or other landscape features (i.e. complex landscapes). Counterintuitively, greater honey bee populations and greater colony mass (a proxy for honey production) resulted when apiaries were adjacent to soybean fields in simple rather than complex landscapes (Dolezal *et al.*, 2019b). Regardless of the surrounding landscape, the primary source of pollen was clover. A positive relationship between honey and wax production and the amount of cultivated cropland in the landscape was also observed in Ohio through surveys of beekeepers (Sponsler and Johnson, 2015).

In the Dolezal *et al.* (2019b) study, regardless of the surrounding landscape, all colonies lost weight beginning in August when soybean and clover ceased flowering. This loss of weight is due to honey bees consuming honey generated up to August. In this landscape, honey bees experience a feast of mass-flowering crops (e.g. soybeans) and a famine after they cease flowering due to a lack of non-cropped flowering plants. This phenomenon has been addressed by midwestern beekeepers through supplemental feeding after an August honey harvest to ensure honey bees survive the winter. These colony-level impacts mirror responses by individual honey bees to increased floral diversity. A more detailed exploration of land use on the health of individual honey bees has revealed that increasing area of row crops is negatively associated with the nutritional state of individual honey bees (Smart *et al.*, 2019).

To address the late-summer famine that occurs after soybeans and clover (a ubiquitous component of non-cropped areas of the Corn Belt) cease flowering, Dolezal *et al.* (2019b) moved colonies to prairies (the biome native to the region of the study) that bloom in the late summer and fall. Not only did colonies stop losing weight, but increased honey storage leading to colony weight gain, suggesting honey bee productivity can be improved by providing floral resources at key times when forage is no longer available in the landscape. Furthermore, forage can be provided through conservation practices that include native plants (Carr-Markell *et al.*, 2020). A loss of land enrolled in programs that keep land out of agricultural production (e.g. agri-environmental schemes such as CRP) is associated with declines in honey bee productivity (Otto *et al.*, 2016) and health (Ricigliano *et al.*, 2019). Reversing declines in native, perennial flowering plants is a central approach to conserve wild, native pollinators in the US, and could contribute to more sustainable beekeeping.

In an Iowa beekeeper survey, Cass *et al.* (2022) found that many beekeepers noted that they do not lack for forage and did not see an immediate need for improving the landscape through conservation, but they identified Varroa mite management and pesticide exposure as more immediate challenges. This is curious as noted above, because honey production in the US is in decline, and some recent experiments have explored how increasing floral abundance and diversity can improve honey bee productivity. These experiments were conducted in Iowa, a state that features the stressors of limited floral resources and pesticide use.

Beekeeping in an Agricultural-dominated Landscape: The Iowa Story

Although pollinators, including honey bees, can be found in corn and soybeans (Gill and O'Neil, 2015; Wheelock and O'Neal, 2016), neither require bee pollination for production. Therefore, there are few incentives in the US Corn Belt where farmers and landowners maintain apiaries for crop pollination. Despite this, approximately 40% of all the managed colonies in the US are kept in Corn Belt states (USDA-NASS, 2024). Many commercial beekeepers maintain their colonies in the Dakotas and Minnesota during the summer after renting them for crop pollination during the spring in California, Oregon, and Washington (Bond *et al.*, 2021). The decline in honey production that has occurred at a national level on a per-state basis (Fig. 12.1) has also occurred for beekeepers in the Dakotas over a 30-year period (Otto *et al.*, 2018).

Although the decline in honey production in Iowa is not as steep as that observed across the US (see regression analysis in Fig. 12.1) there is remarkable variation over a 33-year period. This is coupled with a remarkable fluctuation in the number of colonies kept in Iowa, from a high of 55,000 in 1998 to a low of 24,000 in 2008. As of 2022, colonies rebounded to 48,000 with the majority kept by hobbyists, and several thousand by sideliners and a small community of commercial beekeepers (Cass *et al.*, 2022). The Iowan community faces many of the stressors that all beekeepers in the US face, at a landscape scale.

As the global landscape has changed, the forage available to honey bees has changed as well with much of the natural, unmanaged land converted to agriculture (Foley *et al.*, 2005). Farms were once comprised of multiple crops and pasture for animal production that would have had multiple sources of forage available throughout a growing season for honey bees. As more farmland has been committed to the production of fewer crops with little to no livestock grazing on pasture, the opportunity for honey bees to find forage has declined, a trend observed in the agriculture practiced in Iowa over a more than 100-year period (Brown and Schulte, 2011). By 2020, 75% of Iowa was committed to corn and soybean production, with less than 1% remaining in prairie, the native biome common before European settlers arrived. The current landscape is depauperate of flowering forage, contributing to low diversity and abundance of native bees found in Iowa and the surrounding midwestern states (Koh *et al.*, 2016).

Compounding a decline in forage is the use of broad-spectrum insecticides, toxic to a wide range of insects including honey bees. The amount and variety of insecticides used on Iowa crops has changed dramatically over a 30-year period due to several factors. This includes a transition beginning in the 1990s to genetically modified-based varieties that express toxins, in the form of proteins derived from *Bacillus thuringiensis* (Bt), that target key insect pests (Gassmann and Reisig, 2023; see Chapter 9, this volume). To date, there is no evidence that Bt-expressing crops negatively affect honey bees (Duan *et al.*, 2008). In the US, the adoption of various Bt-producing corn and cotton cultivars resulted in a decline in foliar- and soil-applied broad-spectrum insecticides with reduction in environmental impacts (Yang and Suh, 2015).

The arrival and establishment of the soybean aphid occurred after the transition to genetically modified-based insect pest control. By 2003, the aphid was found throughout the Corn Belt, especially in states producing most of the soybeans: Iowa, Illinois, and Minnesota. Although extreme outbreaks can lower yield by 40%, the average yield loss during a typical untreated outbreak has been estimated, at least for Iowa, at approximately 18% (Dean *et al.*, 2020). Foliar insecticide applications are the main management available to conventional soybean farmers throughout the Corn Belt (Hodgson *et al.*, 2012). Pyrethroids are the most commonly used, with a limited number of other active ingredients due to recent restrictions on chlorpyrifos in the US (Dean *et al.*, 2021). Regardless of the active ingredient, soybean aphid outbreaks when soybeans are flowering are problematic for honey beekeepers. Insecticide labels produced by the Environmental Protection Agency (EPA) in the US restrict the daily timing of foliar insecticide applications to limit exposure to bees.

Neonicotinoids can be applied to both foliage and seed, but the major driver of their prevalence is from the wide adaptation of coated seed treatments in the mid-2000s (USGS, 2017). The change in application was driven by an increase in insect resistance to commonly used insecticides and concerns over the high mammalian toxicity of commercially available active ingredients (Simon-Delso *et al.*, 2015). At times, up to 100% of corn and 50% of soybean have been planted with neonicotinoid-treated seeds in the US (Douglas and Tooker, 2015). Neonicotinoids used on corn and soybean seeds (imidacloprid, clothianidin, and thiamethoxam) move systemically, taken up through the roots and translocated to their leaves, flowers, and pollen (Hladik *et al.*, 2018). Neonicotinoids applied to seeds were considered a way to reduce the impact of insecticides on beneficial insects, while more efficiently targeting pests (Jeschke *et al.*, 2011). However, neonicotinoids have been documented to cause both lethal and sublethal effects on non-target organisms including honey bees (Pisa *et al.*, 2015). For social insects like honey bees, neonicotinoid residues in pollen and nectar of various plants may be brought back to the hive causing adverse effects to the developing brood and queen. To address the impacts this may have on colonies, semi-field studies have been conducted under controlled conditions to address potential effects of chronic exposure. After exposing honey bees through insecticide-contaminated sucrose, numerous parameters were measured, including estimating colony strength via number of eggs, larvae, and brood, as well as sublethal features of exposure like altered foraging behavior. From these studies a "No Observed Adverse Effect Concentration" (NOAEC) was determined, which is the highest concentration that does not result in a negative impact to a colony. Based on 6-week feeding studies of honey bee colonies with imidacloprid, thiamethoxam, and clothianidin,

NOAECs were determined to be 25, 44 and 19 ng/g, respectively (EPA, 2020a, b). These values can be compared to pesticide residues measured in pollen and nectar to characterize potential risk of exposure to honey bees.

There are at least three ways in which bees can be exposed to neonicotinoids used as seed treatments. The first form of exposure occurs when neonicotinoids applied to seeds move off-field in dust produced at planting (Nuyttens *et al.*, 2013; Krupke *et al.*, 2017a). Contaminated dust can land directly on foraging honey bees and/or on the plants they are foraging on, potentially exposing colonies when surviving foragers return with the contaminated provisions (Nuyttens *et al.*, 2013; Long and Krupke, 2016; Mogren and Lundgren, 2016). Field evaluations of honey bee colonies exposed to neonicotinoids from planted corn revealed temporary delays in colony growth (Lin *et al.*, 2021). Efforts to reduce dust at planting have been attempted to limit this exposure route (Krupke and Long, 2015).

Second, honey bees foraging on treated crops can be exposed via contaminated pollen and nectar (Wood and Goulson, 2017). This results in dietary exposure through consumption of contaminated nectar (consumed as honey) and pollen (consumed as bee bread). This exposure varies over time as the concentration of neonicotinoids within treated crops declines post-planting (Whalen *et al.*, 2021; Stewart *et al.*, 2014). This has been studied thoroughly for soybeans, with several field-based experiments revealing lethality to target pests (e.g. soybean aphids). Target pest mortality is limited to the vegetative stages leaving the plant susceptible to pests by around 40 days after planting (Johnson *et al.*, 2009; McCarville and O'Neal, 2013). By the end of the vegetative stage, the concentration of neonicotinoids in soybean leaf tissue between treated and untreated plants is indistinguishable and often undetectable when the plant is producing flowers (Krupke *et al.*, 2017b). We are unaware of any reported detections of neonicotinoids in the nectar or pollen of soybeans. Corn pollen can be contaminated from neonicotinoid seed application alone. The concentration varies by active ingredient from 3.02 to 36.88 ng/g for imidacloprid and clothianidin, respectively (Wood and Goulson, 2017).

An interesting observation from the study of neonicotinoid-treated soybean seeds is the occurrence of neonicotinoids in leaf tissue of soybean grown from untreated seeds due to the persistence of neonicotinoids in the soil from previous growing seasons (Krupke *et al.*, 2017b). This observation reveals that neonicotinoids can be taken up through contaminated soil, revealing a third form of neonicotinoid exposure that is facilitated via overland runoff and subsurface flow post-planting. Imidacloprid, clothianidin, and thiamethoxam have been detected in streams and tributaries within the Great Lakes region of the US (i.e. Michigan and Ohio), due to overland or subsurface flow from treated fields (Hladik and Kolpin, 2015). Such movement likely occurs across the Corn Belt where seed treatment use is common. This movement raises concern that non-target plants bordering crop fields could systemically take up neonicotinoids resulting in exposure to foraging honey bees. Botías *et al.* (2016) reported concentrations of clothianidin, thiamethoxam, and imidacloprid (maximum concentrations of 11.45–106.2 ng/g) in the leaves of wild herbaceous and woody plants collected 1–2 m from a field planted with thiamethoxam-treated seeds indicating that neonicotinoids used on treated seeds can be found in non-target plants close to the crop field. They also reported detection of thiamethoxam in pollen and nectar with a maximum concentration detected of 86.02 and 1.80 ng/g, respectively. Additional research has reported detections of neonicotinoids in non-target wildflowers and flowering shrubs located near row crop fields (Main *et al.*, 2020; Hall *et al.*, 2022). Mullin *et al.* (2010) detected imidacloprid (2.9%) and thiamethoxam (0.3%) in pollen from 13 apiaries across Florida and California. This evidence indicates honey bees could be exposed to neonicotinoids from treated crops and/or habitat established in close proximity to crop fields. Although these exposures could potentially affect individual honey bees, they are substantially below concentrations expected to negatively impact honey bee colonies. It is unclear the extent colony health is affected as susceptibility to neonicotinoids may be buffered by social behaviors that could limit exposure (i.e. grooming and colony hygiene) or diet (Barascou *et al.*, 2021).

The risks and benefits of beekeeping near conventional crop fields where neonicotinoid-treated seeds are used is still an area of investigation. Damage has been observed to honey bee colonies due to exposure from neonicotinoid dust generated during corn planting (Lin *et al.*, 2021). This damage was not consistently biologically significant. Counterintuitively, honey bees located in Ohio landscapes dominated with agricultural fields, with exposure to seed treatment dust were more productive than those kept in landscapes with less agricultural activity (Lin *et al.*, 2021). Foraging on pollen and nectar available in and around crops in the Corn Belt may outweigh the impact of the pesticides used within them. Regardless, when these crops cease flowering, there are limited resources remaining in the landscape (Dolezal *et al.*, 2019b). However, recent efforts to increase availability of native, perennial flowering vegetation to agricultural landscapes in the US is revealing a potential for more sustainable beekeeping into the future.

Benefits of Improving Forage for Honey Bees: An Iowa Perspective

In the US, there are several ongoing efforts to improve forage for pollinators that include activities by non-government organizations (NGOs) (e.g. Bee and Butterfly Habitat Fund, Pheasants Forever) and federal programs, including the USDA's CRP. A number of agri-environmental schemes offered through the European Union include pollinator stewardship (Batáry *et al.*, 2015). The CRP was established in 1985, and since then has amassed over 20 million acres of protected land (USDA-FSA, 2022). The Farm Service Agency (FSA) pays landowners to remove "environmentally sensitive land," from agricultural production, and instead, plant species that can ultimately improve soil quality, reduce topsoil erosion, protect groundwater supply, and provide habitat for wildlife including bees (USDA-FSA, 2022). Within the CRP, there are several conservation practices (USDA-FSA, 2024), with a subset that can help landowners protect bees, including CP42 "Pollinator Habitat" (USDA-FSA, 2013) and CP43 "Prairie Strips" (USDA-FSA, 2023a). Both pay landowners to take land out of production and replace it with native, perennial flowering plants found in prairies. This focus on prairie plants is a promising approach for insect conservation, as several perennial flowering forbs found within them are attractive to natural enemies of pests and pollinators, including honey bees (Tuell *et al.*, 2008; McMinn-Sauder *et al.*, 2020). When a concerted effort is made to add the most attractive of these plants to a conservation tract, the abundance and diversity of pollinators, especially bees, is increased (Gill *et al.*, 2014). Although both CP42 and CP43 provide an opportunity to add more native, perennial flowering plants back into an agricultural landscape, they do so in very different ways.

The CP42 practice, titled "Pollinator Habitat", was created in 2008, due in part to growing concern for declines in native pollinators. This practice requires a landowner to create habitat that includes planting at least nine species of native plants considered pollinator friendly on land that was used for commodity crop production (USDA-FSA, 2013). Additionally, this must include at least three species blooming in each of three periods (April to May, June to July, and August to October) (USDA-FSA, 2013) to provide flowers throughout a growing season. CP42 was initially applicable to a range of parcels with only a minimum size requirement of 0.2 ha. Recently this has been restricted to a maximum of 4 ha per application. The placement of this practice is up to the landowner. If the landowner qualifies, they receive a one-time sign-up bonus on a per acre basis, reimbursement payment of up to 50% of the establishment costs, and a contract for annual payments based on county rental rates.

Unlike the "Pollinator Habitat" practice, the main goal of CP43, the "Prairie Strip" practice (USDA-FSA, 2023a), is to prevent soil erosion and improve water quality, with a secondary goal of providing habitat for pollinators or wildlife. This practice was first introduced in the 2018 US Farm Bill and belongs to the "CLEAR" (Clean Lakes, Estuaries, and Rivers) initiative (USDA-FSA, 2019), which is an attempt to prevent fertilizer contamination from crop fields from contributing to eutrophication caused by algal blooms that thrive on the nitrogen and phosphorus. The "Prairie Strip" practice requires planting a minimum of ten native flowering plants, either on the contour within a field or on the perimeter of a field to achieve the primary purpose of reducing soil erosion and protecting water quality by intercepting water moving. To qualify, the prairie strips must be between 9 and

36.5 m in width and may not account for more than 25% of the cropped area.

A multi-disciplinary team at Iowa State University demonstrated that prairie strips reduce the loss of soil sediment and key nutrients that contribute to watershed eutrophication (Fig. 12.2) (Schulte *et al.*, 2017). Evaluation of CP43 placed in commercial fields where conventional agricultural practices are used has revealed disproportional benefits. Converting only 10% of a field to diverse, native perennial vegetation, reduced sediment movement off the field by 95% and total phosphorous and nitrogen lost through runoff by 90% and 85%, respectively. By reconstructing prairie within or at a field edge, prairie strips also provide habitat and forage for many wildlife species, including pollinators.

Other conservation and agronomic practices can counter the environmental consequence of the production of annual crops in temperate climates like Iowa (i.e. several months of bare soil). For example, cover crops can provide protection from soil erosion and, if flowering species are included, provide forage to pollinators (Lee-Mader *et al.*, 2015). But their capacity to contribute to pollinator conservation in corn and soybean production may be limited as the planting of annual crops occurs and cover crops are terminated before a cover crop flowers. Federal programs (e.g. CRP) in the US financially support cover crops and other practices that can protect soil. Of the practices currently available, CP43 is the most cost effective at achieving multiple benefits on a per acre basis (Tyndall *et al.*, 2013). How widely adopted CP43 will be is not yet clear; however, several states and NGOs have promoted this practice (e.g. Sand County Foundation) because of the additional benefits derived from it.

Several forms of wildlife respond positively to prairie strips, including insects (Cox *et al.*, 2014; Schulte *et al.*, 2017), including several native bees and pollinators (Kordbacheh *et al.*, 2020). The response from these insect taxa includes greater abundance and diversity within the prairie strip than what would be observed if the land had remained in crop production. To date, there is no evidence that the adjacent crop is at greater risk for insect pest attack, and more broadly, no evidence that yield adjacent to the strips is negatively affected (Dutter *et al.*, 2023). There have been limited efforts to measure the conservation of beneficial insects within the prairie strip. To date the most thorough study of prairie strips contributing to suppression of the soybean aphid did not reveal an impact (Cox *et al.*, 2014). Additional studies of the specific impact on other pests have not been done.

Several farmers adopted the "Prairie Strip" practice before it was incorporated within CRP, providing an opportunity to test the impact of these small, often isolated patches of prairie on

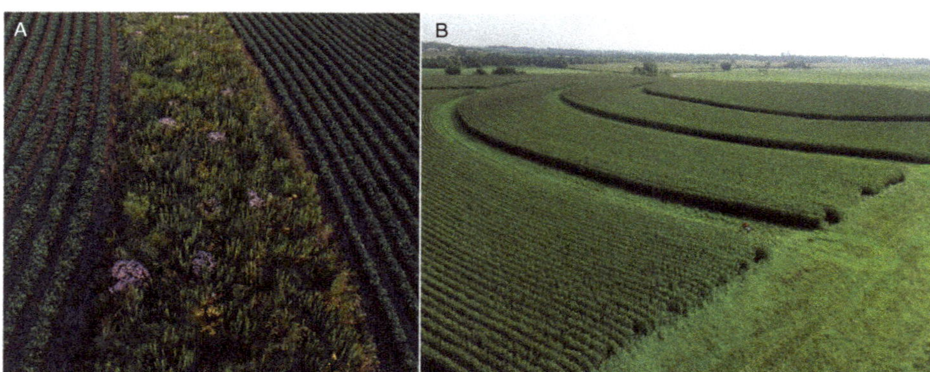

Fig. 12.2. (A) A single prairie strip and (B) a series of strips planted along field contours, both in conventional soybean fields in central Iowa. These strips were established per the guidelines of the US Department of Agriculture's conservation practice (CP43) and include placement along a contour to capture sediment, the primary goal of CP43 (USDA-FSA, 2023a). (Photo credits: Iowa State University (A) and Lynn Betts (B).)

honey bees. Using methodology derived from Dolezal *et al.* (2019b), Zhang *et al.* (2023), observing over a 3-year period, found honey bees kept at farms with prairie strips had larger worker populations and produced 24% more honey than those kept at farms without prairie strips (Fig. 12.3). Beekeeping in farms with prairie strips provided the prairie-rescue effect observed by Dolezal *et al.* (2019b), without moving colonies to a prairie. By following the CP43 guidelines, landowners were able to increase the floral diversity and flower abundance compared to what is found in typical field edges, roadsides, and grassed waterways providing honey bees season-long access to native flowering plants. (Zhang *et al.*, 2023; Murray, 2021). These experiments did not determine the source of nectar for honey bees, but did monitor the plants used for pollen. Honey bees only utilized a subset of plants found within prairie strips, especially, those species that flowered after soybean and clover (*Chamaecrista fasciculata*, *Solidago* spp. and *Silphium* spp.) (Zhang *et al.*, 2021). Even though experimental units (i.e. four colonies) were kept over 3 km apart, colonies kept at farms without prairie strips also used these plants as a source of pollen forage. The greater survival of adult bees in colonies kept at prairie strips is likely due to the increased proximity to resources, reducing stress associated with foraging. Most of the honey produced at farms with or without prairie strips is likely driven by soybean. In a landscape similar to Iowa, honey bees kept in Ohio primarily collected nectar from soybean fields as far as 3 km from their colonies, but primarily within 0.5–1.5 km radius (Lin *et al.*, 2022). These results suggest that if beekeepers utilize farms with prairie strips that produce soybeans, their honey bees can forage on both a mass-flowering crop and a mix of native, perennial flowering vegetation.

The value of prairie stripes as forage for honey bees may be limited if they are immediately adjacent to conventional fields in which pesticide use is common. Participation in the CP43 practice does not require farmers to adjust their farming practices beyond removing a modest amount of land from production. As noted earlier, insecticides can be detected in non-crop habitat adjacent to conventional crop production. This has been observed for both seed-applied neonicotinoids and commonly applied foliar insecticides. The risk of this exposure has been estimated in a series of on-farm experiments involving CP43.

Hladik *et al.* (2018) analyzed samples from fields with and without prairie strips, both planted with clothianidin or imidacloprid-treated

Fig. 12.3. Two apiaries of four honey bee colonies kept at a conventional farm (A) without prairie strips or (B) with prairie strips. Each honey bee colony inhabits the column of boxes that are stacked over the course of the growing season to provide more space for honey storage. Note that added production by each colony in (B) required an extra box. (Photo credit: M.E. O'Neal.)

corn and soybean seeds 3 years prior to sampling. Neonicotinoids were detected in groundwater, surface water, and soil samples taken from the prairie strip sites. However, maximum concentrations in soil were lower at sites with prairie strips (2.8 ng/g) as compared to sites without prairie strips (25.7 ng/g). No detections were recorded in plant root material suggesting that systemic uptake of the neonicotinoids by plants in prairie strips was not occurring at detectable levels. This is likely a result of the low levels present and detected in soil (three detections/27 samples = 9%; concentration range 1–3 ng/g) resulting in low levels present in the root zone and minimal plant uptake. However, samples were taken 3 years after the last neonicotinoid seed treatment application and lower concentrations would be expected compared to samples from sites currently using neonicotinoid-treated seeds. This was addressed in a subsequent study in which soil and plant tissue (including milkweed) from prairie strips established at commercial corn and soybean farms were sampled for neonicotinoids (Hall *et al.*, 2022). Samples were taken from multiple farms in 2 years. At least one neonicotinoid was found in all soil samples and 80% of all plant tissue.

In an additional study, pollen was collected from honey bees, along with collections of Bombus spp. (Hymenoptera: Apidae) visiting flowers in the prairie strips and worker bees from within four colonies kept at fields either with or without prairie strips. Neonicotinoids (thiamethoxam, clothianidin, and imidacloprid) were the least frequently detected insecticides (<15% of pollen samples and <1% of bee samples) (M. Hall *et al.*, unpublished data). No significant differences were observed in the frequency of pesticides detected in honey bees collected from colonies placed at fields with or without prairie strips. Honey bee colonies experienced more frequent exposure to foliar-applied insecticides and fungicides (<60% of pollen samples and <20% of bee samples) (M. Hall *et al.*, unpublished data). The risk to honey bees within the context of prairie strips was considered negligible when residue data collected from these studies was compared to relevant toxicity data using the US EPA screening tool BeeREX (EPA, 2015) as well as higher tier colony-level studies (M. Hall *et al.*, unpublished data).

When combined, these studies suggest beekeeping can be improved in a landscape like Iowa with the addition of prairie strips with limited to no additional risk due to the conventional agricultural practices in the adjacent crop that are typical for Iowa and the US Corn Belt. A similar conclusion has been recently reached for the conservation of the Monarch butterfly, *Danaus plexippus* (Lepidoptera: Nymphalidae), in this same region. A combination of multiple field, laboratory, and model-based studies suggest that a sustained population of Monarch butterflies can be achieved through the establishment of 1.3–1.6 billion new stems of milkweed with a diversity of forbs (Grant *et al.*, 2022). This prediction is possible if the current use of insecticides is limited to a seed application and a single foliar application of insecticide to either corn or soybean. Other soybean-producing regions both in and outside the US typically apply insecticides more frequently (Pozebon *et al.*, 2020). In regions where insecticide use is more frequent, the effectiveness of this approach for conserving flower-visiting insects, Monarch butterflies, or honey bee, will require additional toxicity data and field-relevant pesticide exposure estimates.

Best Practices to Improve Bee Forage and Reduce Pesticide Exposure

The Honey Bee Health Coalition has developed a series of "Best Management Practices" that describe practices for farmers and beekeepers to improve honey bee health (HBHC, 2021b). These include crops that require bee pollination (e.g. apple) and those that do not (e.g. soybean, corn). A shared feature of these practices is the need to coordinate conservation and pest management activities between landowners, farmers, and beekeepers. In this section, we discuss a few key practices highlighted within these recommendations and how this coordination is implemented in Iowa, where much of the previous research used in this review was conducted.

Limiting pesticide exposure

The application of integrated pest management (IPM) practices is considered fundamental to ensuring pollinator protection by both farmers and conservationists. This includes scouting for pests to determine their presence and the need for management if the population exceeds a threshold. For example, applied entomologists developed scouting programs and thresholds for determining when a foliar insecticide should be used against the soybean aphid (Ragsdale *et al.*, 2007; Hodgson *et al.*, 2012). Outbreaks of the soybean aphid in the north central region of the US frequently occur when soybeans are in anthesis. This is remarkable, as insecticides used on soybean including those frequently used for soybean aphid (e.g. lambda-cyhalothrin), are restricted when the crop is blooming. This creates a tension between farmers seeking to achieve optimal yield and the beekeepers who keep bees around these fields. Efforts to reduce exposure of flower-visiting insects to foliar-applied insecticides include additional labelling and state laws that restrict application when flowering to early morning and dusk when honey bee colonies are within a given range of a field to be treated. For example, the state of Iowa's Department of Agriculture and Land Management has expressed a rule that comes with two components (IDALS, 2023): beekeepers can register their colonies and applicators are restricted in the timing of insecticides within a mile of a registered apiary. For this rule to work, a registry has been constructed whereby beekeepers can report the location of their colonies, and applicators can view if a targeted field is within the buffer region. Iowa, like many US states, participate in the DriftWatch program (FieldWatch, 2024), which maintains registries of honey bees and other sensitive crops per the rules of each participating state. Each year the registry is refreshed for new beekeepers to join and existing beekeepers to note if their apiaries moved from year to year.

Finding new forage

Efforts to increase habitat for pollinators and forage for bees have been successful in the US, especially in Iowa. As of 2023, there were more than 100,000 ha of CP42 in Iowa (USDA-FSA, 2023b), which leads the nation in adoption of this practice. However, programs like CP42 and CP43 that create small patches of high-quality habitat for pollinators may not be seen as immediately valuable. Theoretically, ecologists often contrast results from a single large patch with those that create several small patches in an ongoing debate. A recent summary of global data suggests that high-quality, isolated patches of conservation areas have great value for preserving biodiversity (Wintle *et al.*, 2018). This finding has profound implications for flower-visiting insects that forage or migrate great distances. Recent survey efforts in Michigan have demonstrated that several conservation practices (e.g. CP01, CP38E, CP42) that create relatively small patches have increased bee visitations to flowers beyond what is observed in unmanaged fields and roadsides (Quinlan *et al.*, 2021; USDA-FAS, 2024). Honey bee visits were observed to be greater in conservation practice sites than road ditches, but visits were similar to what was observed in unmanaged fields with abundant goldenrod (*Solidago* spp.), a late-summer flowering native perennial.

For beekeepers to take advantage of these patches will require increasing their awareness of land use changes, as US beekeepers typically do not own land. Rather, they practice a form of "free-range" livestock production in which honey bees forage on land owned by others. Applications like Beescape can assist beekeepers in finding and evaluating possible locations that have flowering resources throughout a growing season (Robinson *et al.*, 2021). The capacity to reveal these resources in applications that use spatial data is dependent upon the accuracy of land use estimates. For the continental US, the USDA annually reports the Cropland Data Layer via a publicly accessible website reporting ground-truthed satellite imagery of agricultural regions (USGS, 2017; USDA-NASS, 2022). This produces a raster-based, georeferenced, crop-specific land cover data set with 30-m resolution. Practices like CP42 and especially CP43 that are embedded within larger crop fields may not be observed within these maps if the patch produced is below this resolution. Applications like Beescape that rely upon the Cropland Data Layer will underestimate the value of a landscape in which these practices are utilized. A more fine-grained measurement of cropland cover than what is currently available is needed to account for the presence of features created through programs like CP42 and CP43. To what extent more accurate measurements can improve beekeeper experience of applications like Beescape and lead to subsequent improvements in beekeeping are not clear.

Summary and Conclusions

By sharing the results from Iowa, we hope to reveal opportunities to address declining honey bee productivity by addressing key stressors that affect bees. This includes action by both farmers and policies employed by the US federal government through conservation practices. Similar efforts exist within the European Union, through agri-environmental schemes which include establishing flowering resources at field margins and hedgerows (Batáry *et al.*, 2015). A review of studies from Europe, North America, and New Zealand revealed that the general practice of establishing flowering strips adjacent to crop fields enhances pest control but does not consistently improve pollination and subsequent yield. This was due in part to a decline in pollinator abundance with increasing distance from floral resources (Albrecht *et al.*, 2020). Although specific examples exist where ecosystem services are delivered that result in yield improvement (e.g. Blaauw and Isaacs, 2014), additional study is needed to create recommendations that consistently deliver benefits to the adjacent crop. Overall, adding floral resources aids in pollinator conservation (Haaland *et al.*, 2011), but the benefit for honey beekeeping requires noting a few caveats.

We observed that adding forage into an agricultural landscape with a conservation practice improved honey bee health and productivity. This approach may come with a potential cost to the conservation of native pollinators. A growing body of literature is revealing that honey bees can interact negatively with native bees (Mallinger *et al.*, 2017; Iwasaki and Hogendoorn, 2022; Weaver *et al.*, 2022) via competition and disease spread. This is a concern for native pollinators that have co-evolved with the native vegetation that is explicitly part of practices like CP42 and CP43. A central point of this concern is that honey bees are not native. They are managed as a form of agriculture and best considered livestock. If the goals of a landowner include conserving native bees, then allowing beekeeping within and around patches of native vegetation produced by these practices should be avoided. However, there is evidence that both honey beekeeping and native bee conservation can be practiced simultaneously.

Native pollinators were still present when honey bees were kept within prairie strips established per the CP43 practices (Murray, 2021; Borchardt *et al.*, 2023). A possible mechanism allowing for the sharing of resources in a prairie strip is the use of non-native vegetation by honey bees within the surrounding landscape. A subset of the plants flowering in prairie strips were used by honey bees, with the majority of the pollen collected from clover (Zhang *et al.*, 2023). The phenology of pollen foraging by honey bees reveals native plant species were used as forage after non-native crops ceased flowering (Zhang *et al.*, 2022). This phenomenon extends to the foraging of nectar, as a decline in honey stores was reversed when honey bee colonies had access to prairie after the adjacent crop ceased flowering (Dolezal *et al.*, 2019b). Mass-flowering crops adjacent and surrounding patches of conservation may limit the competition for the native plants that flower within them, allowing for co-occurrence of managed and wild, native bees.

An additional mechanism by which managed bees (both honey bees and bumble bees) negatively affect native bees is through the spread of pathogens. The extent to which pathogen spread is a concern is unclear, as initial studies in the Midwest US suggest that transmission between honey bees and wild *Bombus* spp. may be limited (Pritchard *et al.*, 2021). Similar results were observed in the state of North Carolina (Levenson and Tarpy, 2022). Furthermore, transmission of a disease-causing agent may not result in pathogenicity (Dolezal *et al.*, 2016b). If practices like CP42 and CP43 establish flowering vegetation that attracts beekeepers, this phenomenon will require additional study as it will be of concern to both conservationists and beekeepers, as pathogen spread may occur in both directions.

This need for additional study is especially timely as multiple stakeholders have identified a desire to improve the landscape for pollinators. In a recent survey of Iowa landowners, both farmers and beekeepers were interested in conserving pollinators (Cass *et al.*, 2022).

Landowners expressed an interest in allowing beekeepers access to land with forage created through conservation. This interest is likely not unique to Iowa. Making sure that the optimal use is achieved will require considering the timing at which resources are available, and if this conflicts with conservation goals related to native pollinator conservation.

References

Adgaba, N., Alghamdi, A., Sammoud, R., Shenkute, A., Tadesse, Y. *et al.* (2017) Determining spatio-temporal distribution of bee forage species of Al-Baha region based on ground inventorying supported with GIS applications and remote sensed satellite image analysis. *Saudi Journal of Biological Sciences* 24, 1038–1044. DOI: 10.1016/j.sjbs.2017.01.009

Albrecht, M., Kleijn, D., Williams, N.M., Tschumi, M., Blaauw, B.R. *et al.* (2020) The effectiveness of flower strips and hedgerows on pest control, pollination services and crop yield: a quantitative synthesis. *Ecology Letters* 23, 1488–1498. DOI: 10.1111/ele.13576

Al-Ghamdi, A., Adgaba, N., Getachew, A. and Tadesse, Y. (2016) New approach for determination of an optimum honey bee colony's carrying capacity based on productivity and nectar secretion potential of bee forage species. *Saudi Journal of Biological Sciences* 23, 92–100. DOI: 10.1016/j.sjbs.2014.09.020

Alqarni, A. (2015) Honeybee foraging, nectar secretion, and honey potential of wild jujube trees, *Ziziphus nummularia*. *Neotropical Entomology* 44, 232–241. DOI: 10.1007/s13744-015-0279-4

Barascou, L., Sene, D., Barraud, A., Michez, D., Lefebvre, V. *et al.* (2021) Pollen nutrition fosters honeybee tolerance to pesticides. *Royal Society Open Science* 8, 210818. DOI: 10.1098/rsos.210818

Batáry, P., Dicks, L.V., Kleijn, D. and Sutherland, W.J. (2015) The role of agri-environment schemes in conservation and environmental management. *Conservation Biology* 29, 1006–1016. DOI: 10.1111/cobi.12536

Beekman, M. and Ratnieks, F. (2001) Long-range foraging by the honey bee, *Apis mellifera* L. *Functional Ecology* 14, 490–496. DOI: 10.1046/j.1365-2435.2000.00443.x

Blaauw, B.R. and Isaacs, R. (2014) Flower plantings increase wild bee abundance and the pollination services provided to a pollination-dependent crop. *Journal of Applied Ecology* 51, 890–898. DOI: 10.1111/1365-2664.12257

Bond, J.K., Hitaj, C., Smith, D., Hunt, K., Perez, A. and Ferreira, G. (2021) *Honey Bees on the Move: From Pollination to Honey Production and Back*. ERR-290. Washington, DC: US Department of Agriculture, Economic Research Service. Available at: https://www.ers.usda.gov/publications/pub-details/?pubid=101475 (accessed January 21, 2023).

Borchardt, K., Kadelka, C., Schulte, L. and Toth, A. (2023) An ecological networks approach reveals restored native vegetation benefits wild bees in agriculture. *Biological Conservation* 287, 110300. DOI: 10.1016/j.biocon.2023.110300

Botías, C., David, A., Hill, E. and Goulson, D. (2016) Contamination of wild plants near neonicotinoid seed-treated crops, and implications for non-target insects. *Science of the Total Environment*, 566–567, 269–278. DOI: 10.1016/j.scitotenv.2016.05.065

Brown, P. and Schulte, L. (2011) Agricultural landscape change (1937–2002) in three townships in Iowa, USA. *Landscape and Urban Planning* 100, 202–212. DOI: 10.1016/j.landurbplan.2010.12.007

Calderone, N. (2012) Insect pollinated crops, insect pollinators and US agriculture: trend analysis of aggregate data for the period 1992–2009. *PLoS ONE* 7, e37235. DOI: 10.1371/journal.pone.0037235

Calovi, M., Grozinger, C., Miller, D. and Goslee, S. (2021) Summer weather conditions influence winter survival of honey bees (*Apis mellifera*) in the northeastern United States. *Science Reports* 11, 1553. DOI: 10.1038/s41598-021-81051-8

Carr-Markell, M., Demler, C., Couvillon, M., Schürch, R. and Spivak, M. (2020) Do honey bee (*Apis mellifera*) foragers recruit their nestmates to native forbs in reconstructed prairie habitats? *PLoS ONE* 15, e0228169. DOI: 10.1371/journal.pone.0228169

Cass, R., Hodgson, E., O'Neal, M., Toth, A.L. and Dolezal, A.G. (2022) Attitudes about honey bees and pollinator-friendly practices: a survey of Iowan beekeepers, farmers, and landowners, *Journal of Integrated Pest Management* 13, 30. DOI: 10.1093/jipm/pmac027

CDFA (2020) *California Agricultural Exports 2019–2020*. Sacramento, CA: California Department of Food and Agriculture. Available at: https://www.cdfa.ca.gov/Statistics/PDFs/2020_Exports_Publication.pdf (accessed September 25, 2023).

Cox, R., O'Neal, M., Hessel, R., Schulte, L. and Helmers, M. (2014) The impact of prairie strips on aphidophagous predator abundance and soybean aphid predation in agricultural catchments. *Environmental Entomology* 43, 1185–1197. DOI: 10.1603/EN13129

Dean, A., Pritchard, S., Tyndall, J., Hodgson, E. and O'Neal, M. (2020) Evaluating soybean aphid-resistant varieties in different environments to estimate financial outcomes. *Journal of Economic Entomology* 113, 940–948. DOI: 10.1093/jee/toz309

Dean, A., Niemi, J., Tyndall, J., Hodgson, E. and O'Neal, M. (2021) Developing a decision-making framework for insect pest management: a case study using *Aphis glycines* (Hemiptera: Aphididae). *Pest Management Science* 77, 886–894. DOI: 10.1002/ps.6093

Dolezal, A., Carrillo-Tripp, J., Miller, W., Bonning, C. and Toth, A. (2016a) Intensively cultivated landscape and *Varroa* mite infestation are associated with reduced honey bee nutritional state. *PLoS ONE* 11, e0153531. DOI: 10.1371/journal.pone.0153531

Dolezal, A., Hendrix, S., Scavo, N., Carrillo-Tripp, J., Harris, M. *et al.* (2016b) Honey bee viruses in wild bees: viral prevalence, loads, and experimental inoculation. *PLoS ONE* 11, e0166190. DOI: 10.1371/journal.pone.0166190

Dolezal, A. and Toth, A. (2018) Feedbacks between nutrition and disease in honey bee health. *Current Opinion in Insect Science* 26, 114–119. DOI: 10.1016/j.cois.2018.02.006

Dolezal, A., Carrillo-Tripp, J., Judd, T., Allen, M., Bonning, B. and Toth, A. (2019) Interacting stressors matter: diet quality and virus infection in honeybee health. *Royal Society of Open Science* 6, 181803. DOI: 10.1098/rsos.181803

Dolezal, A., St. Claire, A., Zhang, G., Toth, A. and O'Neal, M. (2019) Native habitat mitigates feast-famine conditions faced by honey bees in an agricultural landscape. *Proceedings of the National Academy of Sciences* 116, 25147–25155. DOI: 10.1073/pnas.1912801116

Douglas, M. and Tooker, J. (2015) Large-scale deployment of seed treatments has driven rapid increase in use of neonicotinoid insecticides and preemptive pest management in U.S. field crops. *Environmental Science and Technology* 49, 5088–5097. DOI: 10.1021/es506141g

Duan, J., Marvier, M., Huesing, J., Dively, G. and Huang, Z. (2008) A meta-analysis of effects of Bt crops on honey bees (Hymenoptera: Apidae). *PLoS ONE* 3, e1415. DOI: 10.1371/journal.pone.0001415

Dutter, C., Damiano, L., Niemi, J., Miller, B., Schulte, L. *et al.* (2023) Contour prairie strips affect adjacent soil but have only slight effects on crops. *Field Crops Research* 296, 108905. DOI: 10.1016/j.fcr.2023.108905

EPA (2015) *BeeREX*. Washington, DC: US Environmental Protection Agency. Available at: https://www.epa.gov/pesticide-science-and-assessing-pesticide-risks/models-pesticide-risk-assessment#beerex (accessed January 21, 2024).

EPA (2020a) *Final Bee Risk Assessment to Support the Registration Review of Clothianidin and Thiamethoxam*. EPA-HQ-OPP-2011-0865-1164. Washington, DC: US Environmental Protection Agency. https://www.regulations.gov/document/EPA-HQ-OPP-2011-0865-1164 (accessed January 21, 2024).

EPA (2020b) *Final Bee Risk Assessment to Support the Registration Review of Imidacloprid*. EPA-HQ-OPP-2008-0844-1611. Washington, DC: US Environmental Protection Agency. Available at: https://www.regulations.gov/document/EPA-HQ-OPP-2008-0844-1611 (accessed January 21, 2024).

FieldWatch (2024) *Welcome to DriftWatch*. Available at: https://driftwatch.org/ (accessed January 21, 2024).

Foley, J., DeFries, R., Asner, G., Barford, C., Bonan, G. *et al.* (2005) Global consequences of land use. *Science* 309, 570–574. DOI: 10.1126/science.1111772

Gallant, A., Euliss, N. and Browning, Z. (2014) Mapping large-area landscape suitability for honey bees to assess the influence of land-use change on sustainability of national pollination services. *PLoS ONE* 9, e99268. DOI: 10.1371/journal.pone.0099268

Garibaldi, L., Schulte, L., Jodar, S., Carella, D. and Kremen, C. (2021) Time to integrate pollinator science into soybean production. *Trends in Ecology and Evolution* 36, 573–575. DOI: 10.1016/j.tree.2021.03.013

Gassmann, A. and Reisig, D. (2023) Management of insect pests with Bt crops in the United States. *Annual Review of Entomology* 68, 31–49. DOI: 10.1146/annurev-ento-120220-105502

Gill, K. and O'Neal, M. (2015) Survey of soybean insect pollinators: community identification and sampling method analysis. *Environmental Entomology* 44, 488–498. DOI: 10.1093/ee/nvv001

Gill, K., Cox, R. and O'Neal, M. (2014) Quality over quantity: buffer strips can be improved with select native plant species. *Environmental Entomology* 43, 298–311. DOI: 10.1603/EN13027

Grant, T., Fisher, K., Krishnan, N., Mullins, N., Hellmich, R. *et al.* (2022) Monarch butterfly ecology, behavior, and vulnerabilities in North Central United States agricultural landscapes. *BioScience* 72, 1176–1203. DOI: 10.1093/biosci/biac094

Haaland, C., Naisbit, R.E. and Bersier, L.F. (2011) Sown wildflower strips for insect conservation: a review. *Insect Conservation and Diversity* 4, 60–80. DOI: 10.1111/j.1752-4598.2010.00098.x

Hall, M., Zhang, G., O'Neal, M., Bradbury, S. and Coats, J. (2022) Quantifying neonicotinoid insecticide residues in wildflowers sampled from prairie strips established in corn and soybean fields planted with treated seeds. *Agriculture Ecosystems & Environment* 325, 107723. DOI: 10.1016/j.agee.2021.107723

Harwood, G. and Dolezal, A. (2020) Pesticide-virus interactions in honey bees: challenges and opportunities for understanding drivers of bee declines. *Viruses* 12, 566. DOI: 10.3390/v12050566

HBHC (2021a) *Varroa Management*. Keystone, CO: Honey Bee Health Coalition. Available at: https://honeybeehealthcoalition.org/resources/varroa-management/ (accessed September 25, 2023).

HBHC (2021b) *Best Management Practices Guides*. Keystone, CO: Honey Bee Health Coalition. Available at: https://honeybeehealthcoalition.org/resource_category/best-management-practices/ (accessed September 25, 2023).

Hladik, M. and Kolpin, D. (2015) First national-scale reconnaissance of neonicotinoid insecticides in streams across the USA. *Environmental Chemistry* 13, 12–20. DOI: 10.1071/EN15061

Hladik, M., Main, A. and Goulson, D. (2018) Environmental risks and challenges associated with neonicotinoid insecticides. *Environmental Science & Technology* 52, 3329–3335. DOI: 10.1021/acs.est.7b06388

Hodgson, E., McCornack, B., Tilmon, K. and Knodel, J. (2012) Management recommendations for soybean aphid (Hemiptera: Aphididae) in the United States. *Journal of Integrated Pest Management* 3(1), E1–E10. DOI: 10.1603/IPM11019

IDALS (2023) *FieldWatch & The Iowa Bee Rule: Frequently Asked Questions*. Des Moines, IA: Department of Agriculture & Land Stewardship. Available at: https://iowaagriculture.gov/sites/default/files/ent/FieldWatch%20and%20Bee%20Rule%202-27-2023.pdf (accessed January 28, 2024).

Insolia, L., Molinari, R., Rogers, S., Williams, G., Chiaromonte, F. *et al.* (2022) Honey bee colony loss linked to parasites, pesticides and extreme weather across the United States. *Science Reports* 12, 20787. DOI: 10.1038/s41598-022-24946-4

Iwasaki, J. and Hogendoorn, K. (2022) Mounting evidence that managed and introduced bees have negative impacts on wild bees: an updated review. *Current Research in Insect Science* 2, 100043. DOI: 10.1016/j.cris.2022.100043

Jeschke, P., Nauen, R., Schindler, M. and Elbert, A. (2011) Overview of the status and global strategy for neonicotinoids. *Journal of Agricultural and Food Chemistry* 59, 2897–2908. DOI: 10.1021/jf101303g

Johnson, K., O'Neal, M., Ragsdale, D., DiFonzo, C., Swinton, S. *et al.* (2009) Probability of economical management of soybean aphid (Hemiptera: Aphididae) in North America. *Journal of Economic Entomology* 102, 2101–2108. DOI: 10.1603/029.102.0613

Kordbacheh, F., Liebman, M. and Harris, M. (2020) Strips of prairie vegetation placed within row crops can sustain native bee communities. *PLoS ONE* 15, e0240354. DOI: 10.1371/journal.pone.0240354

Koh, I., Lonsdorf, E., Williams, N., Brittain, C., Isaacs, R. *et al.* (2016) Modeling the status, trends, and impacts of wild bee abundance in the United States. *Proceedings of the National Academy of Sciences* 113, 140–145. DOI: 10.1073/pnas.1517685113

Krupke, C. and Long, E. (2015) Intersections between neonicotinoid seed treatments and honey bees. *Current Opinion Insect Science* 10, 8–13. DOI: 10.1016/j.cois.2015.04.005

Krupke, C., Holland, J., Long, E. and Eitzer, B. (2017a) Planting of neonicotoind-treated maize poses risk for honey bees and other non-target organisms over a wide area without consistent crop yield benefit. *Journal of Applied Ecology* 54, 1449–1458. DOI: 10.1111/1365-2664.12924

Krupke, C., Alford, A., Cullen, E., Hodgson, E., Knodel, J. *et al.* (2017b) Assessing the value and pest management window provided by neonicotinoid seed treatments for management of soybean aphid (*Aphis glycines* Matsumura) in the Upper Midwestern United States. *Pest Management Science* 73, 2184–2193. DOI: 10.1002/ps.4602

Kulhanek, K., Steinhauer, N., Rennich, K., Caron, D., Sagili, R. *et al.* (2017) A national survey of managed honey bee 2015–2016 annual colony losses in the USA. *Journal of Apicultural Research* 56, 328–340. DOI: 10.1080/00218839.2017.1344496

Lee-Mader, E., Stine, A., Fowler, J., Hopwood, J. and Vaughan, M. (2015) *Cover Cropping for Pollinators and Beneficial Insects*. Washington, DC: Sustainable Agriculture Research and Education. Available at: https://www.sare.org/resources/cover-cropping-for-pollinators-and-beneficial-insects/ (accessed January 21, 2024).

Levenson, H. and Tarpy, D. (2022) Effects of planted pollinator habitat on pathogen prevalence and interspecific detection between bee species. *Science Reports* 12, 7806. DOI: 10.1038/s41598-022-11734-3

Lin, C., Sponsler, D., Richardson, R., Watters, H., Glinski, D. *et al.* (2021) Honey bees and neonicotinoid-treated corn seed: contamination, exposure, and effects. *Environmental Toxicology and Chemistry* 40, 1212–1221. DOI: 10.1002/etc.4957

Lin, C., Suresh, S., Matcham. E., Monagan, P., Curtis, H. *et al.* (2022) Soybean is a common nectar source for honey bees (Hymenoptera: Apidae) in a Midwestern agricultural landscape. *Journal of Economic Entomology* 115, 1846–1851. DOI: 10.1093/jee/toac140

Long, E. and Krupke, C. (2016) Non-cultivated plants present a season-long route of pesticide exposure for honey bees. *Nature Communications* 7, 11629. DOI: 10.1038/ncomms11629

Main, A., Webb, E., Goyne, K. and Mengel, D. (2020) Reduced species richness of native bees in field margins associated with neonicotinoid concentrations in non-target soils. *Agriculture, Ecosystems & Environment* 287, 106693. DOI: 10.1016/j.agee.2019.106693

Mallinger, R., Gaines-Day, H. and Gratton, C. (2017) Do managed bees have negative effects on wild bees? A systematic review of the literature. *PLoS ONE* 12, e0189268. DOI: 10.1371/journal.pone.0189268

Mashilingi, S., Zhang, H., Garibladi, L. and An, J. (2022) Honeybees are far too insufficient to supply optimum pollination services in agricultural systems worldwide. *Agriculture, Ecosystems & Environment* 335, 108003. DOI: 10.1016/j.agee.2022.108003

McCarville, M. and O'Neal, M. (2013) Soybean aphid (Aphididae: Hemiptera) population growth as affected by host plant resistance and insecticidal seed treatments. *Journal of Economic Entomology* 106, 1302–1309. DOI: 10.1603/EC12495

McGregor, S. (1976) *Insect Pollination of Cultivated Crop Plants*. Handbook 496. Washington, DC: US Department of Agriculture, pp. 93–98. Available at: https://www.ars.usda.gov/ARSUserFiles/20220500/OnlinePollinationHandbook.pdf (accessed January 21, 2024).

McMinn-Sauder, H., Richardson, R., Eaton, T., Smith, M. and Johnson, R. (2020) Flowers in Conservation Reserve Program (CRP) pollinator plantings and the upper Midwest agricultural landscape supporting honey bees. *Insects* 11, 405. DOI: 10.3390/insects11070405

Mogren, C. and Lundgren, J. (2016) Neonicotinoid-contaminated pollinator strips adjacent to cropland reduce honey bee nutritional status. *Science Reports* 6, 29608. DOI: 10.1038/srep29608

Mullin, C., Frazier, M., Frazier, J., Ashcraft, S., Simonds, R. *et al.* (2010) High levels of miticides and agrochemicals in North American apiaries: implications for honey bee health. *PLoS ONE* 5, e9754. DOI: 10.1371/journal.pone.0009754

Murray, C. (2021) Pollinator response to prairie strips in the Iowa agricultural landscape. MS thesis. Ames, IA: Iowa State University. DOI: 10.31274/etd-20210609-129

Nekoei, S., Rezvan, M., Khamesipour, F., Mayack, C., Molento, M.B. *et al.* (2023) A systematic review of honey bee (*Apis mellifera*, Linnaeus, 1758) infections and available treatment options. *Veterinary Medicine & Science* 9, 1848–1860. DOI: 10.1002/vms3.1194

Nuyttens, D., Devarrewaere, W., Verboven, P. and Foqué, P. (2013) Pesticide-laden dust emission and drift from treated seeds during seed drilling: a review. *Pest Management Science* 69, 564–575. DOI: 10.1002/ps.3485

Otto, C.R., Roth, C., Carlson, B. and Smart, M. (2016) Land-use change reduces habitat suitability for supporting managed honey bee colonies in the Northern Great Plains. *Proceedings of the National Academy of Sciences* 113, 10430–10435. DOI: 10.1073/pnas.1603481113

Otto, C., Zheng, H., Gallant, A., Iovanna, R., Carlson, B. *et al.* (2018) Past role and future outlook of the Conservation Reserve Program for supporting honey bees in the Great Plains. *Proceedings of the National Academy of Sciences* 115, 7629–7634. DOI: 10.1073/pnas.1800057115

Phiri, B.J., Fèvre, D. and Hidano, A. (2022) Uptrend in global managed honey bee colonies and production based on a six-decade viewpoint, 1961–2017. *Scientific Reports* 12, 21298. DOI: 10.1038/s41598-022-25290-3

Pisa, L., Amaral-Rogers, V., Belzunces, L., Bonmatin, L., Downs, J. *et al.* (2015) Effects of neonicotinoids and fipronil on non-target invertebrates. *Environmental Science and Pollution Research* 22, 68–102. DOI: 10.1007/s11356-014-3471-x

Pozebon, H., Marques, R.P., Padilha, G., O'Neal, M.E., Valmorbida, I. *et al.* (2020) Arthropod invasions versus soybean production in Brazil: a review. *Journal of Economic Entomology* 113, 1591–1608. DOI: 10.1093/jee/toaa108

Pritchard, Z., Hendriksma, H., St. Clair, A., Stein, D., Dolezal, A. *et al.* (2021) Do viruses from managed honey bees (Hymenoptera: Apidiae) endanger wild bees in native prairies? *Environmental Entomology* 50, 455–466. DOI: 10.1093/ee/nvaa181

Quinlan, G., Milbrath, M., Otto, C. and Isaacs, R. (2021) Honey bee (*Apis mellifera*) colonies benefit from grassland/pasture while bumble bee (*Bombus impatiens*) colonies in the same landscapes benefit from non-corn/soybean cropland. *PLoS ONE* 16, e0257701. DOI: 10.1371/journal.pone.0257701

Quinlan, G., Isaacs, R., Otto, C., Smart, A. and Milbrath, M. (2023) Association of excessive precipitation and agricultural land use with honey bee colony performance. *Landscape Ecology* 38, 1555–1569. DOI: 10.1007/s10980-023-01638-6

Ragsdale, D., McCornack, B., Venette, R., Potter, B., Macrae, I. *et al.* (2007) Economic threshold for soybean aphid (Hemiptera: Aphididae). *Journal of Economic Entomology* 100, 1258–1267. DOI: 10.1093/jee/100.4.1258

Ricigliano, V., Mott, B., Maes, P., Floyd, A., Fitz, W. *et al.* (2019) Honey bee colony performance and health are enhanced by apiary proximity to US Conservation Reserve Program (CRP) lands. *Scientific Reports* 9, 4894. DOI: 10.1038/s41598-019-41281-3

Robinson, A., Peeler, J., Prestby, T., Goslee, S., Anton, K. *et al.* (2021) Beescape: characterizing user needs for environmental decision support in beekeeping. *Ecological Informatics* 64, 101366. DOI: 10.1016/j.ecoinf.2021.101366

Schulte, L., Niemi, J., Helmers, M., Liebman, M., Arbuckle, J. *et al.* (2017) Prairie strips improve biodiversity and the delivery of multiple ecosystem services from corn-soybean croplands. *Proceedings of the National Academy of Sciences* 114, 11247–11252. DOI: 10.1073/pnas.1620229114

Simon-Delso, N., Amaral-Rogers, V., Belzunces, L., Bonmatin, J., Chagnon, M. *et al.* (2015) Systemic insecticides (neonicotinoids and fipronil): trends, uses, mode of action and metabolites. *Environmental Science and Pollution Research* 22, 5–34. DOI: 10.1007/s11356-014-3470-y

Siviter, H., Bailes, E., Martin, C., Oliver, T., Koricheva, J. *et al.* (2021) Agrochemicals interact synergistically to increase bee mortality. *Nature* 596, 389–392. DOI: 10.1038/s41586-021-03787-7

Smart, M., Otto, C. and Lundgren, J. (2019) Nutritional status of honey bee (*Apis mellifera* L.) workers across an agricultural land-use gradient. *Science Reports* 9, 16252. DOI: 10.1038/s41598-019-52485-y

Sponsler, D. and Johnson, R. (2015) Honey bee success predicted by landscape composition in Ohio, USA. *PeerJ* 3, e838. DOI: 10.7717/peerj.838

Steinhauer, N., Wilson, M., Aurell, D., Bruckner, S. and Williams, G. (2023) *United States Honey Bee Colony Losses 2022–2023: Preliminary Results from the Bee Informed Partnership*. Bee Informed Partnership, Inc. Available at: https://beeinformed.org/wp-content/uploads/2023/06/BIP-2022-23-Loss-Abstract.pdf (accessed January 21, 2023).

Stewart, S.D., Lorenz, G.M., Catchot, A.L., Gore, J., Cook, D. *et al.* (2014) Potential exposure of pollinators to neonicotinoid insecticides from the use of insecticide seed treatments in the mid-southern United States. *Environmental Science and Technology* 48, 9762–9769. DOI: 10.1021/es501657w

Tuell, J., Fiedler, A., Landis, D. and Isaacs, R. (2008) Visitation by wild and managed bees (Hymenoptera: Apoidea) to eastern U.S. native plants for use in conservation programs. *Environmental Entomology* 37, 707–718. DOI: 10.1603/0046-225X(2008)37[707:VBWAMB]2.0.CO;2

Tyndall, J., Schulte, L., Liebman, M. and Helmers, M. (2013) Field-level financial assessment of contour prairie strips for enhancement of environmental quality. *Environmental Management* 52, 736–747. DOI: 10.1007/s00267-013-0106-9

Urbanowicz, C., Muniz, P.A. and McArt, S.H. (2020) Honey bees and wild pollinators differ in their preference for and use of introduced floral resources. *Ecology and Evolution* 10, 6741–6751. DOI: 10.1002/ece3.6417

Ullah, A., Gajger, I., Majoros, A., Dar, S., Khan, S. *et al.* (2021) Viral impacts on honey bee populations: a review. *Saudi Journal of Biological Sciences* 28, 523–530. DOI: 10.1016/j.sjbs.2020.10.037

USDA-FSA (2013) *CP42 Pollinator Habitat: Establishing and Supporting Diverse Pollinator Friendly Habitat*. Washington, DC: US Department of Agriculture, Farm Service Agency. Available at: https://www.fsa.usda.gov/Internet/FSA_File/cp42_habitat.pdf (accessed September 25, 2023).

USDA-FSA (2019) *Conservation Reserve Program: CP43 Prairie Strips*. Washington, DC: US Department of Agriculture, Farm Service Agency. Available at: https://www.fsa.usda.gov/Assets/USDA-FSA-Public/usdafiles/FactSheets/2019/crp_clear_initiative_prairie_strip_practice-fact_sheet.pdf (accessed January 26, 2024).

USDA-FSA (2022) *Conservation Reserve Program: Fact Sheet*. Washington, DC: US Department of Agriculture, Farm Service Agency. Available at: https://www.fsa.usda.gov/Assets/USDA-FSA-Public/usdafiles/FactSheets/2019/conservation-reserve_program-fact_sheet.pdf%20 (accessed January 26, 2024).

USDA-FSA (2023a) *CP43 Prairie Strips: Erosion Control, Water Quality Enhancement, Wildlife Habitat*. Washington, DC: US Department of Agriculture, Farm Service Agency. Available at: https://www.fsa.usda.gov/Assets/USDA-FSA-Public/usdafiles/Conservation/PDF/CP43.pdf (accessed September 25, 2023).

USDA-FSA (2023b) *Conservation Reserve Program Monthly Summary – October 2023*. Washington, DC: US Department of Agriculture, Farm Service Agency. Available at: https://www.fsa.usda.gov/Assets/

USDA-FSA-Public/usdafiles/Conservation/PDF/CRPMonthlyOctober2023WithPageNumbers.pdf (accessed January 28, 2024).

USDA-FSA (2024) *CRP Practices Library*. Washington, DC: US Department of Agriculture, Farm Service Agency. Available at: https://www.fsa.usda.gov/programs-and-services/conservation-programs/crp-practices-library/index (accessed May 6, 2024).

USDA-NASS (2022) *CropScape – Cropland Data Layer*. Washington, DC: US Department of Agriculture, National Agricultural Statistics Service. Available at: https://nassgeodata.gmu.edu/CropScape/ (accessed August 10, 2023).

USDA-NASS (2024) *Honey*. Washington, DC: US Department of Agriculture, National Agricultural Statistics Service. Available at: https://usda.library.cornell.edu/concern/publications/hd76s004z?locale=en (accessed January 26, 2024).

USGS (2017) *National Water-Quality Assessment Program: Estimated Annual Agricultural Pesticides Use*. Washington, DC: US Geological Survey. Available at: https://water.usgs.gov/nawqa/pnsp/usage/maps/ (accessed January 21, 2023).

Visscher, P. and Seeley, T. (1982) Foraging strategy of honeybee colonies in a temperate deciduous forest. *Ecology* 63, 1790–1801. DOI: 10.2307/1940121

Weaver, J., Ascher, J. and Mallinger, R. (2022) Effects of short-term managed honey bee deployment in a native ecosystem on wild bee foraging and plant-pollinator networks. *Insect Conservation and Diversity* 15, 634–644. DOI: 10.1111/icad.12594

Whalen, A., Catchot, A., Gore, J., Stewart, S., Lorenz, G. *et al.* (2021) Temporal profile of neonicotinoid concentrations in cotton, corn, and soybean resulting from insecticidal seed treatments. *Agronomy* 11, 1200. DOI: 10.3390/agronomy11061200

Winston, M. (1987) *The Biology of the Honey Bee*. Cambridge, MA: Harvard University Press.

Wintle, B., Kujala, H., Whitehead, A., Comeron, A., Veloz, S. *et al.* (2018) Global synthesis of conservation studies reveals the importance of small habitat patches for biodiversity. *Proceedings of the National Academy of Sciences* 116, 909–914. DOI: 10.1073/pnas.1813051115

Wheelock, M.J. and O'Neal, M.E. (2016) Insect pollinators in Iowa cornfields: community identification and trapping method analysis. *PLoS ONE* 11, e0143479. DOI: 10.1371/journal.pone.0143479

Wood, T. and Goulson, D. (2017) The environmental risks of neonicotinoid pesticides: a review of the evidence post 2013. *Environmental Science and Pollution Research* 24, 17285–17325. DOI: 10.1007/s11356-017-9240-x

Yang, Y. and Suh, S. (2015) Changes in environmental impacts of major crops in the US. *Environmental Research Letters* 10, 094016. DOI: 10.1088/1748-9326/10/9/094016

Zhang, G., St. Clair, A., Dolezal, A., Toth, A. and O'Neal, M.E. (2020) Honey bee (Hymenoptera: Apidea) pollen forage in a highly cultivated agroecosystem: limited diet diversity and its relationship to virus resistance. *Journal of Economic Entomology* 113, 1062–1072. DOI: 10.1093/jee/toaa055

Zhang, G., St. Clair, A., Dolezal, A., Toth, A. and O'Neal, M. (2021) North American prairie is a source of pollen for managed honey bees (Hymenoptera: Apidae). *Journal of Insect Science* 21, 17. DOI: 10.1093/jisesa/ieab001

Zhang, G., St. Clair, A., Dolezal, A., Toth, A. and O'Neal, M. (2022) Can native plants mitigate climate-related forage dearth for honey bees (Hymenoptera: Apidae)? *Journal of Economic Entomology* 115, 1–9. DOI: 10.1093/jee/toab202

Zhang, G., Murray, C., St. Clair, A., Cass, R., Dolezal, A. *et al.* (2023) Native vegetation embedded in landscapes dominated by corn and soybean improves honey bee health and productivity. *Journal of Applied Ecology* 60, 1032–1043. DOI: 10.1111/1365-2664.14397

13 Future of Information Systems for Pest Management: Data Acquisition and Integration to Guiding Management Decisions

Mahendra Bhandari*, Pankaj Pal, and Michael J. Brewer
Texas A&M AgriLife Research, Corpus Christi, TX, USA

Abstract

Integrated information systems in agriculture have the potential to increase farm efficiency, reduce environmental impact, and improve overall productivity. This chapter explores information systems in the context of pest management, emphasizing a holistic framework that spans from data acquisition, management, analytics, to informed decision making. Contemporary examples of components of an information system are presented, inclusive of insect pest management, while future developments are partly adopted from examples coming from the allied fields of agronomy, data sciences, and agricultural engineering. The 21st century unveiled unmanned aerial systems (UAS), commonly referred to as drones, revolutionizing high-resolution aerial imaging. Improvements over past sensing include better identification of causes of plant stress and analyzing data facilitated by geographic information systems (GIS) and artificial intelligence (AI). For example, detecting pest-related stress at an early stage may be facilitated by fine-resolution imagery acquired by UAS while continuously tracking the dynamics of plant damage at a larger scale using satellite imagery. Integration of data acquired at the ground (e.g. ground robots, tractor), very low altitude (e.g. UAS), and very high altitude (e.g. satellite) levels, combined with on-farm records, weather, and other data layers, will further improve data quality and integration. Using multiple data sources will open avenues to improve spatial and temporal data analyses and facilitate decision support using advanced analysis and AI tools. Through embracing technological advances, information systems for pest management can become proactive in minimizing crop damage from insect pests from in-field to larger scale pest risk and forecasting assessments. An information system with both plant production and protection features will increase its agricultural appeal and sustainability.

Introduction

Crops are integrators. Their response to environmental conditions, agricultural management, and interactions of these factors throughout their development stages leads to the desired agricultural product output. As such, the impact of biotic stresses inclusive of insect pests and abiotic stresses such as adverse weather is reflected in several forms of plant response. For example, insects with chewing mouth parts (e.g. Coleoptera, larval Lepidoptera, Orthoptera) can reduce plant biomass through their consumption of leaf, stem, roots, and/or reproductive tissue. Changes occur in

*Corresponding author: Mahendra.Bhandari@ag.tamu.edu

 Arthropod Management and Landscape Considerations in Large-Scale Agroecosystems (eds M.J. Brewer and G.L. Hein)
DOI: 10.1079/9781800622777.0013

leaf shape, canopy density, and plant deterioration including discoloration and necrosis. These changes create variations in the visual reflectance spectrum as well as possible changes in spectral reflectance outside of the visual range (Nansen and Elliott, 2016). Insects with sucking and rasping mouth parts (e.g. Hemiptera, Thysanoptera) can remove plant vascular fluids, disrupting photosynthesis, nutrient transport, and other physiological processes. Because these changes often have minimal impact on plant structure, they are more challenging to detect visually, but changes in spectral reflectance within and outside the visual range have been well documented (Nansen and Elliott, 2016). Plant response to specific insect feeding can result in varied spectral reflectance in the visible range, from pale green and yellow, to leaf streaking, and spotting of purple and white shades. Changes in infrared (IR) reflectance and the reflectance pattern also occur. Aphids and other sucking arthropods may introduce plant pathogens that further disrupt plant physiology. These disruptions can result in lesions such as leaf spotting and general discoloration and further spectral reflectance change outside the visual range. For both chewing and sucking insects, the extent of injury varies by crop sensitivity to feeding and pest density and can be exacerbated by other stressors such as plant water stress (e.g. Bhandari *et al.*, 2021). The progression of plant injury in selected parts of the plant or field can expand to more intense injury within the plant canopy, spatially across large sections of a field, and potentially across fields and even regions when conditions are conducive to pest outbreaks.

Regarding spatial detection and evaluation of intensity of plant injury (or insect abundance) in large crop production systems, data collection and information systems (e.g. optical and other sensors, associated data analysis, decision support tools) are needed. Information systems can supplement, authenticate, and improve the effectiveness and efficiency of traditional direct observations by humans (e.g. crop and pest monitoring by direct visual observations). Partial to complete autonomy of data collection and information systems is desirable with the objective of alerting the user to potential pest threats and possible mitigation actions. Research in selected agroecosystems and crop–pest pairings has demonstrated the capabilities of capturing the response of plants to insect stress by remote sensing (for overview, see Nansen and Elliott, 2016; Filho *et al.*, 2020). For example, plant stress in corn caused by the European corn borer (*Ostrinia nubilalis* [Hübner] [Lepidoptera: Crambidae]), has been detected by hyperspectral sensors mounted in fixed-wing piloted aircraft (Carroll *et al.*, 2008). Progress has been made in distinguishing insect-derived plant stress from other forms of plant stress using unmanned aerial systems (UAS)-mounted sensors based on spectral reflectance differences and the spatial form of plant stress. For example, discoloration of foliage following linear rows caused by fertilization or drought issues can be contrasted with irregular patches of discoloration caused by local aphid infestation (Backoulou *et al.*, 2016; Marston *et al.*, 2020). Identification of the cause of plant stress by remote sensing alone remains challenging, and verification by direct observations is advisable. More advanced analytics and artificial intelligence (AI) support (see below) may improve this situation.

However, few applications have addressed large spatial scales and extended time frames appropriate to monitoring crop–pest interactions often found in large-scale crop, forage, and forest systems (Cook *et al.*, 2007; Westbrook *et al.*, 2014). Wulder *et al.* (2008) demonstrated that mountain pine beetle (*Dendroctonus ponderosae* [Hopkins] [Coleoptera: Curculionidae, Scolytinae])-induced tree injury progression over time and space could be monitored by extrapolating tree-level damage to a larger study area by integrating field information, forest inventory data, and high spatial satellite imagery data obtained from QuickBird, an online satellite imagery source (Apollo Mapping, 2023). Such data integration from multiple sources for plant damage monitoring is still in its infancy, but this study serves as a prototype for future development to address pest insects in large-scale systems.

Progress in remote sensing, robotics, data integration, and information delivery may be adapted for insect pest management. New methods may support and improve upon traditional direct sensing techniques, such as visual inspection, that are labor intensive and prone to acquisition (e.g. sampler and spatial) errors, including when aided by a geographic information system (GIS) (e.g. Deleon *et al.*, 2017). For example, entomologists relying on visual inspection of crops to detect and quantify plant injury or insect abundance will likely benefit (in scale and objectivity) from remote sensing of plant canopy density and leaf discoloration (Elliott *et al.*, 2015; Marston *et al.*, 2020). Insect traps and other detection devices (e.g. pitfall traps, light and pheromone traps, suction traps) would benefit from technologies to automate counting and insect identification (Rydhmer *et al.*, 2022). For large crop production systems, these benefits accelerate if the system can be deployed in a time-efficient and scale-appropriate manner.

In this chapter, we discuss current and future technological advancements in data collection, management, and the development of decision support tools. Figure 13.1 provides an overall vision for this chapter in which we propose to discuss data collection from ground, aerial, and satellite systems. The data systems will potentially be complemented with seamless data transfer using wireless connectivity networks and storage using cloud services for further analytics, assessment, and management. The current examples we present are inclusive of research found in the entomological literature, while anticipated developments come in part from allied fields of agronomy, data and computer sciences, and agricultural engineering. Our goal is to showcase current progress and future trends that will better serve pest management from in-field (precision agriculture applications) to larger-scale pest risk and forecasting assessments emphasized in this book.

Evolution of Data Acquisition Technologies

Agriculture is benefitting from new developments in data acquisition methods that can be applied to pest management (e.g. Sharma *et al.*, 2022; Pal *et al.*, 2022; Edan *et al.*, 2023). In the mid 20th century, radar technology that was appropriate for local detection of insect swarms entered the scene and later advanced to use of doppler radar that allowed monitoring of insect movement over large areas. Westbrook *et al.* (2014) explored the use of radar to estimate the migratory flux of corn earworm (*Helicoverpa zea* [Boddie] [Lepidoptera: Noctuidae]). Detecting aerial swarms of insects by radar can be used to provide insights into insect migration and movement patterns and improve interpretation risk assessment of insect pest detection from regional trapping networks (e.g. Lawton *et al.*, 2022; see Chapter 10, this volume).

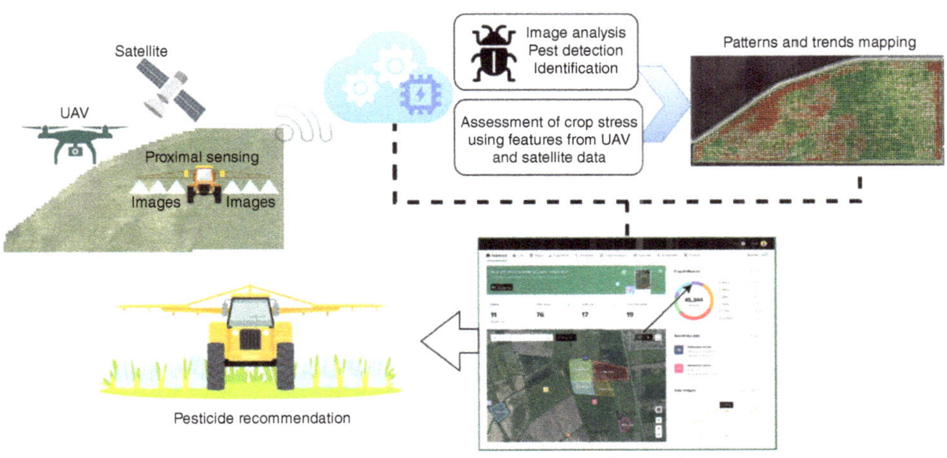

Fig. 13.1. Overview of a workflow for multi-sensing data acquisition, transfer, management, and integration for pest management. This rendering applies to the use of insecticides within or across fields depending on the extent of pest-induced stress. The same process applies for use of other remedial pest management tactics applied at local to regional scales (see Chapters 1 and 2, this volume). (Content is the author's own: M. Bhandari.)

The late 20th century saw advancements in aerial photography and satellite imagery, opening new possibilities for research in crop growth monitoring and assessing pest and disease progression from local to regional scales (e.g. Shahi *et al.*, 2023). Scientists could now map insect and disease infestations as well as study environmental factors influencing insect populations and damage (see Chapters 5, 6, and 7, this volume), but applications were still limited (e.g. Cook *et al.*, 2007). GIS technology enabled overlay of data layers, including spatially referenced insect distribution data, environmental variables, and UAS and other remotely sensed data (e.g. Brewer *et al.*, 2022; see Chapter 10, this volume). This has enabled a deeper understanding of insect habitats and environmental factors and their impact on insect life history and movement, which leads to variation in insect injury and outbreaks across the spatial domain (e.g. Cook *et al.*, 2007).

The 21st century ushered in a new era with the advent of UAS, commonly referred to as drones, revolutionizing high-resolution aerial imaging. The increased quality and efficiency of these technologies make them more useful in collecting an array of data in both the spatial and temporal domains (Quamar *et al.*, 2023). UAS-mounted RGB (red, green, and blue) optical cameras can capture detailed visual information for aerial surveys, aiding in vegetation monitoring in which structural change (e.g. canopy density and height) and color changes may be associated with insect and other plant stressors (Barbedo, 2019; Marston *et al.*, 2020; Bhandari *et al.*, 2021; Zhao *et al.*, 2023). Their increased use in agriculture makes them attractive for multi-stress detection for precision agriculture applications, but the spectral sensing limitation to the visual bands is problematic for insect pest management applications (Filho *et al.*, 2020, Moses-Gonzales and Brewer, 2021). Multispectral and hyperspectral cameras can improve detection and enable the estimation of plant structure and spectral changes associated with insect stress in crops (e.g. Carroll *et al.*, 2008; Elliott *et al.*, 2015; Backoulou *et al.*, 2016). This includes the use of lens filters for RGB cameras that extend detection to IR wavelengths (Stanton *et al.*, 2017). Upon further processing of these imagery data, crop morphological features can be extracted across space and time to better assess crop growth and development as used in yield prediction models (e.g. Feng *et al.*, 2023). These data sets can also provide detailed information about vegetation health and stress, indirectly revealing pest insect presence and severity. Thermal cameras can measure surface temperatures, helping identify variation in crops associated with stress factors including water stress that can be exacerbated by insect pests (Awais *et al.*, 2023). Apart from the passive sensors such as RGB cameras, and the multispectral and hyperspectral sensors that sense reflected light, LiDAR (Light Detection and Ranging) is an active sensor that detects its own emitted laser light pulses to measure distance. LiDAR can play a significant role in studying the structural damage in plants caused by pests. LiDAR can facilitate the creation of accurate three-dimensional maps, enabling the study of vegetation structure, including the extent of plant injury attributable to insect pests (Dwivedi *et al.*, 2020). However, significant research needs to be done to apply LiDAR in real-world environments.

For local to regional insect pest outbreaks, UAS facilitate rapid and targeted assessment of affected areas, allowing for more timely responses to mitigate pest outbreaks that are spatially variable (Stumph *et al.*, 2019). These versatile platforms, whether piloted remotely or flying autonomously, can be equipped with an impressive array of sensors, cameras, and instruments, enabling efficient remote data collection. Although UAS can generate high spatial resolution data that is valuable in pest severity assessment, large area coverage is highly constrained due to limited flight endurance and flight regulations (Moses-Gonzales and Brewer, 2021).

Satellite-based remote sensing provides extensive land coverage and is suitable for large cropping systems. However, it has historically struggled with coarse spatial resolution, which impedes efficient quantification of crop growth and development, including change detection associated with insect pests. Detecting the spatial extent of insect outbreaks by severe pests is feasible (e.g. grasshopper outbreaks causing substantial defoliation of grasslands) (Jonas *et al.*, 2015), but the coarser spatial resolution of satellite images limits more detailed assessment of insect pest infestations. Estimation of outbreak severity at a local scale is limited, as is detection of less intense insect pest outbreaks causing limited change to spectral reflectance and canopy structure.

More recently, significant advancements have been made in the integration of sensors and cameras in farm equipment. For example, one of the latest precision agriculture products is tractor-mounted optical sensors and supporting AI technology that has been applied to weed control (Vijayakumar *et al.*, 2023; John Deere, 2024). This technology utilizes computer vision and AI. It first captures data to detect weeds and activates the specific nozzles to precisely apply herbicide on the detected weed. More generally at the research level, integrating cameras with highly accurate real-time kinematic Global Positioning System (RTK GPS) technology to capture images at plant level and with support from AI has been applied to site-specific pest management applications for weed control (Thorp and Tian, 2004; Batte and Ehsani, 2006; Machleb *et al.*, 2020). Applications to mobile insects such as grasshoppers and planthoppers are more challenging but may be feasible for relatively sessile insects such as wingless aphids and larval stages of moths and beetles (see Chapter 1, this volume).

Looking to the future, we propose that integration of data acquired at the ground (e.g. ground robots, tractor), very low altitude (e.g. UAS), and very high altitude (e.g. satellite) levels will open avenues to improve both spatial and temporal data resolution and acquisition. This will lead to the development of models and AI tools, leading to improved pest identification, forecasting, and decision management tools. Multi-source data integration studies have been conducted for yield predictions of cotton using soil texture, weather conditions, and UAS imagery (Feng *et al.*, 2023), of maize using satellite imagery and crop models (Joshi *et al.*, 2019), and of soybean and maize using different types of satellite imagery (Sagan *et al.*, 2021). Additionally, a crop monitoring framework that integrates satellite-based canopy spectral information with structural information obtained from UAS improved the prediction of soybean leaf area index, aboveground biomass, and leaf nitrogen concentration (Maimaitijiang *et al.*, 2020). Detecting specific diseases and pests or pest-related stress at an early stage may be facilitated by fine-resolution imagery acquired by ground or UAS, while continuously tracking the dynamics of severity and damage at a larger scale may be facilitated by satellite imagery. The satellite resolution challenges can be offset by pairing these broad satellite views with high-resolution "snap-shots" taken by UAS at selection locations of the satellite's coverage. Source data can be integrated, and new information generated with the aid of AI systems.

A Unified Data Ecosystem for Pest Management

Data standardization and harmonization are very important to agricultural applications when data sets are generated from multiple sensors and platforms, as considered here for advanced data integration and pest management decision support. Due to the broad spectrum of disciplines within agriculture, standard best practices for data storage and management are not applied as a common practice at this time (Moore *et al.*, 2022). Nevertheless, considerable attention has been paid from various sectors to address the issue of data management in agriculture. FAIR (Findability, Accessibility, Interoperability, and Reusability) data principles, initially introduced by Wilkinson *et al.* (2016), provide guidelines on enhancing data availability, use, and reuse. Regular reviews and updates ensure that FAIR practices adapt to evolving technologies and research needs. These principles are relevant across applications, particularly in crop production and protection applications that utilize multiple data sources.

In the case of pest management in large-scale cropping systems, vast quantities of data may be collected using different tools such as proximal sensors, *in situ* sensors, weather stations, satellite-based imaging systems, and on-farm scouting (Tonnang *et al.*, 2022). These data sets are in different forms and formats. To facilitate data integration and retain integrity, a user-friendly data system is needed to easily navigate and derive value from the data. The depth of data use and integration is substantial and ranges across spatial scales. At the intrafield and local scales, this includes relatively simple assessment and mapping of in-field and interfield variability of pest occurrence and abundance (e.g. Deleon *et al.*, 2017) and monitoring disease and pest detection across time and space (e.g. Corn ipmPIPE, 2019). At the regional and near-continental scales, predicting routes of insect pest dispersal and outbreak potential (e.g. Westbrook *et al.*, 2011; Wang *et al.*, 2019) and advanced guidance of regional pest management decisions (e.g. Isard *et al.*, 2011; VanKirk *et al.*, 2012) contribute to a farmer's risk assessment and management decision process.

At the early stages of application for agricultural uses, centralized data repositories and data management and analyses systems are valued and needed. Cloud-based systems enhance accessibility and collaborative storage and retrieval for large volumes of structured and unstructured data. Standardized data formats and protocols enable seamless integration across different data sources, while data quality assurance protocols identify and address errors, inconsistencies, and missing information. Metadata documentation enhances transparency and reproducibility, and version control ensures historical data remain accessible. Robust security measures safeguard sensitive information. The question of data ownership and usage rights are topics of debate and concern in the agricultural community that become convoluted with the desire for standardization and accessibility. Clear guidelines on data ownership, privacy, and usage among growers (data generators), technology providers, and data managers need to be established.

Advanced analytics techniques, such as machine learning and statistical modeling, extract valuable insights from large data sets. For example, system-level data allow for predictive modeling of pest outbreaks based on historical and environmental factors, but examples are mainly at the research and prototype stage or have resulted in decision support systems (DSS) that have been difficult to sustain (e.g. VanKirk *et al.*, 2012; Wang *et al.*, 2019; see Chapter 7, this volume). Integration with external data sources further enriches analytical capabilities, while real-time monitoring with sensor data allows for prompt responses to changing conditions. All these processes exist, while the novelty and challenge remain in developing a sustainable information system for pest management. Components of such a system are challenging to maintain (e.g. VanKirk *et al.*, 2012). Some of the databases for integrated pest management (IPM) can be further enhanced by integrating pest monitoring data sets generated from several different platforms to better understand the dynamics and develop management options (VanKirk *et al.*, 2012). By establishing a unified data ecosystem for pest management, stakeholders gain access to timely, accurate, and actionable insights generated by the information system.

Beyond the technology and data integration challenges, implementing these principles and processes requires collaboration with stakeholders and the establishment of community standards. Ideally, this will foster data sharing and promote regional best IPM practices that are broadly beneficial and help address drawbacks envisioned by Hardin's "tragedy of the commons" associated with regional and areawide pest management approaches (see Chapter 1, this volume). We propose that by embracing the FAIR principles and collaborating effectively, a robust and unified data ecosystem can be built, sustained, and empower researchers and insect pest management practitioners to deploy effective IPM strategies, ultimately safeguarding our agricultural systems.

Summary and Conclusions

There is substantial room for development and implementation of effective strategies for identifying, monitoring, and managing insect pests within fields and across broader landscape scales by adapting current and future technologies. Ideally, integrating plant production and crop protection management into a unified information system attractive to agriculturalists, inclusive of both plant production and plant protection components that share databases, will have tremendous value in the advancement of crop pest management systems. The 21st century unveiled UAS, commonly referred to as drones, revolutionizing high-resolution aerial imaging. Improvements included better identification of causes of plant stress and analysis of data facilitated by GIS and AI. For example, ground-based sensors initially identify pests or associated symptoms followed by the detection of pest-related stress at an early stage facilitated by fine-resolution imagery acquired by UAS. This may be further facilitated by adding satellite imagery and AI-aided analysis with products visualized using GIS tools to continuously track the dynamics of damage at a larger scale. Integration of data acquired at the ground (e.g. ground robots, tractors), very low altitude (e.g. UAS), and very high

Case Study 13.1: AI for pest identification and forecasting

The use of AI in agriculture has been explored in several applications (Talaviya *et al.*, 2020) including general crop monitoring and management, irrigation scheduling, weed management, yield predictions, and supply chain systems. It is gaining importance for entomological questions as well (Hartbauer, 2024). Toscano-Miranda *et al.* (2022) conducted a systematic review of AI and sensing techniques for the management of insect pests and diseases in cotton based on the literature published between 2014 and 2021. They found examples of support vector machines (SVMs), fuzzy inference, and convolutional neural networks (CNNs) used in image classification relevant to insect identification and categorization. The data sets were mostly obtained from cameras and *in situ* field sensors, with special attention to whiteflies (Hemiptera: Aleyrodidae). SVM is a supervised machine learning algorithm used for both classification and regression problems (Boser *et al.*, 1992; Xu *et al.*, 2009). CNNs are a class of deep neural networks primarily used for image recognition and computer vision tasks that are effective at capturing patterns within images (Albawi *et al.*, 2017). An SVM algorithm was used to classify and count planthoppers (Hemiptera: Delphacidae) in rice paddy fields using images collected by a handheld digital camera (Yao *et al.*, 2014). Object detection algorithms such as YOLO (You Only Look Once) that use CNN architecture are being explored to detect insect pests using close-up images (Li *et al.*, 2021a; Xu *et al.*, 2023). Liu and Wang (2020) utilized YOLO V3 to detect diseases (i.e. blights, leaf mold, leaf curl virus) and insect pests (i.e. leaf miners, whitefly) in tomatoes.

Most studies currently are focused on training and testing several AI algorithms using close-up images. This creates a challenge to acquire and classify close-up images within crop environments. A deep residual learning model was proposed by Cheng *et al.* (2017) to identify pests in the images captured with complex vegetation in the background. With sufficient imagery data sets, CNNs show promise in categorizing insect types using images obtained under natural settings. This advance may help pave the way to develop more robust models and mobile applications for pest identification (Barbedo, 2014; Li *et al.*, 2020; Høye *et al.*, 2021; Li *et al.*, 2021b; Batz *et al.*, 2023; Tannous *et al.*, 2023). However, the diversity of insects, varying life stages and damage in crops, significant impact of environment in pest occurrence and abundance, intraspecies variability, limited publicly available data sets, and background noise on the images pose significant challenges to train and use CNNs in insect pest identification (Hartbauer, 2024).

In addition to pest identification, a few studies have explored the possibility of using Big Data analytics to develop forecasting models for pest occurrences. One key benefit of using AI algorithms is the ability to process and integrate vast amounts of diverse data, including weather patterns, crop conditions, and historical pest occurrences. This integration of data enables the development of sophisticated models that can identify complex patterns and non-linear relationships, which may lead to more accurate predictions (Hartbauer, 2024). Prabhakar *et al.* (2022) proposed an AI-based ARIMAX-ANN (Artificial Neural Network) model using maximum and minimum temperatures and their deviations, rainfall, and relative humidity. They were able to forecast the incidence of green semilooper (*Chrysodeixis acuta* [Walker] [Lepidoptera: Noctuidae]) using these weather factors 1 week in advance of pest occurrence. Meteorological data sets can be combined with historical data sets of pest occurrence to train machine learning models for better predictions. However, it is important to identify key variables to prevent overfitting of the models (Yang *et al.*, 2009; Bahlai, 2023). In addition to developing robust models, machine learning algorithms can adapt and learn from new data, continuously improving their forecasting accuracy over time which can increase accuracy and reduce variability in pest identification and outbreak predictions.

To date, much of the work has been done by using close-up images for identification and weather parameters for pest occurrence predictions, which are basic aspects of insect pest management. In the future, efforts should include integrating data sets collected from multiple platforms such as UAS and satellite systems to develop a holistic information system inclusive of pest identification, occurrence predictions, and management (Hartbauer, 2024). Additionally, temporal changes in crop growth dynamics can be easily measured by using remote sensing systems. These agronomic models can be integrated into pest forecasting models that may help apply the information system to large-scale cropping systems, which should entice broad and more sustained use and support by agriculturalists. This capacity has a direct link to landscape considerations in pest management (see Chapters 1 and 2, this volume), from intrafield and local landscape scales to regional-scale assessment of pest risk. By employing computer vision, AI, and predictive modeling integrated with expert systems, the future of pest management will not only be reactive to detected pest outbreaks, but also proactive in forecasting pest threats. This will aid in the development and implementation of preventive measures from local to regional scales (see Chapter 2, this volume).

Case Study 13.2: Decision support using apps and web-based interfaces

Mobile applications ("apps") for pest detection and identification are usually equipped with tools that leverage computer vision technologies, and they are becoming popular as stand-alone DSS. These applications provide digital expertise to visually identify pests and assess information on life cycles and behaviors of insect pests (see Chapter 1, this volume). Karar *et al.* (2021) developed an app that uses CNN to recognize crop insects such as aphids, leafhoppers, flea beetles, and spiders. These arguably basic capabilities of an information system are particularly valuable in the developing world where producers have limited access to expert services such as pest management consultants (see Chapter 4, this volume). The Fall Army Worm Monitoring and Early Warning System (FAMEWS), an app for Android phones, was developed for aiding monitoring of fall armyworm (*Spodoptera frugiperda* [J.E. Smith] [Lepidoptera: Noctuidae]) (FAO, 2024). This tool has been developed in multiple languages. Farmers and pest management practitioners can use the app to record fall armyworm incidence in space and time as fields are scouted. The app can be enhanced in a user-friendly web-based interface that integrates real-time data on weather conditions and crop development information obtained from UAS and satellite systems. This step-ladder approach from a basic to more comprehensive system can provide farmers with pest identification and archived pest management options as a base product, with advanced extensions for insect monitoring inputs, pest map production, and pest forecasting capabilities. These interfaces can be equipped with alert systems and notifications to keep farmers informed about potential pest outbreaks, facilitating rapid responses. Educational resources also can be made available through these tools to further empower farmers by providing training modules and information on best practices in insect pest management.

DSS to more advanced information systems can play a pivotal role in promoting sustainable agriculture practices, minimizing the use of chemical inputs, reducing environmental impact, and optimizing crop yield through effective pest management (Tonle *et al.*, 2024). However, there are several factors to be considered when developing a DSS or information system. These include profitability, maintaining a user-friendly interface and adapting to the farm and production system, time requirement for usage, information updating, and level of knowledge of the user. As noted at the beginning of this chapter, the concepts of a pest information system are sound, and technology and process are available; however, application-oriented research and development, social and economic evaluation of grower expectations and needs, and consideration of scale of data and use (e.g. farm-scale to regional pest risk, UAS to satellite data) will be essential to move from basic applications (e.g. pest identification and mapping) to more comprehensive information systems. Ideally, these need to be linked to or part of a unified information system attractive to agriculturalists working in plant production and protection.

altitude (e.g. satellite) levels, combined with on-farm records, weather, and other data layers, will further improve data quality and integration. Using multiple data sources will open avenues to improve spatial and temporal data analyses and facilitate decision support using advanced analyses and AI tools. Through embracing technology advances, information systems for pest management can improve basic pest management support such as pest identification but also advance pest forecasting and pest management decision support from in-field to larger scale applications. They become more sustainable and attractive when linked to or part of a unified information system with both plant production and protection features.

Acknowledgment

The authors would like to thank Ms Stephanie A. Klock for assistance in formatting.

References

Albawi, S., Mohammed, T.A. and Al-Zawi, S. (2017) Understanding of a convolutional neural network. In: *2017 International Conference on Engineering and Technology (ICET), Antalya, Turkey*. Piscataway, NJ: Institute of Electrical and Electronics Engineers, pp. 1–6. DOI: 10.1109/ICEngTechnol.2017.8308186

Apollo Mapping (2023) *Apollo Mapping: The Satellite Imagery Source*. Boulder, CO: Apollo Mapping. Available at: https://apollomapping.com/ (accessed April 22, 2024).

Awais, M., Li, W., Cheema, M.J.M., Zaman, Q.U., Shaheen, A. *et al.* (2023) UAV-based remote sensing in plant stress imagine using high-resolution thermal sensor for digital agriculture practices: a meta-review. *International Journal of Environmental Science and Technology* 20, 1135–1152. DOI: 10.1007/s13762-021-03801-5

Backoulou, G.F., Elliott, N.C. and Giles, K.L. (2016) Using multispectral imagery to compare the spatial pattern of injury to wheat caused by Russian wheat aphid and greenbug. *Southwestern Entomologist* 41, 1–8. DOI: 10.3958/059.041.0101

Bahlai, C.A. (2023) Forecasting insect dynamics in a changing world. *Current Opinion in Insect Science* 60, 101133. DOI: 10.1016/j.cois.2023.101133

Barbedo, J.G.A. (2014) Using digital image processing for counting whiteflies on soybean leaves. *Journal of Asia-Pacific Entomology* 17, 685–694. DOI: 10.1016/j.aspen.2014.06.014

Barbedo, J.G.A. (2019) A review on the use of unmanned aerial vehicles and imaging sensors for monitoring and assessing plant stresses. *Drones* 3, 40. DOI: 10.3390/drones3020040

Batte, M.T. and Ehsani, M.R. (2006) The economics of precision guidance with auto-boom control for farmer-owned agricultural sprayers. *Computers and Electronics in Agriculture* 53, 28–44. DOI: 10.1016/j.compag.2006.03.004

Batz, P., Will, T., Thiel, S., Ziesche, T.M. and Joachim, C. (2023) From identification to forecasting: the potential of image recognition and artificial intelligence for aphid pest monitoring. *Frontiers Plant Science* 14, 1150748. DOI: 10.3389/fpls.2023.1150748

Bhandari, M., Baker, S., Rudd, J.C., Ibrahim, A.M.H., Chang. A. *et al.* (2021) Assessing the effect of drought on winter wheat growth using unmanned aerial system (UAS)-based phenotyping. *Remote Sensing* 13, 1144. DOI: 10.3390/rs13061144

Boser, B.E., Guyon, I.M. and Vapnik, V.N. (1992) A training algorithm for optimal margin classifiers. In: *COLT '92: Proceedings of the Fifth Annual Workshop on Computational Learning Theory*. New York: Association for Computing Machinery, pp. 144–152. DOI: 10.1145/130385.130401

Brewer, M.J., Elliott, N.C., Esquivel, I.L., Jacobson, A.L., Faris, A.M. *et al.* (2022) Natural enemies, mediated by landscape and weather conditions, shape response of the sorghum agroecosystem of North America to the invasive aphid *Melanaphis sorghi*. *Frontiers in Insect Science* 2, 830997. DOI: 10.3389/finsc.2022.830997

Carroll, M.W., Glaser, J.A., Hellmich, R.L., Hunt, T.E., Sappington, T.W. *et al.* (2008) Use of spectral vegetation indices derived from airborne hyperspectral imagery for detection of European corn borer infestation in Iowa corn plots. *Journal of Economic Entomology* 101, 1614–1623. DOI: 10.1093/jee/101.5.1614

Cheng, X., Zhang, Y., Chen, Y., Wu, Y. and Yue, Y. (2017) Pest identification via deep residual learning in complex background. *Computers and Electronics in Agriculture* 141, 351–356. DOI: 10.1016/j.compag.2017.08.005

Cook, S., Cherry, S., Humes, K., Guldin, J. and Williams, C. (2007) Development of a satellite-based hazard rating system for *Dendroctonus frontallis* (Coleoptera: Scolytidae) in the Ouachita Mountains of Arkansas. *Journal of Economic Entomology* 100, 381–388. DOI: 10.1093/jee/100.2.381.

Corn ipmPIPE (2019) *Corn ipmPIPE*. Available at: https://corn.ipmpipe.org/ (accessed January 1, 2024).

Deleon, L., Brewer, M.J., Esquivel, I.L. and Halcomb, J. (2017) Use of a geographic information system to produce pest monitoring maps for south Texas cotton and sorghum land managers. *Crop Protection* 101, 50–57. DOI: 10.1016/j.cropro.2017.07.016

Dwivedi, M., Shadab, M.H. and Santosh, V.R. (2020) Insect pest detection, migration and monitoring using radar and LiDAR systems. In: Chakravarthy, A. (ed.) *Innovative Pest Management Approaches for the 21st Century*. Singapore: Springer, pp. 61–76. DOI: 10.1007/978-981-15-0794-6_4

Edan, Y., Adamides, G. and Oberti, R. (2023) Agriculture automation. In: Nof, S.Y. (ed.) *Springer Handbook of Automation*. Cham, Switzerland: Springer, pp. 1055–1078. DOI: 10.1007/978-3-030-96729-1_49

Elliott, N.C., Backoulou, G.F., Brewer, M.J. and Giles, K.L. (2015) NDVI to detect sugarcane aphid injury to grain sorghum. *Journal of Economic Entomology* 108, 1452–1455. DOI: 10.1093/jee/tov080

FAO (2024) *Global Action for Fall Armyworm Control: FAMEWS Mobile App*. Rome, Italy: Food and Agriculture Organization of the United Nations. Available at: https://www.fao.org/fall-armyworm/monitoring-tools/famews-mobile-app/en/ (accessed April 22, 2024).

Feng, A., Zhou, J., Vories, E. and Sudduth, K.A. (2023) Prediction of cotton yield based on soil texture, weather conditions and UAV imagery using deep learning. *Precision Agriculture* 25, 303–326. DOI: 10.1007/s11119-023-10069-x

Filho, F.H.L., Heldens, W.B., Kong, Z. and de Lange, E.S. (2020) Drones: innovative technology for use in precision pest management. *Journal of Economic Entomology* 113, 1–25. DOI: 10.1093/jee/toz268

Hartbauer, M. (2024) Artificial neuronal networks are revolutionizing entomological research. *Journal of Applied Entomology* 148, 232–251. DOI: 10.1111/jen.13227

Høye, T.T., Ärje, J., Bjerge, K., Hansen, O.L.P., Iosifidis, F. *et al.* (2021) Deep learning and computer vision will transform entomology. *Proceedings of the National Academy of Sciences* 118, e2002545117. DOI: 10.1073/pnas.2002545117

Isard, S.A., Barnes, C.W., Hambleton, S., Ariatti, A., Russo, J.M. *et al.* (2011) Predicting soybean rust incursions into the North American continental interior using crop monitoring, spore trapping, and aerobiological modeling. *Plant Disease* 95, 1346–1357. DOI: 10.1094/PDIS-01-11-0034

John Deere (2024) *See & Spray™ Technology*. Decatur, IL: Deere & Company. Available at: https://www.deere.com/en/sprayers/see-spray/ (accessed February 1, 2024).

Jonas, J.L., Wolesensky, W. and Joern, A. (2015) Weather affects grasshopper population dynamics in continental grassland over annual and decadal periods. *Rangeland Ecology & Management* 68, 29–39. DOI: 10.1016/j.rama.2014.12.011

Joshi, V.R., Thorp, K.R., Coulter, J.A., Johnson, G.A., Porter, P.M. *et al.* (2019) Improving site-specific maize yield estimation by integrating satellite multispectral data into a crop model. *Agronomy* 9, 719. DOI: 10.3390/agronomy9110719

Karar, M.E., Alsunaydi, F., Albusaymi, S. and Alotaibi, S. (2021) A new mobile application of agricultural pests recognition using deep learning in cloud computing system. *Alexandria Engineering Journal* 60, 4423–4432. DOI: 10.1016/j.aej.2021.03.009

Lawton, D., Huseth, A.S., Kennedy, G.G., Morey, A.C., Hutchison, W.D. *et al.* (2022) Pest population dynamics are related to a continental overwintering gradient. *Proceedings of the National Academy of Sciences* 119, e2203230119. DOI: 10.1073/pnas.2203230119

Li, K., Zhu, J. and Li, N. (2021b) Lightweight automatic identification and location detection model of farmland pests. *Wireless Communications and Mobile Computing* 2021, 9937038. DOI: 10.1155/2021/9937038.

Li, W., Zheng, T., Yang, Z., Li, M., Sun, C. and Yang, X. (2021a) Classification and detection of insects from field images using deep learning for smart pest management: a systematic review. *Ecological Informatics* 55 101460. DOI: 10.1016/j.ecoinf.2021.101460

Li, Y., Wang, H., Dang, L.M., Sadeghi-Niaraki, A. and Moon, H. (2020) Crop pest recognition in natural scenes using convolutional neural networks. *Computers and Electronics in Agriculture* 169, 105174. DOI: 10.1016/j.compag.2019.105174

Liu, J. and Wang, X. (2020) Tomato diseases and pests detection based on improved Yolo V3 convolutional neural network. *Frontiers in Plant Science* 11, 898. DOI: 10.3389/fpls.2020.00898

Maimaitijiang, M., Sagan, V., Sidike, P., Daloye, A.M., Erkbol, H. and Fritschi, F.B. (2020) Crop monitoring using satellite/UAV data fusion and machine learning. *Remote Sensing* 12, 1357. DOI: 10.3390/rs12091357

Machleb, J., Peteinatos, G.G., Kollenda, B.L., Andujar, D. and Gerhards, R. (2020) Sensor-based mechanical weed control: present state and prospects. *Computers and Electronics in Agriculture* 176, 1405638. DOI: 10.1016/j.compag.2020.105638

Marston, Z.P.D., Cira, T.M., Hodgson, E.W., Knight, J.F. and Macrae, I.V. (2020) Detection of stress induced by soybean aphid (Hemiptera: Aphididae) using multispectral imagery from unmanned aerial vehicles. *Journal of Economic Entomology* 113, 779–786. DOI: 10.1093/jee/toz306

Moore, E.K., Kriesberg, A., Schroeder, S., Geil, K., Haugen, I. *et al.* (2022) Agricultural data management and sharing: best practices and case study. *Agronomy Journal* 114, 2624–2634. DOI: 10.1002/agj2.20639

Moses-Gonzales, N. and Brewer, M.J. (2021) A special collection: drones to improve insect pest management. *Journal of Economic Entomology* 114, 1853–1856. DOI: 10.1093/jee/toab081

Nansen, C. and Elliott, N. (2016) Remote sensing and reflectance profiling in entomology. *Annual Review of Entomology* 61, 139–158. DOI: 10.1146/annurev-ento-010715-023834

Pal, P., Sharma, R.P., Tripathi, S., Kumar, C. and Ramesh, D. (2022) Machine learning regression for RF path loss estimation over grass vegetation in IoWSN monitoring infrastructure. *IEEE Transactions on Industrial Informatics* 18, 6981–6990. DOI: 10.1109/TII.2022.3142318

Prabhakar, M., Vennila, S., Prasad, Y.G., Kumar, G.S., Paul, R.K. and Yadav, S.K. (2022) ARIMAX – artificial neural network hybrid model for predicting semilooper (*Chrysodeixis acuta*) incidence on soybean. *International Journal of Tropical Insect Science* 42, 3601–3608. DOI: 10.1007/s42690-022-00887-3

Quamar, M.M., Al-Ramadan, B., Khan, K., Shafiullah, M. and El Ferik, S. (2023) Advancements and applications of drone-integrated geographic information system technology – a review. *Remote Sensing* 15, 5039. DOI: 10.3390/rs15205039.

Rydhmer, K., Bick, E., Still, L., Strand, A., Luciano, R. *et al.* (2022) Automating insect monitoring using unsupervised near-infrared sensors. *Scientific Reports* 12, 2603. DOI: 10.1038/s41598-022-06439-6

Sagan, V., Maimaitijiang, M., Bhadra, S., Maimaitiyiming, M., Brown, D.R. *et al.* (2021) Field-scale crop yield prediction using multi-temporal WorldView-3 and PlanetScope satellite data and deep learning. *ISPRS Journal of Photogrammetry and Remote Sensing* 174, 265–281. DOI: 10.1016/j.isprsjprs.2021.02.008

Sharma, R.P., Ramesh, D., Pal, P., Tripathi, S. and Kumar, C., (2022) IoT-enabled IEEE 802.15.4 WSN monitoring infrastructure-driven fuzzy-logic-based crop pest prediction. *IEEE Internet of Things Journal* 9, 3037–3045. DOI: 10.1109/JIOT.2021.3094198

Shahi, T.B., Xu, C.Y., Neupane, A. and Guo, W. (2023) Recent advances in crop disease detection using UAV and deep learning techniques. *Remote Sensing* 15, 2450. DOI: 10.3390/rs15092450

Stanton, C., Starek, M.J., Elliott, N., Brewer, M., Maeda, M.M. and Chu, T. (2017) Unmanned aircraft system-derived crop height and normalized difference vegetation index metrics for sorghum yield and aphid stress assessment. *Journal of Applied Remote Sensing* 11, 026035. DOI: 10.1117/1.JRS.11.026035.

Stumph, B., Virto, M.H., Medeiro, H., Tabb, A., Wolford, S. *et al.* (2019) Detecting invasive insects with unmanned aerial vehicles. In: *2019 International Conference on Robotics and Automation (IRCA)*. Piscataway, NJ: Institute of Electrical and Electronics Engineers, pp. 648–654. DOI: 10.1109/ICRA.2019.8794116

Talaviya, T., Shah, D., Patel, N., Yagnik, H. and Shah, M. (2020) Implementation of artificial intelligence in agriculture for optimisation of irrigation and application of pesticides and herbicides. *Artificial Intelligence in Agriculture* 4, 58–73. DOI: 10.1016/j.aiia.2020.04.002.

Tannous, M., Stefanini, C. and Romano, D. (2023) A deep-learning-based detection approach for the identification of insect species of economic importance. *Insects* 14, 148. DOI: 10.3390/insects14020148

Thorp, K.R. and Tian, L.F. (2004) A review on remote sensing of weeds in agriculture. *Precision Agriculture* 5, 477–508. DOI: 10.1007/s11119-004-5321-1

Tonle, F.B.N., Niassy, S., Ndadji, M.M.Z., Tchendji, M.T., Nzeukou, A. *et al.* (2024) A road map for developing novel decision support system (DSS) for disseminating integrated pest management (IPM) technologies. *Computers and Electronics in Agriculture* 217, 108526. DOI: 10.1016/j.compag.2023.108526

Tonnang, H.E.Z., Salifu, D., Mudereri, B.T., Tanui, J., Espira, A. *et al.* (2022) Advances in data-collection tools and analytics for crop pest and disease management. *Insect Science* 54, 100964. DOI: 10.1016/j.cois.2022.100964.

Toscano-Miranda, R., Toro, M., Aguilar, J., Caro, M., Marulanda, A. and Trebilcok, A. (2022) Artificial-intelligence and sensing techniques for the management of insect pests and diseases in cotton: a systematic literature review. *The Journal of Agricultural Science* 160, 16–31. DOI: 10.1017/S002185962200017X

VanKirk, J.R., Isard, S.A., Cardwell, K.F. and Draper, M.A. (2012) The ipmPIPE: overview, lessons, opportunities, and challenges. *Journal of Integrated Pest Management* 3(2), C1–C7. DOI: 10.1603/IPM11015

Vijayakumar, V., Ampatzidis, Y., Schueller, J.K. and Burks, T. (2023) Smart spraying technologies for precision weed management: a review. *Smart Agricultural Technology* 6, 100337. DOI: 10.1016/j.atech.2023.100337.

Wang, H.-H., Grant, W.E., Elliott, N.C., Brewer, M.J., Koralewski, T.E. *et al.* (2019) Integrated modelling of the life cycle and aeroecology of wind-borne pests in temporally variable spatially heterogeneous environment. *Ecological Modelling* 399, 23–38. DOI: 10.1016/j.ecolmodel.2019.02.014

Westbrook, J.K., Eyster, R.S. and Allen, C.T. (2011) A model for long-distance dispersal of boll weevils. *International Journal of Biometeorology* 55, 585–593. DOI: 10.1007/s00484-010-0359-4

Westbrook, J.K., Eyster, R.S. and Wolf, W.W. (2014) WSR-88D doppler radar detection of corn earworm moth migration. *International Journal of Biometeorology* 58, 931–940. DOI: 10.1007/s00484-013-0676-5

Wilkinson, M., Dumontier, M., Aalbersberg, I., Appleton, G., Axton, M. *et al.* (2016) The FAIR guiding principles for scientific data management and stewardship. *Scientific Data* 3, 160018. DOI: 10.1038/sdata.2016.18

Wulder, M.A., Ortlepp, S.M., White, J.C., Coops, N.C. and Coggins, S.B. (2008) Monitoring tree-level insect population dynamics with multi-scale and multi-source remote sensing. *Journal of Spatial Science* 53, 46–61. DOI: 10.1080/14498596.2008.9635135

Xu, H., Caramanis, C. and Mannor, S. (2009) Robustness and regularization of support vector machines. *Journal of Machine Learning Research* 10, 1485–1510. DOI: 10.48550/arXiv.0803.3490

Xu, L., Shi, X., Tang, Z., He, Y., Yang, N. *et al.* (2023) ASFL-YOLOX: an adaptive spatial feature fusion and lightweight detection method for insect pests of the Papilionidae family. *Frontiers in Plant Science* 14, 1176300. DOI: 10.3389/fpls.2023.1176300

Yang, L., Peng, L., Zhang, L., Zhang, L. and Yang, S. (2009) A prediction model for population occurrence of paddy stem borer (*Scirpophaga incertulas*), based on back propagation artificial neural network and principal components analysis. *Computers and Electronics in Agriculture* 68, 200–206. DOI: 10.1016/j.compag.2009.06.003

Yao, Q., Xian, D., Liu, Q., Yang, B., Diao, G. and Tang, J. (2014) Automated counting of rice planthoppers in paddy fields based on image processing. *Journal of Integrative Agriculture* 13, 1736–1745. DOI: 10.1016/S2095-3119(14)60799-1

Zhao, G., Zhang, Y., Lan, Y., Deng, J., Zhang, Q. *et al.* (2023) Application progress of UAV-LARS in identification of crop diseases and pests. *Agronomy* 13, 2232. DOI: 10.3390/agronomy13092232

14 Past, Present, and Future Applications of Molecular Tools in Landscape-scale IPM: Fall Armyworm as a Case Study

Robert L. Meagher, Jr,[1,*] Ashley E. Tessnow,[2] and Gregory A. Sword[2]
[1]US Department of Agriculture, Agricultural Research Service, Insect Behavior and Biocontrol Research Unit, Gainesville, FL, USA; [2]Department of Entomology, Texas A&M University, College Station, TX, USA

Abstract
Molecular tools can have many uses in pest management systems across landscapes of various scales. Molecular identification of indistinguishable life stages or closely related species, either in the field or intercepted at ports of entry, plays a key role in agricultural biosecurity. Identification of within-species types, such as biotypes, host forms, host strains, or cryptic species, can help growers and consultants determine crop varieties to plant and the best management strategies to use once pests are present. Molecular tools can identify dispersing, migrating, or invasive insect pests which have types that originate from different geographic sources. Molecular tools can also be used to determine the population source if types have different pathways. Finally, molecular approaches can provide important insights into the evolution of insect resistance to management approaches such as *Bacillus thuringiensis*-engineered plants or chemical insecticides. These approaches range from detecting the genetic variance resulting from the mechanisms underlying resistance to monitoring the frequency and movement of resistance alleles across the landscape. The fall armyworm, *Spodoptera frugiperda* (J.E. Smith), is a worldwide noctuid pest of maize and many other crops that has been shown to have two genetically distinct host strains. Genetic analysis of seasonal migrations in the United States (US) led to the development of diagnostic markers for one of the strains that could be used to identify migratory pathways. The source of migrants that establish new populations annually in the northern US can be identified as originating earlier in the season from populations in either southern Texas or southern Florida. These same tools have been used to assess the recent invasion of this species in the Eastern Hemisphere. Future applications of molecular tools in pest identification and management will be discussed.

Introduction

Genetic and molecular tools have been used in the management of pest insects for many decades. At different landscape scales, pest species require accurate identification so that control measures can be targeted, and the damage level for certain crops can be anticipated. Many pest species are morphologically indistinguishable during immature (nymphal, larval, and/or pupal) life stages, so molecular tools are often required to identify these immature collections to species. This has become especially routine at border ports of entry where insect hitchhikers must be intercepted and identified prior to importing goods (Madden *et al.*, 2019). In these cases, misidentification

*Corresponding author: Rob.Meagher@usda.gov

 Arthropod Management and Landscape Considerations in Large-Scale Agroecosystems (eds M.J. Brewer and G.L. Hein)
DOI: 10.1079/9781800622777.0014

can result in invasive pests entering a novel landscape and causing economic damage to that region. In addition to species-level identification, many insect groups such as aphids (Harris-Shultz *et al.*, 2018), whiteflies (Crossley and Snyder, 2020; De Marchi *et al.*, 2023), beetles (Crossley *et al.*, 2013), and lepidopterans (Coates *et al.*, 2018) are comprised of distinct within-species types ("biotypes," "host forms," "host strains," and "sibling species") that can only be identified using molecular tools. Many of these types have sympatric distributions that feed on different but related plant species and/or different crop varieties at field, regional, or global landscape scales. Identification of these species' types is important for integrated pest management (IPM) practices. For example, knowing the species' biotype infesting cultivated and non-cultivated hosts in a given area can provide information on the crop variety that should or should not be planted (Anstead *et al.*, 2003). Agricultural landscapes can be manipulated within or between seasons using crop rotation and expansion or compression of non-crop areas. The dispersal of pest species or genetically identified types will be affected at different spatial levels by these changes in landscapes. Therefore, pest management can be greatly improved by using genetic variation among these populations to monitor and track pest dispersal in agroecosystems.

In this chapter, we highlight the use of different molecular tools to determine the genetic variation, population structure, and migratory trajectories of a serious global pest of corn (maize) and other crops, the fall armyworm, *Spodoptera frugiperda* (J.E. Smith) (Lepidoptera: Noctuidae). Migration can be narrowly defined as movements between the natal breeding site to another breeding site, and then return to the original site (ecological definition) (Clobert *et al.*, 2009). For this chapter, we define insect migration where it, "...is characterized by persistent, straightened-out movements that are undistracted by the cues which would arrest other types of movements" (behavioral definition) (Chapman *et al.*, 2015). Therefore, these definitions describe migration as either unidirectional or with a return during another season, but both include dispersal across the landscape to suitable habitats. We first describe how molecular tools have been used to characterize two genetically distinct subtypes or strains that exist in sympatry within the species in its native range in the US. This is followed by a discussion of how the implementation of molecular tools over the years has enabled researchers to answer questions about the overwintering source of migrant infestations. Although fall armyworm originated in the neotropics, it recently became invasive across the tropical and subtropical areas of Africa, Asia, Australia, and Europe. As such, we will discuss how molecular tools initially developed in the US have been improved and are being used by researchers to better understand fall armyworm movements across spatial landscape scales all over the world. Finally, we will conclude by describing future opportunities for using molecular tools in landscape IPM.

Fall Armyworm

The fall armyworm is a migratory neotropical pest of many agricultural crops (Sparks, 1979) and has hundreds of plant hosts (Montezano *et al.*, 2018) (Fig. 14.1). In the southeastern US, fall armyworm damages corn in spring (Fig. 14.1) and then infests peanuts, cotton, and forage grasses in the summer. Migration of fall armyworm in North America has been studied since the early 1900s by using appearance of larvae as a way to map the distribution of populations moving north in the spring (Sparks, 1979) and possibly limited movement south in the fall (Pair *et al.*, 1987). In the last few decades, pheromone-baited traps have monitored movement of male moths, and information collected by extensive trapping networks can be used to follow the northward movement (Fig. 14.2). Because this species does not diapause, they can only overwinter in locations that are warm enough year-round to support continuous generations. In the US, this usually occurs in the subtropics of southern Texas and southern Florida, but internationally this can include Mexico, Central America, and the Caribbean (Luginbill, 1928). Populations then move north from these overwintering zones into the southern temperate areas of the US and even into Canada by the end of the season. Therefore, studies in the early 20th century focused on populations moving from the central and eastern US through migratory pathways that were undetermined at the time.

Fig. 14.1. Damage to a corn plant from larval feeding of fall armyworm, Homestead, FL. (Photo credit: R.L. Meagher, Jr, USDA-ARS.). Inset: fall armyworm larva on cowpea, Gainesville, FL. (Photo credit: C. Stuhl, USDA-ARS.)

Fig. 14.2. Right: universal moth trap ("bucket trap") used with commercial pheromone lures to attract and capture male fall armyworm, Hague, FL. Left: application of pesticides to young sweet corn using a ground sprayer, Homestead, FL. (Photo credits: R.L. Meagher, Jr, USDA-ARS.)

Detection of Host Strains

Researchers in Louisiana tested the hypothesis that an assessment of genetic differences among populations could provide clues to identify the location of the source population(s) that migrate into the state each spring (Pashley *et al.*, 1985). Larval samples from Puerto Rico, Florida, Georgia, Louisiana, Texas, and Mexico were characterized using protein electrophoresis to test for variation in isozymes, but the genetic differences found could not be explained by geographic collections. Rather, the genetic differences found were due to a divergent population obtained from Puerto

Rico that had been collected from rice, compared to other populations collected from corn (Pashley *et al.*, 1985). Further, larvae collected from bermudagrass in Louisiana were shown to be genetically identical at 11 polymorphic enzyme loci to the population collected in rice from Puerto Rico. This foundational work established the existence of two genetically distinct fall armyworm subtypes, which will henceforth be referred to as host strains (corn or C-strain and rice or R-strain) (Pashley, 1986).

In the years since the initial isozyme studies, other genetic markers have been used to separate host strains, including amplified fragment length polymorphisms (AFLPs), restriction fragment length polymorphisms (RFLPs) (Pashley Prowell *et al.*, 2004; Cano-Calle *et al.*, 2015), and genomic microsatellite markers (tandem repeats) (Machado *et al.*, 2008). Polymorphisms in the mitochondrial cytochrome oxidase subunit 1 (*COI*) gene provided several restriction sites that have been used to separate host strains (Nagoshi *et al.*, 2006), and it is also clear that variation in the mitochondria genome plays a significant role in strain divergence (Orsucci *et al.*, 2022). Comparisons of the polymorphisms of the intron sequences from the triosephosphate isomerase (*Tpi*) gene on the Z-chromosome in the nuclear genome have also been used to identify host strains (Nagoshi and Meagher, 2016). Because polymorphisms in *COI* and *Tpi* result in unique restriction enzyme cut sites that are easy and inexpensive to assess in the laboratory, these have been the most used markers to identify host strains. More recently, double digest restriction site associated DNA sequencing (ddRADSeq) was used to identify genome-wide single nucleotide polymorphisms (SNPs) that vary between strains (Tessnow *et al.*, 2022). From the ddRADSeq data, four real-time polymerase chain reaction (PCR)-based TaqMan diagnostic assays were developed that use genomic SNP markers to differentiate between strains (Tessnow *et al.*, 2021). Interestingly, most SNPs that were highly divergent between strains were located on the Z-chromosome, the same chromosome where *Tpi* is located. This indicates that the sex chromosome may play an important role in strain identity and differentiation (Tessnow *et al.*, 2022).

A few studies have directly compared the efficacy of strain markers within the same set of collections. In one such study from Louisiana, the esterase (*Est3*) gene, which is located on the Z-chromosome, seemed to be a slightly less reliable predictor of host strain than known AFLP markers, or strain-specific polymorphisms in the mitochondrial *nd1* gene (Pashley Prowell *et al.*, 2004). Nagoshi and Meagher (2022) compared 26 collections from multiple locations and times that were analyzed with both the *COI* and *Tpi* markers. The *Tpi* marker detected a significantly higher percentage of C-strain than *COI* in 24 of the 26 collections in data from C-strain preferred habitats, but not in collections from R-strain hosts (Nagoshi and Meagher, 2022). Tessnow *et al.* (2021) compared *COI*, *Tpi*, and four genomic SNP markers for host strain identity and found that accuracy did not significantly differ among any of these six markers.

One interesting result of these studies was the number of individuals in the fall armyworm population that were discordant in their host strain markers, i.e. one marker identified an individual as one host strain while the other marker identified it as the other host strain. For the *COI–Tpi* study, the marker combination $R_{COI}C_{TPI}$ was a substantial fraction (over 35%) of the population. Two explanations have been proposed to explain the appearance of this combination (Pashley Prowell *et al.*, 2004; Nagoshi and Meagher, 2022). The first is that one or more of the markers did not achieve fixation at the time of host strain divergence and as such may not be completely accurate in their identification. The second explanation involves hybridization between host strains (R-strain female × C-strain male). This latter explanation was further supported by combining *COI* strain assignments with ddRADSeq data that identified host strain by using nuclear SNP markers. In this study, several individuals that exhibited a mismatch between nuclear and mitochondrial host strain assignments ($R_{COI}C_{Genome}$) showed significant evidence of interstrain admixture or hybridization (Tessnow *et al.*, 2022). Importantly, these same data also provided clear evidence for unidirectional gene flow following hybridization between the strains. Individuals with predominately C-strain nuclear genomes can have either C- or R-strain mitochondria, whereas the opposite pairing of R-strain nuclear genome with C-strain mitochondrial DNA (mtDNA) does not occur (Tessnow *et al.*, 2022).

Distribution of Host Strains

The identification of host strains questions the dogma that fall armyworm shifts from corn to forages grasses as a single population in the southern US. C-strain larvae can be found infesting corn as early as April in Louisiana, whereas R-strain larvae can be found infesting corn in June and then later infesting pasture grasses. This observation led to speculation that host strain subgroups migrate separately from overwintering areas (Pashley Prowell *et al.*, 2004). Host fidelity is not absolute, as C-strain genotypes are less common on the R-strain hosts, but the R-strain has a wider range of host use (Pashley Prowell *et al.*, 2004). In southern Florida, strong habitat to habitat differences were found between host strains as R-strain moths were the primary population observed in naturalized pastures and wetlands, whereas a higher proportion of C-strain moths were observed in areas associated with golf courses, agriculture, or urban development (Meagher and Nagoshi, 2004). There was also a large seasonal difference in sweet corn habitat, where moths of both host strains were captured in the spring and summer, but R-strain moths dominated captures in the fall and winter (Nagoshi and Meagher, 2004) (Fig. 14.3). This pattern was observed using 16 months of pheromone trap sampling data, so it remains possible that these patterns of host strain abundance fluctuate annually. The opposite pattern was found in northcentral Florida where C-strain larvae infested a smaller patch of turfgrass bordering a much larger area of field corn (R.L. Meagher, Jr, unpublished data). In Texas, both C- and R-strain larvae and moths are readily recovered from corn and sorghum, but the R-strain tends to be dominant in pasture grasses (A.E. Tessnow and G.A. Sword, unpublished data). Host use of strains in the Caribbean and South America is variable. Collections in Puerto Rico indicated that there is almost no overlap of crops used by the two strains, whereas in Guadeloupe, there was some overlap (Pashley Prowell *et al.*, 2004).

Landscape-scale variation in agroecosystem composition is also a likely factor affecting where host strains are found, as R-strain larvae infested a corn field in French Guiana that was surrounded by a much larger rice field (Pashley Prowell *et al.*, 2004). In Colombia, field collections of both host strains were found in corn, but C-strain larvae were found predominately in cotton and sorghum while R-strain larvae were found predominately in rice (Saldamando and Vélez-Arango, 2010). Studies in Argentina have provided mixed results. In two studies, there was no consistent pattern of host association between the two sympatric strains and their respective host plants, either using *COI* markers (Juárez *et al.*, 2012) or *COI* and *Tpi* markers (Juárez *et al.*, 2014). However, C-strain larvae were collected from corn and sorghum and R-strain larvae from grasses and alfalfa in another study that used both markers with field-collected specimens (Murúa *et al.*, 2015). The disagreement among these studies can possibly be explained by the use of field-collected larvae rather than laboratory-reared material and the greater use of larvae from pasture grass sites rather than rice fields (Murúa *et al.*, 2015). An important take away from these studies is that host strain cannot be assumed solely based on the host plant or field where an individual is collected. Rather, researchers must employ molecular diagnostic assays to verify the host strain before conducting strain-specific research or implementing strain-specific management.

Relationship of Host Strains to Fall Armyworm Management

The management of fall armyworm populations depends on the crop plant, host strain, and environmental conditions/landscape that are present. Biological differences in feeding, mating behavior, and susceptibility to insecticides and *Bacillus thuringiensis* (Bt)-engineered crops are all factors that can affect the efficacy of management strategies. Therefore, knowledge of the host strains that are present and are being managed in the field is important. Physiological differences between strains in response to feeding on different plants were found to be related to differences in food consumption and utilization (Veenstra *et al.*, 1995), although host plant defensive responses to strain-specific elicitors in saliva can also affect larval growth, and therefore, may influence crop plant associations in the field (Acevedo *et al.*, 2018). Feeding studies using susceptible and resistant bermudagrass germplasm resulted in different responses between host strains. "Tifton 292" bermudagrass was found

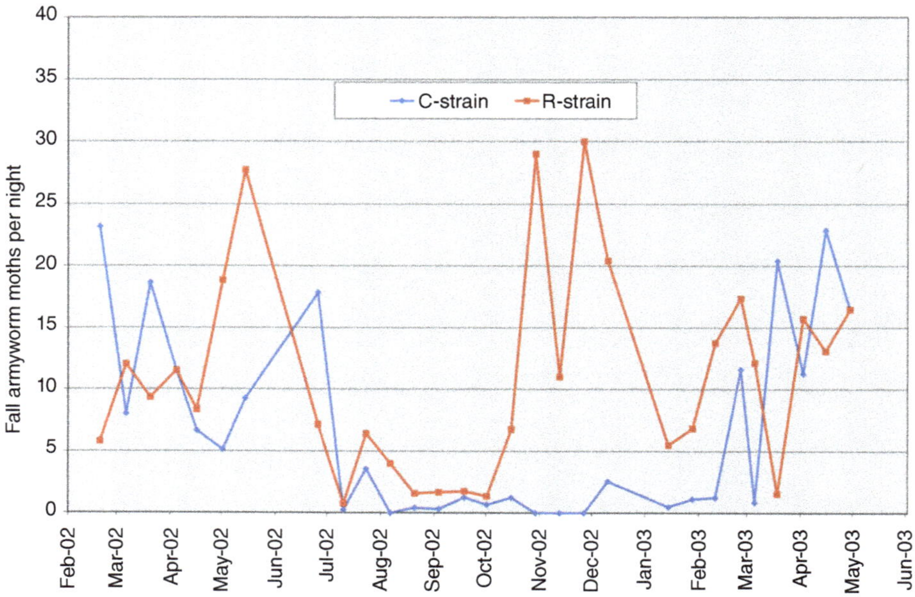

Fig. 14.3. Pheromone trap data from Redland, FL, with traps placed next to sweet corn fields. Samples in the spring and summer 2002 (February–July) and spring 2003 (February–May) were a mixture of corn (C) and rice (R) strains, whereas the samples in the fall 2002 (October–January) were predominantly R-strain (Meagher and Nagoshi, 2004; Nagoshi and Meagher, 2004). (Original artwork by R.L. Meagher, Jr, USDA-ARS.)

to be resistant to fall armyworm feeding when tested with C-strain larvae, but it was determined to be susceptible when fed with R-strain larvae (Pashley *et al.*, 1987). The conclusion was a clear indication that plant resistance bioassays should be conducted using the appropriate host strain expected for the crop.

Growers and public agency personnel occasionally use pheromone traps to monitor the arrival of fall armyworm in locations where migrating populations occur seasonally. In many field situations, moths arriving in traps trigger field scouting to determine if control measures are appropriate. Depending on the crop being managed, there was a concern that commercial pheromone lures differentially attracted moths from host strain populations that were not a threat to the crop. Sympatric host strains require different mating behaviors to maintain their status. One possible pre-mating behavior for fall armyworm is divergence of host strains in pheromone production or selective male attraction to pheromone blends. Several studies suggest that pheromone composition between C- and R-strain females is different (Unbehend *et al.*, 2013), although other studies suggest that geographical variation and time of season may be more important in male attraction (Meagher *et al.*, 2013; Unbehend *et al.*, 2014). Without a study comparing other adult monitoring techniques (light traps, floral attractants, etc.) with pheromone trapping, there is no evidence that commercial pheromone lures (i.e. those commercially available in the US) selectively capture either host strain.

Chemical control is often used to manage fall armyworm populations and several studies have shown statistical differences in susceptibility of host strains to various classes of insecticides (Ríos-Díez and Saldamando-Benjumea, 2011) (Fig. 14.2). Generally, C-strain larvae are less susceptible to insecticides than R-strain larvae. As was shown above, at some times of the year R-strain larvae may predominate in corn plantings (Meagher and Nagoshi, 2004), providing growers with a wider choice of insecticides to use since they are generally more susceptible to certain insecticide classes. Knowledge of the host strain to which the insecticides are being applied is important information for the grower. Bt proteins have also been shown to cause differences in mortality between host strains, again with C-strain larvae generally being less susceptible than R-strain larvae (Ríos-Díez *et al.*, 2012; Ingber *et al.*, 2018). Molecular tools have been used to detect resistance to Bt insecticidal proteins, and analysis of the genes involved in resistance can help predict the spread of these alleles between populations across landscapes (Banerjee *et al.*, 2022).

Mapping Overwintering Source Populations

Sequence analysis of C-strain individuals identified variation in a locus on the *COI* gene that results in four different haplotype subgroups (Nagoshi *et al.*, 2007) (Fig. 14.4). Two of these haplotypes are rare (known as h1 and h3), and account for less than 10% of all individual moths collected (Nagoshi *et al.*, 2012). The other two haplotypes are common in both North and South America (known as h2 and h4). Although all four haplotypes can be found across all locations, the frequency of haplotypes differs geographically. For example, the dominant haplotype in Florida and Puerto Rico (h4; >70%) is different from the dominant haplotype in Texas and Brazil (h2; >75%) (Nagoshi *et al.*, 2012).

Because of consistent geographic differences in haplotype frequency across multiple years (Nagoshi *et al.*, 2012), a ratio among the C-strain haplotypes (h4/h2), or more recently a modified ratio of these two haplotypes (h4 – h2)/(h4 + h2), has been effectively used to define the natal origins of a population collected outside the overwintering range. Using primarily pheromone trap specimens, collections from several southern, central, and northeastern US locations provided evidence that the Florida type (h4) was primarily found along the Atlantic coastal plain, whereas the Texas type (h2) was dominant in the central and northeastern states as well as two eastern provinces of Canada (Nagoshi *et al.*, 2012; Tessnow *et al.*, 2023) (Fig. 14.5). Mixing of individuals from different haplotype groups was found near the Alabama/Georgia border and in parts of the northeast, east of the Appalachian range.

Internationally, samples from the Caribbean and several South American countries suggest limited genetic exchange between those two areas (north to south migration). Haplotype ratios were different between Trinidad and Tobago and Barbados and Dominica (Nagoshi *et al.*, 2017) (Fig. 14.5); however, regional differences in haplotype frequencies have been stable over time.

These patterns are contrasted with other genomic analyses which suggest that fall armyworm shows little genetic differentiation and high

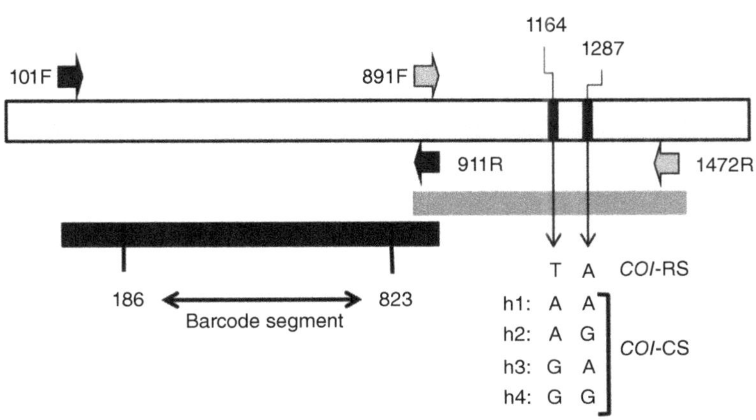

Fig. 14.4. Diagram of fall armyworm mitochondrial cytochrome oxidase I gene (*COI*), with two diagnostic segments. Segment A (PCR 101F/911R) is used for species and host strain identification; segment B (PCR 891F/1472R) is also used for host strain identification, but also for geographical subgroups. (Original artwork by R.L. Meagher, Jr, R. Nagoshi, USDA-ARS.)

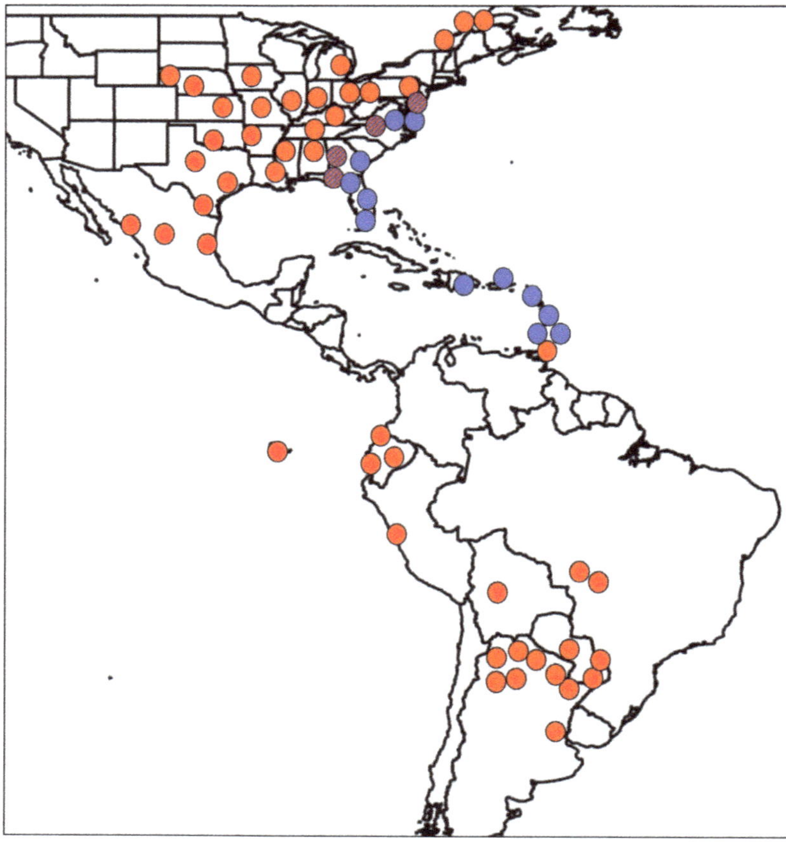

Fig. 14.5. Geographic regions with h4-dominant "Florida" (blue circles) and h2-dominant "Texas" (red circles) C-strain migratory haplotypes in the Western Hemisphere. Purple hatched circles represent areas of mixing between h2 and h4 haplotypes. (Original artwork by R.L. Meagher, Jr, R. Nagoshi, USDA-ARS.)

genetic diversity across populations from several countries in the Western Hemisphere (Belay *et al.*, 2012). Their results showed that within-population variability was higher than among-population variability, with the conclusion that there was modest gene flow across large-area populations. Whole-genome nuclear diversity analysis also concluded that fall armyworm is a panmictic population with no separation at the nuclear level (Schlum *et al.*, 2021) and that the host strains exhibit a low level of differentiation across the whole genome (Nam *et al.*, 2020). It should be noted that the original studies using haplotype ratios to define migratory origin were done exclusively with individuals of the C-strain because no consistent polymorphic markers were found in the *COI* gene of R-strain individuals. However, more recent genomic studies have pooled C- and R-strain individuals together to assess geographic patterns of population structuring. Because genetic variation between strains is higher than genetic variation between populations, when strains are pooled together, we would expect that equal or even more genetic variation would be present within a location than between locations, consistent with the findings from these studies. Ongoing whole-genome sequencing studies are exploring geographic structuring as a nested variable within host strain (A.E. Tessnow, unpublished data), and these data show similar C-strain migratory populations in North America as those that had previously been observed using haplotype data (as in Fig. 14.5).

Modeling Migration in North America

Observations in local landscapes show that fall armyworm appears to be a species active at dusk, and its movement within the atmosphere is determined by light intensity thresholds and meteorological parameters, including temperature, wind speed and direction, atmospheric pressure, and humidity (Sparks *et al.*, 1986; Chapman *et al.*, 2015). If conditions for long-range transport predominate, moth flight and wind direction will carry moths out of the flight boundary layer and up to the high altitude nocturnal aeroecological boundary layer which may contain low-level jet stream winds (Westbrook, 2008). As a result, these moths disperse out of the local landscape (Fig. 14.6) where they were produced. These wind-borne migrants redistribute vertically and become dispersed within the migratory "cloud" of moths. As flight terminates, they descend and are distributed across the landscape to search for suitable habitats (Gatehouse, 1997). If atmospheric conditions are not favorable for long-distant transport or if the wind direction is unfavorable, then moth movement is short range and within or between local habitats.

Meteorological approaches, especially synoptic weather analysis (i.e. back-track analysis), showed how fall armyworm moths moved out of local landscapes and roughly 1600 km in two nights from the southcentral US to southern Ontario (Rose *et al.*, 1975). Ecological, entomological, and meteorological studies improved atmospheric transport modeling of noctuid populations moving north from southern Texas and northern Mexico to the southcentral US (Westbrook *et al.*, 1995). The use of radar, including weather doppler radar, detected insect flight as high as 500 m above ground level and identified characteristic patterns of migratory flight that were largely associated with vertical profiles of temperature and wind speed (Westbrook, 2008). Weather tools such as radar can provide information for 1 or 2 nights of activity for distances up to 200 km away from the source when large numbers of moths move through the aeroecological boundary layer, but because individual radar units are in one spot, further movement of the moths is not charted.

Genetic markers have provided a unique tool to map fall armyworm migration in North America, support modeling of its northward migration, and provide a means to test model predictions. Two important studies simulated migratory flight from distant winter-breeding source areas by using the combination of genetic analyses that characterized migratory pathways (Nagoshi *et al.*, 2012) and a biophysical air transport model (Hybrid Single-Particle Lagrangian Integrated Trajectory [HYSPLIT]). The single-year model simulated moth movement into regions of the country dominated by the Florida or Texas source populations plus two regions of overlapping populations with ambiguous haplotype profiles where there are similar numbers of moths with h4 and h2 haplotypes (Westbrook *et al.*, 2016). The follow up study coupled corn plant distribution and moth development with the physical and atmospheric processes (HYSPLIT) in a multigenerational model of fall armyworm's seasonal migration, and compared simulated migrations to observed pheromone trap captures and population genetic markers at a continental scale (Westbrook *et al.*, 2019) (Fig. 14.7).

Biophysical models showed that migratory flights were primarily influenced by wind speed and direction, as suggested by Sparks *et al.* (1986). The inclusion in the models of other variables showed that moth flights were also influenced by biotic and abiotic factors such as host plant development and distribution across the landscape, moth development and behavior, and meteorological conditions (Westbrook *et al.*, 2016, 2019). The advantages of estimating the future potential of pest populations that distribute through the landscape are numerous and include analysis of host-resistant crop germplasm and implementation of IPM strategies. Additionally, determining the genetic background of populations that are included in the models provides the ability to identify geographically leeward landscapes that may be at risk for migratory populations possessing such attributes as insecticidal and/or transgenic resistant alleles (Westbrook *et al.*, 2019). Large numbers of these pest populations (e.g. 600 million fall armyworm moths from

Fig. 14.6. (A) Sweet corn field in Miami-Dade County, FL. (B) Cotton field soon to be harvested in southern Alabama. (Photo credits: R.L. Meagher, Jr, USDA-ARS.)

south Texas) (Pair *et al.*, 1991) influence crop yields and pest control in the advanced landscapes (Chapman *et al.*, 2015).

Invasive Populations

Although originally from the Western Hemisphere, fall armyworm has been recorded at various times in the Eastern Hemisphere since the late 1960s, but was never determined to be established (Tay *et al.*, 2023). However, the invasion of fall armyworm in the Eastern Hemisphere was confirmed in 2016 using DNA barcoding of specimens collected in West Africa (Benin, Nigeria, São Tomé and Príncipe, and Togo) (Goergen *et al.*, 2016). Since then, the species has been documented in most sub-Saharan countries in Africa, India,

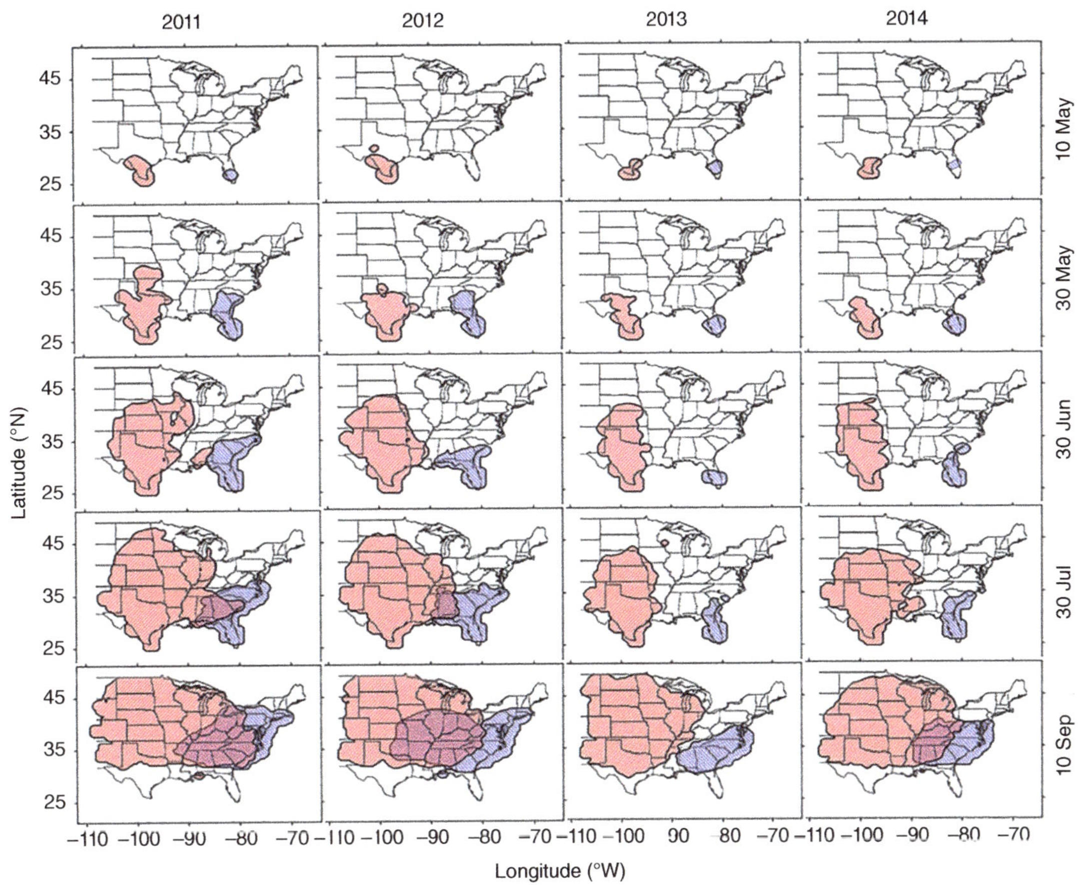

Fig. 14.7. Simulated fall armyworm moth migration (>500 moths per 1600 km²) from 10 May to 10 September 2011–2014. Colored contours represent Texas (red), Florida (blue), and mixed (purple) sources of immigrant fall armyworms (Westbrook *et al.*, 2019). (USDA-ARS publication by Westbrook *et al.*, not copyrighted.)

southeast Asia, China, Korea, Australia, and New Zealand (FAO, 2020; see Chapter 4, this volume). This new global distribution provides the opportunity to genetically compare populations that have undergone genetic isolation, hybridization, and potentially new selection pressures in novel agricultural landscapes.

Molecular analysis of populations from the West African nation of Togo indicated that specimens collected were predominately C-strain (*COI*-CS and *Tpi*C), and that most *COI*-CS specimens displayed h4, which is the allele of the Florida migratory haplotype (Nagoshi *et al.*, 2018). The h2 allele (Texas migratory haplotype) was found in a 2018 collection from Ghana and Togo, possibly indicating a second introduction into West Africa (Nagoshi *et al.*, 2022). Unfortunately, given that both haplotypes are present across all North American populations, albeit in varying ratios, these haplotype ratios in the invasive populations could have changed due to a founder's effect making them an unreliable indicator of the initial population's origin. A more exhaustive collection of specimens from West, Central, and South Africa suggested that the individuals with R-strain nuclear DNA are rare, because over 95% of the specimens tested expressed the C-strain diagnostic *Tpi*C marker. However, the majority of the *Tpi*C specimens showed the R-strain *COI*-RS haplotype indicating they likely descended from a R × C hybrid that then backcrossed to the C-strain (Nagoshi, 2019). Therefore, the *COI* strain markers in the Eastern Hemisphere are in substantial disagreement with the *Tpi* marker and are an unreliable indicator of strain identity across the invasive region. Although fall armyworm collections have been made in West African rice and pasture fields, these collections are rare, and it remains unclear if the R-strain is present in the Eastern Hemisphere (Nagoshi *et al.*, 2021). *COI* and *Tpi* analysis of samples from South Africa, eastern India, and southeast Asia indicated these populations were closely related, suggesting a common origin and a single recent introduction into the Eastern Hemisphere followed by rapid dispersion (Nagoshi *et al.*, 2020).

Researchers using other molecular tools have come to different conclusions concerning the genetic makeup of fall armyworm populations in the Eastern Hemisphere. Phylogenetic analysis of larvae collected from maize in Ghana showed both host strains ("species") present (Cock *et al.*, 2017), whereas specimens collected in Uganda and analyzed using partial mtDNA *COI* gene sequences found mtDNA haplotypes representative of both host strains ("sister species") (Otim *et al.*, 2018). Population genomic analysis identified 12 maternal mtDNA genome lineages from Africa, India, and China that suggests that there were multiple fall armyworm introductions, and the directional gene flow was from Asia and/or Southeast Asia into Eastern Africa (Tay *et al.*, 2022b). Characterization of insecticide and Bt resistance alleles from southern Asian and Australian populations provide further evidence of multiple introductions from the Western Hemisphere into Asia (Tay *et al.*, 2022a). SNP and mtDNA markers from populations sampled in southern Asia, Eastern Asia, southeastern Asia, and Pacific/Australia do not support the "African origin spread" hypothesis (Rane *et al.*, 2023).

Future Use of Molecular Tools in Pest Management

In this chapter we have reviewed how molecular tools have been used to gain a better understanding of fall armyworm ecology to increase the effectiveness and sustainability of management strategies. The work was started in the 1980s using allozyme or protein markers (Pashley *et al.*, 1985; Pashley, 1986) and has continued to adopt newer technologies including the recent use of ddRADSeq (Tessnow *et al.*, 2022) and whole-genome sequencing technologies (Gimenez *et al.*, 2020; Schlum *et al.*, 2021). With increases in the accuracy, speed, and affordability of sequencing, we expect molecular tools and molecular data to continue to play a critical role in informing and assisting IPM into the future. These advances will primarily take two interrelated forms: increased understanding of insect pest ecology and the active deployment of genetic pest management strategies.

A major component to advancing IPM has been to gain a broader understanding of insect pests in the field. Specifically, there is interest in better understanding physiology (i.e. responses to toxins, pesticides, and pathogens), ecology (i.e. movement across landscapes and interactions with other species including host plants and natural enemies), and evolution (i.e. genetic differentiation and resistance to management strategies). Sequencing projects take many forms and provide insights across all three of these disciplines. For example, molecular tools have allowed us to characterize and track the development of Bt resistance in several lepidopteran pests (Banerjee *et al.*, 2022) and recommend alternative control strategies in regions where current pest management strategies are expected to be ineffective. The original tracking of resistance was slow, costly, and labor intensive, but targeted sequencing approaches have shown promise in more efficiently detecting and tracking known resistance mutations (Tandy *et al.*, 2023). Additional molecular tools such as molecular gut content analysis (MGCA) of natural enemies have also been effectively used to tease apart food webs and assess efficacy of biological control agents (Schmidt *et al.*, 2021). Identification of endoparasitoids using mtDNA can be used to estimate parasitism of pest species before the parasitoid emerges, providing time to assess the potential level of control before application of insecticides. Molecular identification of parasitoids has become a widespread technique across many different orders of pests, including cereal aphids (Jones *et al.*, 2005), Lepidoptera (Nam *et al.*, 2022), and chrysomelid beetles (Wilson *et al.*, 2022). This information, in turn, can feed into conservation biological control strategies by enabling dynamic spraying thresholds that account for current levels of predator–prey activity in the field.

Population genomic approaches have also allowed us to understand broad scale patterns of evolution and gene flow in insect pest populations. For example, genome-wide SNP markers revealed that populations of the infamous pest, the cotton boll weevil, *Anthonomus grandis* (Boheman) (Coleoptera: Curculionidae), are highly structured and associated with geography (Raszick *et al.*, 2021). This knowledge then enabled determination of the geographic source of weevil populations that re-infested previously eradicated regions in the southern US (Raszick *et al.*, 2023; see Chapter 1, this volume). Other molecular tools have provided identification and information about the phylogenetics and landscape movements (local, regional, and continental) of many smaller pest species, including *Bemisia tabaci* (Gennadius) (Hemiptera: Aleyrodidae) (McKenzie *et al.*, 2012; Crossley and Snyder, 2020) and various aphid pests (Shufran and Puterka, 2011; Kati *et al.*, 2013; Liu *et al.*, 2010).

Using similar population genomic and molecular evolution approaches, it has also become possible to assess patterns of selection across the genome to identify specific genes that may be undergoing selection in field populations, particularly in response to management practices. Whole-genome resequencing analysis showed that there was adaptive introgression of a single region from the invasive Old World bollworm, *Helicoverpa armigera* (Hübner), into local corn earworm, *H. zea* (Boddie) (Lepidoptera: Noctuidae), in Brazil, including an insecticide resistance allele that increased in frequency over time (Valencia-Montoya *et al.*, 2020). These newer analyses may open the door to prediction and early detection of the onset of resistance to newly deployed control strategies before it becomes widespread and problematic.

Molecular tools are also facilitating the development of novel genetic pest management strategies. Genetic pest management strategies are highly attractive because they are designed to target a specific pest, and therefore, have limited off-target and environmental impacts, unlike many synthetic insecticides. Genetic management strategies include plant-incorporated protectants (PIPs) or genes that are expressed directly by a plant that can kill insect herbivores. The most common PIPs are Bt transgenic traits, originally derived from the bacterium *B. thuringiensis*, that are expressed within the plant to produce proteins that directly target lepidopteran, coleopteran, and thysanopteran insect pests. Another commercially available PIP is RNA interference (RNAi) that causes plants to produce double-stranded RNA that specifically targets an insect herbivore by down regulating the expression genes that are essential for its survival. Development of RNAi strategies relies extensively on molecular tools to understand the genome and gene expression patterns within the target insect pests. Unfortunately, the regulatory processes to approve PIPs often take years. Nanoparticle delivery of RNAi is also being explored as an option for future pest management. If these efforts are successful, the hope is that RNAi could be applied to crops as a spray formulation, similar to a chemical insecticide (Christiaens *et al.*, 2020).

Another promising genetic management technique for pest population suppression involves releasing sterile insects. The traditional sterile insect technique (SIT) entails mass rearing pests that are exposed to high levels of radiation to render them sterile. They are then released into the environment to mate with the wild population. Since these insects are sterile, their offspring do not survive, resulting in reduced population numbers. This strategy has been instrumental in the management of a variety of agricultural and livestock pests across the world, including species of Diptera, Lepidoptera, and Hemiptera (Bourtzis and Vreysen, 2021). Although this technique involves genetic manipulation, it does not require advanced knowledge of the insect pest's genetics. An alternative genetics-based approach to producing sterile insect lines is the development and release of male insects with a dominant self-limiting gene. In this case, genomic information about the pest is required to engineer a lethal genetic construct. For example, insects may be engineered to overexpress a gene that results in death of females when reared in the absence of the antibiotic tetracycline (Leftwich *et al.*, 2014; Zhao *et al.*, 2020). When reared in the laboratory, these insects (Mediterranean fruit fly, *Ceratitis capitata* [Wiedemann] [Diptera: Tephritidae], and *Drosophila melanogaster* [Meigen] [Diptera: Drosophilidae]) are provided with the tetracycline antidote, but when they are released into the wild, all female offspring have the same lethal trait and would not have access to the tetracycline antidote, resulting in population suppression. This genetically engineered population suppression technique has been marketed by the company Oxitec as "Friendly" insect release and has been shown to reduce the populations of some crop pests including Mediterranean fruit fly (Leftwich *et al.*, 2014) and fall armyworm (Reavey *et al.*, 2022). While this technology shows promise in pest population suppression in caged studies and is currently undergoing testing in the field, it also highlights the need for molecular genetics tools to understand and monitor population structure and geneflow of insect pests, as this information will be critical in evaluating the spatial dynamics of spread and efficacy of these novel control techniques in the field.

Although genomic data is now easier than ever to generate and has promise for advancing our agricultural pest management strategies, there are limitations to the current implementation of these data sets. First, genomic data sets are very large, with single files commonly exceeding 1 TB in size. As a result, analyzing these data sets requires both advanced computational skills and an advanced understanding of molecular ecology. This has made genomics inaccessible for many biological or entomological researchers and has occasionally resulted in erroneous publications being submitted by researchers who did not fully understand their computational analyses. Therefore, the development and implementation of molecular tools in pest management is currently limited by the number of researchers trained to analyze and interpret these large-scale genomic data sets. Additionally, analyzing these large genomic data sets is also currently limited by access to computational resources. Expensive supercomputing networks are generally required to analyze large genomic data sets, and gaining access to these resources can be challenging. Despite these current limitations, the future of molecular tools in agricultural pest management is bright, and implementing these technologies has the potential to drastically increase the efficiency and sustainability of pest management strategies in the future.

Summary and Conclusions

Historically pest management strategies involved applying control measures to localized agroecosystems (fields, orchards, pastures, vineyards, stored grain facilities, etc.). Scientific advancements have led to considering areawide pest management on much larger agroecosystem scales. Pest population dynamics within local, regional, and continental landscapes need to be identified so that management tactics can be developed and implemented. The tools used to identify a pest's population growth and movement have also evolved over time. This chapter focused on a migratory pest, the fall armyworm, and how the use of molecular tools and genetic analysis provided information on within-species variation and its interactions with movement and management.

Being only identifiable using molecular tools, fall armyworm host strains were documented to possess different physiological, behavioral, and habitat characteristics that affect its management.

Host strain determination of field populations across the country led to the discovery of genetic markers (haplotypes) in one of the host strains that were geographically stable in the central and eastern part of the US. Therefore, overwintering source populations could be determined by sampling moths during their migration in the spring. Seasonal invasion of these northern landscapes via two different migratory pathways allowed researchers to follow populations that could damage crops in distant habitats. Modeling these movements at a continental scale using meteorological, crop plant, and genetic processes could aid strategic implementation of crop protection measures. Additional genetic analysis showed the same haplotypes were present throughout the Western Hemisphere.

In the last decade, fall armyworm has invaded every continent except Antarctica. Molecular tools used in the Western Hemisphere and new genetic analytical techniques are now being used throughout the world to monitor fall armyworm migration, determine crops that could be damaged, and follow the movement of insecticide resistance genes across large-scale agroecosystems. Many of these tools are being used with pests in other insect orders, including Hemiptera, Coleoptera, Diptera, and Lepidoptera, to identify sibling species or biotypes, to understand broad scale patterns of evolution and gene flow in insect pest populations, and to develop novel genetic pest management strategies. These future advanced research topics will add valuable information to address spatial and landscape questions that will improve pest management in many cropping systems.

Acknowledgements

Mention of trade names or commercial products in this publication is solely for the purpose of providing specific information and does not imply recommendation or endorsement by the USDA. The USDA is an equal opportunity provider and employer.

References

Acevedo, F.E., Peiffer, M., Ray, S., Meagher, R., Luthe, D.S. *et al.* (2018) Intraspecific differences in plant defense induction by fall armyworm strains. *New Phytologist* 218, 310–321. DOI: 10.1111/nph.14981

Anstead, J.A., Burd, J.D. and Shufran, K.A. (2003) Over-summering and biotypic diversity of *Schizaphis graminum* (Homoptera: Aphididae) populations on noncultivated grass hosts. *Environmental Entomology* 32, 662–667. DOI: 10.1603/0046-225X-32.3.662

Banerjee, R., De Bortoli, C.P., Huang, F., Lamour, K., Meagher, R. *et al.* (2022) Large genomic deletion linked to field-evolved resistance to Cry1F corn in fall armyworm (*Spodoptera frugiperda*) from Florida. *Scientific Reports* 12, 13580. DOI: 10.1038/s41598-022-17603-3

Belay, D.K., Clark, P.L., Skoda, S.R., Isenhour, D.J., Molina-Ochoa, J. *et al.* (2012) Spatial genetic variation among *Spodoptera frugiperda* (Lepidoptera: Noctuidae) sampled from the United States, Puerto Rico, Panama, and Argentina. *Annals of the Entomological Society of America* 105, 359–367. DOI: 10.1603/an11111

Bourtzis, K. and Vreysen, M.J.B. (2021) Sterile insect technique (SIT) and its applications. *Insects* 12, 638. DOI: 10.3390/insects12070638

Cano-Calle, D., Arango-Isaza, R.E. and Saldamando-Benjumea, C.I. (2015) Molecular identification of *Spodoptera frugiperda* (Lepidoptera: Noctuidae) corn and rice strains in Colombia by using a PCR-RFLP of the mitochondrial gene cytochrome oxydase I (COI) and a PCR of the gene FR (for rice). *Annals of the Entomological Society of America* 108, 172–180. DOI: 10.1093/aesa/sav001

Chapman, J.W., Reynolds, D.R. and Wilson, K. (2015) Long-range seasonal migration in insects: mechanisms, evolutionary drivers and ecological consequences. *Ecology Letters* 18, 287–302. DOI: 10.1111/ele.12407

Christiaens, O., Whyard, S., Vélez, A.M. and Smagghe, G. (2020) Double-stranded RNA technology to control insect pests: current status and challenges. *Frontiers in Plant Science* 11, 451. DOI: 10.3389/fpls.2020.00451

Clobert, J., Le Galliard, J-F., Cote, J., Meylan, S. and Massot, M. (2009) Informed dispersal, heterogeneity in animal dispersal syndromes and the dynamics of spatially structured populations. *Ecology Letters* 12, 197–209. DOI: 10.1111/j.1461-0248.2008.01267.x

Coates, B.S., Dopman, E.B., Wanner, K.W. and Sappington, T.W. (2018) Genomic mechanisms of sympatric ecological and sexual divergence in a model agricultural pest, the European corn borer. *Current Opinion in Insect Science* 26, 50–56. DOI: 10.1016/j.cois.2018.01.005

Cock, M.J.W., Beseh, P.K., Buddie, A.G., Cafa, G. and Crozier, J. (2017) Molecular methods to detect *Spodoptera frugiperda* in Ghana, and implications for monitoring the spread of invasive species in developing countries. *Scientific Reports* 7, 4103. DOI: 10.1038/s41598-017-04238-y

Crossley, M.S. and Snyder, W.E. (2020) What is the spatial extent of a *Bemisia tabaci* population? *Insects* 11, 813. DOI: 10.3390/insects11110813

Crossley, M.S., Rondon, S.I. and Schoville, S.D. (2013) Effects of contemporary agricultural land cover on Colorado potato beetle genetic differentiation in the Columbia Basin and Central Sands. *Ecology and Evolution* 9, 9385–9394. DOI: 10.1002/ece3.5489

De Marchi, B.R., Gama, A.B. and Smith, H.A. (2023) Evidence of the association between the Q2 mitochondrial group of *Bemisia tabaci* MED species (Hemiptera: Aleyrodidae) and low competitive displacement capability. *PLoS ONE* 18, e0280002. DOI: 10.1371/journal.pone.0280002

FAO (2020) *The Global Action for Fall Armyworm Control: Action Framework 2020–2022. Working Together to Tame the Global Threat.* Rome, Italy: United Nations, FAO. Available at: https://www.fao.org/documents/card/en/c/ca9252en (accessed April 22, 2024).

Gatehouse, A.G. (1997) Behavior and ecological genetics of wind-borne migration by insects. *Annual Review of Entomology* 42, 475–502. DOI: 10.1146/annurev.ento.42.1.475

Gimenez, S., Abdelgaffar, H., Le Goff, G., Hilliou, F., Blanco, C.A. *et al.* (2020) Adaptation by copy number variation increases insecticide resistance in the fall armyworm. *Communications Biology* 3, 664. DOI: 10.1038/s42003-020-01382-6

Goergen, G., Kumar, P.L., Sankung, S.B., Togola, A. and Tamò, M. (2016) First report of outbreaks of the fall armyworm *Spodoptera frugiperda* (J E Smith) (Lepidoptera, Noctuidae), a new alien invasive pest in West and Central Africa. *PLoS ONE* 11, e0165632. DOI: 10.1371/journal.pone.0165632

Harris-Shultz, K.R., Brewer, M.J., Wadl, P.A., Ni, X. and Wang, H. (2018) A sugarcane aphid "super-clone" predominates on sorghum and johnsongrass from four US states. *Crop Science* 58, 2533–2541. DOI: 10.2135/cropsci2018.03.0151

Ingber, D.A., Mason, C.E. and Flexner, L. (2018) Cry1 Bt susceptibilities of fall armyworm (Lepidoptera: Noctuidae) host strains. *Journal of Economic Entomology* 111, 361–368. DOI: 10.1093/jee/tox311

Jones, D.B., Giles, K.I., Chen, Y. and Shufran, K.A. (2005) Estimation of hymenopteran parasitism in cereal aphids by using molecular markers. *Journal of Economic Entomology* 98, 217–221. DOI: 10.1093/jee/98.1.217

Juárez, M.L., Murúa, M.G., Garcia, M.G., Ontivero, M., Vera, M.T. *et al.* (2012) Host association of *Spodoptera frugiperda* (Lepidoptera: Noctuidae) corn and rice strains in Argentina, Brazil, and Paraguay. *Journal of Economic Entomology* 105, 573–582. DOI: 10.1603/ec11184

Juárez, M.L., Schöfl, G., Vera, M.T., Vilardi, J.C., Murúa, M.G. *et al.* (2014) Population structure of *Spodoptera frugiperda* maize and rice host forms in South America: are they host strains? *Entomologia Experimentalis et Applicata* 152, 182–199. DOI: 10.1111/eea.12215

Kati, A., Shufran, K.A., Taylor, M.S., Barjadze, S., Eastop, V.F. *et al.* (2013) Identity of *Schizaphis* species (Hemiptera: Aphididae) in the United Kingdom: are they a threat to crops? *Bulletin of Entomological Research* 103, 425–440. DOI: 10.1017/S0007485312000909

Leftwich, P.T., Koukidou, M., Rempoulakis, P., Gong, H-F., Zacharopoulou, A. *et al.* (2014) Genetic elimination of field-cage populations of Mediterranean fruit flies. *Proceedings of the Royal Society B: Biological Sciences* 281, 20141372. DOI: 10.1098/rspb.2014.1372

Liu, X., Marshall, J.I., Stary, P., Edwards, O., Puterka, G. *et al.* (2010) Global phylogenetics of *Diuraphis noxia* (Hemiptera: Aphididae), an invasive aphid species: evidence for multiple invasions into North America. *Journal of Economic Entomology* 103, 958–965. DOI: 10.1603/EC09376

Luginbill, P. (1928) *The Fall Army Worm.* Technical Bulletin 34. Washington, DC: US Department of Agriculture.

Machado, V., Wunder, M., Baldissera, V.D., Oliveira, J.V., Fiúza, L.M. *et al.* (2008) Molecular characterization of host strains of *Spodoptera frugiperda* (Lepidoptera: Noctuidae) in Brazil. *Annals of the Entomological Society of America* 101, 619–626. DOI: 10.1603/0013-8746(2008)101[619:MCOHSO]2.0.CO;2

Madden, M.J.L., Young, R.G., Brown, J.W., Miller, S.E., Frewin, A.J. *et al.* (2019) Using DNA barcoding to improve invasive pest identification at U.S. ports-of-entry. *PLoS ONE* 14, e0222291. DOI: 10.1371/journal.pone.0222291

McKenzie, C.I., Bethke, J.A., Byrne, F.J., Chamberlin, J.R., Dennehy, T.J. *et al.* (2012) Distribution of *Bemisia tabaci* (Hemiptera: Aleyrodidae) biotypes in North America after the Q invasion. *Journal of Economic Entomology* 105, 753–766. DOI: 10.1603/EC11337

Meagher, R.L. and Nagoshi, R.N. (2004) Population dynamics and occurrence of *Spodoptera frugiperda* host strains in southern Florida. *Ecological Entomology* 29, 614–620. DOI: 10.1111/j.0307-6946.2004.00629.x

Meagher, R.L., Nagoshi, R.N., Armstrong, J.S., Niogret, J., Epsky, N.D. *et al.* (2013) Captures and host strains of fall armyworm (Lepidoptera: Noctuidae) males in traps baited with different commercial pheromone blends. *Florida Entomologist* 96, 729–740. DOI: 10.1653/024.096.0305

Montezano, D.G., Specht, A., Sosa-Gómez, D.R., Roque-Specht, V.F., Sousa-Silva, J.C. *et al.* (2018) Host plants of *Spodoptera frugiperda* (Lepidoptera: Noctuidae) in the Americas. *African Entomology* 26, 286–300. DOI: 10.4001/003.026.0286

Murúa, M.G., Nagoshi, R.N., Dos Santos, D.A., Hay-Roe, M.M., Meagher, R.L. *et al.* (2015) Demonstration using field collections that Argentina fall armyworm populations exhibit strain-specific host plant preferences. *Journal of Economic Entomology* 108, 2305–2315. DOI: 10.1093/jee/tov203

Nagoshi, R.N. (2019) Evidence that a major subpopulation of fall armyworm found in the Western Hemisphere is rare or absent in Africa, which may limit the range of crops at risk of infestation. *PLoS ONE* 14, e0208966. DOI: 10.1371/journal.pone.0208966

Nagoshi, R.N. and Meagher, R.L. (2004) Seasonal distribution of fall armyworm (Lepidoptera: Noctuidae) host strains in agricultural and turf grass habitats. *Environmental Entomology* 33, 881–889. DOI: 10.1603/0046-225x-33.4.881

Nagoshi, R.N. and Meagher, R.L. (2016) Using intron sequence comparisons in the *triose-phosphate isomerase* gene to study the divergence of the fall armyworm host strains. *Insect Molecular Biology* 25, 324–337. DOI: 10.1111/imb.12223

Nagoshi, R.N. and Meagher, R.L. (2022) The *Spodoptera frugiperda* host strains: what they are and why they matter for understanding and controlling this global agricultural pest. *Journal of Economic Entomology* 115, 1729–1743. DOI: 10.1093/jee/toac050

Nagoshi, R.N., Meagher, R.L., Adamczyk, J.J., Jr, Braman, S.K., Brandenburg, R.L. *et al.* (2006) New restriction fragment length polymorphisms in the cytochrome oxidase I gene facilitate host strain identification of fall armyworm (Lepidoptera: Noctuidae) populations in the southeastern United States. *Journal of Economic Entomology* 99, 671–677. DOI: 10.1093/jee/99.3.671

Nagoshi, R.N., Silvie, P. and Meagher, R. (2007) Comparison of haplotype frequencies differentiate fall armyworm (Lepidoptera: Noctuidae) corn-strain populations from Florida and Brazil. *Journal of Economic Entomology* 100, 954–961. DOI: 10.1603/0022-0493(2007)100[954:cohfdf]2.0.co;2

Nagoshi, R.N., Meagher, R.L. and Hay-Roe, M. (2012) Inferring the annual migration patterns of fall armyworm (Lepidoptera: Noctuidae) in the United States from mitochondrial haplotypes. *Ecology and Evolution* 2, 1458–1467. DOI: 10.1002/ece3.268

Nagoshi, R.N., Fleischer, S., Meagher, R.L., Hay-Roe, M., Khan, A. *et al.* (2017) Fall armyworm migration across the Lesser Antilles and the potential for genetic exchanges between North and South American populations. *PLoS ONE* 12, e0171743. DOI: 10.1371/journal.pone.0171743

Nagoshi, R.N., Goergen, G., Tounou, K.A., Agboka, K., Koffi, D. *et al.* (2018) Analysis of strain distribution, migratory potential, and invasion history of fall armyworm populations in northern sub-Saharan Africa. *Scientific Reports* 8, 3710. DOI: 10.1038/s41598-018-21954-1

Nagoshi, R.N., Htain, N.N., Boughton, D., Zhang, L., Xiao, Y.T. *et al.* (2020) Southeastern Asia fall armyworms are closely related to populations in Africa and India, consistent with common origin and recent migration. *Scientific Reports* 10, 1421. DOI: 10.1038/s41598-020-58249-3

Nagoshi, R.N., Koffi, D., Agboka, K., Adjevi, A.K.M., Meagher, R.L. *et al.* (2021) The fall armyworm strain associated with most rice, millet, and pasture infestations in the Western Hemisphere is rare or absent in Ghana and Togo. *PLoS ONE* 16, e53528. DOI: 10.1371/journal.pone.0053528

Nagoshi, R.N., Goergen, G., Koffi, D., Agboka, K., Adjevi, A.K.M. *et al.* (2022) Genetic studies of fall armyworm indicate a new introduction into Africa and identify limits to its migratory behavior. *Scientific Reports* 12, 1941. DOI: 10.1038/s41598-022-05781-z

Nam, H., Kwon, M., Ramasamy, S. and Kim, J. (2022) Identification of two diamondback moth parasitoids, *Diadegma fenestrale* and *Diadegma semiclausum*, using LAMP for application in biological control. *Horticulturae* 8, 366. DOI: 10.3390/horticulturae8050366

Nam, K., Nhim, S., Robin, S., Bretaudeau, A., Nègre, N. *et al.* (2020) Positive selection alone is sufficient for whole genome differentiation at the early stage of speciation process in the fall armyworm. *BMC Evolutionary Biology* 20, 152. DOI: 10.1186/s12862-020-01715-3

Orsucci, M., Moné, Y., Audiot, P., Gimenez, S., Nhim, S. *et al.* (2022) Transcriptional differences between the two host strains of *Spodoptera frugiperda* (Lepidoptera: Noctuidae). *Peer Community Journal* 2, e1. DOI: 10.24072/pcjournal.77

Otim, M.H., Tay, W.T., Walsh, T.K., Kanyesigye, D., Adumo, S. *et al.* (2018) Detection of sister-species in invasive populations of the fall armyworm *Spodoptera frugiperda* (Lepidoptera: Noctuidae) from Uganda. *PLoS ONE* 13, e0194571. DOI: 10.1371/journal.pone.0194571

Pair, S.D., Raulston, J.R., Rummel, D.R., Westbrook, J.K., Wolf, W.W. *et al.* (1987) Development and production of corn earworm and fall armyworm in the Texas High Plains: evidence for reverse fall migration. *Southwestern Entomologist* 12, 89–99.

Pair, S.D., Raulston, J.R., Westbrook, J.K., Wolf, W.W. and Adams, S.D. (1991) Fall armyworm (Lepidoptera: Noctuidae) outbreak originating in the lower Rio Grande valley, 1989. *Florida Entomologist* 74, 200–213. DOI: 10.2307/3495298

Pashley, D.P. (1986) Host-associated genetic differentiation in fall armyworm (Lepidoptera: Noctuidae): a sibling species complex? *Annals of the Entomological Society of America* 79, 898–904. DOI: 10.1093/aesa/79.6.898

Pashley, D.P., Johnson, S.J. and Sparks, A.N. (1985) Genetic population structure of migratory moths: the fall armyworm (Lepidoptera: Noctuidae). *Annals of the Entomological Society of America* 78, 756–762. DOI: 10.1093/aesa/78.6.756

Pashley, D.P., Quisenberry, S.S. and Jamjanya, T. (1987) Impact of fall armyworm (Lepidoptera: Noctuidae) host strains on the evaluation of Bermuda grass resistance. *Journal of Economic Entomology* 80, 1127–1130. DOI: 10.1093/jee/80.6.1127

Pashley Prowell, D., McMichael, M. and Silvain, J.F. (2004) Multilocus genetic analysis of host use, introgression, and speciation in host strains of fall armyworm (Lepidoptera: Noctuidae). *Annals of the Entomological Society of America* 97, 1034–1044. DOI: 10.1603/0013-8746(2004)097[1034:MGAOHU]2.0.CO;2

Rane, R., Walsh, T.K., Lenancker, P., Gock, A., Dao, T.H. *et al.* (2023) Complex multiple introductions drive fall armyworm invasions into Asia and Australia. *Scientific Reports* 13, 660. DOI: 10.1038/s41598-023-27501-x

Raszick, T.J., Dickens, C.M., Perkin, L.C., Tessnow, A.E., Suh, C.P-C. *et al.* (2021) Population genomics and phylogeography of the boll weevil, *Anthonomus grandis* Boheman (Coleoptera: Curculionidae), in the United States, northern Mexico, and Argentina. *Evolutionary Applications* 14, 1778–1793. DOI: 10.1111/eva.13238

Raszick, T.J., Perkin, L.C., Shirley, X.A., Ruiz-Arce, R., Kramer, Z.A. *et al.* (2023) Source tracing of boll weevils (*Anthonomus grandis*) captured in areas of the United States where the species had previously been eradicated. *Journal of Pest Science* 97, 585–595. DOI: 10.1007/s10340-023-01656-y

Reavey, C.E., Walker, A.S., Joyce, S.P., Broom, L., Willse, A. *et al.* (2022) Self-limiting fall armyworm: a new approach in development for sustainable crop protection and resistance management. *BMC Biotechnology* 22, 5. DOI: 10.1186/s12896-022-00735-9

Ríos-Díez, J.D. and Saldamando-Benjumea, C.I. (2011) Susceptibility of *Spodoptera frugiperda* (Lepidoptera: Noctuidae) strains from central Colombia to two insecticides, methomyl and lambda-cyhalothrin: a study of the genetic basis of resistance. *Journal of Economic Entomology* 104, 1698–1705. DOI: 10.1603/EC11079

Ríos-Díez, J.D., Siegfried, B. and Saldamando-Benjumea, C.I. (2012) Susceptibility of *Spodoptera frugiperda* (Lepidoptera: Noctuidae) strains from Central Colombia to Cry1Ab and Cry1Ac entotoxins of *Bacillus thuringiensis*. *Southwestern Entomologist* 37, 281–293. DOI: 10.3958/059.037.0304

Rose, A.H., Silversides, R.H. and Lindquist, O.H. (1975) Migration flight by an aphid, *Rhopalosiphum maidis* (Hemiptera: Aphididae), and a noctuid, *Spodoptera frugiperda* (Lepidoptera: Noctuidae). *Canadian Entomologist* 107, 567–576. DOI: 10.4039/Ent107567-6

Saldamando, C.I. and Vélez-Arango, A.M. (2010) Host plant association and genetic differentiation of corn and rice strains of *Spodoptera frugiperda* Smith (Lepidoptera: Noctuidae) in Colombia. *Neotropical Entomology* 39, 921–929. DOI: 10.1590/s1519-566x2010000600012

Schlum, K.A., Lamour, K., de Bortoli, C.P., Banerjee, R., Meagher, R. *et al.* (2021) Whole genome comparisons reveal panmixia among fall armyworm (*Spodoptera frugiperda*) from diverse locations. *BMC Genomics* 22, 179. DOI: 10.1186/s12864-021-07492-7

Schmidt, J.M., Acebes-Doria, A., Blaauw, B., Kheirodin, A., Pandey, S. *et al.* (2021) Identifying molecular-based trophic interactions as a resource for advanced integrated pest management. *Insects* 12, 358. DOI: 10.3390/insects12040358

Shufran, K.A. and Puterka, G.J. (2011) DNA barcoding to identify all life stages of holocyclic cereal aphids (Hemiptera: Aphididae) on wheat and other Poaceae. *Annals of the Entomological Society of America* 104, 39–42. DOI: 10.1603/AN10129

Sparks, A.N. (1979) A review of the biology of the fall armyworm. *Florida Entomologist* 62, 82–87. DOI: 10.2307/3494083

Sparks, A.N., Jackson, R.D., Carpenter, J.E. and Muller, R.A. (1986) Insects captured in light traps in the Gulf of Mexico. *Annals of the Entomological Society of America* 79, 132–139. DOI: 10.1093/aesa/79.1.132

Tandy, P., Lamour, K., de Bortoli, C.P., Nagoshi, R., Emrich, S.J. *et al.* (2023) Screening for resistance alleles to Cry1 proteins through targeted sequencing in the native and invasive range of *Spodoptera frugiperda* (Lepidoptera: Noctuidae). *Journal of Economic Entomology* 116, 935–944. DOI: 10.1093/jee/toad061

Tay, W.T., Rane, R.V., James, W., Gordon, K.H.J., Downes, S. *et al.* (2022a) Resistance bioassays and allele characterization inform analysis of *Spodoptera frugiperda* (Lepidoptera: Noctuidae) introduction pathways in Asia and Australia. *Journal of Economic Entomology* 115, 1790–1805. DOI: 10.1093/jee/toac151

Tay, W.T., Rane, R.V., Padovan, A., Walsh, T.K., Elfekih, S. *et al.* (2022b) Global population genomic signature of *Spodoptera frugiperda* (fall armyworm) supports complex introduction events across the Old World. *Communications Biology* 5, 297. DOI: 10.1038/s42003-022-03230-1

Tay, W.T., Meagher, R.L., Jr, Czepak, C. and Groot, A.T. (2023) *Spodoptera frugiperda*: ecology, evolution, and management options of an invasive species. *Annual Review of Entomology* 68, 299–317. DOI: 10.1146/annurev-ento-120220-102548

Tessnow, A.E., Gilligan, T.M., Burkness, E., De Bortoli, C.P., Jurat-Fuentes, J.L. *et al.* (2021) Novel real-time PCR based assays for differentiating fall armyworm strains using four single nucleotide polymorphisms. *PeerJ* 9, e12195. DOI: 10.7717/peerj.12195

Tessnow, A.E., Raszick, T.J., Porter, P. and Sword, G.A. (2022) Patterns of genomic and allochronic strain divergence in the fall armyworm, *Spodoptera frugiperda* (J.E. Smith). *Ecology and Evolution* 12, e8706. DOI: 10.1002/ece3.8706

Tessnow, A.E., Nagoshi, R.N., Meagher, R.L. and Fleischer, S.J. (2023) Revisiting fall armyworm population movement in the United States and Canada. *Frontiers in Insect Science* 3, 1104793. DOI: 10.3389/finsc.2023.1104793

Unbehend, M., Hänniger, S., Meagher, R.L., Heckel, D.G. and Groot, A.T. (2013) Pheromonal divergence between two strains of *Spodoptera frugiperda*. *Journal of Chemical Ecology* 39, 364–376. DOI: 10.1007/s10886-013-0263-6

Unbehend, M., Hänniger, S., Vásquez, G.M., Juárez, M.L., Reisig, D. *et al.* (2014) Geographic variation in sexual attraction of *Spodoptera frugiperda* corn- and rice-strain males to pheromone lures. *PLoS ONE* 9, e89255. DOI: 10.1371/journal.pone.0089255

Valencia-Montoya, W.A., Elfekih, S., North, H.L., Meier, J.I., Warren, I.A. *et al.* (2020) Adaptive introgression across semipermeable species boundaries between local *Helicoverpa zea* and invasive *Helicoverpa armigera* moths. *Molecular Biology and Evolution* 37, 2568–2583. DOI: 10.1093/molbev/msaa108

Veenstra, K.H., Pashley, D.P. and Ottea, J.A. (1995) Host-plant adaptation in fall armyworm host strains: comparison of food consumption, utilization, and detoxication enzyme activities. *Annals of the Entomological Society of America* 88, 80–91. DOI: 10.1093/aesa/88.1.80

Westbrook, J., Fleischer, S., Jairam, S., Meagher, R. and Nagoshi, R. (2019) Multigenerational migration of fall armyworm, a pest insect. *Ecosphere* 10, e02919. DOI: 10.1002/ecs2.2919

Westbrook, J.K. (2008) Noctuid migration in Texas within the nocturnal aeroecological boundary layer. *Integrative and Comparative Biology* 48, 99–106. DOI: 10.1093/icb/icn040

Westbrook, J.K., Raulston, J.R., Wolf, W.W., Pair, S.D., Eyster, R.S. *et al.* (1995) Field observations and simulations of atmospheric transport of noctuids from northeastern Mexico and the south-central U.S. *Southwestern Entomologist* Supplement 18, 25–44.

Westbrook, J.K., Nagoshi, R.N., Meagher, R.L., Fleischer, S.J. and Jairam, S. (2016) Modeling seasonal migration of fall armyworm moths. *International Journal of Biometeorology* 60, 255–267. DOI: 10.1007/s00484-015-1022-x

Wilson, C.R., Cooke-McEwen, C., Gilligan, T.M. and Tembrock, L.R. (2022) Detection of *Centistes* sp. (Hymenoptera, Braconidae) from intercepted *Diabrotica undecimpunctata* (Coleoptera, Chrysomelidae) using CO1 DNA barcodes and larval morphology. *Journal of Hymenoptera Research* 91, 69–81. DOI: 10.3897/jhr.91.84139

Zhao, Y., Schetelig, M.F. and Handler, A.M. (2020) Genetic breakdown of a Tet-off conditional lethality system for insect population control. *Nature Communications* 11, 3095. DOI: 10.1038/s41467-020-16807-3

Index

www.ingramcontent.com/pod-product-compliance
Lightning Source LLC
LaVergne TN
LVHW071337210926
848486LV00003B/161